STIFF COMPUTATION

Stiff Computation

Edited by
RICHARD C. AIKEN

New York Oxford
OXFORD UNIVERSITY PRESS
1985

Library of Congress Cataloging in Publication Data
Main entry under title:
Stiff computation.
Bibliography: p. Includes index.
1. Stiff computation (Differential equations)
I. Aiken, Richard C.
QA371.S8474 1984 519.4 83-26737
ISBN 0-19-503453-8

Printing (last digit): 9 8 7 6 5 4 3 2 1

Printed in Great Britain

In memoriam Gladys Aiken matri,
Catherine Delonchaw aviae hunc
librum pie dedit filius neposque

Preface

Dynamic models are becoming ever more detailed as our experimental methods improve and our objectives expand. Some computational penalties that result from this detail, that we classify here as *stiff computation,* were first recognized in the early 1950s. The number of occurrences in practice were numerous enough by the mid-60s to encourage special study. In the last several years stiff method and algorithm development have reached at least a local peak, and it now seems appropriate to examine whether or not this peak is a global one.

This book grew out of the *International Conference on Stiff Computation,* held April 12–14, 1982, at Park City, Utah. The purpose of that meeting was to *review* the state of the art and practice of stiff computation, rather than to present latest research. Speakers represented the spectrum of individuals involved in stiff computation from theoretician to software developer to end user. The participant selection was consistent with a systems philosophy that would strive to organize rather diverse activity in stiff computation toward a common objective. Some excellent contributions to the meeting do not appear here because of space limitations or because they were not consistent with our objective of providing an overview. Reference to these papers are to the informal proceedings made available at the time of the meeting. Contributions included here that were not presented at the conference are those of Burrage; Deuflhard, Bader and Nowak; Lomax; Pratt; and those of my own. In total, these contributions comprise roughly half of this book. Contributions presented at the conference have been edited into a consistent format. Thus the printed text is not the conference proceedings per se, but is intended to be a unified review of stiff computation.

In chapter one, Shampine defines stiffness. It is very important that the user of stiff software know for what problem class the software is intended. We thus try to respect Shampine's classification in this book. Other difficult computational problems with differential models, some related to stiffness, are classified in chapter one and touched upon later in the book. If not mathematically related to stiffness, these newer difficulties share with stiffness an ancestry of model detail.

Typical application areas from which stiffness results are found in chapter two. Application areas include: the atmosphere (Hindmarsh), heat transfer (Churchill), chemical kinetics (Edsberg), nuclear kinetics (Devooght), and processing (Seider and coworkers; Byrne; Rodrigues and coworkers). I contend that very often stiffness need not be in the model, so that its physical identification is important for model simplification. Physical features leading to stiffness are identified in chapter two.

A review of some of the newer methods and formulae for stiffness comprises chapter three. Cash provides a thorough look at Runge-Kutta methods; Burrage has additional thoughts. Partitioning is examined by Krogh, Mattheij, and O'Malley. The highly oscillatory problem is covered by Miranker. Finlayson and Schiesser both review techniques for handling ordinary differential equations arising from discretization of partial differential equations.

Packages are in chapter four. Gear gives an overview of our present condition; Hindmarsh traces his involvement in taking early general purpose GEAR codes to their present form; Enright indicates how codes can be compared. Thompson and Tuttle provide a unique insight into how a code evolves in an industrial environment. Chen and Schiesser demonstrate a user's point of view in choosing software for large applications.

Software (or hardware) tailored for a specific application should out-perform that intended for general purpose, and this topic is taken up in chapter five. The extra effort to tailor a package seems justified if the application area is common, has a characteristic structure, or cannot be solved otherwise. Chemical kinetics provides such an area: Edsberg looks at some generalities; Deuflhard and coworkers concentrate on the largeness aspect. Pratt; Wendt and Hahn attack combustion kinetics. Seider and Bailey and coworkers offer approaches to processing applications.

"There is nothing more practical than a good theory," Dahlquist would remind us. He provides a summary of his work and that of others on stability and error analysis for stiff nonlinear problems in chapter six. Liniger reviews aspects of contractivity.

Chapter seven addresses the stiff computation problem as it relates to experimentation, modelling, parameter estimation, and to the overall objective for this long train of activity. Aspects of the "inverse problem" for stiff computation are formalized for the first time.

Finally, in chapter eight, we look more qualitatively at where stiff computation is and where it is going; Cellier provides his opinions. Excerpts from a panel discussion held at the Park City meeting and from a questionnaire that followed the meeting yield candid answers—everything you always wanted to ask about stiffness but were afraid to know. The impact of next-generation computers on stiff computation is the topic of Lomax's contribution.

Contributing authors' names appear on their section headings or on subsections if more than one source has contributed to a section. Full names and affiliations of the authors appear in the list of contributors.

I would like to recognize my former advisor, the late Leon Lapidus of Princeton, and Germund Dahlquist of the Swedish Institute of Technology (whom I visited in 1975) for stimulating my interest in stiff computation. My appreciation goes to Major C. E. Oliver and the Air Force Office of Scientific Research for funding the Park City meeting; also to Celanese Corporation for support in the preparation of this book; and to the University of Utah for the David P. Gardner Award providing time for this project. Thanks also to my graduate student S. Venkateshwaran, who coauthored some material with me. My deepest gratitude goes to Mary Schmid Aiken for her outstanding work on this project from literature searching to organizing the Park City conference to preparing this manuscript.

Salt Lake City
April 1983

R.C.A.

Contents

5. SPECIAL NUMERICAL APPROACHES TO SPECIFIC APPLICATIONS

6. THEORY

7. ANALYSIS OF DETAILED DYNAMIC SYSTEMS

8. THE FUTURE

Contributors

Dr. Richard C. Aiken
Dept. of Chemical Engineering
University of Utah
Salt Lake City, UT 84112

Dr. Patrick G. Bailey
Electric Power Research Institute
Nuclear Power Division
P. O. Box 10412
Palo Alto, CA 94303

Dr. Kevin Burrage
Dept. of Computer Science
University of Auckland
Auckland, New Zealand

Dr. George D. Byrne
Computing Technology and Services Division
EXXON Research & Engineering Co.
P.O. Box 45
Linden, NJ 07036

Dr. J. R. Cash
Dept. of Mathematics
Huxley Building
Imperial College of Science and Technology
Queen's Gate, London SW 7 2BZ
England

Dr. Francois E. Cellier
Dept. of Electrical and Computer Engineering
University of Arizona
Tucson, AZ 85721

Dr. Katherine L. Chen
Raw Materials & Chemical Processes
Homer Research Laboratories
Bethlehem Steel Corporation
Bethlehem, PA

Dr. Stuart W. Churchill
Dept. of Chemical Engineering
392 Towne Building
University of Pennsylvania
Philadelphia, PA 19174

Dr. Germund Dahlquist
Dept. of Numerical Analysis
and Computing Science
Royal Institute of Technology
S-100 44 Stockholm 70, Sweden

Drs. Peter Deuflhard,
G. Bader, and U. Nowak
Universitat Heidelberg
Inst.fur Angewandte Mathematik
Im Neuenheimer Feld 293
D-6900 Heidelberg 1, West Germany

Dr. J. Devooght
Universite Libre de Bruxelles
Faculte des Sciences Appliquees
Service de Metrologie Nucleaire
Avenue F.-D. Roosevelt 50
1050 Bruxelles, Belgium

Drs. M. Q. Dias, J. C. Lopes,
C. A. Costa, and A. E. Rodriques
Instituto Nacional de Investigacao Cientifica
Centro de Engenharia Quimica
Faculdade de Engenharia
Rua Dos Bragas
4099 Porto-Codex Portugal

Dr. Lennart Edsberg
Dept. of Numerical Analysis
and Computing Science
Royal Institute of Technology
S-100 44 Stockholm 70, Sweden

Dr. Wayne H. Enright
Dept. of Computer Science
University of Toronto
Toronto, Canada M5S 1A7

Dr. Bruce A. Finlayson
Dept. of Chemical Engineering
105 Benson Hall, BF-10
University of Washington
Seattle, WA 98195

Dr. Charles W. Gear
Dept. of Computer Science
University of Illinois at Urbana-Champaign
1304 West Springfield Avenue
Urbana, IL 61801-2987

Drs. P. V. Girijashankar,
D. L. Hetrick, W. N. Keepin,
and O. A. Palusinski
Dept. of Nuclear Engineering
University of Arizona
Tucson, AZ 85721

Dr. Alan C. Hindmarsh
Lawrence Livermore National Laboratory
L-316 Mathematics & Statistics Division
University of California
P. O. Box 808
Livermore, CA 94550

Dr. Fred T. Krogh
Jet Propulsion Laboratory
125-128
4800 Oak Grove Drive
Pasadena, CA 91103

Dr. Werner Liniger
Mathematical Sciences Dept.
IBM T. J. Watson Research Center
P. O. Box 218
Yorktown Heights, NY 10598

Dr. Harvard Lomax
Computational Fluid Dynamics Branch
Ames Research Center
Moffett Field, CA

Dr. Robert M. M. Mattheij
Mathematisch Instituut
Katholieke Universiteit
Toernooiveld, Nymegen
The Netherlands

Dr. Willard L. Miranker
IBM T.J. Watson Research Center
P. O. Box 218
Yorktown Heights, NY 10598

Dr. Robert E. O'Malley, Jr.
Dept. of Mathematical Sciences
Rensselaer Polytechnic Institute
Troy, NY 12181

Dr. David Pratt
Dept. of Mechanical Engineering
Mail Stop FU-10
University of Washington
Seattle, WA 98195

Dr. William E. Schiesser
Dept. of Chemical Engineering
Lehigh University
Whitaker Laboratory 5
Bethlehem, PA 18015

Drs. Warren D. Seider,
C. W. White, III,
G. J. Prokopakis
Dept. of Chemical Engineering
Towne Building D3
220 South 33rd Street
University of Pennsylvania
Phialdelphia, PA 19104

Dr. Larry F. Shampine
Sandia National Laboratories
Numerical Mathematics Division 5642
Albuquerque, NM 87185

Drs. S. Thompson,
P. G. Tuttle
Babcock & Wilcox
Nuclear Power Generation Division
P. O. Box 1260
Lynchburg, VA 24505

S. Venkateshwaran
Dept. of Chemical Engineering
University of Utah
Salt Lake City, UT 84112

Drs. Jost O. L. Wendt
and W. A. Hahn
Dept. of Chemical Engineering
The University of Arizona
Tucson, AZ 85721

1

INTRODUCTION TO STIFFNESS

1.1. What is stiffness? (Shampine)

1.1.1. *Introduction*

The numerical solution of the initial value problem for a system of N first-order ordinary differential equations is considered:

$$y' = f(x, y) \qquad a \leqslant x \leqslant b \tag{1.1}$$

$$y(a) \text{ given.} \tag{1.2}$$

Modern codes based on step-by-step methods begin with $y(a) = y_0$ at $x_0 = a$ and step from a to b successively producing approximations y_n to $y(x_n)$ on a mesh $x_0 < x_1 < \ldots < x_M = b$. At the step from x_n the code selects a step size h_{n+1} so that the resulting approximation at $x_{n+1} = x_n + h_{n+1}$ is to satisfy a certain accuracy requirement.

The standard assumption about $f(x, y)$ is that it has as many continuous derivatives as needed and that it satisfies a Lipschitz condition,

$$\| f(x, u) - f(x, v) \| \leqslant L \| u - v \|, \tag{1.3}$$

for $a \leqslant x \leqslant b$ and all y. This guarantees (1.1, 1.2) has a unique solution $y(x)$. The "classical situation" is that $L(b - a)$ is not "large." In this situation classical numerical methods such as Runge-Kutta and Adams are quite satisfactory with one major exception—too frequent output can drastically reduce the step size. There are a variety of other reasons why the step size might have to be restricted, but in the classical situation, none can lead to a *severe* restriction of the step size.

When $L(b - a)$ is "large," the matter is quite different. In the first place, the class of mathematical problems must be restricted in order for step-by-step methods to have a chance of solving the problem adequately. It develops that in some extremely important circumstances, classical methods suffer a variety of restrictions on the step size which are so severe that such methods are impractical. This is what is usually termed "stiffness."

Many authors have sought a definition of stiffness involving only the mathematical problem (1.1, 1.2). Unfortunately the situation is far more complex than that. In this section we shall explore those factors determining the step size in a sequence like that of the usual treatment of Runge-Kutta methods. All the seeds of step size restrictions are present in the classical situation. We shall see how a large Lipschitz constant affects the algorithms and shall describe ways to circumvent or overcome the difficulties. We aim to provide some feeling as to the kinds of problems leading to severe step size restrictions for classical methods and to show how the computational problem, formula, and implementation are related to "stiffness."

1.1.2. *Mathematical problem*

We are interested in approximating a vector function $y(x)$ which satisfies (1.1) as an identity for given function $f(x, y)$ and interval $[a, b]$. The solution $y(x)$ is to have the specified initial value $y(a)$ at $x = a$. The assumption that $f(x, y)$ is continuous and satisfies a Lipschitz condition (1.3) for $a \leqslant x \leqslant b$ and all y guarantees that (1.1) has a unique solution for any given initial value y_n at any point x_n in $[a, b]$. It is easy to reduce the requirements on f to holding only in a neighborhood of the solution $y(x)$. Although this reduction is of great theoretical and practical value, we do not want to complicate matters by going into details.

One-step methods exemplify step-by-step methods. They have the form

$$y_{n+1} = y_n + h_{n+1}\,\Phi(x_n, y_n; h_{n+1}). \tag{1.4}$$

Having reached x_n, such methods are supplied only the values x_n, y_n, h_{n+1}, and the ability to evaluate f. It is clear that the best one can hope to do is to approximate $u(x_{n+1})$ where $u(x)$ is the "local solution" of (1.1) with initial value y_n:

$$u' = f(x, u), \quad u(x_n) = y_n. \tag{1.5}$$

Some of the classical methods use a small number of the most recent previously computed approximations, but the situation is not really any different for them.

The question which immediately arises is how well does $u(x)$ approximate $y(x)$ for $x \geqslant x_n$. This is a question of the stability of the solution $y(x)$. A classical result states that if $u(x)$, $v(x)$ are two solutions of (1.1), then

$$\|u(x + \Delta) - v(x + \Delta)\| \leqslant e^{\Delta L}\,\|u(x) - v(x)\|, \tag{1.6}$$

where L is the Lipschitz constant of (1.3). The result is sharp, as the single equation $y' = Ly$ shows.

We simply cannot solve unstable problems in practice by the kinds of methods we consider. A single error of ϵ at x_n moves us from $y(x)$ to the $u(x)$ of (1.5). If $u(x)$ diverges strongly from $y(x)$, the error of ϵ is amplified accordingly. Of course what "too" much means depends on the accuracy desired and the computing budget. In the classical situation, (1.6) implies that $y(x)$ is reasonably stable.

A large Lipschitz constant tells us that some solutions of (1.1) either diverge or converge rapidly in a relative sense. Suppose equality holds in (1.3) for x_0, u_0, v_0 near $y(x_0)$. Let $u(x)$ be the local solution of (1.1) with $u(x_0) = u_0$ and similarly define $v(x)$. Further let $\delta(x) = u(x) - v(x)$. Then (1.3) states that

$$\frac{\|\delta'(x_0)\|}{\|\delta(x_0)\|} = L \gg 1.$$

There are problems of great practical importance for which the Lipschitz constant is "large." Here and later we say the Lipschitz constant is large as an abbreviation for the statement that $(b - a)L$ is large; the reader should remember that the length of the interval is also important. As noted, a problem with a large Lipschitz constant may be unstable. For the kinds of numerical methods we consider, we must assume that the solution is stable. In order to formulate quantitative results, we must supplement the Lipschitz condition with some other condition guaranteeing stability.

The most studied class of problems with large Lipschitz constants consists of linear problems with constant Jacobians:

$$y' = Jy + F(x). \tag{1.7}$$

It is often assumed for convenience, and we do so, that J has a complete set of eigenvectors $\{v_i\}$ and associated eigenvalues $\{\lambda_i\}$. The general solution can then be written in terms of a particular solution $p(x)$ as

$$y(x) = p(x) + \sum_{i=1}^{n} \alpha_i v_i \exp[\lambda_i(x-a)]. \tag{1.8}$$

This class is so simple that many numerical procedures can be analyzed in considerable detail and so provides insight as to more general problems. The representation (1.8) makes it clear that stability is equivalent to the requirement that for all i, either $\mathrm{Re}(\lambda_i) \leqslant 0$ or if $\mathrm{Re}(\lambda_i) > 0$, then $(b-a)\mathrm{Re}(\lambda_i)$ is not large. Notice that $\|J\| = L \geqslant |\lambda_i|$ for all i, so the presence of some λ_j with $(b-a)\mathrm{Re}(\lambda_j) \ll -1$ implies a large Lipschitz constant.

It is often suggested that the general problem (1.1) be modelled near a point $x_n, y(x_n)$ by a problem of the form (1.7):

$$z' = f[x_n, y(x_n)] + f_x[x_n, y(x_n)](x - x_n) + f_y[x_n, y(x_n)][z - y(x_n)],$$

where f_y is the Jacobian matrix of first partial derivatives of f. This is a time-honored tactic of applied mathematics. We say that our numerical methods should perform well on problems of the form (1.7) and that we hope the linearization stated will provide a useful guide for more general problems.

Recent work has considered f which satisfy

$$\langle f(x,u) - f(x,v), u - v\rangle \leqslant l\|u - v\|^2 \tag{1.9}$$

for a suitable inner product $\langle\cdot\cdot\cdot,\cdot\cdot\cdot\rangle$. This is a one-sided Lipschitz condition because in general

$$-L\|u - v\|^2 \leqslant \langle f(x,u) - f(x,v), u - v\rangle \leqslant L\|u - v\|^2.$$

The new thing is that l might be a lot less than L and even negative. By differentiating $D(x) = \exp(-2lx)\|u(x) - v(x)\|^2$, one can show that

$$\|u(x+\Delta) - v(x+\Delta)\| \leqslant e^{l\Delta}\|u(x) - v(x)\|, \tag{1.10}$$

which can be a great improvement over (1.6).

A variant is to consider problems of the form

$$y' = Jy + g(x, y), \tag{1.11}$$

where g satisfies a Lipschitz condition

$$\|g(x,u) - g(x,v)\| \leqslant \rho\|u - v\|, \tag{1.12}$$

and μ is the smallest constant such that

$$\langle Jz, z \rangle \leqslant \mu \| z \|^2 \qquad \text{for all } z$$

for the constant matrix J. The quantity $\mu = \mu[J]$ is called the logarithmic "norm" of J and it may be negative. One can then prove

$$\|u(x+\Delta) - v(x+\Delta)\| \leqslant e^{(\mu+\rho)\Delta} \|u(x) - v(x)\|.$$

Notice that this class is included in (1.9) with $l = \rho + \mu$. This class is a natural one for the investigation of several kinds of methods and, as we shall see later, is natural for certain practical reasons as well.

In summary, continuity of f and a Lipschitz condition guarantee existence and uniqueness of the solution $y(x)$ of (1.1, 1.2). We cannot solve (1.1, 1.2) in practice using step-by-step methods if $y(x)$ is not moderately stable. In the classical situation of $(b - a)L$ not large, the Lipschitz condition alone guarantees adequate stability. Otherwise we must assume stability, or supplement the Lipschitz condition with other hypotheses about the problem which guarantee stability. We have mentioned three such subsets of Lipschitzian problems for which stability can be demonstrated even when the Lipschitz constant is large.

A qualitative way to describe the problems with large L which interest us is that $y(x)$ is stable in the direction of integration and some solution of (1.1) approaches $y(x)$ very rapidly. A way we prefer for describing the latter condition is that $y(x)$ is very unstable in the opposite direction. Notice that the stability result (1.1) shows that in the classical situation, $y(x)$ is stable in both directions. For stiffness to occur, there must be a strong directional effect.

1.1.3. *Classical formulae*

The Adams family of formulae for approximating the solution of (1.1) arises from an equivalent integral form:

$$y(x_{n+1}) = y(x_n) + \int_{x_n}^{x_{n+1}} y'(t)\mathrm{d}t. \tag{1.13}$$

Approximations to $y'(t)$ arise naturally from approximations y_i to $y(x_i)$ by

$$y'(x_i) = f[x_i, y(x_i)] \approx f(x_i, y_i). \tag{1.14}$$

The Adams formulae approximate $y'(t)$ in (1.13) by interpolating values $f_i = f(x_i, y_i)$ previously computed. Thus an Adams-Bashforth formula of order $p + 1$ forms the (unique) polynomial $P(x)$ of degree p interpolating to f_{n-i} for $i = 0, 1, \ldots, p$ and then defines

$$y_{n+1} = y_n + \int_{x_n}^{x_{n+1}} P(t)\mathrm{d}t.$$

This explicit formula is defined for any set of mesh points but simplifies in the case of constant step size h to a formula of the form

$$y_{n+1} = y_n + h \sum_{i=0}^{p} \alpha_i f_{n-i}.$$

The simplest example is the (forward) Euler method

$$y_{n+1} = y_n + hf_n. \tag{1.15}$$

The Adams-Moulton formula on $p + 1$ points interpolates to the unknown value $f(x_{n+1}, y_{n+1})$, too:

$$P(x_{n+1-i}) = f_{n+1-i}, \qquad i = 0, 1, \ldots, p.$$

This formula defines y_{n+1} implicitly. For constant step size it has the form

$$y_{n+1} = y_n + h\beta_0 f(x_{n+1}, y_{n+1}) + h \sum_{i=1}^{p} \beta_i f_{n+1-i}.$$

The simplest example is the backward Euler method

$$y_{n+1} = y_n + hf(x_{n+1}, y_{n+1}).$$

The backward differentiation formulae (BDF) have a similar origin. Now the polynomial $P(x)$ interpolates to solution values:

$$P(x_{n+1-i}) = y_{n+1-i} \qquad i = 0, 1, \ldots, p,$$

and the differential, rather than the integral, form of the equation is used:

$$P'(x_{n+1}) = f[x_{n+1}, P(x_{n+1})] = f(x_{n+1}, y_{n+1}).$$

For constant step size h this results in a formula of the form

$$y_{n+1} = h\gamma_0 f(x_{n+1}, y_{n+1}) + \sum_{i=1}^{p} \gamma_i y_{n+1-i}.$$

The backward Euler formula happens to be one of the BDF.

These formulae are representative of classical methods using previously computed solution values. In the case of implicit methods such as Adams-Moulton and BDF there are the immediate questions as to whether they are well-defined, and how they are to be evaluated in practice. In the classical use this is always done by simple iteration. A predicted value $y_{n+1}^{(0)}$ is first formed with an explicit formula. The iteration method will be exemplified for the backward Euler method:

$$y_{n+1}^{(m+1)} = y_n + hf(x_{n+1}, y_{n+1}^{(m)}) \qquad m = 0, 1, \ldots. \tag{1.16}$$

It is easy to see that a sufficient condition for convergence is that

$$hL < 1.$$

It is necessary that

$$h\rho[f_y(x_{n+1}, y_{n+1})] < 1$$

where f_y is the Jacobian matrix of f and $\rho(M)$ is the spectral radius of the matrix M. The sufficient condition guarantees contraction in the norm being used. It is virtually the necessary condition in practice because only a very few iterations are allowed in the codes and convergence must be observed.

There is a variation on the use of the Adams-Moulton methods which does a fixed number of iterations, a common choice being one. An example is the use of the Euler method as predictor and one "correction" with the backward Euler method:

$$y_{n+1}^{(0)} = y_n + hf(x_n, y_n),$$
$$y_{n+1} = y_n + hf(x_{n+1}, y_{n+1}^{(0)}).$$

This variation results in explicit methods which behave similarly, but by no means identically, to the implicit methods.

The one-step methods were mentioned earlier. The most important classical examples are the Runge–Kutta methods. An s-stage method has the form

$$y_{n+1} = y_n + \sum_{i=1}^{s} c_i k_i,$$

where

$$k_i = f\left(x_n + ha_i, y_n + h \sum_{j=1}^{s} b_{ij} k_j\right), \qquad i = 1, \ldots, s.$$

The constants a_i satisfy

$$a_i = \sum_{j=1}^{s} b_{ij} \qquad i = 1, \ldots, s,$$

and the constants b_{ij}, c_i define the method. If $b_{ij} = 0$ for all $j \geqslant i$, the k_i are evaluated explicitly in the order $i = 1, \ldots, s$ and the method is explicit. Otherwise it is implicit. The Euler method (1.15) is also an example of an explicit Runge–Kutta formula and the backward Euler method is an example of an implicit Runge–Kutta formula. In the classical use of implicit Runge–Kutta formulae, they are also evaluated by simple iteration:

$$k_i^{(m+1)} = f\left(x_n + ha_i, y_n + h \sum_{j=1}^{s} b_{ij} k_j^{(m)}\right) \qquad m = 0, 1, \ldots.$$

A sufficient condition for convergence is $h\gamma L < 1$, where γ is a constant depending only on the formula.

The classical formulae, as exemplified above, are all explicit, or are evaluated by simple iteration. In every case, simple iteration converges if the step size satisfies $hL < \tau$ for a constant τ which depends only on the method and which is not particularly small. In the classical situation $hL \leqslant (b - a)L$ and L is not large, hence simple iteration does not pose a *severe* restriction on the step size. Obviously the situation is quite different if the Lipschitz constant is large. It is important to appreciate that a severe restriction on the step size arises *only* when $hL \gg 1$. Thus the problem must have a large Lipschitz constant, the solution must be stable, and the solution must be easy to approximate in the sense that the desired accuracy can be achieved with a step size h such that $hL \gg 1$. Later we shall explore circumstances leading to this situation. The restriction is one manifestation of the complex phenomena called stiffness. The difficulty of defining stiffness is evident here. For some kinds of methods there is no step size restriction of this kind at all. The ones that do have such restrictions suffer in varying degrees. Clearly stiffness depends on the method as well as the problem.

The key to solving problems with large Lipschitz constants is to resort to a more powerful iteration method for evaluating implicit formulae. The only popular way is to linearize the algebraic equations by the simplified Newton's method. In the case of the backward Euler method this is

$$y_{n+1}^{(m+1)} = y_n + h[f(x_{n+1}, y_{n+1}^{(m)}) + J(y_{n+1}^{(m+1)} - y_n^{(m)})]. \tag{1.17}$$

Here $J \approx f_y(x_{n+1}, y_{n+1})$. Newton's method does not use a fixed matrix J, rather it uses $f_y(x_{n+1}, y_{n+1}^{(m)})$ to calculate $y_{n+1}^{(m+1)}$. Generally the formation of approximate Jacobians is very expensive. For example, each iteration of (1.16) requires only one evaluation of f. In the common case of forming J by differences for a system of N equations, N evaluations of f are made just to form J. For this reason it is considered impractical to use Newton's method itself. In (1.17) one must repeatedly solve linear systems with matrix $I - hJ$. Simply decomposing this matrix into triangular factors is rather expensive, and the repeated substitution processes to solve for the $y_{n+1}^{(m+1)}$ are a cost by no means negligible.

The simplified Newton iteration is so much more expensive than simple iteration that it can be worthwhile only if the step size must be severely restricted to get convergence with simple iteration. Of course a great deal of research has been, and is being, devoted to minimizing the costs of this iteration. For example, the s-stage Runge–Kutta process involves a system of sN algebraic equations at each step. This can be reduced in various ways to dealing with systems of size N. The structure of the Jacobians can be used to reduce the costs of forming these matrices and of solving the resulting linear systems.

An important cost-saving device is to use an approximate Jacobian J for as many steps as possible. Convergence of the iteration and the rate of convergence depend on how well J approximates $f_y(x_{n+1}, y_{n+1})$ for the various x_{n+1}. As long as convergence is adequate, one should continue to use J. It is a fundamental practical assumption that the Jacobian is roughly constant for distances much larger than $1/L$. This is one reason why the class of problems of (1.11) is particularly significant for analyzing practical computation.

Codes intended for stiff problems are based on implicit methods for reasons we shall examine below. Until recently the simplified Newton iteration was used exclusively during an integration. Whenever hL is not large, this is grossly inefficient. The inefficiency is not as bad as it might seem at first because efficient reuse of an approximate Jacobian reduces the waste. Still, the inefficiency is significant when solving a problem which, e.g., does not have a large Lipschitz constant. Codes intended for non-stiff problems use simple iteration exclusively if they are based on implicit methods. This is grossly inefficient if the problem has a large Lipschitz constant and its solution allows a step size h such that $hL \gg 1$. It is worth remarking that the same formula, e.g., the backward Euler, might be used in both kinds of code. Recent research has considered how to recognize when simple iteration is feasible and efficient. There has been considerable success at recognizing and coping automatically with this manifestation of stiffness.

1.1.4. *Output*

So far we have considered approximate solutions only at mesh points. If the user wants results at specific points, the code will have to reduce its step size as necessary so that these points are in the mesh. A more efficient alternative is natural with the Adams formulae and the BDF. They are based on polynomial interpolants which can simply be evaluated at the places specified by the user. In this way the step size of the code is not affected by the output requirements. (Actually it is affected weakly in the popular codes for various software reasons.)

One might not think that output could *severely* restrict the step size, but for some methods it certainly can. Runge-Kutta methods of many stages must take "large" steps to compensate for the expense of a step. The classical extrapolated mid-point rule can be viewed in typical implementations as a family of Runge-Kutta methods of a *great* many stages. Obviously the methods described for problems with large Lipschitz constants do a great deal of work at each step. This pays off only when the step size is large. The fact that the BDF are not impacted by output is one reason for their popularity. Recent research has considered ways to "interpolate," or at least to weaken the effect of output on methods which produce values only at mesh points.

The effects of output depend on the formula and how it is implemented. Excessive output for certain methods has exactly the same effect as what is called stiffness. However, it is easier to see the origin of the difficulty and there is usually not a severe restriction on the step size. Furthermore, alternative methods not seriously affected by output have been widely available. We shall refer no more to this issue, but the reader should keep in mind that the *computational* problem (which includes specification of output points) may determine whether the step size possible is severely restricted and further that this depends on the formula and how it is implemented.

1.1.5. *Stability*

In our description of convergence we shall restrict our attention to one-step methods, more particularly to Runge-Kutta methods, in order to avoid technical details arising from use of previously computed solution values. Suppose the integration has reached (x_n, y_n). A one-step method produces

$$y_{n+1} = y_n + h_{n+1}\Phi(x_n, y_n; h_{n+1}).$$

Let us define

$$z_{n+1} = y(x_n) + h_{n+1}\Phi[x_n, y(x_n); h_{n+1}],$$

the value the formula would produce if it had available the correct solution value at x_n. A fundamental decomposition of the error is

$$y(x_{n+1}) - y_{n+1} = [y(x_{n+1}) - z_{n+1}] + (z_{n+1} - y_{n+1}). \tag{1.18}$$

The difference $y(x_{n+1}) - z_{n+1}$ in (1.18) is the local truncation error, lte, which measures how well the formula approximates the behavior of the differential equation. Notice that

$$y(x_{n+1}) = y(x_n) + h_{n+1}\Phi[x_n, y(x_n); h_{n+1}] + \text{lte}. \tag{1.18}$$

The difference $z_{n+1} - y_{n+1}$ measures the stability of the formula. We need to relate it to $y(x_n) - y_n$ in a way analogous to the stability results for the differential equation itself.

The step size h_{n+1} is chosen so that

$$\|\text{lte}\| \leqslant h_{n+1}\epsilon \tag{1.19}$$

for a tolerance ϵ provided by the user of the code. We shall see how this can be accomplished in the next section. Let $h_{\min}$ and $h_{\max}$ be the smallest and largest step sizes allowed by (1.19) for any x in $[a, b]$.

It is easy to establish the stability result for Runge-Kutta methods that

$$\|z_{n+1} - y_{n+1}\| \leqslant (1 + h_{n+1}£)\|y(x_n) - y_n\| \tag{1.20}$$

which comes directly from the Lipschitz continuity of the function Φ:

$$\|\Phi(x, u; h) - \Phi(x, v; h)\| \leqslant £\|u - v\|. \tag{1.21}$$

These results hold only for $hL \leqslant c$. When c is small, $£ \approx L$ and the factor $(1 + h_{n+1}£)$ in (1.20) approximates the analogous factor $\exp(h_{n+1}L)$ of the differential equation.

Combining (1.19) and (1.20) leads to

$$\begin{aligned}\|y(x_{n+1}) - y_{n+1}\| &\leqslant (1 + h_{n+1}£)\|y(x_n) - y_n\| + h_{n+1}\epsilon \\ &\leqslant (1 + h_{\max}£)\|y(x_n) - y_n\| + h_{\max}\epsilon,\end{aligned}$$

which then implies

$$\max_k \|y(x_k) - y_k\| \leqslant \frac{\epsilon}{£}\left[\exp\left(£(b-a)\frac{h_{\max}}{h_{\min}}\right) - 1\right]. \tag{1.22}$$

In the classical situation the stability of all Runge-Kutta methods is adequate and the convergence result (1.22) satisfactory. Let us consider the stability of the formulae when L is large. The constraint that hL be not large is not artificial as the simplest examples show. (It is also not due to implicit formulae as we shall see in a moment.)

In §3.2.2 the stability of solutions of (1.7)

$$y' = Jy + F(x),$$

when J is a (constant) matrix with a linearly independent set of eigenvectors is discussed. The difference of two solutions, $w = u - v$, satisfies $w' = Jw$. The (constant) change of variables $w = M\delta$, where the columns of M are the eigenvectors of J, uncouple the equations into the set

$$\delta_i'(x) = \lambda_i \delta_i(x) \qquad i = 1, \ldots, N,$$

where λ_i is a (possibly complex) eigenvalue of J. It is easy to verify that the same scheme works for the Runge-Kutta methods with the result that

$$\delta_{i,n+1} = R(h\lambda_i)\delta_{i,n},$$

where $\delta_{i,n} \approx \delta_i(x_n)$. Here $R(z)$ is a polynomial for explicit Runge-Kutta methods and is a rational function for implicit methods.

For the sake of simplicity we consider the case of $\mathrm{Re}(\lambda_i) \leqslant 0$ for each i so that $\delta_i(x)$ is non-increasing. It is noteworthy that this means that in a suitable norm

$$\|u(x+\Delta) - v(x+\Delta)\| \leqslant \|u(x) - v(x)\|. \tag{1.23}$$

The formula preserves this stability property of the differential equation only if $|R(h\lambda_i)| \leqslant 1$ for each λ_i. The region of absolute stability S of the method is the set of z such that $\mathrm{Re}(z) \leqslant 0$ and $|R(z)| \leqslant 1$. The step size h must be constrained so that $h\lambda_i \in S$ for each eigenvalue of $\boldsymbol{J}$.

All the explicit Runge-Kutta methods have finite regions of absolute stability. This follows because $|R(z)| \to \infty$ as $|z| \to \infty$ for these non-trivial polynomials. Indeed, all the classical explicit methods have finite regions of absolute stability. We point out that $L = \|\boldsymbol{J}\| \geqslant |\lambda_i|$ so that in the classical situation this stability constraint on h is not severe, as follows already from the general result for Lipschitzian problems. Notice that in the norm of (1.23), if the formula is absolutely stable with the step size h, the convergence results are much stronger than in (1.22):

$$\begin{aligned}\|y(x_{n+1}) - y_{n+1}\| &\leqslant \|y(x_n) - y_n\| + h_{n+1}\epsilon \\ &\leqslant \ldots \leqslant (h_1 + h_2 + \ldots + h_{n+1})\epsilon \\ &= (x_{n+1} - a)\epsilon,\end{aligned}$$

using $y(a) = y_0$. This reflects the additional information available about the stability of the differential equation.

Obviously, even to solve the very special class of problems (1.7) with $hL \gg 1$, we need methods with infinite stability regions. As already noted, we *must* resort to implicit methods. Very stable methods exist, e.g., the backward Euler method has

$$|R(z)| = \left|\frac{1}{1-z}\right| \leqslant 1 \qquad \text{for all } \mathrm{Re}(z) \leqslant 0,$$

but we then must pay the price of implicitness discussed in the preceding section. A great deal of effort has been devoted to finding formulae with good stability properties which can also be evaluated relatively cheaply. It should be appreciated that the formulae in common use do not have ideal stability even for the class (1.7). For example, all the popular BDF of orders at most 6 have infinite absolute stability regions. Still, some are finite for eigenvalues near the imaginary axis. It is quite easy, for example, to write down a problem for which the BDF of order 3 has no stability restriction, but the one of order 4 does, and the restriction suffered is *just* as severe as that of an explicit Runge-Kutta formula. Evidently stiffness depends on the formula as well as the problem.

In our view these stability constraints arising from the simple class (1.7) are necessary requirements for a reasonable numerical procedure. Of course we hope that the behavior for (1.7) is indicative of the behavior of the procedure for more general problems. A considerable amount of practical experience suggests that it *is*, provided that one is a little cautious. For example, the trapezoidal rule has $|R(z)| \leqslant 1$ for all z with $\mathrm{Re}(z) \leqslant 0$. However, $R(z) \to -1$ as $|z| \to \infty$ so that the formula is barely stable for $|h\lambda_i| \gg 1$. One expects, and sees, slowly damped oscillations in the numerical solutions. Furthermore, it is not difficult to write down problems only a little different from those of (1.7) for which this rule is unstable.

Naturally, stability results for classes broader than (1.7), e.g. (1.9) or (1.11), are of great interest. We mention one such result. Suppose we consider the class (1.9) with $l = 0$ so that no two solution curves spread apart. A Runge-Kutta method is called BN-stable if the approximate solutions also do not spread apart. The backward Euler method is an example of a BN-stable formula.

We have said that the step size should be restrained so as to lie in the region of absolute stability, but codes do not impose this restraint directly. It is interesting that the classical methods, as implemented in good codes, do bring this about. Some details are provided in §3.2.7. The commonly used procedures with infinite stability regions seem not to have stability difficulties often, but typical codes do not handle them well when they do occur. Providing that a method with adequate stability is used, the principal restraint on the step size is the iteration restraint discussed in §3.2.4. We do not wish to minimize the importance of a better understanding of stability or of the need for more stable formulae which are more easily evaluated, but very stable formulae which are apparently adequate for most practical problems are already known.

By resorting to implicit methods, one can avoid stability restrictions in a practically useful way. There are costs involved. One is the implicitness already discussed. Another is accuracy. Restricting the choice of parameters defining a Runge-Kutta method so as to assure good stability properties leaves less freedom to develop very accurate formulae. As a rule, the classical methods with poor stability are much more accurate than those at present being used because of their good stability. As a consequence, the classical formulae permit significantly larger step sizes (and hence are more efficient) when accuracy dominates the selection of the step size.

1.1.6. *Accuracy*

In §1.1.5 the local truncation error, lte, of a one-step method was defined by the relation

$$y(x_{n+1}) = y(x_n) + h\,\Phi[x_n, y(x_n); h] + \text{lte}.$$

By Taylor series expansion one finds that

$$\text{lte} = h^{p+1}\tau[x_n, y(x_n)] + O(h^{p+1}) \tag{1.24}$$

for sufficiently smooth f. In such a case the formula is said to be of order p.

For methods of order $p \geq 1$ it is always possible to choose h so that

$$\|\text{lte}\| \leq h\epsilon. \tag{1.25}$$

Indeed, for small h, or equivalently small tolerances ϵ (1.24) shows that the largest h satisfying (1.25) is approximately given by

$$h_{n+1} \approx \{\epsilon/\|\tau[x_n, y(x_n)]\|\}^{1/p}.$$

The smallest step size needed in the integration is approximately given by

$$h_{\min} \approx \left(\frac{\epsilon}{\max\limits_{[a,\,b]} \|\tau[x, y(x)]\|} \right)^{1/p}$$

and the largest,

$$h_{\max} \approx \min \left[(b-a), \left(\frac{\epsilon}{\min\limits_{[a,\,b]} \|\tau[x, y(x)]\|} \right)^{1/p} \right].$$

By way of example, both the forward and backward Euler methods are seen to have

$$\text{lte} = \frac{h^2}{2} y''(x_n) + O(h^3). \tag{1.26}$$

The general case of two-stage, second-order explicit Runge-Kutta formulae form a one-parameter family of formulae. In terms of the parameter $\alpha \neq 0$, they are

$$\Phi(x, y; h) = (1-\alpha) f(x, y) + \alpha f\left(x + \frac{h}{2\alpha}, y + \frac{h}{2\alpha} f(x, y) \right).$$

The local truncation error is

$$\text{lte} = h^3 \left[\left(\frac{1}{8\alpha} - \frac{1}{6} \right) y'''(x) - \frac{1}{8\alpha} f_y[x, y(x)]\, y''(x) \right] + O(h^4). \tag{1.27}$$

With the exception of output, none of the restrictions considered in this paper can be *severe* in the classical situation when $L(b-a)$ is not large. It is true that, for example, stability might restrict the step size some so that one needs to consider stability regions when designing a code for non-stiff problems, but there cannot be the kind of restriction that we call stiffness.

Some valuable information can be gleaned easily from the expression for the lte of a method. We are now interested in problems with solutions $y(x)$ which are smooth, but $f_y[x, y(x)]$ is "large." If the term τ in (1.24) involves f_y, it is often easy to see that the step size must be restricted so that hL is not large in order to yield the desired accuracy. This excludes some methods immediately from their use for problems with large L. For example, the second term in (1.27) will impose an unwelcome restriction on the step size when f_y is large.

The Adams-Moulton methods are implicit methods which use past solution values. When the step size is a constant h, the lte for the method of order k has

$$\tau = \gamma_k^* y^{(k+1)}(x),$$

where γ_k^* is a constant characteristic of the method. A popular variant of these formulae which is explicit was mentioned earlier. This variant has

$$\tau = \alpha_{k,0}^* f_y[x, y(x)]\, \gamma_k y^{(k)}(x) + \gamma_k^* y^{(k+1)}(x).$$

The coefficients $\alpha_{k,0}^*$, γ_k are constants. It is evident that this variant is unsuitable for problems with large $f_y[x, y(x)]$ despite whatever merits it might have in the classical situation.

These observations do not exclude some formulae which are, in fact, useless. For example, in the asymptotic expression (1.26) we do not see a difference between the explicit and implicit Euler methods. Still this is an easy way to see that some formulae cannot be helpful when solving problems with large Lipschitz constants.

Considerable insight can be attained from a study of the class (1.7) of problems

$$\mathbf{y}' = J\mathbf{y} + \mathbf{F}(x).$$

The representation (1.8) of the solution,

$$\mathbf{y}(x) = \mathbf{p}(x) + \sum_{i=1}^{N} \alpha_i \mathbf{v}_i \exp[\lambda_i(x-a)],$$

tells us how all integral curves behave qualitatively. Suppose the eigenvalues are ordered so that

$$\operatorname{Re}(\lambda_N) \leqslant \ldots \leqslant \operatorname{Re}(\lambda_i) \leqslant \ldots \leqslant \operatorname{Re}(\lambda_1),$$

where $(b-a)\operatorname{Re}(\lambda_N) \ll -1$ and if $\operatorname{Re}(\lambda_1) > 0$, then $(b-a)\operatorname{Re}(\lambda_1)$ is not large. The constants α_i are determined by the initial conditions. The terms corresponding to "large" $\operatorname{Re}(\lambda_i)$ vanish rapidly as x increases. We might say that the rapidly varying components decay out quickly so that the solution looks smoother and smoother. Of course, just when some term is computationally negligible depends on the norm and tolerance used.

Generally, we expect to need a small step size to get accuracy when the solution is changing rapidly. This arises from the fact that τ in (1.24) consists of various derivatives of the solution and of $\mathbf{f}$. When the solution changes rapidly, one expects τ to be relatively large, although this is not necessarily so. Here we see that

$$\mathbf{y}^{(m)}(x) = \mathbf{p}^{(m)}(x) + \sum_{i=1}^{N} \alpha_i \lambda^m \mathbf{v}_i \exp[\lambda_i(x-a)].$$

There is a region of rapid change near $x = a$ in which all derivatives of $\mathbf{y}^{(m)}(x)$ are large and a small step size is necessary. As x increases, the exponential terms dominate the λ_i^m factors so that the solution looks smoother and a larger step size is permissible. It must be understood that if $\mathbf{p}(x)$ itself is not easy to approximate, e.g., it is not smooth, then neither is $\mathbf{y}(x)$ and one cannot expect $hL \gg 1$ will be possible.

If $\mathbf{p}_n$ is the numerical result of solving (1.7) with initial value $\mathbf{p}(a)$ and $\mathbf{y}_n$ the result with $\mathbf{y}(a)$, the difference

$$\mathbf{y}_n - \mathbf{p}_n = \mathbf{w}_n$$

is the result of solving the homogeneous problem with the Runge-Kutta method. Uncoupling the equations again leads us to considering how well $\delta_i'(x) = \lambda_i \delta(x)$ is approximated by our Runge-Kutta method. We saw earlier that

$$\delta_i(x_{n+1}) = \exp(\lambda_i h_{n+1})\delta_i(x_n)$$

and

$$\delta_{i,n+1} = R(\lambda_i h_{n+1})\delta_{i,n}.$$

We have already considered when $|R(\lambda_i h_{n+1})| \leqslant 1$ in the stability analysis. We are led to the same consideration again on grounds of accuracy. When a solution component is negligible in the norm used, we want it to stay negligible in the numerical solution, too. The asymptotic expression (1.26) for the truncation error did not distinguish the forward and backward Euler methods. They certainly differ dramatically in the present situation.

In qualitative terms, $hL \gg 1$ is possible only when the solution $y(x)$ is easy to approximate for the method at hand. There is typically a region of rapid change after which the solution smooths out so that it becomes easier to approximate. Just how fast this happens and how smooth the solution becomes depends on the stability properties of the differential equation and on the norm and tolerance supplied by the user. With nonlinear problems the solution may go through other regions of rapid change. Van der Pol's equation undergoing relaxation oscillations is a familiar example.

1.1.7. *Error estimator*

There are a number of related ways in which large Lipschitz constants pose additional restrictions on the step size. Fundamental to them is the question of approximating $y'(x)$. With some methods this is a very natural task. Indeed, the Adams methods are best viewed as methods for approximating $y'(x)$, rather than $y(x)$. Traditionally the approximation

$$y'_n = f(x_n, y_n) \approx y'(x_n)$$

is used. The function f is usually evaluated at (x_n, y_n) anyway, so this approximation is often returned by codes as an approximate derivative. The error of the approximation is easily analyzed. For dimensional reasons, the error of the scaled derivative $hy'(x_n)$ is more relevant. Now

$$\begin{aligned} \|hy'_n - hy'(x_n)\| &= h\|f(x_n, y_n) - f[x_n, y(x_n)]\| \\ &\leqslant hL\|y_n - y(x_n)\|. \end{aligned}$$

If we know only that

$$\|y_n - y(x_n)\| \leqslant \mu,$$

the best we can say is that

$$\|hy'_n - hy'(x_n)\| \leqslant hL\mu.$$

Obviously, this approximation is satisfactory only when hL is of modest size. How can we generate a reasonable approximation when the Lipschitz constant is large? One way is to use only approximations to the smooth solution $y(x)$ and so avoid the large derivatives associated with local solution. For example, a Taylor expansion results in

$$hy'(x_n) = h\left[\frac{y(x_{n+1}) - y(x_n)}{h}\right] - \frac{h^2}{2}y''(x_n) + O(h^3),$$

which suggests the finite difference approximation

$$hy'(x_n) \approx h\left(\frac{y_{n+1} - y_n}{h}\right).$$

What is the error of this approximation?

$$\left\| h\left(\frac{y_{n+1} - y_n}{h}\right) - hy'(x_n) \right\| \leq \|(y_{n+1} - y_n) - [y(x_{n+1}) - y(x_n)]\|$$

$$+ \|y(x_{n+1}) - y(x_n) - hy'(x_n)\|$$

$$\leq 2\mu + \frac{h^2}{2}\|y''(x_n)\| + O(h^3),$$

if

$$\|y_{n+1} - y(x_{n+1})\| \leq \mu \text{ and } \|y_n - y(x_n)\| \leq \mu.$$

This is a perfectly reasonable approximation even when the Lipschitz constant is large.

The truncation error expressions for the Adams-Moulton formulae and the BDF involve only derivatives of the solution. In typical Adams codes these derivatives are approximated by divided differences of the $f(x_n, y_n)$ values. As we have seen, when the Lipschitz constant is large these values do not give a good approximation to $y'(x_n)$ and differencing them results in even worse approximations for higher derivatives. If the code should try step sizes too large for stability, these local truncation error estimates are affected; the higher the order of the formula, the more the estimator is affected. Indeed, propagated error due to instability "looks" to the code like a solution which is not smooth. Modern Adams codes, like ODE, consider several orders at a time and select that which appears most efficient. The result is that the order is lowered and the step size reduced until the computation is stable. This is used in ODE to diagnose stiffness when the code does a lot of work. The typical BDF code is intended for problems with large Lipschitz constants, hence must difference the y_n values to obtain reasonable estimates of the truncation error. In some codes this is not obvious because of the representation of the formula, but they are all doing much the same thing.

Although described in a variety of ways, all the popular procedures for estimating local truncation error in explicit Runge-Kutta codes can be viewed as taking each step with two formulae of different orders and estimating the error of the lower order formula by comparison. Obviously, this estimate is disturbed if the step size is such that either formula is unstable. The interesting behavior can be used to recognize stiffness. Basically, the situation is the same as with the Adams codes. Instability looks like inaccuracy arising from a solution which is not smooth, so that step size is reduced until the computation is stable.

1.1.8. *Summary*

It is assumed that the function f of (1.1) is continuous and satisfies the Lipschitz condition (1.3). This guarantees the initial value problem (1.1, 1.2) has a unique solution $y(x)$. In the classical situation when $L(b-a)$ is not large, this also implies that $y(x)$ is moderately stable. Step-by-step methods for approximating (1.1) numerically select at each step a step size h_{n+1} intended to satisfy the local truncation error requirement (1.19). There are other restrictions on the step size, too, but in the classical situation they cannot be a *severe* restriction except in the case of very frequent output for some methods and some implementations.

To consider problems with $L(b-a)$ large, we must assume $y(x)$ is stable. Some mathematical conditions guaranteeing this have been stated. Some solution curves must approach $y(x)$ very rapidly so as to cause L to be large, or, as we prefer to put it, in the reverse direction $y(x)$ is very unstable. There is really nothing new unless $h_{acc}L \gg 1$ for the step size h_{acc} yielding the desired accuracy. For this to happen $y(x)$ must be "easy" to approximate. Typically $y(x)$ has a region of rapid change in which $h_{acc}L$ is not large, but h_{acc} increases rapidly as $y(x)$ smoothes out in the course of the integration. All the classical methods must restrict the step size h actually used so that hL is not large in order to integrate even very simple problems stably. There are formulae which seem to have satisfactory stability properties, but they are all implicit. The classical way of evaluating implicit formulae also must restrict h so that hL is not large. This can be avoided by resorting to the expensive simplified Newton iteration. Implicit is the assumption that the Jacobian changes slowly along $y(x)$ so that it need not be evaluated "often." The classical ways the local truncation error is estimated also impose a restriction that hL be not large. However, there are estimation procedures which do not suffer such restrictions.

Except for the case of too frequent output, a problem is "stiff" in a region for a given code if the step size must be severely reduced from that value which would yield the requested accuracy. The classical situation has no stiff problems. We have described certain characteristics of the mathematical problem which must be present to have stiffness. Unfortunately stiffness also depends on the computational problem, e.g., norm and tolerance. Furthermore, whether a given code exhibits stiffness depends on the particular formula, how it is implemented, and how it relates to the particular mathematical problem. Apparently minor changes can change a severe restriction into none at all!

1.2. Classifications of other "stubborn" problem types (Aiken)

In addition to the stiff ODE problem class defined in the last section, there are a number of other related ODE problem classes which contain features that can severely restrict numerical computation:

(i) differential-algebraic
(ii) differential-difference
(iii) discontinuous
(iv) highly oscillatory
(v) partial differential (discretized)
(vi) large sparse
(vii) unstable (including two-point boundary value problems with sharp transients)
(viii) real time.

These are discussed briefly in the following sections. A ninth category:

(ix) reoccurring stiffness

is discussed in §2.2, associated with diurnal kinetics, and in §2.11, associated with lasing relaxations. See also Gear's §4.2 for a further discussion of most of these stubborn problem classes.

1.2.1. *Differential-algebraic*

Differential-algebraic equations occur in models of power systems, control systems and from application of the steady-state approximation or perturbation methods. They present particular problems in determining initial conditions, error estimation and step size selection (Gear, Hsu, and Petzold 1981). If the algebraic equations resulted from setting a derivative to zero, they can sometimes be explicitly solved individually and back-substituted; for example, in chemical kinetics of the mass action form this is often possible. If this elimination cannot be effected, the differential-algebraic set can exhibit infinite stiffness (in a sense). Some of these problems cannot be solved with stiff methods without extensive modifications in the error estimates and other strategies of the code; and sometimes they apparently cannot be solved at all by stiff methods (Petzold 1981b). See too her package DASSL (Petzold 1982a).

Liniger (1979) gives us multistep and one-leg methods for implicit mixed differential-algebraic systems. Soderlind (1980) has written DASP3, a program for partitioned stiff ODEs and differential-algebraic sets. Chua and Dew (1982) attack these mixed systems that also include discontinuities. Gross (1976) presents a method that makes special use of the structure in the differential-algebraic set; the nonlinear system is split into a stiff part with a sparse Jacobian and a nonstiff part. Datta and Martens (1974) investigate automatic step-size-selection techniques for a method tailored for this combination of equation types. Refer also to the algorithm of Starner (1976).

1.2.2. *Differential-difference*

Delay terms arise in lossless transmission line modelling, economic modelling, and ecological modelling, to name only a few areas. Brayton (1974) develops conditions for numerical A-stability for these systems. Bickart (1981) offers a program package for differential-difference systems. Van der Staay (1982) explores composite integration-interpolation methods. Bickart (1982b) determines F-stable and F(alpha, beta)-stable integration-interpolation methods. Weiderholt (1976) studies the stability of multistep methods for this class of mixed equation forms. Carver (1977) studies the efficient handling of discontinuities and time delays in ordinary differential equations. See also Roth (1980), Watanabe and Roth (1982), and Moore (1974).

1.2.3. *Discontinuous*

Discontinuities occur when certain variables pass through known critical points, but the time for this passage is usually unknown. The critical point may be just a switch when some criterion is satisfied. Nakhle and Rouz (1982) look at stiff problems that are also discontinuous; they solve a model of taxiing-induced vibrations in aircraft by a track runway.

Eillison (1981) classifies events that cause discontinuities as either a time event or a state event. Automatic detection of time events is straightforward; detection of state events is not, but is achievable on examples tried with a special program. Hedstrom, Rodrigue, Berger, and Oliger (1982) work on an adaptive mesh refinement to handle discontinuities in hyperbolic systems. Halin (1976) points out shortcomings of popular stiff software on discontinuities: he applies a "quasi-analytic" integration technique.

Halin and Benz (1982) use a variant of this technique in the formation of a new simulation language called RSCSP (Power Series Continuous-Systems Simulation Program). Their approach overcomes some of the disadvantages of numerical integration programs providing the solution at discrete points. Analytic manipulations of the model description at run time and a variable-order, variable step-size, semianalytic integration method allow efficient, accurate handling of special functions, such as those containing discontinuities.

De Boncker (1978) presents an automatic integration algorithm (in QUADPACK) that makes use of a nonlinear extrapolation technique to jump discontinuities. Hay, Crosbie, and Chaplin (1974) also have a means for detection of a discontinuity and readjustment of the step so that the discontinuity is at the step's end. Dew and Chua (1982) discuss the occurrence of discontinuities in a model of compressors loading in pipeline networks; here the problem results from directly including in the model discretely measured user loads. Li, Yang, and Ye (1982) present a method for ODEs with discontinuities that is reported to achieve high accuracy.

If the location of a singularity is known, it seems best to integrate toward it, rather than away from it. Brackbill and Saltzman (1982) of Los Alamos construct an adaptive mesh generator for singular problems in two dimensions.

1.2.4. *Highly oscillatory*

Gaffney (1982a) has completed a critical survey and testing of software (STRIDE, BLEND, STINT, and DIRK) for solving stiff highly oscillatory ordinary differential equations; none of these packages is given a very high mark on the test problem. Some highly oscillatory equations do not qualify as being stiff in the usual sense as there may not be any large negative real eigenvalues, although they can be "stubborn".

Petzold (1981a) presents a numerical method for this (non-stiff) highly oscillatory problem as does Fatunla (1980). Gear and Gallivan (1981) address automatic detection of highly oscillatory behavior, period determination, and efficient integration. See also Gallivan (1980), Gear (1980e), Miranker (1981), and §3.5.

1.2.5. *Large sparse*

Large systems are often sparse because there is usually direct interaction among only a relatively few variables in the set. This is true, for example, in large kinetic rate equations. Large systems resulting from discretization of PDEs are sparse with special structure. Special handling of the sparsity can both reduce storage and computation time. Curtis (1978) reviews the state of the art.

The efficiency of Gear's method depends heavily on the efficiency of matrix operations, particularly for larger systems. Hindmarsh (1974) provided a package, called GEAR, that allowed for the chord method with either a full Jacobian matrix or a diagonal approximation. Later, he added an option for banded matrices, as from PDEs, in GEARB (Hindmarsh 1977b). Both codes have been superceded by

LSODE (Hindmarsh 1980a). The banded structure also occurs in models of stage-wise processing (Tyreus, Luyben, and Schiesser 1975). Carver and Baudouin (1976) used GEARB to solve a stiff set of 242 ODEs modelling neutron kinetics and transport. They found that only the chord method allowed the solution to be reached in reasonable time, but storage was near to machine capacity and twenty seconds were required for each Jacobian evaluation and decomposition. They therefore added to the package a method from the Harwell subroutine library for solution of large linear equations, which stores only non-zero entries and uses a pivotal scheme optimal in some sense. The resulting package is called FORSIM. The greatest saving was probably in the numerical approximation to the Jacobian: the matrix is evaluated first to find non-zero elements, then the Jacobian is evaluated by perturbing as many of the individual variables as affect ony one derivative.

Sherman and Hindmarsh (1980) present GEARS, a package for stiff sparse ODEs, using YSMP (Yales Sparse Matrix Package, Eisenstat, Ursky, Schultz, and Sherman 1981); this was superceded by LSODES (Hindmarsh 1982a). The two main sparse techniques here are a special method for computing finite difference approximations to the Jacobian and YSMP non-pivoting Gaussian elimination linear equation solver.

Schaumberg, Wasniewski, and Zlatev (1980) and Zlatev, Schaumberg, and Wasniewski (1980) analyze implementation of a Gustavson storage scheme and a generalized Markowitz pivotal strategy for large stiff linear ODEs. Enright (1979) also examines stiff sparse linear ODEs and suggests modifications to GEAR for three classes of linear equations and four levels of structure. Rabbat, Sangiovanni-Vincente, and Hsieh (1979) mention that sparse matrix techniques have allowed time domain analysis of circuits with hundreds of elements, but large-scale integrated circuits present the challenge of solving thousands of active devices. Johnson and Barney (1976) look at several sparse techniques (SIMULT, IMP, and LINEQ4).

For strategies solving applications resulting in PDEs-turned-ODEs, see Iserles (1981), Sincovec and Madsen (1975), Melgaard and Sincovec (1981), Hunding (1980) and Karasalo and Kurylo (1981). The many other example implementations of sparse techniques include: Franke (1980), transient field problems; Watson (1976) for CSMP III; Gross (1976) for power systems; Carver and Boyd (1979) for mass-action kinetics; Dove and Raynor (1979, 1982), molecular dynamics; Enright (1980), structural mechanics; Sincovec, Erisman, Yip, and Epton (1981) for describer systems; Prasad and Huntress (1980), interstellar clouds; Atkinson, Carter, Darnael, Winer, and Pitts (1980), atmospheric pollution.

1.2.6. *Partial differential*

The most popular method for the numerical solution of parabolic and hyperbolic partial differential equations is the method of lines. In this method the spatial coordinates are finite differenced, resulting in a set of ordinary differential equations. For parabolic equations particularly, the ODE set is often large and very stiff. Some PDE packages that use the method of lines and a variant of GEAR are: FORSIM VI (Carver 1979), MOL1D (Hyman 1976), and DISPEL (Leaf, Minkoff, Byrne, Sorensen, Bleaknery, and Sullivan 1977). One-dimensional reaction-transport problems, for example, once discretized in space yield a set of ODEs with a banded Jacobian. EPISODEB recognizes this feature for reduced storage and faster execution. DISPL2 frees the user from discretization, requiring only a specification of the problem (Byrne 1981). It uses Galerkin's procedure together with B-spines. PDECOL (Madsen and Sincovec 1976) discretizes with the collocation method using B-splines also; it then uses GEAR. For a review of software for PDEs see Karplus (1982) and Machura and Sweet (1980) and §4.1.

Gaffney (1982b) identifies some pitfalls with the method of lines; he states that most packages in use with this method are for pure parabolic equations. If the equations are hyperbolic or mixed parabolic-hyperbolic, there can be difficulties with the discretization or in the numerical solution of the ODEs. Gaffney recognizes that PDEs tend to be even less classifiable than ODE sets, so that general software may not be a practical route to take.

Adaptive mesh refinement techniques are becoming ever more effective. They can be quite useful when there are regions on a spatial grid where components of the exact solution are discontinuous or have large gradients. Hedstrom, Rodrigue, Berger, and Oliger (1982) offer a technique for a one-dimensional gas dynamics problem. Bank, Sherman, and Wiser (1982) discuss the use of local mesh refinement for elliptic equations by the finite element method; they say it is effective if, among other requirements, the resulting linear system remains sparse. An arbitrary local mesh refinement scheme can lead to density; their scheme can preserve sparsity. See Rice (1982) and Boisvert and Sweet (1982) for a summary of elliptic software.

1.2.7. *Unstable*

Lindberg (1974b) discusses the fact that many stiff methods fail to detect inherent instability of an equation, particularly when large negative eigenvalues turn positive. Aiken (1982b) notes automatic methods can skip over a narrow interior region where positive eigenvalues become suddenly active, as in models for explosions. Hoppensteadt, Alfeld, and Aiken (1981) propose a numerical method that focuses on the positive eigenvalues. Brown (1978) examines the error behavior of multistep methods applied to unstable differential equations.

An obvious trick is to perform the integration in the opposite direction, particularly useful for two-point boundary value and cyclical problems (Aiken and Lapidus 1975); positive eigenvalues then become negative. This is not helpful when there are positive and negative eigenvalues present together, as in counter diffusion (Aiken 1982c). Also, note that reversing direction could change an unstable problem into a stiff problem.

1.2.8. *Real time*

There are some advantages for use of an analog computer for time-critical computation, that is for some real-time needs or for stiff computation. There has not been a widely accepted recent evaluation of this usefulness, however. Reasons that analogs are used rather infrequently include the basic equipment expense (more than a thousand dollars an integrator) for an analog in a laboratory that typically already has a digital computer. Also there does not yet exist software in the digital sense. Therefore one must "patch" an analog manually, although this could be based on a diagram written by someone else. Another huge inconvenience is the need to scale the problem so that all variables vary on the same normalized interval a normalized amount. For stiff ODEs this is equivalent to requiring that the user set up his problem in a dimensionless singular perturbation form. However, once this is done, the ease of parameter variation and the continuous graphical availability of the solution on an oscilloscope makes for an excellent environment to explore sensitivities of parameters and interactions of variables nonlinearly related.

Our laboratory owns a hybrid EAI PACER 100/580 system and an EAI 2000 analog connected serially to an Apple. We have on-going work in the evaluation of the analog for stiff computation, as occurs from mass-action kinetic models of certain combustion reactions. We have noted an upper limit on the degree of stiffness that the analog can handle (for example, eigenvalue spreads greater than about six orders of magnitude seem impossible to solve). However, there is likely to be a relation between a variable not solvable on the analog and one that is really not that important to the overall solution. This is often evident once the equations have been properly scaled, before being patched.

For solutions of stiff equations on a hybrid computer, reference is given to Kogan, Zykov, and Petrov (1980), Karba, Bremsak, and Kozjek (1980), Neundorf (1981), and El-Zorkany (1981); for stiff problems on an analog alone: Bernard-Weil and Duvelleroy (1978).

More recently the digital has further cut into the analog market by offering real-time digital computers, like the Applied Dynamics AD-10. Amoco, for example, now uses them for digital real-time control of processes; time constants as different as 10 000 to 1 can run ten times faster than real-time using fixed-step nonstiff integration techniques like the Euler or Adams–Bashforth second-order method. This type of computer appears to be particularly attractive for time-critical problems that are at most moderately stiff, very nonlinear and large.

Severe time-critical requirements for the conventional digital can lead to numerical difficulties as can the output requirements for real-time simulation. Refer to Gear (1977c) for comments on the use of the digital for real-time dynamics. See Shinohara and Ricci of C-E Lummus for experiences with three real-time digital process simulators.

Tzafestas (1982) has written a state-of-the-art review on the use of hybrid computers for solution of distributed parameter systems.

2

OCCURRENCE IN PRACTICE

2.1. Stiffness: who needs it? (Aiken)

How often does the stiff component of a model really need to be included? Usually it need *not* be present for the modelling objective (§7.1). However, the modeller may not know a priori what level of model detail is necessary to achieve his objectives, nor what aspects of the model lead to severe computational requirements. Once a full model has been formulated, sensitivity analysis or perturbation analysis can help determine how to reformulate the model with less detail. Even if the decision to remove the detail is not made, knowledge of the stiffness origin can aid the numerical solution (see §3.4 on partitioning).

Examples can be cited in which attention is *focused* on the stiff variables but they may no longer be part of a stiff problem. Consider, for example, steady-state mass transfer to a solid particle reacting with a component of a fluid flowing past it. If the particle is soluble in the fluid, it may occur that essentially all of the reaction takes place in the hydrodynamic boundary layer. Concentration change with respect to distance is stiff on the scale of the particle size but if interest is entirely focused on the boundary-layer scale, the problem is not stiff.

However, there are situations where inclusion of stiff variables is mandatory. For example, in chemical kinetics this may be necessary in order to provide mechanistic information (§7.3); this fundamental information could then be used, for example, to engineer increased production or suppression of some chemical species. In pollutant chemistry, it has been found (§7.2) that fast-acting low-concentration free radicals greatly influence the production of pollutants. In some control problems, the fast controller response (compared to the plant response) cannot always be neglected since the controller dynamics can greatly influence or govern the inherent stability of the plant–controller complex (see §2.5). Numerical difficulties arising from discretizing the spatial dimensions in partial differential equations may be unavoidable. One reason for this is related to the boundary-value nature of these problems: integration over large spatial dimensions may be necessary even if most changes occur over relatively short distances. See §8.4 (Question No. 6 on questionnaire) for further discussion.

2.2 Atmosphere

2.2.1 *Introduction* (*Aiken*)

Description of atmospheric phenomena involves transport with chemical reaction; thus stiffness can occur because the time scales of the reactions are much smaller than times for movement over distances. The transport equations themselves can be stiff because these distances can be very long indeed; and the chemical kinetic rate equations are usually large and usually stiff. Pollutant formation models naturally are stiff because highly reactive free radical transients are included in the model as they are intimately related to formation of trace quantities of pollutants. McRae, Goodin, and Seinfeld (1982) provided an excellent review of the numerical solution of the atmospheric diffusion equations for chemically reacting flows.

Brown (1980) used EPISODE (§4.3) for solution of a diurnal kinetics example and judges it too inefficient. Here some concentrations are small during the night, suddenly increase by orders of magnitude with dawn, vary slowly during the day with maximum values around noon, and then drop sharply at sunset. With the cycle time well known, Brown transformed the time coordinate to stretch it during the times of rapid change; in the transformed time the step size is more constant.

Miller, Meakin, Franks, and Jesson (1978) presented a one-dimensional model of atmospheric fluorocarbon-ozone photochemistry with transport. Chang, Hindmarsh, and Madsen (1974) solved over 14,000 ODEs to describe the effect of the supersonic transport on the ozone layer, using GEARBI (§4.3).

Bottenheim and Strausz (1980) modelled gas-phase chemistry in clean air as a prelude to including pollutants. Gelinas and Skewes-Cox (1977) explored tropospheric photochemical mechanisms. Baldwin, Barker, Golden, and Hendry (1977) studied rate parameter estimates in a photochemical smog kinetics model. Atkinson, Carter, Darnall, Winer, and Pitts (1980) did a modelling study for the gas phase NO_X-air photo-oxidation of toluene and the cresols. Pitts and Finlayson (1975) proposed various mechanisms of photochemical air pollution. Preussner and Brand (1981) applied a semi-implicit Euler method to photochemical smog kinetics. Carmichael, Kitada, and Peters (1980) applied the Galerkin finite element method to the movement of pollutants in the atmosphere. Kuhlman, Fox, and Jefferies (1978) studied the effect of CO on sulfate aerosol formation. Wallace, Springer, and Stedman (1980) modelled photochemical ozone and NO formation. Watkins (1981) solved an ionospheric model with several unknown initial conditions; these are chosen so to avoid the initial steep transient (see also §7.2.9 for problem approximations like this). Isaacson (1981) predicted extremely high wind speeds at ground level on the downstream side of a mountain range.

Difficulties with the use of steady-state approximation on the stiff reaction rate equations have been pointed out by Farrow and Edeslon (1974). Reasons could include: the necessity of having in the model radicals which directly affect the trace variables of interest, the reoccurring nature of the stiffness, or that the typically large size of the equation set makes choice of stiff variables more difficult. Dickinson and Gelinas (1976) performed sensitivity analysis on these types of equations to better understand what reactions are important in one specific example. Farrow and Graedel (1977) found that the steady-state approximations applied to some species but not to others for which they would have expected it to work; some species rates can have this approximation hold at various diurnal times.

Several very effective packages have been developed at Lawrence Livermore National Laboratory for the solution of differential equations modelling atmospheric phenomena. The motivation for these packages is described in the next section; the packages are discussed in §4.3.

2.2.2. *Applications to atmospheric models at LLNL (Hindmarsh)*

Initial value problems for systems of ordinary differential equations have long been a topic of great interest at Lawrence Livermore National Laboratory (LLNL). Stiff systems are particularly prevalent and are, of course, much more challenging. Applications giving rise to stiff ODE systems vary widely. One area that typifies the difficulties encountered is that of atmospheric computer models, on which a great deal of effort has been spent at LLNL since about 1971. These problems, in most cases, take the form of systems of partial differential equations in space and time, involving chemical kinetics and transport processes. A discretization process leads to large stiff ODEs.

The problem features discussed here are not at all unique to this particular application, nor to problems at LLNL, and the software developed for their solution was designed with full awareness of that fact. Thus, the ODE solvers used were designed to be as much general purpose as possible. However, a great deal has been learned in the intervening years about methods, algorithms, and software design for general ODE solvers. As a result, a new collection of initial value solvers has recently evolved at Livermore—the ODEPACK collection. There are currently five solvers in the collection. They are based on Adams methods (nonstiff case) and on the backward differention formula (BDF) methods (stiff case), and also on an inter-laboratory effort to set user interface standards for initial value solvers.

The computer modelling of various atmospheric chemistry and transport processes has been of great interest in the context of (a) ozone depletion from supersonic transport exhausts in the stratosphere (Chang, Hindmarsh, and Madsen 1974; MacCracken 1975), (b) stratospheric ozone depletion from terrestial fluorocarbon sources (Chang, Wuebbles, and Duewer 1976), and (c) regional air pollution in the lower atmosphere (MacCracken, Wuebbles, and Walton 1978), among others. In all cases, the mathematical model can be put in the form of a set of time-dependent PDEs in space (or, as a special case, as ODEs without space effects), in the concentrations c^i of the various chemical species of interest. These PDEs can be written

$$\partial c^i/\partial t = \nabla \cdot (\boldsymbol{D}\nabla c^i + \boldsymbol{V}c^i) + K^i + S^i, \qquad i = 1, 2, \ldots, p, \tag{2.01}$$

where $\boldsymbol{D}$ is a matrix of diffusion coefficients, $\boldsymbol{V}$ is the vector of mean atmospheric motion, K^i is the kinetic rate for species i, and S^i is its external source rate. All of these quantities can in general depend on time t, on the point in space, and on the dependent variable vector $\boldsymbol{c} = (c^1, \ldots, c^p)^{\mathrm{T}}$. However, in the applications of interest, $\boldsymbol{D}$ and $\boldsymbol{V}$ and the S^i depend only on space and possibly time. The number of space dimensions is usually 2, but sometimes 1 or 0. The kinetics rates K^i usually involve diurnally varying rate coefficients, corresponding to photochemical reactions in the system.

The particular kinetics system involved varies from one application to another, but in all cases there is a mixture of fast and slow reactions (large and small rate coefficients), and the rates are nonlinear. The fast reactions correspond to strong damping effects with short time constants, i.e., they cause stiffness. This means that the ODE problem will be stiff regardless of the transport processes in the model, although they can also contribute to stiffness. The small time constants from the kinetics are usually in the microsecond range or smaller, while the time range of interest is usually measured in years. Thus stiffness ratios exceeding 10^{12} are common in these problems.

Historically, the spatial operator in (2.01) has usually been treated by finite difference approximations, because of their simplicity and their long history of use. In 2-D, a rectangular mesh is used. However, there is nothing inherent in the problem to prevent the use of a finite element, collocation, or Galerkin treatment, although in the latter case the local integrals pose some difficulty.

To illustrate the spatial differencing process, consider the one-dimensional operator

$$Lc = \frac{\partial}{\partial x}\left(D(x)\frac{\partial c}{\partial x} + V(x)c\right). \tag{2.02}$$

On a mesh $x_1 < x_2 < \ldots < x_M$, the standard central differencing of (2.02), in terms of discrete values $c_i \approx c(x_i)$, is

$$Lc|_{x_i} \approx \frac{D(x_{i+1/2})\,c_x(x_{i+1/2}) - D(x_{i-1/2})c_x(x_{i-1/2})}{\frac{1}{2}(x_{i+1} - x_{i-1})} + \frac{V(x_{i+1})c_{i+1} - V(x_{i-1})c_{i-1}}{x_{i+1} - x_{i-1}}, \tag{2.03}$$

where, by definition,

$$x_{i+1/2} = \tfrac{1}{2}(x_i + x_{i+1}),$$
$$c_x(x_{i+1/2}) = (c_{i+1} - c_i)/(x_{i+1} - x_i),$$

and similarly at $x_{i-1/2}$. The differencing process is handled in a similar way for two-dimensional operators. Boundary conditions are similarly approximated in discrete form. For example, a zero normal derivative boundary condition, say

$$\left.\frac{\partial c}{\partial x}\right|_{x_1} = 0,$$

is approximated by setting $c_0 = c_2$ in the ODEs corresponding to $x = x_1$.

In this way, a set of p species PDEs becomes a coupled stiff system of ODEs whose size is

$N = p$	in 0-D (space-independent),
$N = pM$	in 1-D (mesh size M),
$N = pM_xM_y$	in 2-D (mesh size M_x by M_y).

It is clear that for a reasonably realistic model in 2-D, the system can easily exceed 10 000.

In 1971, the only reliable stiff system solver available (to us) was Gear's DIFSUB. This routine was installed, modified and improved, and used as the GEAR package (Hindmarsh 1974a). For 0-D problems or small 1-D problems, it worked well, and produced accurate answers with efficiencies that were quite impressive when compared with anything else we tried (including a number of nonstiff methods, before the nature of stiffness was recognized). The GEAR package (like DIFSUB), also had a nonstiff method option (using Adams methods), so that it was widely usable (and used) as a general purpose ODE solver. Still, the GEAR package gave little hope of solving the full atmospheric models, because it had to construct, and perform LU factorizations on, full $N \times N$ matrices, followed by backsolve operations for solutions of linear systems. If the ODE system is written

$$\mathrm{d}y/\mathrm{d}t = f(t, y), \tag{2.04}$$

and has a system Jacobian matrix

$$\boldsymbol{J} = \partial f/\partial y, \tag{2.05}$$

then the linear systems take the form

$$\boldsymbol{P}\boldsymbol{x} = (\text{BDF residual vector}), \tag{2.06}$$

where $\boldsymbol{x}$ is a correction vector and $\boldsymbol{P}$ is an approximation to $\boldsymbol{I} - h\beta_0\boldsymbol{J}$. Here $\boldsymbol{I}$ denotes the identity matrix, h is a time step size, and β_0 is a scalar depending on the current method order. The $N \times N$ system (2.06) occurs within a modified Newton iteration to solve the implicit BDF relation.

The next step was the realization that our problems had a very sparse Jacobian matrix, and that the LU method was extendable to sparse matrices. This was clearest if $\boldsymbol{J}$ was thought of as banded, and so a variant solver, with $\boldsymbol{J}$ treated as banded, was written and called GEARB (Hindmarsh 1977b). This approach could be applied to the atmospheric problems if the dependent variables in the ODE system were correctly ordered. The appropriate ordering is to group together all p values c^i at one node, then all the c^i at the next node, and so on, with a natural linear (in 1-D) or rectangular (in 2-D) ordering of the nodes. Ordering in the reverse manner (by nodes, then by species) may seem natural, but produces much larger bandwidths, as long as p is small compared to either of the mesh dimensions. In 2-D, the bandwidth is also minimized by numbering the nodes in the shorter direction first (if there is one). GEARB was used successfully in solving the 1-D models of interest, and small 2-D models.

The full 2-D models still seemed out of reach. Combining GEARB with reordering algorithms to reduce bandwidth failed because of the regular structure of the problem. A general sparse LU approach seemed inappropriate (though it was not thoroughly pursued) because matrix fill-in would result in much the same storage costs as for the banded treatment. Inasmuch as the matrix elements in (2.06) are rather easily generated, the idea of iterative linear system methods was a natural next choice. For example, successive overrelaxation (SOR), with careful attention to the choice of relaxation parameter, was known both theoretically and experimentally to do well on linear systems based on transport PDEs of the type (2.01) with a single species and no kinetics. The implementation of SOR for (2.06) requires that $\boldsymbol{P}$ be written as

$$\boldsymbol{P} = \boldsymbol{L} + \boldsymbol{D} + \boldsymbol{U}, \tag{2.07}$$

with

L strictly lower triangular,

D diagonal and nonsingular,

U strictly upper triangular.

so that one can generate (or access from memory) the elements of D, and that one can easily generate matrix-vector products $(L + U)x$. Specifically, for a given relaxation parameter ω, the νth SOR iteration in the solution of $Px = r$ is given by

$$(D + \omega L)^{\nu+1} = \omega r + (1 - \omega)Dx^\nu - \omega Ux^\nu. \tag{2.08}$$

This can be rephrased as

$$x^{\nu+1} = (1 - \omega)x^\nu + \omega z,$$
$$Dz = r - Ux^\nu - Lx^{\nu+1}. \tag{2.09}$$

This pair of equations appears circular, but is not; z is computed and x^ν updated to $x^{\nu+1}$ one component at a time according to (2.09).

For the atmospheric problems, the coupling induced by the kinetics does not even come close to satisfying the conditions needed for ordinary SOR. Thus, to have any hope for convergence, it is necessary to treat the c vector (of length p) as a unit, and use a block-SOR algorithm. This results from replacing scalars in SOR by matrix or vector blocks of size p. Thus one must write

$$P = L + D + U, \tag{2.10}$$

with L strictly lower block-triangular,

D block diagonal and nonsingular,

U strictly upper block-triangular.

One must again generate products $(L + U)x$, and one must generate D and be able to solve $p \times p$ systems with the diagonal blocks in D. This can easily be done by forming and using LU factorizations of those blocks. Equations (2.09) also provide an algorithm for the block-SOR iteration, by updating x^ν in blocks (from first to last).

For a 2-D atmospheric model, a special block-SOR variant of GEAR was written and later made into a general-purpose solver, called GEARBI (Hindmarsh 1976c), BI denoting Block-Iterative. This was soon modified to take advantage of Large Core Memory (LCM) storage on the CDC-7600 computer, resulting in a package called GEARBIL. The latter stores most of the large data arrays in LCM, the largest being the one containing the matrix D in (2.10) (and later its LU decomposition), of size $p^2M_xM_y$. These two solvers, as they stand, are still somewhat specialized to the atmospheric models, in that L and U matrices in (2.10) have a regular block structure (in $p \times p$ blocks), with each block being a scalar multiple of the $p \times p$ identity matrix. (The latter is due to the fact that D and V are independent of c in (2.01). However, it is a straightforward matter to modify the codes to accommodate a more general coupling, and this has been done in at least two instances.

The GEARBIL package was used for a variety of studies of kinetics in the stratosphere, including ozone depletion from SSTs (Chang, Hindmarsh, and Madsen

1974; MacCracken 1975) and from fluorocarbons (Chang, Wuebbles, and Duewer 1976), with the two dimensions being altitude and latitude. In most cases, the mesh sizes were 37 × 44 (1628 spatial nodes), and the number of chemical species p was 9, giving an ODE system size of $N = 14\,652$. GEARBIL was subsequently used in several similar models developed for regional air quality calculations (MacCracken, Wuebbles, Walton, Duewer, and Grant 1978), with the third dimension (altitude) accounted for by assuming uniformity between the ground and the temperature inversion layer. One of these models, LIRAQ2, is currently used to model the San Francisco Bay Area air layer as an ongoing pollution control tool of regional government.

In the stratosphere studies, one of the lessons learned concerns the tolerance on the part of the ODE solver for errors in the Jacobian matrix, when supplied by the user. For some time, routines to supply $\boldsymbol{J}$ were written by hand, and therefore subject to error, especially for these complicated systems. When plots of step-size history were generated, and in one case published (Chang, Hindmarsh, and Madsen 1974, p. 58), they often showed great irregularities, as if frequent instabilities or problem discontinuities were forcing drastic reductions in step size periodically. Subsequently, with the aid of automated Jacobian generators, errors in $\boldsymbol{J}$ were found and corrected, and reruns of these problems showed a remarkably smooth, nearly monotone, growth in step size.

Independently of the development of complex transport models in 1-D and 2-D, the kinetics mechanisms of the lower and upper atmosphere were studied. In particular, the detailed kinetics processes of the stratosphere, with diurnal effects included, required much effort in both the model-building and the numerical solution phases, even in the absence of spatial transport, i.e., in 0-D. The stiff ODE systems that arise are further complicated by the fact that rate coefficients for photochemical reactions follow a nearly square-wave pattern in response to sunlight (Gelinas 1974).

In many cases, these diurnal kinetics problems were found to cause great difficulty for the GEAR package, sometimes causing it to crash irrecoverably. It was found that this is due to the buildup of errors associated with the use of fixed-step BDFs with interpolatory step changing. As a result, variable-step forms of the BDFs were developed, jointly with Byrne (Byrne and Hindmarsh 1975). These were implemented in a general solver called EPISODE (Hindmarsh and Byrne 1977; Hindmarsh and Byrne 1976a), along with variable-step Adams methods for nonstiff problems. EPISODE resembles GEAR externally, but differs internally in all details associated with integration coefficients, error estimation, and step selection. EPISODE was found to be able to handle the diurnal problems quite reliably, although it was usually somewhat less efficient than GEAR on problems with smooth solutions (Byrne, Hindmarsh, Jackson, and Brown 1977). A banded Jacobian variant EPISODEB (Byrne and Hindmarsh 1976) was also written to accommodate diurnal kinetics–transport problems.

After most of the model building for these atmospheric problems was completed, the question of alternative space discretizations was nevertheless studied to some extent. Madsen and Sincovec (1979) showed that collocation methods could be used quite effectively on 1-D problems of this type, and developed a general-purpose PDE package, called PDECOL, from that idea. In the process, it was clear that a different type of ODE solver was needed, namely one which would treat linearly implicit systems,

$$A\dot{y} = g(t, y) \tag{2.11}$$

(A being a square matrix), in a direct and efficient manner. To this end, another GEAR variant, GEARIB, was written (Hindmarsh 1976a), IB standing for Implicit systems with Banded matrix treatment, and a modified form of this is used in PDECOL. Later, an analogous EPISODE variant, EPISODEIB, was also written (Byrne 1979). These two ODE solvers are intended mainly for the case of a non-singular A matrix in (2.11), which is the most common situation in PDE-based problems, but can also be used in the singular case, if used with caution.

See §4.3 for the further evolution of these packages (the ODEPACK collection) and §4.3.7 for a simple atmospheric model that illustrates the use of the various packages.

2.3 Biology (Aiken)

Most chemical reactions occurring in living species involve a catalyst that facilitates proper orientation of the big biological molecules. This catalyst is called an enzyme. These reactions are stiff for one of two reasons: (i) there is typically a large scale difference between the concentration of the reactant, called the substrate, and the enzyme, or (ii) there is a large difference in the rate constants (see §7.3).

It is rare to find enzyme models where the Michaelis–Menten approximation has not been used. The reason for this seems to be primarily an experimental limitation: the enzyme–substrate complex, typically the stiff intermediate variable, is like the enzyme itself present in very small quantities and may not differ greatly from measurable characteristics of the enzyme or the substrate (§7.3.1). It was measured spectroscopically in the mid-twentieth-century and allowed the first complete model of enzyme kinetics with all parameters specified. There seems to be a growing interest in obtaining this (stiff) model detail today (Hiromi 1982; Kondo, Nakatani, Matsuno, and Hiromi 1980). Kinderlehrer and Ainsworth (1976) have written a program to predict detailed enzyme mechanisms.

In addition to enzyme kinetics, stiffness appears in pharmocological compartment models of drug response. Gehring and Blau (1977), of Dow Chemical, modelled dose response to suspected carcinogens and noted the initial large transients. Bloch, Ingram, Sweeney, and Dickinson (1980) found that the processes of reversible binding of drug to protein occur rapidly on the time scale of the drug disposition in the body. Jackson (1980) used a version of GEAR for the kinetic simulation of anticancer drug interactions. Perelson and DeLisi (1980) encountered stiffness with receptor clustering on a cell surface. Perelson (1979) found that an antigen will attach to and detach from the surface of a cell many times before cross-linking occurs. Karba, Bremsak, Kozjek, Mrhar, and Matko (1980) used hybrid computer simulation to overcome stiffness in drug pharmacokinetics.

Loomis, Rabbinge, and Ng (1979), in modelling crop physiology, found that while they were interested in crop growth over a period of many days, they had to use a time scale of hours to include diurnal events or a scale of minutes or seconds if cellular processes were to be included. Chu and Berman (1974) developed a program using an exponential method for modelling and simulation of complex biological systems. Gottwald and Wanner (1982) compared various methods for stiff differential equations occurring in biology. Hunding (1980) came across stiffness and chemical hysteresis in simulating a biological cell or early blastula.

2.4 Combustion (Aiken)

This is a special area of reaction with transport and of chemical kinetics. Stiffness can come from the time scale differences between reaction and transport or from the stiff kinetic system or both.

Combustion kinetics could be defined simply as reactions of the oxidative type, but the primary reactions of interest are initiated by combination of oxygen with hydrocarbons. From this class the most important are the chain branching type or autocatalytic reactions that proceed ever faster as fuel is consumed: explosions. Most combustion applications involve explosive reactions confined in space; the spatial location of most intense reactivity is termed the flame heart. Aiken (1982b) explores the definition of the term "explosion" and develops criteria for the degree of explosive activity for the oxygen–hydrogen problem (see also §5.2.1).

Stiffness in explosive reactions is different from that in most other kinetic systems in that fast transients do not occur only for initial times but usually appear later in the transient. Thus codes that are taking giant time-integration steps can jump over the explosive zone where most of the action is. Another difference is that the stiff variables change during the transient. Typically, at least three major time scales are important in these systems: initiation, explosion, and termination. For these reasons stiffness in combustion kinetics can be particularly severe; only recently has the steady-state approximation been successfully and automatically applied to these reactions (Aiken, 1982b). Hoppensteadt, Alfeld, and Aiken (1981) present a projection method that focuses on the positive eigenvalue during the explosion. Pratt (1979) has investigated methods that make use of the special structure of the kinetic equations. See also White and Seider (1981).

Addition of the spatial variable further complicates the potential numerical problems. The flow in which combustion occurs is usually turbulent with intermittent changes on time scales different from the kinetic time scales. Stochastic methods are promising for overcoming this stiffness (Chorin 1980). Kansa (1981), of the LLNL, has combined some aspects of block implicit PDE schemes with stiff ODE solvers for this problem; this was partially motivated by the experience that block implicit methods, although unconditionally stable for linear equations, were found to have inadequate stability properties for the severe nonlinear Arrhenius temperature dependence of the reaction rate.

Operator splitting techniques are often used for the numerical solution of multicomponent gas mixtures undergoing rapid reaction. They offer better storage economy than fully implicit schemes and better stability properties than explicit schemes; their low-order accuracy is appropriate, and they allow flexibility and modularity in the overall numerical attack. Karasalo and Kurylo (1981), also of LLNL, discuss in this context ways to reduce overall computation time by improving the efficiency of the stiff kinetics step at each grid point, where by far the most time is spent. They employ a stiff ODE package (like GEAR) but with three modifications. First, they avoid repeated evaluations of the Jacobian at each mesh point by storing selected historical values from adjacent mesh points. Second, they allow step sizes and order to vary more frequently since this is not the expensive feature. Third, they artificially impose that concentrations remain nonnegative during prediction and corrector iterations. This last point is an interesting one as it was found that small negative concentrations that can result from rounding or truncation cause stability problems that require a local error bound smaller than that otherwise sufficient for a given global accuracy.

Otey (1978), of Sandia, formulated a test problem with combustive reaction and diffusion to make comparisons among solutions by the method of lines, linearized block tridiagonal procedures, and other techniques. He found that the block diagonal implicit procedure was by far the best in all variants of the test problem, including ones for which the system was quite stiff due to stiffness in the kinetic equations. Wendt, Martinez, Lilley, and Corley (1979; see also §5.2.2) present methods for solving stiff boundary-value problems with combustion and diffusion in a pore; the reaction occurs on the surface of the pore. A variable grid mesh is used so as to be finer at the pore mouth where concentrations change fastest. Sundaresan and Amundson (1980) also report the stiff nature of this problem.

Oran and Boris (1981) present a comprehensive examination of modelling and simulation of combustive flow problems. They also mention that kinetic ODE stiffness within a spatial model cannot be handled adequately for realistic models because the usual matrix inversions would be too expensive for a large number of chemical species and many grid points; storage is also an obvious problem. These authors further review the complex nature of modelling and simulating turbulence for reactive environments.

Sandusky, Bellan, and Ohlemiller (1979) advocate finite element techniques for combustive transport. Westbrock (1978) offers an improvement in the operator splitting method. George and Harris (1977) report stiffness from a model of in situ oil shale retorting. See also Scaccia and Kennedy (1974), McDonald (1979), Choi and Churchill (1979), and Lowe, Wall, and Stewart (1977).

2.5 Control (Aiken)

Many engineering control systems can be modelled by the form:

$$\mathrm{d}\boldsymbol{x}/\mathrm{d}t = \boldsymbol{A}\boldsymbol{x} + \boldsymbol{B}\boldsymbol{z},$$

$$\epsilon \mathrm{d}\boldsymbol{z}/\mathrm{d}t = \boldsymbol{C}\boldsymbol{z} + \boldsymbol{B}\boldsymbol{u},$$

where the first equation represents a large linear multivariable plant, and the second represents a multivariable actuator. ϵ is a small parameter that indicates a fast controller response in comparison to the plant-variable time-scale. The controller might be electrical and the plant mechanical or involving fluid transport. In general ϵ cannot be neglected (set equal to zero) since the presence of the controller dynamics can govern the inherent stability of the plant–controller complex; see Porter (1976), however, for some conditions for which this simplification is permissible. Shimizu, Yoshihara, and Watanabe (1980) describe some stiff nonlinear control problems.

This singularly perturbed simple linear form has been exploited by Khalil and Kokotovic (1980) in the design of decentralized feedback controllers. Anderson (1980) offers a time-varying transformation to separate the fast from the slow modes. Dontchev (1974) explores by sensitivity analysis optimal control systems with changes in order. Womble (1976) looks at some further approximations to Ricatti equations having fast and slow modes.

Any differential equation can be considered stiff if a solution is required in a short enough computing time. Real-time aerospace control applications can have this feature. Bulirsch and Branca (1974) mention that for an optimal deceleration maneuver an Apollo-type vehicle would need information in about one second and would allow up to one per cent error. Gear (1977) discusses the conflict between real-time and software; he says that real-time implies that implicit methods cannot be used in the usual sense and presents some of the semi-implicit methods. See Hiestand and George (1976) for other stiff aerospace applications.

Ojika, Nishikawa, and Okudaira (1979) give a "time-decomposition" algorithm for a stiff two-point boundary-value problem applied to nonlinear optimal control problems.

2.6 Dispersed phases (Aiken)

Consider a liquid or solid phase dispersed as droplets or particles in another gas, liquid or solid phase. If the size distribution of the dispersed phase is broad, stiffness can result from models that include heat or mass transfer, particularly complex when reaction is also occurring in the dispersed phase. This is because it is much easier to transport mass and heat to smaller sizes, roughly proportional to the reciprocal of the effective diameter. For example, Kayihan (1980) describes a model and solution for coal devolatilization in which heat is transferred preferentially to the smaller particles that therefore devolatilize relatively rapidly and cause severe numerical simulation problems; he estimated GEAR would require 10 hours on a CDC CYBER 73 computer and used an algebraic approximation that preserved the expected solution features. As usual, a modelling choice must be made as to how small a size to include. The size distribution functions typically have a skewed maximum with long tails—particularly in the direction of the smaller sizes. The modeller must often determine a cut-off point for small particles; smaller than the cut-off would cause numerical problems but ideally would not contribute significantly to the solution variables of interest.

Wall and Anlansson (1980) use a version of the GEAR package to solve a model of stepwise micelle association. Lahey, Cheng, Drew, and Flaherty (1980) also use this package for modelling bubbles flowing through a nozzle.

See also the sections in this review on heat transfer and reactions. Bubble columns and spray reactions can exhibit this type of stiffness as can processing crushed shale or coal. The problem can be particularly severe in the case of in-situ solids processing for which the particle sizing from underground explosions, for example, is rather uncontrollable.

2.7 Electronics (Aiken)

The time domain analysis of electronic circuits requires the solution of nonlinear algebraic–differential equations. Implicit integration methods and sparse matrix techniques have made possible analysis of circuits containing hundreds of active devices. Advances in large-scale integrated circuits have indicated potential for analysis of thousands of active elements. Hybrid simulation is an interesting concept that applies different methods to sections of the circuit that require different accuracy, but effects of interaction among the subsystems can be difficult to access a priori. The concept of latency, rather like a temporary steady-state approximation, has also received attention in this area. The relationship between latency and the numerical method has been explored by Rabbat, Sangiovann-Vincente, and Hsieh (1979), of IBM Data Systems Division. See also Tadeusiewicz (1981).

Power system dynamic response involves the solution of large differential-algebraic equations. The differential equations model the dynamics of the machines and their control systems while the algebraic equations model the network steady-state relationships. Gross and Bergen (1977) pursue this combination by partitioning the set into a nonstiff part and a stiff part with a sparse Jacobian matrix.

Resonant circuits, time variant and time invariant, have been studied by Ruehli, Brennan, and Liniger (1980), of IBM T. J. Watson Research Center. A-contractive arc methods were shown to perform well for both types of circuits. Oscillatory nonlinear circuits with a finite number of continuous derivatives has been the subject of work by Hajj and Skelboe (1979). Zein (1980), of IBM Data Systems Division, discusses the use of "sensitivity circuits" for the transient analysis of periodic circuit behavior.

A range of integration algorithms has been tested on some model problems of large ODE sets for power system dynamics by Humpage, Wong, and Lee (1974). Methods for fast contingency analysis at a power systems control center are advanced by Chamorro (1981).

Alvarado (1979) reports some results on stiff transient stability analysis of circuits. Weaver (1977) gives a stiff model for radiation-induced bulk electrical conductivity in insulators. Covello and White (1977), of the U.S. Air Force Weapons Laboratory, discuss stiffness in investigating radiation response characteristics of networks. Charge transfer in a nonlinear stiff model of charge-coupled devices has been simulated by McKenna and Schryer (1975), from Bell Labs. Warner and Wilson (1980), also working at Bell Labs, use some analytic transformations to help lessen the stiffness in their equations related to the fabrication of narrow-channel MOS transistors. Gambart–Ducros and Maral (1980) concerned themselves with the stiff differential equations that arise from some computer-aided design techniques; see also Dietze and Reibiger (1978). Von Pragenau (1981) reported his own method for greatly reducing the computation time of stiff digital filters. Stiff nonlinear switching circuits are attacked by Boness (1979) and specifically switching surges by Tripathy and Rao (1978).

See §2.5 on control systems for further stiff applications involving electronics.

2.8 Fluids (Aiken)

Stiffness occurs in spatial coordinates within homogeneous fluids with sharp changes in physical properties or from abrupt obstructions in the flow path. Compressible flow with compression and rarefaction (shock) waves, reflection, flow reversal and choked flow all can lead to numerical problems.

The method of characteristics is "characteristically" used on problems of inviscid flow because it naturally handles discontinuous derivatives as it follows waves but cannot be used on viscous shock layer equations. Srivastava, Werle, and Davis (1979) present a finite differencing scheme for viscous flow past blunted cones, where derivative discontinuities are encountered at the sphere-cone juncture point. To avoid large truncation errors associated with these points, differencing across the discontinuity is carefully avoided. The method of characteristics can become too expensive on inviscid flow problems to follow long term transients involving shock waves; Carver (1980) gives a spatial discretization which utilizes the directional aspects of the method of characteristics.

Blottner (1980), from Sandia National Laboratory, uses a variable grid approach to solve turbulent boundary-layer flows that involve jumps in viscosity and density. Blottner has used both coordinates stretching with uniform grid in the stretched coordinate as well as discontinuous grid spacing that is effective for discontinuous changes in variables. MacCormack and Paullay (1974), of NASA Ames, provide a study on the effect of the mesh spacing on inviscid supersonic shock flows. Bell, Shubin, and Solomon (1982), of the Naval Surface Weapons Center, develop a fully implicit shock tracing method that uses an implicit form for both interior and boundary points; the location of the shock (free boundary) is also treated implicitly and coupled with all the other unknowns. Stewart (1979), of the Atomic Energy Laboratory of France, examines a model-oriented numerical method for solving flow with sharp changes in phase as occurs in cooling water superheated locally. Other variable mesh approaches are shown by Yanenko, Lisseikin, and Kovenia (1979) for boundary layer shear flows.

The PDEs that model unsteady flow in one, two, or three dimensions can yield stiff ODEs when spatially discretized; this technique is called the method of lines (see §3.7). This will be the case when there are many spatial grid points compared to the time steps that one would like to use. A large number of spatial mesh points can result from tight spatial coupling or simply from "long" distances to be covered. The method of lines can be attractive because of its programming simplicity but may not be as efficient as finite differencing for problems without tight spatial coupling (Kurtz, Smith, Parks, and Boney 1978).

Madala and Piacsek (1977), of the U.S. Naval Laboratory, have studied numerically the response of oceans to weather changes. They avoid small time steps associated with fast moving surface gravity waves by dividing the flow into baroclinic and barotropic vertically averaged modes; the baroclinic waves are treated explicitly and the barotropic waves implicitly (even so, computation times reach over 60 hours on a large computer).

See also Coleman, Mesick, and Darby (1977) for averaging methods applied to stiff circulatory flow and Gersting (1980) for the Orr–Somerfield flow approached as an initial-value problem. Leventhal (1982), of Gulf Science and Technology constructs an implicit method of exponential type for diffusion-convection problems with steep gradients, as occur, for example, when the cell Reynolds number is large. Issacson (1981) looks at the mountain wind problem and suggests a filtering scheme and a hybrid method for handling shocks in the atmosphere. Oey (1982) examines some implicit schemes for hyperbolic equations with a large Courant number.

Chua and Dew (1981 and 1982) designed a variable-step integrator for the simulation of gas transmission networks. Rapid changes in the solution were due to the incorporation of a discontinuous model of consumer demand and from the on–off control of compressors. The performance of a stiff integrator is compared with that of the British Gas simulation program PAN on a large system of equations.

2.9 Heat Transfer (Churchill)

2.9.1 *Introduction*†

Stiffness in heat transfer originates physically in one of two ways: sharp changes in the thermal environment or large differences in the rates which components of the system can transfer heat. The first class could also occur in a boundary-value problem with the sharp changes represented in the boundary conditions. A realistic model for these boundary conditions is a tough problem in itself since discontinuities do not exist in nature, or at least have not been measured; for example, step changes in temperature can result in infinite heat fluxes, certainly not observed. The other class of stiff heat transfer problems originates from differences in heat capacities or by size differences among the components. For this category see also §2.6 on dispersed phase transport where, for example, heating of coal particles for pyrolysis is discussed.

Krishnan and Sastri (1978) solved the thermal entry length problem for high Prandtl numbers, that is from large differences in heat flow by convective versus conductive means. Mazhukin, Uglov, and Chetverushkih (1980) dealt with stiffness that occurred through large spatial temperature differences created through laser irradiation of targets and interaction with the resulting plasma above the surface. This type of stiff heat problem commonly comes from nonisothermal chemical kinetics coupled in a model with transport.

Mention also should be made of the stiff set of ODEs that results from discretization of the PDEs that describe unsteady heat transport. See, for example, Bushard (1976), who solved the heat conduction equation with the method of lines. Wood (1977) discussed the solution of the stiff equations that result from a finite element discretization of the heat conduction equation.

Distillation is a chemical engineering heat transport process that can result in stiff numerical models because of, for example, differences in the liquid hold-up in the boiler at the bottom of the column compared to the much smaller hold-up on the plates. If dynamics of the vapor traffic is included together with that of the liquid, stiffness may occur through the great difference in heat capacities between the liquid and the gas. Tyreus, Luyben, and Schiesser (1975) examined stiffness in a specific model of a distillation tower and found it became more

† Introduction by Aiken.

severe the more difficult the separation (from high purity requirements or from the components to be separated being similar in their volatilities). Billingsley (1982) reported that stiffness in a packed distillation example is the result of "boundary-layer behavior in the mole fractions at the top and bottom of the column." An adaptive semi-implicit Runge–Kutta algorithm was used by Prokopakis and Seider (1982) in a model in which the rapidly changing liquid flow rates were decoupled in a sense from the relatively slowly changing mole fractions. Gallun and Holland (1982) reported use of a GEAR variant for stiff differential–algebraic sets of distillation equations. See also §2.5.

Stiffness resulting from heat capacity differences commonly occurs in reactors which contain two or more phases. For example, a reactor tube containing packing that is processing a gas will lead to stiffness if the dynamics of the temperature changes in the solid and gaseous phases are included in the model.

2.9.2 *Heat exchangers*

The interchange of energy between two or more confined streams can be represented by the corresponding number of ODEs for most geometrical configurations in so far as the effect of variable transport properties can be lumped. An example of an exception is rectangular cross-flow, which requires partial-differential modelling.

Simple analytical solutions have been obtained for concurrent and countercurrent flow with constant heat capacities and constant or linearly varying heat transfer coefficients. Other geometries, developing flow, developing heat transfer and variations in the physical properties with temperature may give rise to combinations of differential equations which are difficult to integrate numerically. Most practical problems of geometrical complexity have already been solved, but only a few of the simpler cases of developing flow and developing heat transfer have been considered. The effects of variable physical properties on heat exchange are still not well resolved because of the lack of general models of the temperature dependence and because of the strong influence of variable viscosity and density on momentum transfer.

2.9.3 *Laminar thermal boundary layers*

Heat transfer in laminar boundary layers, although a very old and worked-over problem, still offers some challenges. The boundary layer equations for free convection are numerically intractable for low Prandtl numbers, apparently because of instability. Although asymptotic solutions are readily derived for a Prandtl number of zero, numerical results have not been obtained for any boundary condition or geometry for finite values less that 0.001. Such low values of the Prandtl number only occur in galactic applications, but solutions would be very useful in developing correlations for terrestrial applications in the generalized form proposed by Churchill and Usagi (1972).

The coupled differential equations for the conservation of momentum and energy in dissipative flow over a flat plate have apparently never been solved directly despite the many practical applications, including thermometry. For some particular Prandtl numbers, numerical solutions have been obtained from an integral formulation (see, for example, Churchill and Char 1983).

2.9.4 *Shock waves*

Supersonic compression waves can be generated by the motion of a strong pressure pulse or the rapid motion of a piston. Normal reflection of a wave off a solid surface results in an even stronger wave. After their full development, shock waves and reflected shock waves can be modelled as stationary, in so far as their decay due to dissipative processes can be neglected, by choosing the wave front as a frame of reference. Solution of the resulting set of nonlinear algebraic equations yields a step function in pressure, velocity, and temperature. The effects of diffusion and viscous dissipation on a pseudo-stationary wave can be modelled by a set of coupled ODEs representing the conservation of mass, momentum, and energy. Shock waves may be generated in a tube, adding the complication of the drag of the wall. Spherical and cylindrical waves are usually modelled by the same equations as planar waves in pseudostationary terms.

2.9.5 *Detonation waves*

Supersonic compression waves can also be generated by a rapid chemical reaction. Again, if diffusion and dissipation are neglected and chemical equilibrium is assumed behind the wave, the fully developed behavior can be modelled algebraically as a step function. This idealized model provides excellent predictions for the detonation velocity (see, for example, Moyle, Churchill, and Morrison 1960). Even so, computations of the effect of finite reaction rates and diffusion are of interest. For example, Nicholls (1960) established experimentally a stationary detonation wave for a mixture of hydrogen and oxygen, and thereby determined the reaction rate.

The development phase of a detonation wave following ignition is another unresolved and challenging problem involving stiffness and partial differential equations.

2.9.6 *Deflagration waves*

Unconfined laminar flames produce truly stationary expansion waves, but the postulates of chemical equilibrium and negligible diffusion are not applicable even as first-order approximations. On the other hand, the general model for this process is only moderately stiff. Hahn and Wendt (1982) have developed a numerical solution for an opposed flat flame, taking into account the unique flow field, diffusion and free-radical reactions.

The transient problems of ignition and flame development provide a further challenge. In addition, axial symmetry is of more direct interest than planar.

2.9.7 *Radiantly stabilized combustion*

The stationary combustion of premixed air and fuel (either gas or an atomized volatile liquid) in a refractory tube has been successfully modelled by an ordinary integro-differential equation and a large set of coupled ODEs. This complex model is characterized by split boundary conditions as well as by sharpness in the transient; the process of solution by instability; and the solutions by multiplicity.

The physical behavior can be described as follows. In passage down the inlet region of the tube the fuel–air mixture is heated to the point of ignition by convection from the tube wall. Combustion then occurs with a rapid rise in temperature.

Downstream from this combustion front the burned gas heats the colder tube wall by convection. The resulting hot downstream wall radiates and conducts energy to the colder upstream wall, thereby providing the thermal feedback required to preheat the fuel–air mixture. For a sufficiently large tube diameter, the stable range of operation falls in the turbulent regime. Hence the process can be modelled with reasonable accuracy as plug flow. The model consists of separate energy balances for the gas, liquid (if any) and solid (the wall), and separate material balances for each of the chemical species in the gas phase, including free radicals. The energy balance for the wall is an ordinary integro-differential equation. The other balances are ODEs. The energy balances for the gas and wall are coupled through a convective boundary condition, and the energy and material balances for the gas by the temperature- and composition-dependence of the reaction mechanisms. In the case of liquid droplets, the mass balances as well as the energy balances for the gas and liquid streams are also coupled through convective boundary conditions. Secondary couplings occur due to physical property dependence on temperature and composition, and the dependence of the convective coefficients on the reaction-generated energy. The inlet conditions are known for the fuel and air, but the wall temperature at the inlet is a dependent variable, thereby requiring a shooting method for solution.

We have not yet been able to obtain a convergent solution without making approximations in the model. The general problem involves a double iteration, one for the postulated inlet temperature which satisfies the boundary condition of overall conservation of energy, and the other for the temperature distribution in the integral. Our initial numerical solution (Chen and Churchill 1972), although not convergent, did yield the correct behavior semiquantitatively, and unexpectedly predicted the existence of seven stable solutions. These multiple stationary states arise from differences in the eighth significant figure of the trial value of the inlet wall temperature. The inlet wall temperatures which lead to stable solutions are thus equivalent to eigenvalues. The existence of the multiple stationary states has subsequently been confirmed experimentally for both gaseous and liquid fuels.

We have subsequently (Choi and Churchill 1979) obtained a convergent solution by approximating and analytically evaluating the integral term. However, in so doing we lost five of the stable solutions. We have also carried out thermal and chemical-kinetic calculations for the fluid stream using an experimental wall-temperature profile, thereby avoiding the computational difficulties associated with the integro-differential component of the model. Such calculations, although stiff in the sense of a 25 to 1 ratio in the eigenvalues, have converged satisfactorily. The classical postulates of quasisteady states or local equilibrium for the free radicals were not found to be valid or useful for these conditions.

Recently we have been calculating spontaneous ignition for various fuels at various specified rates of heating, using over one hundred free-radical mechanisms for thermal cracking, preliminary oxidation, primary oxidation, secondary oxidation and post-flame reactions. We have surmounted problems related to stiffness, and particularly the occurrence of negative concentrations for reasonable time steps, by local self-scaling.

It appears from this experience that an integral term, such as that arising from radiant interchange between surfaces, may be a more serious source of computational difficulty than widely separated eigenvalues or split boundary conditions.

The principal idealization in the above modelling is that of plug flow. Two-dimensional modelling with eddy diffusivities for the radial transport of momentum, energy and species would undoubtedly be more realistic. The principal advantage would be the elimination of the present uncertain prediction of the reaction-dependent heat transfer coefficient for the wall. Calculation of the Eulerian transient behavior would also be of practical value in connection with ignition, quenching and physical perturbations. Both radial and transient modelling require the use of PDEs without any relaxation of the previously mentioned difficulties.

2.9.8 *Simultaneous radiation and conduction in dispersed media*

Radiative transfer through a dispersed medium depends on scattering, absorption and reradiation. This behavior can be modelled by an integro-differential equation involving one to three dimensions and two angles. For the idealized case of isotropic scattering, the integro-differential equation can be approximated by the Poisson equation. However for anisotropic scattering this approximation breaks down near the boundaries, precisely where the results are of most interest. The parameters of the problem all vary with the wavelength of the radiation, but mean values can be postulated as a further approximation. The transmission of solar radiation through clouds to the surface of the earth is one application. In a thermal insulation or packed bed, radiative transfer may also be coupled with thermal conduction. To be rigorous, conduction in the solid and vapor phases must be considered individually, but as an approximation, a combined, effective conductivity may be used to represent both phases.

These simplified models for radiative transfer and combined radiative and conductive transfer may still pose stiff problems, particularly if the reflectivity of one or both of the boundaries is near unity (Larkin and Churchill 1959; Chen and Churchill 1963).

2.9.9 *Thermal regenerators*

The fully developed temperature profile in a thermal regenerator can be modelled as stationary. If the solid and fluid are postulated to be at thermal equilibrium locally, and if diffusion is neglected in both phases, the process can be modelled algebraically as a step function. Chen and Schiesser (1980) solved the transient model for this limiting behavior numerically, but stiffness resulted in considerable inaccuracy. The superimposed effect of diffusion can be modelled by one ODE and finite convective resistance by two.

This stationary model is not valid for the fundamentally transient behavior near the inlet and outlet of the regenerator, which requires modelling by one PDE in the case of diffusion and two in the case of convection. Cyclic operation introduces the further complication of a different initial temperature distribution for each cycle, although an asymptotic initial distribution is approached after many cycles.

Abbrecht, Churchill, and Chu (1957) have reviewed some of the analytical solutions in series form and in terms of high-order functions. Interestingly, the stiffness of the various models appears to have a counterpart in the numerical evaluation of these analytical solutions.

2.9.10 *Wave-like mass transfer*

Ion exchange and gas-phase chromatography bear some analogy to thermal regeneration but are generally more complicated.

The dispersion of CO_2 and O_2 in the process of respiration, as described by Boy (1981), is also somewhat analogous, although the conical shape of the bronchial tubes, etc., introduces a different set of complications.

2.9.11 *Thermoacoustic waves*

A sonic compression wave is generated in an unconfined gas by a step increase in the temperature of a bounding surface. If the gas is confined, the wave reflects repeatedly. Oxoe, Sato, and Churchill (1980) failed to obtain a convergent numerical solution for this problem owing to stiffness, although their computed pseudosteady-state pressures, velocities, and temperatures are self-consistent.

2.9.12 *Forced convection*

Developing heat transfer in forced convection poses a fundamentally two-dimensional problem. Graetz derived a series solution for the temperature field and heat transfer coefficient in fully developed laminar flow in a round tube due to a step in wall temperature. However, this series does not converge for short distances from the onset of heating. Lévêque used a similarity transformation to derive an approximate solution for the inlet region, which is correct in the limit but not for finite distances. Worsøe-Schmidt developed a series solution in the form of a perturbation on the Lévêque model which is valid for small distances but not for large. A number of numerical solutions have been attempted but none are accurate as the inlet is approached, owing to the ever-decreasing thickness of the boundary layer.

Many closely related problems have been considered for other geometries, other boundary conditions and developing flow. They all have the same limitations. More detailed description of these problems are given by Churchill and Usagi (1972) and Churchill and Ozoe (1973a, b).

2.9.13 *Transient conduction*

Transient thermal conduction in multiple solid media, such as an insulated semi-infinite region subjected to a step in surface temperature (Churchill 1965), may pose computational difficulties associated with the nearly discrete transition from a transient state to a pseudosteady state followed by another nearly discrete transition to another transient state.

The nonlinear boundary condition provided by a phase transition, such as in the freezing of wet soil, particularly in two or three dimensions, may introduce computational problems related to the location of the phase boundary (Churchill and Gupta 1977).

2.9.14 *Transient heating of a dispersion*

The transient heating of a dispersion of suspended particles by radiation, with conduction to the surrounding air, is difficult computationally because of the nearly discrete shifts from transient single-particle heating, to a pseudosteady state, to transient interparticle effects.

2.9.15 *Thermohydrodynamic instability*

Physical instability may cause computational difficulties which are similar to those associated with stiffness. For example, the mode of circulation for convection in an enclosure heated from below changes from multiple roll cells with axes parallel to the shorter horizontal dimension to a single circulation with its axis parallel to the axis of inclination. For inclinations near the critical value the computed values oscillate chaotically instead of converging (Ozoe, Yamamoto, and Churchill 1979).

2.9.16 *Summary and conclusions*

I have now completed my physical description of a representative set of problems in heat transfer whose numerical solution is known or suspected to encounter computational difficulties.

Most important unsolved problems in heat transfer (as well as in mass transfer, momentum transfer, and chemical kinetics) are found to be either transient or multidimensional, and hence must be modelled by partial differential equations.

Shock, detonation, deflagration, and thermoacoustic waves, thermal regeneration, ion exchange, gas-phase chromatography, respiration, and radiantly stabilized combustion, with dispersion due to diffusion, chemical reactions and/or interphase transfer, whether stationary or transient, pose classical problems of stiffness. That is, the stiffness is associated with different terms having widely varying time constants or the equivalent.

The simplified model for radiative and conductive transfer through dispersed material and the illustrative model for transient radiant heating of a dispersion of particles are similarly stiff owing to discordant mechanisms of heat transfer. This same effect occurs in the transient heating of an insulated semi-infinite medium and in the freezing of a semi-infinite region of wet soil (locating the freezing front further complicates the latter problem, particularly in two or three dimensions).

On the other hand, the original integro-differential model for radiative transfer through a dispersion and the integro-differential equation in the model for radiantly stabilized combustion generate even more severe computational difficulties.

Split boundary conditions are a further source of difficulty in some boundary layer and heat exchanger problems, and also in solving the model for radiantly stabilized combustion.

The generation of a thermoacoustic wave by a step function in temperature at the boundary appears to pose a uniquely stiff problem.

Physical instability, such as that associated with low values of the Prandtl number in free and natural convection and with changes in the mode of circulation in natural convection, is another severe source of computational difficulty.

Mathematical attention to the solution of stiff differential equations and related problems has been limited to trivial problems relative to those encountered in realistic models for heat transfer, mass transfer, flow and chemical kinetics, separately or combined. Hence, in the future as well as in the past, it is probable that the significant advances will be made heuristically by practitioners rather than by mathematicians. This is particularly true of packaged programs which, because of their objective of generality for a broad class of problems, are seldom useful for realistic problems unless modified. Also, they are seldom as efficient as special-purpose programs.

2.10 Chemical kinetics

2.10.1 *Introduction* (*Aiken*)

This is by far the largest stiff application area on the basis of numbers of problems or numbers of users. Stiffness is caused in the vast majority of cases simply by a large difference among the reaction rate constants. The larger the system or the more detailed the model, the more likely that stiffness will occur. If the elementary reactions are known, the law of mass action dictates the form of the rate expressions: either linear in a concentration variable or quadratic. If temperature is also considered to vary, severe exponential nonlinearities are introduced through the Arrhenius temperature term.

Several investigators have made use of the special simple structure of isothermal mass-action kinetics. Edsberg (1974, 1976) simplifies the problem set-up and Jacobian evaluation for the user and efficiently handles the Jacobian, but Dahlquist Edsberg, Skollermo, and Soderlind (1980) feel more can be done to make use of structure as well as the users' knowledge of the stiffness. This knowledge often consists of: a partitioning of variables into stiff variables and nonstiff variables; a fast transient that occurs only initially; and inherent tight stability of the stiff variables whose concentrations must not be negative. Karasalo and Kurylo (1980) point out an advantage in keeping these concentrations artifically non-negative when using a version of GEAR. Robertson (1975, 1976) also suggests some structure-related handling of the Jacobian for faster convergence. See Pratt (1979) for a special numerical method tailored for this equation type.

Packages exist specifically for the mass-action kinetics form. Uhlen (1979) describes KINRATE and KINBOX. Edelson (1976), of Bell Laboratory, presents a simulation language and compiler for mass-action kinetics; he used a version of GEAR in forming a package called BELLCHEM. Rider (1977) offers CAKE, a user-friendly version of GEAR that makes use of the typical sparse structure of large mass-action kinetic equations. Deuflhard, Bader, and Nowak (1981) describe LARKIN to handle large systems of kinetic equations. Gottwald (1981) gives us KISS for coupled chemical reactions, Stabler and Chesick (1978) have written a program for reaction rate equations using a version of GEAR. David (1977) describes a FORMAC program for direct integration using formula manipulation and a tailor-made numerical method; Kennedy and Moore (1977) also recognized the virtue in using a Taylor-series expansion as the basis for a numerical method with such simple functions.

Enright and Hull (1976) compare numerical methods for stiff kinetic problems and found that the backward differentiation methods were superior to most other methods, including an implicit Runge–Kutta technique; pitfalls in generalizing such conclusions are explained by Enright (1982).

The steady-state approximation has been used extensively to eliminate stiffness in chemical kinetic systems (Aiken and Lapidus 1975), see §7.2.5. Noyes (1978) discusses the importance in including reversible reaction when the approximation is made.

Sensitivity analysis is becoming an effective means of determining appropriate model detail (see §7.6 and §7.7). Koda, McRae, and Seinfeld (1979) studied automatic sensitivity analysis of kinetic mechanisms and developed FAST (Fourier Amplitude Sensitivity Test). Dougherty and Rabitz (1980) looked at the sensitivity of hydrogen combustion. Rabitz (1981) gave an overview of this area applied to chemical kinetics. See also Kuchel (1980), Sundaresan and Amundson (1980), Dougherty, Hwang, and Rabitz (1979), Dove and Raynor (1979), Dickinson and Gelinas (1976), and Lowe, Wall, and Stewart (1977).

A specific kinetic application area not covered in any of the other sections of this review is pyrolysis. Hautman, Santoro, Dryer, and Glassman (1981) stated that at low conversions the primary reactions govern the dynamics, but at lower conversion the secondary reactions do. Layokun and Slater (1979) modelled a free-radical mechanism of propane pyrolysis and solved it with a semi-implicit trapezoidal rule. A number of thermal cracking models were solved in detail by Sundaram and Froment (1978). Liquid phase pyrolysis of 1,2 diphenylethane was studied by Miller and Stein (1981).

An overview on the computational techniques for the study of reaction processes is available from Edelson (1981). A stiff model for chemistry in interstellar clouds is advanced by Prasad and Huntress (1980). Ross (1977) mentions the problem of loss of detailed balance when applying the steady-state approximation and uses instead a Markov matrix method. Rosenbaum (1976) classifies certain numerical methods as "conservative" (if they satisfy the detailed balance). Ong and Mason (1976) discuss a computational problem that could be different from the type we have been implicitly focusing on: one in which a right-hand side is the difference of two large terms; they convert the initial-value problem into a two-point boundary-value problem for the case that the right-hand side passes through zero. See also the entire volume 81, number 25, pp. 2309–2559 of the Journal of Physical Chemistry (1977).

2.10.2 *Some mathematical properties of chemical kinetics* (*Edsberg*)

Consider a closed system where several chemical reactions are taking place simultaneously. The system is held at constant temperature and the volume does not change. It is also assumed that the system is operating under homogeneous conditions, i.e. no spatial gradients occur. Under these idealized conditions we have an isothermal batch reactor.

Assume that n chemical species A_i, $i = 1, 2, \ldots, n$ simultaneously interact in m reactions

$$\sum_{i=1}^{n} p_{ij} A_i \xrightarrow{k_j} \sum_{i=1}^{n} q_{ij} A_i, \qquad j = 1, 2, \ldots, m, \tag{2.12}$$

where p_{ij} and q_{ij} are integers (stoichiometric coefficients) and k_j are parameters called the rate constants of the reactions. Under the physical conditions mentioned above and the assumption that the mass action law is applied for the rate functions, the following mathematical model consisting of a set of ODEs can be set up:

$$\dot{c}_i = \sum_{j=1}^{m} (q_{ij} - p_{ij})k_j \prod_{k=1}^{n} c_k^{p_{kj}}, \qquad c_i(0) = c_{i0}, \tag{2.12}$$

where c_i are the concentrations of A_i and c_{i0} are the concentrations of A_i at time $t = 0$. The system (2.12) can be written in compact form:

$$\dot{\boldsymbol{c}} = (\boldsymbol{Q} - \boldsymbol{P})\boldsymbol{K}\boldsymbol{g}(\boldsymbol{c}), \qquad \boldsymbol{c}(0) \text{ given}, \tag{2.13}$$

where $\boldsymbol{Q}$ and $\boldsymbol{P}$ are $n \times m$ matrices, $\boldsymbol{K}$ a diagonal matrix = diag(k_i) and $\boldsymbol{g}$ is a vector with m components containing the polynomial expressions in (2.12). The vector $\boldsymbol{Kg}$ is denoted by $\boldsymbol{r}$ and is a vector of m components containing the rate-functions (formed by the mass action law). The matrix $\boldsymbol{S} = \boldsymbol{Q} - \boldsymbol{P}$ is referred to as the stoichiometric matrix. Thus a chemical kinetics problem is specified by the two matrices $\boldsymbol{S}$ and $\boldsymbol{P}$ (usually being very sparse with integers as entries), the diagonal matrix $\boldsymbol{K}$ (with only positive entries), and the initial concentration vector $\boldsymbol{c}(0)$ (with nonnegative entries).

Consider the following example of chemical pyrolysis, originating from Datta (1967); this set of ODEs is also one of the test problems in the stiff integrator comparison by Enright, Hull and Lindberg (1975):

$$\begin{aligned} A_1 &\to A_2 + A_3 \\ A_2 + A_3 &\to A_5 \\ A_1 + A_3 &\to A_4 \\ A_4 &\to A_3 + A_6 \end{aligned} \tag{2.14}$$

The corresponding set of ODEs is, according to mass action kinetics:

$$\begin{aligned} \dot{c}_1 &= -k_1c_1 - k_3c_1c_3, \\ \dot{c}_2 &= k_1c_1 - k_2c_2c_3, \\ \dot{c}_3 &= k_1c_1 - k_2c_2c_3 - k_3c_1c_3 + k_4c_4, \\ \dot{c}_4 &= k_3c_1c_3 - k_4c_4, \\ \dot{c}_5 &= k_2c_2c_3, \\ \dot{c}_6 &= k_4c_4, \end{aligned} \tag{2.15}$$

with

$$\boldsymbol{S} = \begin{bmatrix} -1 & 0 & -1 & 0 \\ 1 & -1 & 0 & 0 \\ 1 & -1 & -1 & 1 \\ 0 & 0 & 1 & -1 \\ 0 & 1 & 0 & 0 \\ 0 & 0 & 0 & 1 \end{bmatrix}, \qquad \boldsymbol{P} = \begin{bmatrix} 1 & 0 & 1 & 0 \\ 0 & 1 & 0 & 0 \\ 0 & 1 & 1 & 0 \\ 0 & 0 & 0 & 1 \\ 0 & 0 & 0 & 0 \\ 0 & 0 & 0 & 0 \end{bmatrix}.$$

For this system $n = 6$ and $m = 4$. The rate constants used in the test problem were $k_1 = 7.89 \times 10^{-10}$, $k_2 = 1.13 \times 10^9$, $k_3 = 1.1 \times 10^7$, $k_4 = 1.13 \times 10^3$ and the initial value $c_1(0) = 1.76 \times 10^{-3}$; the others were set to zero.

Note that $\dot{c}_2 = \dot{c}_3 + \dot{c}_4$ and $2\dot{c}_1 + \dot{c}_2 + \dot{c}_3 + 3\dot{c}_4 + 2\dot{c}_5 + 2\dot{c}_6 = 0$, i.e. there are linear relations between the variables due to rank deficiency in $\boldsymbol{S}$. In fact, for this example, rank($\boldsymbol{S}$) = 3. The wide spread of size in the rate constants is the reason for stiffness.

Examples of reactors where transport and energy-processes must be taken into account is the continuous-flow stirred tank reactor (CSTR) and the plug-flow tubular reactor (PFTR) which, however, can be modelled by sets of ODEs obtained by making small modifications to (2.13); See Edsberg (1976) for further discussion. For the CSTR, the equations are

$$\dot{\boldsymbol{c}} = \frac{1}{\theta}(\boldsymbol{c}_f - \boldsymbol{c}) + (\boldsymbol{Q} - \boldsymbol{P})\boldsymbol{K}\boldsymbol{g}(\boldsymbol{c}), \tag{2.16}$$

where θ is a constant and $\boldsymbol{c}_f$ is a constant vector.

For systems of ODEs coming from chemical kinetics it is obvious that the variation of the variables c_i is restricted to certain bounds. Returning to the model (2.13) which is valid under closed conditions the total mass must be preserved. This can be expressed by

$$\boldsymbol{m}^{\mathsf{T}}\boldsymbol{c} = \boldsymbol{m}^{\mathsf{T}}\boldsymbol{c}(0) = \alpha, \tag{2.17}$$

where $\boldsymbol{m}$ is a vector of molecular weights of the species A_i. Thus

$$\boldsymbol{m}^{\mathsf{T}}\dot{\boldsymbol{c}} = 0, \tag{2.18}$$

which means that $\dot{\boldsymbol{c}}$ belongs to a subspace L being orthogonal to the vector $\boldsymbol{m}$, which is directed into the positive orthant $\mathbb{R}^n_+$. The concentration vector $\boldsymbol{c}$ then belongs to a hyperplane L_α being parallel with L and intersecting the coordinate axis at intercepts $\gamma_i > 0$.

2.10.3 *The use of scaling in chemical kinetics* (*Edsberg*)

Various ways of performing automatic scaling of the local error estimate were discussed by Curtis (1978a) and Deuflhard (1981). In some numerical experiments reported by Edsberg (1976) it was found that for certain chemical systems some scaling device seemed to be necessary in order to obtain reliable approximations of the solution trajectories at least when the machine accuracy and/or the tolerance used in the numerical integrator was not small enough. The scaling factors were chosen from a priori knowledge of the size of the variables $\boldsymbol{c}$, and scaling was tested both by transforming (2.13) to a scaled set and by using different weights in the norm of the local error estimate used to control the step size. In both cases the results were satisfactory when tried on some stiff ODE-solvers if scaling was used but unreliable if no scaling was used (corresponding to using an absolute error tolerance). Similar results are reported by Curtis (1978a).

A more systematic approach to the scaling problem is given by Dahlquist, Edsberg, Sköllermo, and Söderlind (1980), who presented a semiautomatic algorithm for scaling the variables. The algorithm is tested on some chemical kinetics problems and the scaling procedure emerges in a singular perturbation formulation of (2.13):

$$\begin{aligned} \dot{x}_1 &= S_1 f_1(x_1, x_2, \epsilon), \qquad x_1(0), \\ \epsilon\dot{x}_2 &= S_2 f_2(x_1, x_2, \epsilon), \qquad x_2(0), \end{aligned} \tag{2.19}$$

where

$$x = \begin{bmatrix} x_1 \\ x_2 \end{bmatrix}, \; S = \begin{bmatrix} S_1 \\ S_2 \end{bmatrix},$$

and x is a vector of the variables c_i scaled by certain scaling factors. x and S are then partitioned according to the non-stiff variables x_1 and the stiff variables x_2; ϵ is a small parameter. The system (2.19) is then treated with a stiff ODE-solver.

The scaling procedure is simply based on a diagonal scaling of the dependant variables c_i and the time t:

$$\begin{aligned} c &= Ex, \\ t &= e_0\tau, \end{aligned} \tag{2.20}$$

where E is a diagonal matrix with e_i in the diagonal, $i = 1, 2, \ldots, n$. With this transformation inserted into (2.13) we obtain

$$Ex' = S\tilde{K}g(x), \qquad x(0) = E^{-1}c(0), \tag{2.21}$$

where

$$\tilde{K} = e_0 KG(e), \qquad G(e) = \operatorname{diag}(g_i(e)), \tag{2.22}$$

and e is a vector of the e_i, $i = 1, 2, \ldots, n$, and $'$ denotes differentiation with respect to τ. Thus only the rate constants k_i and the initial values $c_i(0)$ are affected by this transformation; neither S nor g is changed (provided E is kept on the left-hand side in front of the derivatives). In this transformed system (2.21) we now try to use the scaling factors to identify which variables are stiff and which are nonstiff.

The way of finding the scaling factors e_i, $i = 0, 1, \ldots, n$ is outlined by Dahlquist and Mao-Zu-Fan (1982) and is based on the following assumptions:

(i) In a time-interval $[\tau_1, \tau_2]$ where the scaling is performed, the scaled variables x_i and their derivatives x_i' are of the same order of magnitude, $O(1)$.

(ii) The vector x is partitioned into two vectors x_1 and x_2. For the vector x_1 (the nonstiff variables), the derivatives x_i' balance at least one term on the right-hand side. For the vector x_2 (the stiff variables), there is balance between terms of opposite sign in the left-hand side. (Balance here means that two terms are of the same order of magnitude.)

If the scaling procedure succeeds in balancing terms the system will be transformed into singular perturbation form. In Dahlquist and Mao-Zu-Fan (1982) there is outlined an algorithm which gives the scaling factors e_i provided the "user" indicates which terms balance each other in all equations. The problem can then be formulated as a linear program which tries to find the e_i that match the balance conditions as well as possible. Work is underway in our department which aims at designing an automatic scaling procedure. We illustrate the scaling procedure here on the problem (2.15) and show the advantages of finding the transformed system (2.19).

With rate constants and initial values inserted and after using one of the linear relations between the c_i, the system can be written:

$$\begin{aligned} \dot{c}_1 &= -7.89 \times 10^{-10} c_1 - 1.1 \times 10^7 c_1 c_3, & c_1(0) &= 1.76 \times 10^{-3}, \\ \dot{c}_2 &= 7.89 \times 10^{-10} c_1 - 1.13 \times 10^9 c_2 c_3, & c_2(0) &= 0, \\ \dot{c}_4 &= 1.1 \times 10^7 c_1 c_3 - 1.13 \times 10^3 c_4, & c_4(0) &= 0, \\ c_2 &= c_3 + c_4. \end{aligned} \tag{2.23}$$

Let the scaling factors e_i be expressed in powers of ϵ, being a small positive given number

$$e_i = \epsilon^{\alpha_i}. \tag{2.24}$$

The problem now is to make a hypothesis in each equation concerning the leading terms which balance each other. In the first equation, since both terms on the right-hand side have the same sign, one of them (or both) balance the derivative. Let's assume the second term does. Assume also that c_2 and c_4 are stiff variables; hence the terms on the right-hand sides balance each other in the last two ODEs. In the algebraic equation assume that all terms are of the same magnitude, i.e. both terms on the right-hand side balance the left-hand side. Finally, guided by the magnitude of the initial value, let $\epsilon = 10^{-3}$.

The balance assumptions now give the following relations:

$$\begin{aligned} \epsilon^{\alpha_1 - \alpha_0} &= k_3 \epsilon^{\alpha_1 + \alpha_3}, \\ k_1 \epsilon^{\alpha_1} &= k_2 \epsilon^{\alpha_2 + \alpha_3}, \\ k_3 \epsilon^{\alpha_1 + \alpha_3} &= k_4 \epsilon^{\alpha_4}, \\ \alpha_2 = \alpha_3 &= \alpha_4. \end{aligned} \tag{2.25}$$

With the known numerical values inserted into (2.25) and rounding every rate constant to the closest power of ten, we get the following set of linear equations:

$$\begin{aligned} 3\alpha_0 &= 7 - 3\alpha_3, \\ -10 - 3\alpha_1 &= 9 - 3\alpha_2 - 3\alpha_3, \\ 7 - 3\alpha_1 - 3\alpha_3 &= 3 - 3\alpha_4, \\ \alpha_2 = \alpha_3 &= \alpha_4. \end{aligned} \tag{2.26}$$

The closest integer solution of (2.26) is $\alpha_0 = -1$, $\alpha_1 = 1$, $\alpha_2 = \alpha_3 = \alpha_4 = 4$, which gives the following set of scaled ODEs:

$$\begin{aligned} x_1' &= 0.789\epsilon^2 x_1 - 0.011 x_1 x_2, & x_1(0) &= 1.76 \\ \epsilon x_2' &= 0.787 x_1 - 0.011 x_2 x_3, & x_2(0) &= 0, \\ \epsilon^2 x_4' &= 11 x_1 x_3 - 1.13 x_4, & x_4(0) &= 0, \\ x_2 &= x_3 + x_4, \end{aligned} \tag{2.27}$$

where

$$t = 10^3 \tau, \; c_1 = 10^{-3} x_1, \; c_i = 10^{-12} x_i, \qquad i = 2, 3, 4.$$

From the equations (2.27) we now have good information on the qualitative behavior of the solution, and we have also obtained a natural scaling of each variable where 1 is a proper measure of each variable in a time interval where the variables x_2 and x_4 have reached a nearly steady state. We see that the transient phase consists of one very quick phase determined by ϵ^2 and one slower interval determined by ϵ. Hence we get a priori information on what can be expected of the step-size regulation of a stiff ODE-solver.

So after this "analytical preprocessing" of the system the problem is now well-suited for being treated by a stiff method, and with a very moderate value given to the tolerance parameter, we obtain the solution with good relative accuracy.

Thus, if it is possible to find a scaling that transforms the set of ODEs (2.13) into the form (2.19), we have gained the following advantages:

(i) We have better insight into the dynamics of the problem.
(ii) The problem is better suited for numerical treatment with a stiff numerical integrator; we have variables that are scaled so that 1 is appropriate measure of the size of all variables.
(iii) The partitioned form (2.19) makes it possible to use numerical methods that utilize this form (Söderlind 1981).

2.10.4 *Chemical reaction systems* (*Seider, White III, and Prokopakis*)

The dynamics of many chemical reaction systems can be adequately described with reactions involving molecular species. The intrinsic rates of forward and back reactions are relatively slow (as compared with reactions involving free-radicals and ionic species) and the mass balances are usually not stiff. Hence, with integrators having a stability bound (so-called "nonstiff" integrators—all explicit and some implicit methods), the step size is not reduced below that to give desired accuracy.

For some systems, such as in pyrolysis, reforming, and combustion, reactions involving free radicals are necessary to give an adequate representation of conversion in time. These fast reactions introduce large $|\mathrm{Re}(\lambda)|$. At times when the concentrations of the molecules and radicals respond slowly, the mass balances are usually stiff; with widely-spread response times they may not be stiff.

Pyrolysis and reforming systems are endothermic and hence nearly isothermal in industrial furnaces. In contrast, combustion systems are highly exothermic with rapid rises of temperature and conversion in the flame front. The high sensitivity of reaction rates to temperature, usually described by the Arrhenius equation

$$k = k_0 e^{-E/RT},$$

causes all reactions to be accelerated. The $|\mathrm{Re}(\lambda)|_{max}$ increases, and stiffness of the differential equations depends upon the largest rate of change in concentration.

Shampine and Gear (1979) refer to the Van der Pol equation as an example of a limit cycle that is intermittently stiff and nonstiff. Another example is the

Belousov reaction system, studied by Prokopakis and Seider (1981) with the kinetic model of Field and Noyes (1974). For a mixture of 1.25M H_2SO_4, 0.0125M $KBrO_3$, 0.001M $Ce(NH_4)_2(NO_3)_5$, and 0.025M $CH_2(COOH)_2$, Field and Noyes derive the mass balances for the intermediates BrO_2, Br^-, and Ce^{4+}, with dimensionless concentration, y_1, y_2, y_3, respectively:

$$\frac{dy_1}{dt} = 77.27(y_2 - y_1y_2 + y_1 - 8.375 \times 10^{-6}y_1^2), \qquad y_1(0) = 4,$$

$$\frac{dy_2}{dt} = \frac{-y_2 - y_1y_2 + y_3}{77.27}, \qquad y_2(0) = 1.1,$$

$$\frac{dy_3}{dt} = 0.161(y_1 - y_3), \qquad y_3(0) = 4.$$

The period of the limit cycle is 300 seconds as illustrated in Figure 2.1. Note that the bold curves in Figure 2.1b denote the fastest changing variable, the stiff variable. Table 2.1 shows the eigenvalues of the system which vary by at most four orders of magnitude (and usually far less) during the rapid transients, or steep concentration fronts. In this region, the system is not very stiff; probably a nonstiff integrator could be used without unduly small step sizes. Then the system gains inherent stability, with $|\lambda|_{max} = 1.33 \times 10^5$. Since the concentrations change more slowly, the system becomes stiff and the stiffness increases as it stabilizes. At longer times, the system slows further and $|\lambda|_{max}$ decreases as instability sets in, just prior to the rapid transient. Consequently, stiffness decreases with destabilization.

Another limit cycle, of environmental interest, involves the reactions of atomic oxygen, O, oxygen, O_2, and ozone, O_3, in the atmosphere. In the mechanism of Chapman (Dickenson and Gelinas 1976):

$$O + O_2 \xrightarrow{k_1} O_3$$

$$O + O_3 \xrightarrow{k_2} 2O_2$$

$$O_2 \xrightarrow{k_3(t)} 2O$$

$$O_3 \xrightarrow{k_4(t)} O + O_2,$$

the last two reactions require photochemical energy and have rate constants that vary diurnally:

$$k_i(t) = \begin{cases} \exp(-c_i/\sin \omega t) & \sin \omega t > 0 \\ 0 & \sin \omega t \leqslant 0 \end{cases} \qquad i = 3, 4,$$

where half the frequency of oscillation, $\omega = \pi/43200\ s^{-1}$ $(=\pi/12\ hr^{-1})$ and $c_3 = 22.62$ and $c_4 = 7.601$. These rate constants rise rapidly beginning at dawn ($t = 0, 24, 48$ hr, . . .), reaching a peak at noon ($t = 6, 30, 54$ hr, . . .), and decreasing to zero at sunset ($t = 12, 36, 60$ hr, . . .). The mass balances for O and O_3 are:

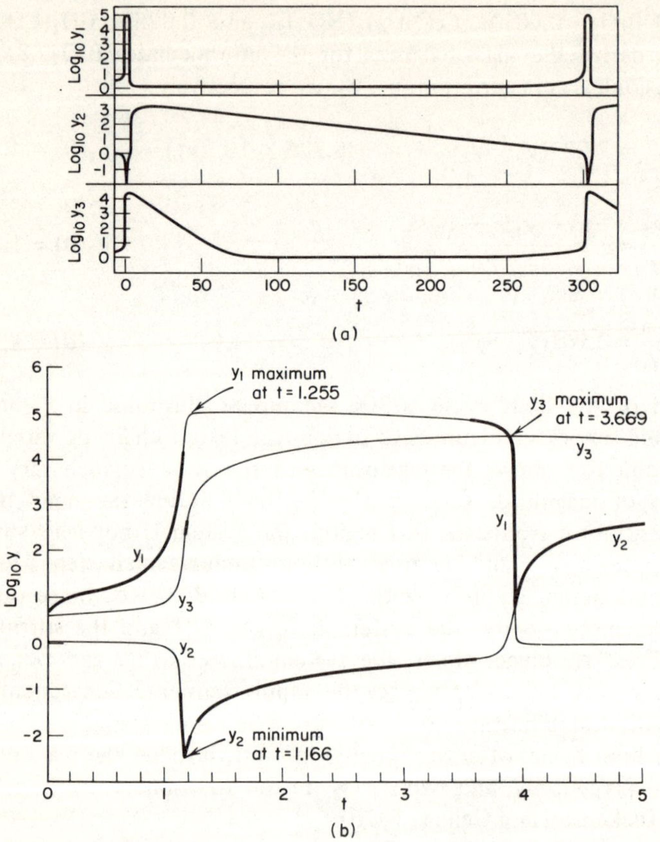

Fig. 2.1. Profiles of dimensionless concentration for the Belousov reaction system.

$$\frac{dy_1}{dt} = -k_1 y_1 y_3 - k_2 y_1 y_2 + 2k_3(t) y_3 + k_4(t) y_2, \qquad y_1(0) = 10^6 \text{cm}^{-3},$$

$$\frac{dy_2}{dt} = k_1 y_1 y_3 - k_2 y_1 y_2 - k_4(t) y_2, \qquad y_2(0) = 10^{12} \text{cm}^{-3},$$

where y_1, y_2, y_3 are concentrations of O, O_3, and O_2, respectively, and $k_1 = 1.63 \times 10^{-16}$ and $k_2 = 4.66 \times 10^{-16}$. When integrated with $y_3 = 3.7 \times 10^{16} \text{cm}^{-3}$, O traverses a limit cycle while O_3 builds just prior to noon and remains constant at other times, as illustrated in Fig. 2.2. Note that y_1 drops from 10^6cm^{-3} initially to virtually zero ($< 10^{-30} \text{cm}^{-3}$) within one minute and builds to 2.92×10^{-2} in one hour. Similarly, y_1 drops from $3.15 \times 10^{-2} \text{cm}^{-3}$ after 11 hours to virtually zero after 12 hours, where it remains until 24 hours. At 24 hours, the rapid rise begins. The rates of change and eigenvalues are given in Table 2.2. λ_1 is nearly constant and λ_2 reflects the rate of change of the system. $|\lambda_2|$ increases toward noon as the system accelerates and decreases toward sundown as the system decelerates. Since the rates of change are very slow at night, the system is very

TABLE 2.1

Concentration derivatives and eigenvalues for the Belousov reaction system

t, sec	dy_1/dt	dy_2/dt	dy_3/dt	λ_1	λ_2	λ_3	Stiffness ratio
0.1969	17.7	−0.100	0.935	−84.082	$-0.0933 \pm 0.1172i$		901
1.057	7.75×10^3	−3.19	55.5	−60.565	$-0.1020 \pm 0.9684i$		594
1.137	5.53×10^5	−10.5	1.28×10^3	−7.949	−112.65	−1.6311	69
1.145	1.00×10^6	−6.01	2.17×10^3	−16.339	−178.20	−0.9162	195
1.152	1.53×10^6	−2.43	3.23×10^3	−25.381	−261.43	−0.6524	401
1.164	3.11×10^6	-4.84×10^{-2}	6.48×10^3	−51.985	−523.04	−0.4010	1304
1.168	3.85×10^6	1.51×10^{-2}	8.02×10^3	−64.401	−646.66	−0.3546	1824
1.243	8.99×10^6	0.368	1.87×10^4	−152.20	−1524.8	−0.2428	6280
2.990	4.43×10^6	−17.5	9.24×10^3	−104.70	−1100.5	−0.2772	3970
3.864	2.40×10^5	11.8	3.53×10^2	−36.480	−509.61	−0.4521	1127
3.916	-2.09×10^4	3.99×10^2	-5.01×10^3	−612.06	$-0.0934 \pm 0.0759i$		6553
3.921	-1.15×10^3	4.02×10^2	-5.01×10^3	−771.13	−0.0287	−0.1582	2.69×10^4
3.994	-1.39×10^{-2}	3.97×10^2	-4.95×10^3	-3.02×10^3	−0.0259	−0.1610	1.16×10^5
14.24	-1.92×10^3	31.8	-9.51×10^2	-1.33×10^5	−0.0259	−0.1610	5.14×10^7
45.92	-4.21×10^2	−12.5	−5.80	-7.76×10^4	−0.0259	−0.1610	3×10^6
64.41	-2.52×10^2	−16.1	−0.296	-4.82×10^4	−0.0259	−0.1610	1.86×10^6
115.7	−18.2	−4.29	6.44×10^{-5}	-1.28×10^4	−0.0259	−0.1610	4.94×10^5
219.6	-2.21×10^{-2}	−0.291	2.30×10^{-3}	−368.36	−0.0260	−0.1610	3.34×10^4
286.1	5.37×10^{-2}	-5.40×10^{-2}	3.80×10^{-2}	−152.62	−0.0344	−0.1525	4437
292.8	0.120	-4.79×10^{-2}	7.08×10^{-2}	−126.25	−0.0429	−0.1440	2943
298.9	0.431	-4.86×10^{-2}	0.172	−104.03	−0.0825	−0.1043	1261
300.0	0.606	-5.13×10^{-2}	0.224	−99.744	$-0.0934 \pm 0.0759i$		1068

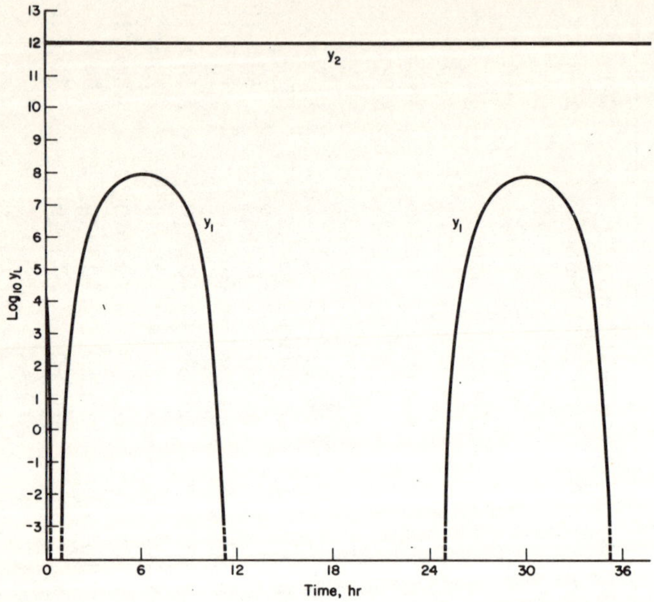

Fig. 2.2. Profiles of atomic oxygen and ozone concentration for atmospheric reaction system (computed using LSODE (Hindmarsh 1980)).

stiff. However, throughout most of the daytime, with rapid rates of change, the system is nonstiff.

2.11 Lasers (Aiken)

Lasing results from creating a highly excited vibrational state in a group of molecules ("pumping" to an inverted energy state). Then a remarkable fact of nature dictates that whatever relaxations to lower energy states that begin, the resulting photons will stimulate other of the still excited energy states to relax in the same way, creating an autocatalytic effect—and coherent radiation. These relaxations are very fast and one source of stiffness in the modelling of lasers. Shampine and Gear (1979) point out that the fast pump–emission cycle of the various energy levels is a re-occurring stiffness that cannot be dealt with effectively with typically available automatic stiff packages. This is because fast transients occur throughout the solution, so that accuracy in handling these transients can be continually important. Cukier and Levine (1978) mention use of a steady-state approximation during the lasing action, but details on this apparently *ad hoc* approach are not given. There is also an initial fast transient from the onset of pumping (Milonni 1977), i.e. at $t = 0$.

Cukier and Levine also examine the sensitivity of the full solution for a model of an HF chemical laser and find that only a few of the rate constants are responsible for the computed gain. More detailed models of an HF laser are given by Ben-Shaul, Feliks, and Kafri (1979) and by Kerber (1977). It should also be mentioned here that the rate constants, particularly the fastest ones, are known only approximately and errors of several orders of magnitude are not uncommon.

TABLE 2.2
*Concentration derivatives and eigenvalues for the atmospheric reaction system. Computed using LSODE**

t, hr	dy_1/dt	dy_2/dt	λ_1	λ_2
0	-6.0315×10^{6}	6.0305×10^{6}	-6.031	-9.319×10^{-10}
1	2.324×10^{-4}	-2.595×10^{-4}	-6.031	≈ 0
2	79.331	-117.94	-6.031	-7.723×10^{-11}
4	9.495×10^{3}	3.022×10^{5}	-6.032	-4.773×10^{-8}
6	9.178×10^{2}	1.102×10^{7}	-6.032	-1.622×10^{-7}
8	-1.021×10^{4}	3.180×10^{5}	-6.032	-5.140×10^{-8}
10	-85.518	40.671	-6.032	-8.324×10^{-11}
12	≈ 0	≈ 0	-6.032	≈ 0
18	≈ 0	≈ 0	-6.032	≈ 0
24	≈ 0	≈ 0	-6.032	≈ 0
30	9.115×10^{2}	1.101×10^{7}	-6.032	-1.741×10^{-7}
42	≈ 0	≈ 0	-6.032	≈ 0

*Relative error tolerance = 10^{-8}. Absolute error tolerance = 10^{-45}. UNIVAC 1100—Double precision. Values ≈ 0 are $<10^{-17}$.

High-energy lasers usually have a gas at the lasing medium for best efficiency, and as a flowing medium to remove heat. The addition of flow can require the laser model to include hydrodynamics coupled to the chemistry. The time scales of transport versus reaction are much different here. Inclusion of hydrodynamics, translational and rotational energy interactions, wall effects and the like are necessary to predict the performance of the laser—to compute how much power output can be extracted and its nature. Additional sources of stiffness can result from the variety of time scales among the three very different types of energy transitions for a molecule: vibrational, rotational, translational (see also §2.13). A molecule in a particular energy state of the many potential combinations is considered a distinct species, so that a large number of highly interactive species can result from only a few different molecules.

A meeting held at the U.S. Air Force Army Weapons Laboratory brought in stiffness experts Shampine, Gear, Liniger, Hindmarsh, and Byrne to convene with laser modellers. Some of these latter have been referenced above; others are: Franklin (1977), who spoke on modelling general kinetic processes in lasers; Lundstrom (1977), of the Naval Weapons Laboratory, modelling the CO laser; Holmes (1977) on CO2/N2 vibrational kinetic equations; Hines (1977), on CO2 electric discharge modelling; and Young and Boris (1977), of the Naval Research Laboratory, on general numerical techniques for chemical kinetics with reactive flow.

Plasma chemistry has been studied as it relates to laser discharge and target interactions. The chemical reactions can include neutral molecule–neutral molecule collisions as well as electron–ion, electron–molecule, and molecule–ion collisions. Roberts (1979) presented his program PLASKEM for this problem. Pert (1978) calculated ionization in rapidly changing plasmas in a model that includes hydrodynamics but integrates these two regimes separately in a steady-state type of approximation. Mazhukin, Uglor, and Cetveruskin (1980), in a numerical investigation of laser breakdown of a dense gas, rejected a version of GEAR and the method of lines for that of Samarskii (1971); his method also is a type of steady-state-approximate decoupling of kinetics from transport. Christiansen and Winsor (1980) studied a numerical model for laser targets, essential to the feasibility investigation of laser fusion.

Refer to Lawton, Richards, Newman, Specht, and DeTemple (1979) for numerical work on the high-pressure infrared xenon laser; Greene and Brau (1978) for KrF and ArF lasers; Barker (1980) for infrared multiphoton decomposition; Pirkle, Poehler, and Sigillito (1974) for pulsed DF–CO_2 transfer lasers; and Bui (1979, 1980, 1981) for model design and analysis of a new type of blast-wave induced laser.

2.12 Mechanics (Aiken)

The term "stiffness" is commonly used in structural mechanics in a much different sense than the present context. The stiffness matrix results from a linear model

$$f = Ky,$$

where f is force and y is displacement. The denser K, the stiffer the problem (finite elements used) in the physical sense that there are more interactions among the components. This would typically imply a more stable system, but a stiff system in this sense is not necessarily associated with fast transients.

Large second-order systems of the form

$$My'' + Cy' + Ky = f(t),$$
$$y(0), y'(0) \text{ given},$$

occur frequently in the transient analysis of dynamic structures (Enright 1980). M, C, and K are the mass, damping, and stiffness matrices, respectively. In the case of large deflections, the problem becomes nonlinear through K depending on y. Here the forcing function $f(t)$ can make accuracy an important consideration for any component of y, including the stiff ones, at any time in the transient; the numerical stability of all components is essential. This special type of stiffness results in computational costs highly dependent on requested accuracy; once a step size has been selected it should remain relatively constant throughout the solution, there being no boundary layers. Thus fixed step size, low-order multistep methods have been commonly used, although Enright (1980) has pointed out the advantages of second-order variable step-size approaches. Jensen (1976, 1974) of the Palo Alto Structural Mechanics Laboratory examines stiffly stable third-order methods for this same problem. See also Wright (1979), von Pragenau (1981), and Addison (1980) for the linear case and Park (1975) for the nonlinear case. This last work is particularly interesting as it demonstrates that methods unconditionally stable for linear problems are not so for the nonlinear case.

Jain and Jain (1981) develop hybrid P-stable methods that improve efficiency for solving periodic problems in celestial mechanics. De Silva and Grant (1978) of

NASA Ames describe research into the development of automatic structural synthesis methods for turbine disk and blade assemblies. Second-variation methods resulted in systems of stiff inhomogeneous matrix Riccatti equations.

2.13 Molecular dynamics (Aiken)

Mathematical models of atomic and molecular dynamic interaction have successfully predicted macroscopic physical properties of fluids. Only a very small quantity of the fluid can be modelled in order to limit the variables to hundreds or thousands of particles. Each particle can theoretically exert a force on all the other particles, but the simplification is made that beyond some cut-off distance the force is too small to consider in the model. This causes a discontinuity in the interaction potential that is a source of error in many molecular dynamic simulations.

The strength of the interaction is a very strong function of distance. Nearest neighbors are thus led by a rapidly varying primary force, while particles farther apart change more slowly with time. This natural partitioning by distance of the stiff from nonstiff variables has been used by Street and Tildesley (1978) with a second-order Taylor-series method and extrapolation to decrease computation time. Their application was to 108 methane molecules with five sites of interaction per molecule, or 540 potential interactions. Adequate accuracy was established arbitrarily on the basis of satisfaction of conservation of energy to within 0.05 per cent per 1000 time steps.

The mathematical model allows "computer experiments" to be performed that could not be done in the laboratory or would be expensive to do. Broughton and Abraham (1980) illustrate this in their study of crystal–melt interfaces. They use a variety of GEAR.

Heinzinger, Riede, Schaefer, and Szasz (1978) investigated simulations of liquids with ionic interactions and found that higher-order integration schemes were necessary for the faster rotational motion compared to the translational motion. Rossky and Karplus (1979) studied solvation; Karplus, Gelin, and McCannon (1980) studied internal dynamics of proteins; and Rossky and Karplus (1979) further explored solvent-solute interactions; all using a version of GEAR with typical time steps on the order of E −16 seconds. In this last study, the limitation of step size was attributed to rapid liberational motion of the water molecules and the correspondingly rapid change in the interaction energy.

Dove and Raynor (1982), Dove and Teitelbaum (1979), and Dove and Raynor (1979) offer an interesting approach to the study of vibrational relaxation in hydrogen. The relaxation was treated as a chemical kinetics problem, with each vibrational/rotational level considered a distinct species. Reference also Gerlich, Nowotny, Schlier, and Telay (1980), Haile and Graben (1980), and Powles, Evans, McGrath, Gubbins, and Murad (1979) for other stiff molecular dynamic simulations.

Microscopic simulations of reacting systems has evolved separately from molecular or atomic dynamics without reaction. This area has, in addition to the time-consuming potential surface evaluations, multiple time scales from the mechanical versus the chemical: while some modest time frame may be enough to model motion, a reactive interaction is typically rare on that time scale—and fast. See

Turner (1978) for a review of reactive molecular dynamics and consideration of how the interplay of the physical and chemical on the molecular level can influence macroscopic physical properties.

2.14 Nuclear

2.14.1 *Introduction* (*Aiken*)

Safety considerations in this field encourage detailed dynamic models for worst-case numerical experiments, training, design, and control. Thus stiffness can be identified at the atomic stage, mechanical fuel-handling stage, spent-fuel disposal stage, or the overall process stage.

Hetrick, Girijashankar, and Palusinski (1981) of the Electric Power Research Institute (see also §5.4.7) presented a model of a pressurized water reactor system and a partitioning technique (together with EPISODE) to handle numerical trouble that occurs after the initial phase because of increasing imaginary contributions to the eigenvalues. MMS (Modular Simulation System) is a code also written at EPRI for nuclear plant dynamics simulation (Divakaruni, Long, and Armor 1982).

The radiolytic decomposition of water is important to both moderator and coolant chemistry in nuclear reactors and is stiff because of disparity of rate constants. Carver and Boyd (1979) and Boyd, Carver, and Dixon (1980) presented a model and solve it by their own stiff sparse integration method for mass action kinetics called MACKSIM. Examples of stiff nuclear reactions are given in Ward and Fowler (1980), handled by a GEAR program.

Lawrence and Dorning (1977) used smoothing and extrapolation techniques on equations that are stiff because they include the greatly differing time constants associated with prompt and delayed neutrons in subcritical and delayed supercritical transients in a fast reactor. The so-called "prompt-jump" approximation is the same as the commonly used steady-state approximation; Blenski, Gadomski, and Mika (1978) gave higher-order singular perturbations for this application. Thermal reactor transients are moderately stiff and prompt supercritical reactivities non-stiff.

Devooght (1980a, b) has developed a more general steady-state method for nuclear reactor space–time kinetics as used for description of power transients in fast and thermal reactors. He gives a review of stiffness in these models in §2.14.2.

Rapid ejection of a control rod from the center of a thermal reactor is solved by Carver and Baudouin (1976) using a version of GEAR with their own sparse matrix solver. They also solve a stiff test model of a control rod withdrawn in a direction parallel to a reactor channel. The transient is a long 200 seconds.

The training of control center operators is often done on analog computers. The real-time critical nature of this problem and the curse of dimensionality for digital simulation explain the attractiveness of the analog approach here.

Patterson and Rozsa (1980), of Lawrence Livermore Laboratory, describe a nuclear process simulator called DYNSYL, also usable for more general chemical processes. Chambers (1978) documents use of his AGR (Advanced Gas-cooled Reactor) digital simulator for real-time solution. Thompson and Tuttle (§5.4.8) present recent software developed at Babcock and Wilcox with an interesting explanation of its historical development in this industrial environment. Halin (1976) discussed performance of conventional stiff methods on nuclear problems with discontinuities. Earles, Wagner, and Foster (1982) described FRST, a Fast Running Space–Time code for nuclear power plant simulations; they report that coarse-mesh finite differencing allows a three-dimensional problem to be solved in real-time on a minicomputer!

Borgonovi, Hammelman, and Miller (1980) investigated solution of stiff models for predicting plutonium inventory on a continuous basis. Carver (1981) discussed numerical aspects of thermohydraulics. Gaffney (1982), of Union Carbide Nuclear Division, surveys methods for solution of stiff oscillatory problems as arise from magnetohydrodynamic equations.

2.14.2 *Nuclear reactor kinectics* (*Devooght*)†

The integration of the reactor kinetics equations is a time-consuming operation whenever it is performed in the full space–energy–time frame. Moreover when coupling of the neutronic parameters with thermohydraulic parameters is present, the problem is even more taxing for the computer. A large number of approximate integration schemes have therefore been studied to reduce the numerical problem to a manageable size. However, neutron kinetics, despite its large spread of eigenvalues, is not typically a difficult stiff problem.

Special techniques have been devised to cope with the particular structure of the kinetics equations (see §5.4.1) owing to two facts:

(i) Neutron kinetics is a linear evolution problem, which becomes nonlinear in so far as it becomes a neutron dynamics problem through coupling with other variables.

(ii) Stiffness can more or less be "factored out," i.e. if we write the neutron density

$$N(r, v, t) = T(t)\psi(r, v, t), \tag{3.28}$$

a factorization used in "quasistatic approximation", a suitable function ψ can be obtained such that the stiff behavior is more or less confined to $T(t)$.

One way to cope with stiffness problems is to use the steady-state approximation (see §7.2.5) which has been known in reactor kinetics as the prompt-jump approximation. Goldstein and Shotkin (1969, 1970) tried to extend the prompt-jump approximation to higher orders but their results based upon intuitive derivation are valid only for the first order. The systematic application of the singular perturbation method to reactor kinetics was first proposed by Hendry and Bell (1969) who derived the algorithm including the intermediate expansion by a purely heuristic argument. Recent developments have been given by Mika (1975, 1976), Blenski, Gadomski, and Mika (1978), and Blenski (1978). The method seems attractive for point kinetics and some nonlinear kinetics problems. So far, limited numerical experience is available, and we shall not develop this point here.

Various approximations have been devised: at one end of the spectrum we have point kinetics (a system of seven linear differential equations) where the reactor is acting as a whole, i.e. a single point. At the other end of the spectrum

† See Devooght (1982) for a more detailed treatment.

we have a full-fledged three dimensional space and energy problem which could involve 10^5 unknowns. Intermediate options involve coarse mesh methods or flux-synthesis methods which we group under the heading multipoint reactor kinetics.

Advanced work in numerical reactor dynamics is mostly related to safety. The need for fast and reliable computing schemes for neutron transients stems from the fact that neutron fluxes yield the thermal power that drives a set of thermo-hydraulic equations. By far, the latter are the more difficult and more time consuming, and neutron kinetics is no longer the main problem. However, two points remain pressing: how to lengthen the time step and how to bound the error. Generally speaking, development has been outside the main stream of ODE stiff solvers, perhaps due to lack of familiarity. Some progress has been made for the first point, but adequate schemes are sorely lacking for the second point. Benchmark problems allow a common point of comparison, but so far the lack of analytical solution for complicated problems does not allow a distinction between convergence from a possible bias common to various methods. It is sometimes implied that the need for a very accurate solution of the reactor kinetics problems is not obvious because of the very large uncertainties in the thermohydraulics. However time-step lengthening is of utmost importance and is limited by the fact that kinetic operators are not constant.

The lack of memory space for large problems makes multistep methods rather unsuitable, and therefore one-step methods are almost the rule. As it may be surmised, almost all methods make implicit or explicit use of collocation, Galerkin or interpolatory quadrature, and therefore of an assumed time-dependent behavior of the solution, usually linear combinations of exponentials and polynomials. Polynomials, however, are not very suitable for fast transients of exponential type, and some methods place increasing reliance on spectral matching, i.e. the one-step growth function $W(hA)$ which approximates $\exp(hA)$ is chosen with the property that

$$W(h\lambda_i) = \exp(h\lambda_i),$$

where the λ_i are some eigenvalues (or their approximations) of A.

2.15 Process industries

2.15.1 *Introduction* (*Aiken*)

Barney and Johnson (1975) explain the incorporation of a GEAR version ODE-solver into modular simulation framework of DYNSYS; the latest version of this package, DYNSYL, developed at Lawrence Livermore, is described in Patterson and Rozsa (1978). Nilsen and Karplus (1974) give a review of continuous-system simulation languages. In particular, DYFLO2 (Franks, 1982) interfaces easily with the user interested in simulating typical dynamic chemical engineering unit operations; a stiff option is available. Also see Fagley and Carnahan (1982) for chemical-process equipment-grouping strategies for minimizing the size of the matrices manipulated by implicit stiff methods. Ockendon (1980) gives a survey of dynamic simulation of Oxford industry problems.

Dissinger and Shiesser (1982) point out that in the chemical industry, the ODEs often have complicated physical and chemical functionalities or table look-ups so that the computational effort to integrate the set can be strongly related to minimization of derivative evaluations.

Refer also to §4.7 for tips on the selection of large-scale integrators in process industries.

2.15.2 *Heat and mass transfer with reaction* (*Seider, White III and Prokopakis*)†

Models that couple heat and mass transfer between a bulk fluid phase and a solid catalyst are justified when the resistances to heat and mass transfer are important. Such a model is presented by Luss and Amundson (1968) to represent the dynamics of a batch fluidized-bed reactor where external resistances are significant. Initially the catalyst particles are heated to temperature, T_p^0, and have a partial pressure of reacting species A, p_p^0 (normally zero). Then the bulk phase (gas or liquid is circulated through the bed with initial temperature, $T_e (< T_p^0)$ and partial pressure, $p_e (> p_p^0)$. A is transferred from the bulk phase to the particles where the first-order reaction $A \to B$ occurs while the particles are cooled. The mass and heat balances for the bulk phase are:

$$\frac{dp}{d\tau} = p_e - p + H_g(p_p - p), \qquad p(0) = p_e,$$

$$\frac{dT}{d\tau} = T_e - T + H_T(T_p - T) + H_w(T_w - T), \qquad T(0) = T_e,$$

and for the reacting catalyst:

$$A\frac{dp_p}{d\tau} = -H_g K k p_p + H_g(p - p_p), \qquad p_p(0) = p_p^0,$$

$$C\frac{dT_p}{d\tau} = H_T F K k p_p + H_T(T - T_p), \qquad T_p(0) = T_p^0,$$

where T and p are the temperature and partial pressure of species A in the bulk phase; T_p and p_p, the particle properties; and k, the rate constant for a first-order reaction, A → B. The results of Luss and Amundson are illustrated in Figure 2.3. Using the explicit Runge-Kutta-Gill integration formula, the step size is small initially and decreases as $T_p \to T$ and $p_p \to p$, with decreasing rates of change. $|\lambda|_{max}$ increases and, hence, stiffness increases with stabilization. When the temperatures and partial pressures differ by less than 1 percent, stiffness sets in ($|\lambda|_{max}$ = 2187). To permit efficient integration with an explicit integrator, Luss and Amundson set $T_p = T$ and $p_p = p$ and add equations (12) and (14) and equations (13) and (15), giving two less-stiff ODEs ($|\lambda|_{max}$ = 1.242),

$$(1 + A)\frac{dp}{d\tau} = p_e - p - H_g K k p$$

† §2.15.2, §2.15.3, and §2.15.4 appeared after the Park City meeting in *Proceedings of the AICHE–CIESC Conference, Beijing, China, September (1982)*; published here with permission.

$$(1+c)\frac{dT}{d\tau} = T_e - T + H_w(T_w - T) + H_T FKkp$$

which are integrated with a step-size increased by a factor of 100.

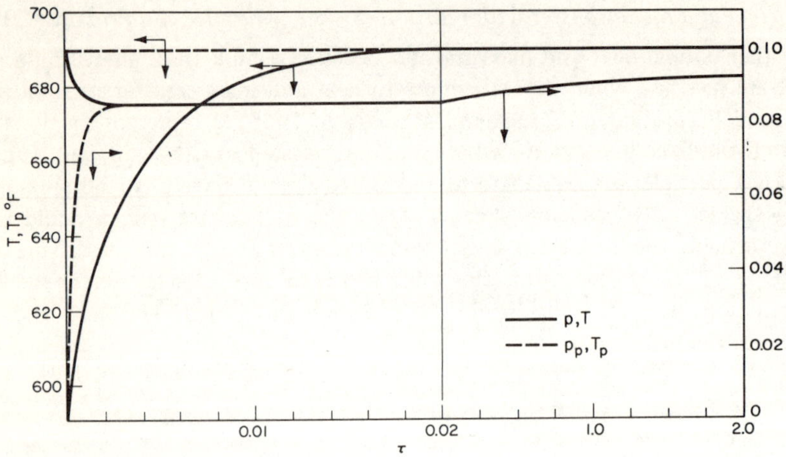

Fig. 2.3. Profiles of bulk and particle temperature and partial pressure for the batch fluidized-bed reactor. From Luss and Amundson (1968).

2.15.3 *Transport* (*Seider, White III, and Prokopakis*)

For weak shock waves, involving a perfect gas, Burgers (32) combined the momentum and energy balances, and continuity equation to give:

$$\frac{\partial u}{\partial t} = -u\frac{\partial u}{\partial x} + \mu\frac{\partial^2 u}{\partial x^2}, \tag{2.29}$$

where μ is an effective viscosity which depends on viscosity as well as thermal conductivity. This equation was subsequently found to apply for accoustical dissipation in sound waves of low frequency. With the boundary conditions,

$$u(t, 0) = \phi(t, 0),$$
$$u(t, 1) = \phi(t, 1),$$
$$u(0, x) = \phi(0, x),$$

where

$$\phi(x, t) = \frac{0.1e^{-A} + 0.5e^{-B} + e^{-C}}{e^{-A} + e^{-B} + e^{-C}},$$

$$A = \frac{0.5}{\mu}(x - 0.5 + 4.95t),$$

$$B = \frac{2.5}{\mu}(x - 0.5 + 0.075t),$$

$$C = \frac{5}{\mu}(x - 0.375).$$

The analytical solution is $u\{x, t\} = \phi\{x, t\}$. Figure 2.4 shows that $\phi\{x, t\}$ represents two shock waves that propagate in the x-direction.

Equation (2.29) has been integrated with the method of lines, using a five-point central difference approximation for $\partial^2 u/\partial x^2$ and a third-order four-point, up-wind finite-difference approximation for $\partial u/\partial x$. Such an approximation is found to improve the tracking of steep waves when the viscosity $\mu = 0$ (Carver 1976, and Hu and Schiesser 1981).

With the grid spacing $\Delta x = 0.01$, we integrated the ODEs to give the solution in Figure 2.4, which compares favorably with the analytical solution. Of course, as μ increases, viscous dissipation spreads the fronts as shown.

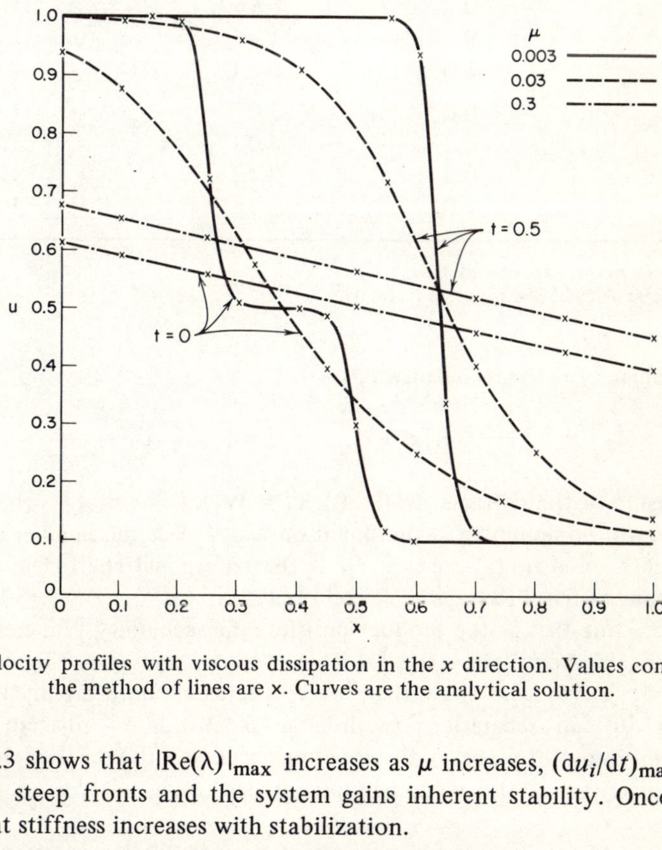

Fig. 2.4. Velocity profiles with viscous dissipation in the x direction. Values computed using the method of lines are x. Curves are the analytical solution.

Table 2.3 shows that $|\text{Re}(\lambda)|_{max}$ increases as μ increases, $(du_i/dt)_{max}$ decreases in the less steep fronts and the system gains inherent stability. Once again, we observe that stiffness increases with stabilization.

2.15.4 *Distillation (Seider, White III, and Prokopakis)*

It is well known that feedback of mass and energy more closely couples slow and fast process units in recycle loops. This is exemplified in the countercurrent cascade of vapor–liquid separators (or trays) of a distillation tower. The trays (small hold-up) respond rapidly to changes in flow rate, but the large reboiler slows the coupled response. After the initial rapid transients, the ODEs become stiff as compositions and temperatures respond slowly.

TABLE 2.3

Eigenvalues for momentum balance with viscous dissipation†

Dimensionless viscosity, μ	t	$\lvert Re(\lambda)\rvert_{min}$	$\lvert Re(\lambda)\rvert_{max}$
0.003*	0.0	0.937	224
	0.2	0.964	238
	0.5	1.04	254
	1.0	2.65	263
0.03	0.0	0.721	1.69×10^3
	0.2	0.900	1.71×10^3
	0.5	1.44	1.72×10^3
	1.0	5.39	1.72×10^3
0.3	0.0	3.07	1.61×10^4
	0.2	3.08	1.61×10^4
	0.5	3.12	1.61×10^4
	1.0	3.18	1.61×10^4

* Most eigenvalues are complex.

† All eigenvalues have a negative real part.

The Jacobian of the mass balances:

$$\frac{d\boldsymbol{x}_j}{dt} = \boldsymbol{G}_j\boldsymbol{x}_j + \boldsymbol{d}_{x_j}, \qquad j = 1, \ldots, C, \tag{2.30}$$

gives a measure of the stiffness. In (2.30), $\boldsymbol{x}_j = [x_{1j}, \ldots, x_{Nj}]^T$, where x_{ij} is the mole fraction of compound j in the liquid on tray i. N is the number of trays and C the number of chemical species. $\boldsymbol{G}_j$ is the tridiagonal coefficient matrix due to the coupling of tray i to trays $i+1$ and $i-1$.

It is interesting that as the product purities approach unity, N increases and the mole fractions respond more slowly. This is demonstrated by Tyreus, Luyben, and Schiesser (1975) who show that $|\lambda|_{min}$ decreases more rapidly than $|\lambda|_{max}$. Hence, for difficult separations (with large pinch zones—adjacent trays with small concentration differences), the equations become stiffer as the response times increase. This effect is also demonstrated for decreasing relative volatility, $\alpha_{jr} = K_j/K_r$.

Typical operating conditions for an azeotropic distillation tower to dehydrate alcohol are shown in Figure 2.5 (Prokopakis and Seider 1983). The trays in the upper portion of the stripping section have very small changes in alcohol and entrainer concentration while water is removed. Then, in a few trays the alcohol mole fraction is increased to near unity as the entrainer is eliminated. These trays respond more rapidly than others in the stripping section, as illustrated in Figure 2.6, which shows the profiles of liquid mole fractions after a thirty percent increase in the feed flow rate.

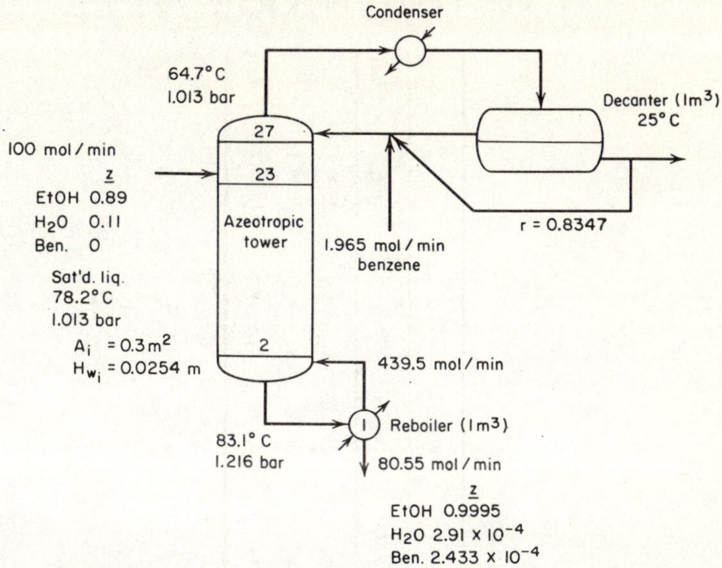

Fig. 2.5. Azeotropic distillation tower for dehydration of ethanol with benzene. From Prokopakis and Seider (1983).

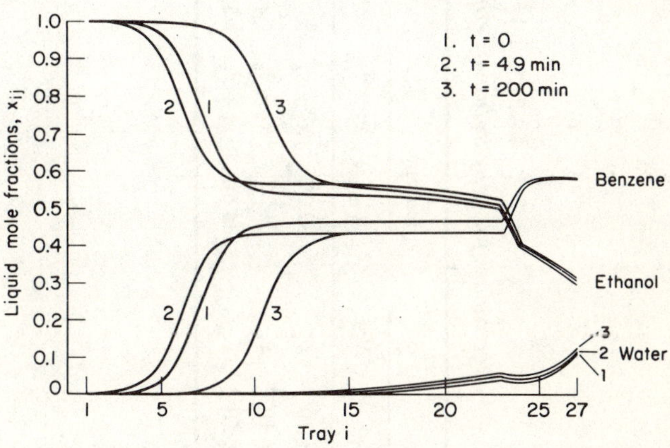

Fig. 2.6. Profiles of liquid mole fractions after a 30 percent increase in feed flow rate. From Prokopakis and Seider (1983).

Table 2.4 shows $|\lambda|_{max}$ and $|\lambda|_{min}$ for ethanol, benzene, and water. Note that $|\lambda|_{max}$ is not associated with liquid mole fractions on trays in the concentration fronts because the bordering trays are more stable. Furthermore, $|\lambda|_{max}$ increases with stabilization as the steady-state is approached. $|\lambda|_{min}$ is associated with Tray 1 because it has the largest holdup.

2.15.5 *Experiences at Exxon* (*Byrne*)

Stiff differential systems are fairly common at Exxon Research and Engineering (ER&E). Chemical kinetics represents a fairly routine set of problems, which

TABLE 2.4

*Eigenvalues for dehydration of ethanol with benzene**

	Ethanol				Benzene				Water			
t,min	$\|\lambda\|_{max}$	Var.	$\|\lambda\|_{min}$	Var.	$\|\lambda\|_{max}$	Var.	$\|\lambda\|_{min}$	Var.	$\|\lambda\|_{max}$	Var.	$\|\lambda\|_{min}$	Var.
0	19.65	$x_{10}-x_{18}$	6.31×10^{-3}	x_1	32.77	x_2-x_6	2.08×10^{-2}	x_1	26.87	$x_{11}-x_{18}$	7.05×10^{-3}	x_1
3.7	19.07	x_9-x_{16}	6.10×10^{-3}	x_1	31.77	x_2-x_5	2.91×10^{-2}	x_1	25.62	x_9-x_{15}	6.96×10^{-3}	x_1
127	18.06	$x_{13}-x_{18}$	5.94×10^{-3}	x_1	34.60	x_2-x_9	5.15×10^{-2}	x_1	23.92	$x_{14}-x_{18}$	6.17×10^{-3}	x_1

* All eigenvalues are real and negative.

are generally solved directly by the user. So far, most kinetics models we have seen involve from ten to thirty ODEs and reaction rates that frequently vary over eight orders of magnitude. Usually, the transient and steady-state results are both of interest. The kineticists sometimes use CHEMKIN (Kee, Miller, and Jefferson 1980) and LSODE (Hindmarsh 1981) to reduce a stoichiometric model to FORTRAN-usable data and to integrate the system with LSODE. The most frequent difficulties we see are related to inappropriate error control. For example, absolute error control or error control relative to the largest component value seen so far does not generally work well. This caveat has been issued by others (Curtis 1978a) but not often heard, it seems.

A variant of the kinetics model was described by Gear (1971a). Suppose some data are collected for one or two chemical species in a reactive system at time t^*. Similar sets of data may correspond to different initial data but to the original chemical kinetics model. The idea is to estimate the members of the components of the generalized Arrhenius reaction rates for specified reactions. The problem is badly scaled because Arrhenius parameters may be gigantic when compared to the relative differences between the computed and observed data. Of course, the function evaluations for the optimizer require the solution of one or more systems of ODEs. Further, the point t^* at which the data are collected may lie in the steady regime of a process involving very fast reactions. Some current work on the ER&E's Thermal DeNox process involves a problem of this type. This process reduces nitric oxide levels in flue gases and is used by industrial and utility companies. The determination of correct reaction rate coefficients influences the numerical model and hence the physical installation and operation of the scrubbing system itself.

The numerical method of lines is also a common source of ODEs at ER&E. Typical partial differential equations have the form

$$\partial_t \boldsymbol{u} - \nabla \cdot (\mathbf{v}\boldsymbol{u}) = \nabla \cdot D\nabla \boldsymbol{u} + \boldsymbol{p}$$

where $\boldsymbol{u}$ is vector-valued, the domain may be one or two-dimensional, and $\boldsymbol{p}$ is a reactive source whose components are of the kinetics form. The PDE may be solved with a package such as DSS2 (Schiesser 1977) or DISPL2 (Leaf and Minkoff 1981), or the user may develop his own discretization.

One such problem involved no convection ($\mathbf{v} = 0$), very little diffusion ($D \ll 1$), reactive source/sink terms, and a few species (about half a dozen). The reaction rates varied over nine orders of magnitude, with the largest rate around 10^{10}. I used B-splines and a very uneven grid to handle the treatment of the initial profile, since the initial and boundary conditions were not compatible and since sharp fronts were anticipated and observed. For this one-dimensional problem, the gradient $\nabla \boldsymbol{u}$ was of particular interest. The problem was very stiff owing to the fast reactions and an interest in the solution through the transient regime and into the steady state regime. The boundary conditions were of Neumann and Dirichlet types. For this work I used DISPL2. Further research in this area is expected to be carried out both computationally and in physical experiments.

Other problems involve flow of reactive fluids in baffled reactors, absorbers, and so on. Thus, unusual domains appear to be common. However, my involvement in these areas is limited.

A general observation is this: many models do not match up well with the off-the-shelf PDE packages. Hence, the evolution of codes like DISPL2 is quite natural. That is, more general coordinate systems, cross-diffusion, and certain flux terms have been accommodated in DISPL2 as a consequence of its applications. By the way, the error control caveat is very important here, for temporal tolerances that are too loose or too tight have dramatic impact on the quality of results, as well as cost.

Dynamic grid strategies, in the setting of the numerical method of lines, appear to have great potential in the simulation of transport phenomena. Thus, we are beginning a study of this area.

Some time ago Hindmarsh and I started an investigation of linear and nonlinear systems solvers in a numerical method of lines setting. We used a combustion model as the basic problem, finite differences for the discretization, and GEAR (Hindmarsh 1974a) for the study. We tried several things—using only the diagonal blocks of the Jacobian, Newton-block SOR and nonlinear block SOR, to name a few. We are now concluding this work, which led to some unexpected results.

It may be somewhat surprising that we also see nonstiff problems in a variety of settings, including small differential–algebraic systems and two-point boundary value problems, although not frequently. Various strategies are used to solve these problems.

2.16 Process vessels

2.16.1 *Introduction* (*Aiken*)

Naturally a reactor model will be stiff if its kinetic equations are stiff or if there is a difference in the characteristic transport time from the reaction time, but we are concerned here with other types of stiffness originating from the nature of the reactor model. A reactor is a vessel through which reactant continuously flows in and product out; there may be multiphase flow or the reactor may contain a fixed solid phase on which reaction takes place. Introduction of each new phase increases the potential for stiffness, depending on the model detail, as different phases may have much different physical, chemical, thermodynamic, and transport properties. For example, a tubular reactor containing a solid particle phase changes its temperature much more slowly than a gas phase flowing through it.

The strong exponential temperature dependence of reaction rates can cause sharp temperature spikes as one marches down an exothermic tubular reactor. This nonlinearity is also responsible for the existence of multiple steady states. Aiken and Lapidus (1974) gave an example of a nonisothermal fluidized bed that is very stiff and possesses three possible steady states. See also Fan, Tojo, and Chang (1979) and Michelsen (1976), who solved this problem with a semi-implicit Runge–Kutta method.

Inclusion of a diffusive or dispersion term under conditions where convection very much dominates the flow description (high Peclet numbers) leads to stiff computation (see, for example, Shah and Parakos 1975, and Serth 1975). Smith (1980) used a finite element approach to this problem. Discontinuous boundary conditions for tubular reactors, often used under some controversy, cause stiffness.

Interphase mass transport with reaction can produce stiff boundary-value problems inherently unstable in any direction. Aiken (1982c) solved a gas purification model of simultaneous transport of two gases into a liquid where each reacts with a third species; one of the two reactions is much faster than the other.

Karanth and Hughes (1974) used orthogonal collocation to solve a detailed model of an adiabatic packed bed reactor, including interphase transfer to the particles and intraphase particle transfer. Billingsley (1982) reported a collocation method for solving problems with packed columns; for a distillation example, the stiffness manifests itself as boundary layer behavior in the mole fractions at the top and bottom of the column. Cavendish and Oh (1979) solved the equations for diffusion and reaction in a bed of poisoned automotive catalyst pellets by applying first Galerkin's method, then a version of GEAR. Guertin, Sorensen, and Stewart (1977) used exponential collocation on some stiff reactor models. Eigenberger and Butt (1976) explained a technique for automatic non-equidistant grid-size space for finite differences on reactors with steep gradients. Edelson and Schryer (1978) compared finite difference with finite elements for one-dimensional reactive flow. Varma, Georgakis, Amundson, and Aris (1976) explored a number of computational methods for tubular reactors. Ramshaw (1980) discussed the use of the steady-state approximation in reactive three-dimensional flow. Cho and Joesph (1981) solved a heterogeneous model for moving-bed coal gasification reactors and removed stiffness with the steady-state approximation. Chin and Braun (1980) solved a model of reacting flow in a porous medium, George and Harris (1977) of in situ oil shale retorting—all quite stiff problems.

See also §2.4 on combustion where reactive flow in the confines of a combustor is discussed.

2.16.2 *Batch adsorption* (*Rodrigues, Costa, Dias, Lopes*)

A nonsteady-state mass balance of solute in a volume element of an adsorbent particle, assuming homogeneous structure and Fick's law for diffusion is:

$$\frac{\partial y}{\partial \theta} = \frac{\partial^2 y}{\partial r^{*2}} + \frac{2}{r^*}\frac{\partial y}{\partial r^*}. \tag{2.31}$$

This is a parabolic PDE where y is the normalized solid concentration, $y = q/q_0$, $r^* = r/r_0$ is the reduced radial position and $\theta = D_e t/r_0^2$ is the reduced time (D_e is the effective diffusivity of the solute in the particle), and q is the concentration of solute within the particle. The subscripted zero refers to the particle surface.

If no solute is present on the solid at the beginning of the operation, and the solid is contacted with a solution containing solute at concentration c_0, the initial condition is:

$$y(r^*, 0) = 0, \qquad r^* < 1. \tag{2.32}$$

The boundary conditions to (2.31) are the symmetry condition for the radial concentration profile, i.e.,

$$\left.\frac{\partial y(r^*, \theta)}{\partial r^*}\right|_{r^*=0} = 0, \tag{2.33}$$

and an overall mass balance at the periphery of the particles:

$$1 - x(\theta) = \xi\bar{y}(\theta), \tag{2.34}$$

where $x = c/c_0$ is the reduced fluid phase solute concentration, $\xi = (1 - \epsilon)q_0/\epsilon c_0$ is the capacity factor (ϵ is void fraction), and the average solute concentration in the solid is

$$\bar{y}(\theta) \equiv 3\int_0^1 y(r^*, \theta)r^{*2}\mathrm{d}r^*. \tag{2.35}$$

Assuming solid-fluid equilibrium at the interface (no film mass transfer resistance) we get:

$$y(1, \theta) = f[x(\theta)]. \tag{2.36}$$

Combining (2.34), (2.35), and (2.36), we obtain for the second boundary condition:

$$\frac{\mathrm{d}f^{-1}\{y(r^*, \theta)\}}{\mathrm{d}y} \cdot \frac{\partial y(r^*, \theta)}{\partial \theta}\bigg|_{r^*=1} = -3\xi \frac{\partial y(r^*, \theta)}{\partial r^*}\bigg|_{r^*=1}. \tag{2.37}$$

This problem has an analytical solution for the linear case, $f(x, \theta) = x(\theta)$, i.e.,

$$\bar{y}(\theta) = \frac{1}{1 + \xi} \quad 1 - \sum_{n=1}^{\infty} \frac{6\alpha(\alpha + 1)\exp(-q_n^2\theta)}{9 + 9\alpha + q_n^2\alpha^2}, \tag{2.38}$$

where $\alpha = 1/\xi$ and q_n are tabulated values (Crank 1956).

We used this analytical expression to examine two numerical methods for this type of problem: the method of lines with GEAR, and the discretization from global orthogonal collocation or B-spline collocation. In both cases $u = r^{*2}$ is used as an independent variable.

Applying global collocation with Lagrange polynomials (Villadsen and Michelsen 1978) and using N interior points plus one exterior point we get a system of $N + 1$ stiff ordinary differential equations:

$$\frac{\mathrm{d}y_j(\theta)}{\mathrm{d}\theta} = 6\sum_{i=1}^{N+1} A_{ji}y_i + 4u_j \sum_{i+1}^{N+1} B_{ji}y_i, \qquad j = 1, 2, \ldots, N, \tag{2.39a}$$

$$\frac{\mathrm{d}y_{N+1}(\theta)}{\mathrm{d}\theta} = -6\xi\left(\frac{\mathrm{d}f^{-1}|y_{N+1}(\theta)|^{-1}}{\mathrm{d}y}\right)\sum_{i=1}^{N+1} A_{N+1,i}y_i, \qquad y_j(0) = 0. \tag{2.39b}$$

This system was integrated for various N using GEAR. Table 2.5 compares the results from the analytical solution with this numerical computation. For a precision criterion $|x_{N+1}(\theta) - x_N(\theta)| \leqslant 10^{-3}$ we can see that more than four collocation points are needed (for Langmuir isotherms, $f(x) = (1 + Kc_0)x/(1 + Kc_0x)$, we needed $N = 8$ to satisfy the same precision criterion).

Some oscillations are observed along the radius, particularly at early times, although the results in terms of fluid concentration at the surface, $x(\theta)$, are good.

TABLE 2.5
Comparison of orthogonal collocation and analytical solutions

θ	Analytical solution, x	Numerical solution				
		$N = 1$	$N = 2$	$N = 4$	$N = 6$	$N = 8$
0.001	0.319*	−1.50	−0.219	0.388	0.448	0.453
0.02	0.161	0.071	0.185	0.165	0.161	0.161
0.04	0.128	0.091	0.159	0.126	0.128	0.128
0.06	0.115	0.093	−0.140	0.114	0.115	0.115
0.08	0.107	0.093	0.126	0.107	0.107	0.107
0.10	0.102	0.093	0.116	0.102	0.102	0.102
0.12	0.099	0.093	0.109	0.099	0.099	0.099
0.14	0.097	0.093	0.105	0.097	0.097	0.097
0.16	0.096	0.093	0.101	0.096	0.096	0.096
0.18	0.095	0.093	0.099	0.095	0.095	0.095
0.20	0.094	0.093	0.097	0.094	0.094	0.094

* For a very low times we should consider more terms in the series development of the analytical solution.

Radial profiles without oscillations required twenty or more collocation points. This is due to the very steep initial profile near $u = 1$.

The second approach can recognize sharp gradients since the collocation interval is divided into several subintervals; in each subinterval collocation is used, which allows a better distribution of collocation points.

2.17 Listing review (Aiken)

For a listing of some available work on stiff methods (1974–82), refer to Appendix I. This is meant to be only a guide to the literature and is not exhaustive or definitive.

3

STIFF METHODS

3.1 Runge-Kutta

3.1.1 *Introduction* (*Cash*)

A q-stage Runge-Kutta formula for the numerical integration of the initial value problem

$$\frac{dy}{dx} = f(x, y), \qquad y(x_0) = y_0, \tag{3.1}$$

can be written in the general form

$$y_{n+1} - y_n = h \sum_{i=1}^{q} b_i k_i, \tag{3.2a}$$

$$k_i = f\left(x_n + c_i h, y_n + h \sum_{j=1}^{q} a_{ij} k_j\right), \qquad 1 \leqslant i \leqslant q, \tag{3.2b}$$

where $h \equiv x_{n+1} - x_n$.

Such formulae can be represented conveniently by the arrays

$$\begin{array}{c|cccc} c_1 & a_{11} & a_{12} & \cdots & a_{1q} \\ \cdot & \cdot & & & \cdot \\ \cdot & \cdot & & & \cdot \\ \cdot & \cdot & & & \cdot \\ c_q & a_{q1} & a_{q2} & & a_{qq} \\ \hline & b_1 & b_2 & \cdots & b_q \end{array} \equiv \begin{array}{c|c} \mathbf{c} & A \\ \hline & b^{\mathrm{T}} \end{array} . \tag{3.3}$$

Classically all Runge-Kutta formulae were explicit, i.e. $a_{ij} = 0$ for $j \geqslant i$, but they became interesting for the numerical integration of stiff differential systems with the introduction of fully implicit Runge-Kutta formulae by Butcher (1964). Butcher showed that for all q there exist fully implicit q-stage Runge-Kutta formulae of order $2q$, and Ehle (1969) has shown that all of these maximal order formulae are A-stable. Thus, among the class of implicit Runge-Kutta formulae we have the possibility of deriving A-stable methods of arbitrarily high order whereas it is well known from a celebrated result of Dahlquist (1963) that the highest attainable order of an A-stable linear multistep method is limited to two. In addition to this, the pioneering work of J. C. Butcher (1963, 1965) has led to the development of a general algebraic framework which allows the investigation of quite high-order Runge–Kutta formulae in a comparatively straightforward manner.

One of the main arguments against the use of fully implicit Runge–Kutta formulae was on the grounds of the amount of computational effort required to solve (3.2b) for the k_i. If (3.2) is applied to a stiff system it is necessary to solve the resulting system of algebraic equations, defining the k_i, using some form of Newton iteration. The system of algebraic equations to be solved is of size qs, so that a modified Newton iteration scheme, if applied in the most obvious way, will for large s require the order of $\frac{1}{3}q^3s^3$ multiplications to obtain the required solution. This is about q^3 times as much work as is required to solve the algebraic equations arising from the use of a linear multistep method. It follows that if these two classes of methods are compared on this basis, the immediate conclusion is to reject fully implicit Runge–Kutta formulae in favor of linear multistep methods. However, because of the inherent stability of some classes of implicit Runge–Kutta formulae, coupled with the desirability of having high-order single-step integration methods, there has been a large amount of research into the possibility of overcoming these computational difficulties. Broadly speaking, research into finding efficient high-order implicit Runge–Kutta methods has followed two main directions:

(i) the investigation of the use of transformation methods to obtain a solution of algebraic equations defining the k_i in an efficient manner;
(ii) the derivation of different classes of implicit Runge–Kutta formulae which do not call for the solution of a system of qs simultaneous algebraic equations.

In what follows we shall survey some of the recent results concerned with Runge–Kutta formulae. The literature on implicit Runge–Kutta formulae is extensive, and an attempt to cover it fully in a single chapter has proved impossible. Instead, what we have done is to gather those results which we think will have a significant effect on how Runge–Kutta codes will be developed in the near future.

3.1.2 *Stability requirements for implicit Runge–Kutta formulae* (*Cash*)

If the Runge–Kutta formula (3.2a, b) is applied with a fixed step-size h to the scalar test equation

$$\frac{dy}{dx} = \lambda y, \qquad \lambda \in \mathbb{C}, \ \mathrm{Re}(\lambda) < 0, \tag{3.4}$$

we obtain the relation

$$y_{n+1} = R(z) y_n, \qquad z = h\lambda, \tag{3.5}$$

where

$$R(z) = 1 + z\boldsymbol{b}^{\mathsf{T}}(\boldsymbol{I} - z\boldsymbol{A})^{-1}\boldsymbol{e}, \qquad \boldsymbol{e} = \underbrace{[1, 1, \ldots, 1]}_{q}{}^{\mathsf{T}}, \tag{3.6}$$

and $\boldsymbol{I}$ is the $q \times q$ unit matrix.

Here $R(z)$ is a rational function of z with both the numerator and the denominator having maximum degree q. The complex number z lies inside the region of absolute stability of (3.2) if $|R(z)| \leqslant 1$. The first, and still most widely used, stability criterion for Runge–Kutta formulae applied to stiff systems is that of A-stability due to Dahlquist (1963).

Definition 3.1. A numerical method is said to be A-stable if its region of absolute stability contains the whole of the left-hand half-plane $\text{Re}(z) < 0$.

A stronger condition, which ensures that the method has the correct damping at $z = -\infty$, is that of L-stability due to Ehle (1969).

Definition 3.2. A Runge–Kutta formula is said to be L-stable if it is A-stable and, in addition,

$$\lim_{\text{Re}(z)\to-\infty} |R(z)| = 0. \tag{3.7}$$

The A-stability criterion was originally developed for linear multistep methods but has been widely applied to Runge–Kutta formulae. Many results are known concerning the conditions under which the rational function $R(z)$ has modulus less than unity for all z in the complex left-hand half-plane. (Ehle (1969) calls such a property A-acceptability and if, in addition, (3.7) holds, then $R(z)$ is said to be L-acceptable.) In particular, it is known that if $R(z)$ is a diagonal of the Padé table for e^z, then $R(z)$ is A-acceptable (Birkhoff and Varga 1965). Furthermore, if $R(z)$ lies on one of the two subdiagonals then $R(z)$ is L-acceptable (Ehle 1969). Runge–Kutta formulae which, when applied to (3.4), give rational approximations of this type, have been investigated by Axelsson (1969), Chipman (1971) and Ehle (1969). However, the investigation of the A-acceptability of rational approximations to e^z remains a difficult task although recently there have been some important advances in this area (Nørsett 1975, Iserles 1979, Wanner, Hairer, and Nørsett 1978). In particular, we mention the concept of order stars developed by Wanner *et al.* which is fast becoming an important tool in the theory of approximation.

Although the criterion of A-stability has proved to be a valuable one, particularly in helping us to discard those methods which are not promising for integrating stiff systems, it has the obvious limitations that the model test equation (3.4) is linear, homogeneous and has constant coefficients. Several authors have given examples of stiff equations for which a particular A-stable method with a certain variable step-size gives a violently unstable solution (Stetter 1973, p. 181). A quite widely used generalization of A-stability is that of S-stability due to Prothero and Robinson (1974). This is associated with the inhomogeneous test equation

$$y' = g'(x) + \lambda[y - g(x)]. \tag{3.8}$$

Prothero and Robinson sought to categorize those one-step methods which give a stable numerical solution when applied with a step length h to any equation of the form (3.8) where λ is a complex constant with negative real part and $g'(x)$ is any function that is defined and bounded in some interval $x \in [x_0, \bar{x}]$. They define the concept of S-stability in the following way.

Definition 3.3 A one-step numerical method is said to be S-stable if, for a differential equation of the form (3.8) and for any real positive constant λ_0, there exists a real positive constant h_0 such that

$$\frac{\|y_{n+1} - g(x_{n+1})\|}{\|y_n - g(x_n)\|} < 1,$$

providing $y_n \neq g(x_n)$, for all $0 < h < h_0$ and all complex λ with $\text{Re}(-\lambda) \geqslant \lambda_0$, and $x_n, x_{n+1} \in [x_0, \bar{x}]$.

In addition an S-stable one-step method is said to be strongly S-stable if

$$\frac{\|y_{n+1} - g(x_{n+1})\|}{\|y_n - g(x_n)\|} \to 0 \text{ as } \text{Re}(-\lambda) \to \infty$$

for all $h > 0$ such that $x_n, x_{n+1} \in [x_0, \bar{x}]$. Clearly, the concept of S-stability reduces to that of A-stability in the case $g(x) \equiv 0$. For further comments on S-stability see Stetter (1975).

Although many of the Runge–Kutta formulae that have recently been proposed for the integration of stiff systems still seek to satisfy the conditions of A-stability or S-stability, the test equation (3.8) has the major limitation that it is linear. A significant step in overcoming this deficiency was made with the G-stability criterion of Dahlquist (1975c) for multistep methods and the B-stability criterion of Butcher (1975) for Runge–Kutta methods. The concept of B-stability is concerned with the nonlinear test equation

$$\frac{dy}{dx} = f[y(x)] \tag{3.9}$$

and is defined by Butcher (1975) in the following way.

Definition 3.4 Let $y_{n-1}, y_n, \ldots, z_{n-1}, z_n, \ldots$ be two sequences of approximate solutions of (3.9) computed by an implicit Runge–Kutta method using the same step size h. Let $\langle .,. \rangle$ denote a scalar product on $\mathbb{R}^N$ and $\|\cdot\|$ the corresponding norm. The method is defined to be B-stable if, for any f satisfying

$$\langle f(u) - f(v), u - v \rangle \leqslant 0 \text{ for all } u, v \in \mathbb{R}^N, \tag{3.10}$$

it holds that $\|y_n - z_n\| \leqslant \|y_{n-1} - z_{n-1}\|$.

This condition was further extended by Burrage and Butcher (1979) to the nonlinear system

$$y' = f[x, y(x)], \quad f\colon \mathbb{R}^{N+1} \to \mathbb{R}^N, \tag{3.11}$$

satisfying the monotonicity condition

$$\langle f(x, y) - f(x, z), \quad y - z \rangle \leqslant 0 \text{ for all } y, z \in \mathbb{R}^N \text{ and all } x \in \mathbb{R}. \tag{3.12}$$

Associated with the test equation (3.11) Burrage and Butcher (1979) give the following definition of BN-stability.

Definition 3.5 Let $[y_n]$, $[z_n]$ be two sequences of approximations to the solution of (3.11) computed by (3.2a, b) using a fixed step h. If

$$\|y_n - z_n\| \leqslant \|y_{n-1} - z_{n-1}\|, \tag{3.13}$$

then the method is said to be BN-stable.

Condition (3.13), which with slight modifications is equivalent to the concept of contractivity used by Nevalinna and Liniger (1978), is stronger than that of

A-stability since A-stability requires only that $\|y_n - z_n\|$ is bounded as $n \to \infty$. Burrage and Butcher give the following sufficient condition for BN-stability:

Consider the quadratic form

$$Q(\theta_1, \theta_2, \ldots, \theta_q) \equiv \sum_{i,j=1}^{q} m_{ij}\theta_i\theta_j, \text{ where } m_{ij} = b_i a_{ij} + b_j a_{ji} - b_i b_j.$$

If the method (3.2a, b) is such that $b_i \geqslant 0$ for $1 \leqslant i \leqslant q$ and Q is non-negative, then (3.2) is BN-stable. This leads to the following definition of algebraic stability due to Burrage and Butcher (1979).

Definition 3.6 A Runge–Kutta method which is such that $b_i \geqslant 0$ and $\zeta^{\mathrm{T}} M \zeta$, where $M \equiv [m_{ij}]$, is a non-negative form, is said to be algebraically stable.

This condition has been given independently by Crouzeix (1979). The condition of BN-stability is equivalent to the A-contractivity condition of Nevalinna *et al.* (1978). Dahlquist and Jehsch (1979) also observed that it is necessary to have $b_i > 0$ for all i since if one of the b_i is zero we have $m_{ii} = 0$, and the non-negative definiteness of M implies that the ith row of M must be zero and the method becomes reducible. Burrage and Butcher (1979) have shown that if the Runge–Kutta formula is nonconfluent (i.e., $c_1, c_2, \ldots, c_q$ are distinct) then

$$\text{algebraic stability} \Leftrightarrow \text{BN-stability.}$$

A further interesting result has been proved by Hundsdorfer and Spijker (1981), who showed that for any irreducible Runge–Kutta formula (see, for example, Stetter 1973) which is not necessarily nonconfluent, the concepts of B-, BN- and algebraic stability are equivalent.

An interesting way of proving the B-stability of certain classes of Runge–Kutta formulae has been given by Wanner (1976), who showed that the implicit Runge–Kutta formulae based on Gaussian points and Ehle's IIA methods based on Radau points are B-stable by using the well known fact that these particular Runge–Kutta formulae are equivalent to collocation methods (Wright 1970). This technique has recently been extended by Nørsett and Wanner (1981) to include other classes of Runge–Kutta formulae. The problem of classifying classes of algebraically stable Runge–Kutta formulae has been considered by Burrage (1978b). He showed that the q-stage methods of order $2q$ are all algebraically stable and has given a classification of all algebraically stable q-stage Runge–Kutta formulae of order $2q - 2$ which are nonconfluent. However, Burrage made the point that such a classification is much harder for lower-order formulae.

In conclusion we can say that in the past few years there has been considerable development in the stability analysis of Runge–Kutta formulae applied to nonlinear stiff systems. The condition of algebraic stability has two major advantages over A-stability. First, it seems likely that an algebraically stable Runge–Kutta method will be superior for nonlinear stiff systems to one which is only A-stable. Second, it is possible to give an algebraic condition for algebraic stability (i.e. one depending only on the coefficients of the Runge–Kutta method). Because of this, algebraic stability, and the closely related concept of BN-stability, should play an important role in the future development of new Runge–Kutta formulae.

3.1.3 *Fully implicit Runge–Kutta formulae* (*Cash*)

It was mentioned in the introduction that the main objection to implementing fully implicit Runge–Kutta formulae has traditionally been on the gounds of the amount of computational effort required to solve the nonlinear equations resulting from a modified Newton iteration. Algorithms to overcome this problem for a special class of Runge–Kutta formulae have been given by Butcher (1976b) and for more general formulae, including the maximal-order ones based on Gaussian points, by Bickart (1977) and Varah (1979). These algorithms are all based on transformation methods and in what follows we shall describe Butcher's approach since this has already been implemented in a successful computer package.

Consider the solution of the differential system

$$y'(x) = f[y(x)] \tag{3.14}$$

using the q-stage formula (3.2). The numerical solution at the point $x_n = x_{n-1} + h$ is computed using

$$y_n = y_{n-1} + h \sum_{j=1}^{q} b_j f(Y_j), \tag{3.15}$$

where

$$Y_i = y_{n-1} + h \sum_{j=1}^{q} a_{ij} f(Y_j) \qquad 1 \leqslant i \leqslant q. \tag{3.16}$$

To evaluate $Y_1, Y_2, \ldots, Y_q$ satisfying (3.16) we use a modified Newton iteration scheme so that at the end of a correction iteration Y_i is replaced by $Y_i + \omega_i$. The increments ω_i satisfy

$$\omega_i - h \sum_{j=1}^{q} a_{ij} J \omega_j - Z_i = 0 \qquad 1 \leqslant i \leqslant q, \tag{3.17}$$

where J is the $s \times s$ Jacobian matrix evaluated at an appropriate point and

$$Z_i = -Y_i + y_{n-1} + h \sum_{j=1}^{q} a_{ij} f(Y_j), \qquad 1 \leqslant i \leqslant q. \tag{3.18}$$

Let $M = \bar{I} \otimes I - hA \otimes J$ be the matrix of coefficients in (3.15) and (3.16) where $\bar{I}$ is $q \times q$, I is $s \times s$ and $\otimes$ denotes a tensor product. Relation (3.17) can then be written as

$$M\omega - Z = 0 \tag{3.19}$$

for the appropriate vectors ω, Z, and the matrix M.

We assume that system (3.19) is to be solved for ω in the usual way by first LU decomposing the matrix M. The number of multiplications required to carry out this decomposition is $Cs^3/3 + O(s^2)$ for large s and for the back substitution we require $Ds^2 + O(s)$ multiplications where $C = q^3$ and $D = q^2$. The idea behind Butcher's approach is to solve a suitably transformed system of equations so that the factors C and D are considerably lowered.

Let $\boldsymbol{P}$ and $\boldsymbol{Q}$ be nonsingular $q \times q$ matrices so that

$$\tilde{\omega} = (\boldsymbol{Q}^{-1} \otimes \boldsymbol{I})\omega, \tilde{z} = (\boldsymbol{P} \otimes \boldsymbol{I})\boldsymbol{Z}.$$

Then $\tilde{\boldsymbol{M}} = (\boldsymbol{P} \otimes \boldsymbol{I})\boldsymbol{M}(\boldsymbol{Q} \otimes \boldsymbol{I}) = \boldsymbol{PQ} \otimes \boldsymbol{I} - h\tilde{\boldsymbol{A}} \otimes \boldsymbol{J}$, where $\tilde{\boldsymbol{A}} = \boldsymbol{PAQ}$. It immediately follows that (3.19) is equivalent to the transformed system

$$\tilde{\boldsymbol{M}}\tilde{\omega} - \tilde{\boldsymbol{Z}} = 0. \tag{3.20}$$

Suppose now that the Jordan canonical form of $\boldsymbol{A}^{-1}$ is

$$\boldsymbol{T}^{-1}\boldsymbol{A}^{-1}\boldsymbol{T} = \begin{bmatrix} \lambda_1^{-1} & & & & & \\ \mu_1 & \lambda_2^{-1} & & & & \\ & \mu_2 & \lambda_3^{-1} & & & \\ & & \cdot & \cdot & & \\ & & & \cdot & \cdot & \\ & & & & \cdot & \cdot \\ & & & & \mu_{q-1} & \lambda_q^{-1} \end{bmatrix}, \mu_i = \begin{cases} 0 \text{ if } \lambda_i \neq \lambda_{i+1} \\ \\ \dfrac{1}{\lambda_i} \text{ if } \lambda_i = \lambda_{i+1} \end{cases},$$

and set $\boldsymbol{D} = \operatorname{diag}(\lambda_1, \lambda_2, \ldots, \lambda_q)$, $\boldsymbol{P} = \boldsymbol{D}\boldsymbol{T}^{-1}\boldsymbol{A}^{-1}$, $\boldsymbol{Q} = \boldsymbol{T}$. Butcher (1976) shows that in the case where all the λ_i are real and equal we have $C = 1$, $D = q$. This is a result of great practical significance because it says that for a q-stage implicit Runge–Kutta formula whose defining matrix has just one real q-fold eigenvalue, the computational effort required to solve the algebraic equations arising from the application of a modified Newton iteration scheme decreases from $q^3s^3/3 + O(s^2)$ to $s^3/3 + O(s^2)$ multiplications for large s.

The obvious question now is whether there exist implicit Runge–Kutta formulae having this special property. This question was answered in the affirmative by Nørsett (1976) who showed that formulae of this type can be constructed by the method of collocation using the ratio between zeros of certain Laguerre polynomials as collocation points. The class of methods derived by Nørsett was extended by Burrage (1978a). The family of Runge–Kutta formulae of order q, or greater, given by Burrage is

$$
\begin{array}{c|cccc}
c_1 & & & & \\
c_2 & & & & \\
\vdots & & V_q A_q V_q^{-1} & & \\
c_q & & & & \\
\hline
 & b_1 & b_2 & \dots & b_q
\end{array}
$$

where

$$[b_1, b_2, \dots, b_q] = \left[1, \tfrac{1}{2}, \dots, \frac{1}{q}\right] V_q^{-1}, \qquad V_q = \begin{bmatrix} 1 \dots c_1^{q-1} \\ \vdots \quad \vdots \\ 1 \dots c_q^{q-1} \end{bmatrix},$$

$$A_q = \begin{bmatrix} 0 & 0 & 0 & \dots & 0 & \alpha_{1q} \\ 1 & 0 & 0 & \dots & 0 & \alpha_{2q} \\ 0 & \frac{1}{2} & 0 & \dots & 0 & \alpha_{3q} \\ \vdots & & & & & \vdots \\ 0 & 0 & & \dots & \dfrac{1}{q-1} & \alpha_{qq} \end{bmatrix},$$

$\alpha_{kq} = (-1)^{q-k}(k^q - 1)(q-1)!\lambda^{q-k+1}/(k-1)!$, $1 \leqslant k \leqslant q$ and λ is real and non-zero. Burrage (1978) defines the concept of a transformed Runge-Kutta method and then gives the following definition.

Definition 3.7 A transformed method of order q or greater whose Runge-Kutta matrix has just one real q-fold eigenvalue is called a singly implicit method.

Burrage also proved that the maximum attainable order of a singly implicit Runge-Kutta method with q stages is $q + 1$. The connection between these formulae and the formulae of Nørsett (1976) comes from the observations that if z_1, $z_2, \dots, z_q$ are the zeros of the qth degree Laguerre polynomial,

$$L_q(z) = \sum_{j=0}^{q} (-1)^j \binom{q}{j} z^j/j!,$$

then the singly implicit Runge-Kutta method of Burrage is such that $c_i = \lambda z_i$, $1 \leqslant i \leqslant q$, where λ is the single q-fold eigenvalue of A_q,

$$\sum_{j=1}^{q} a_{ij} c_j^{k-1} = \frac{1}{k} c_i^k, \qquad i, k = 1, 2, \dots, q,$$

$$\sum_{j=1}^{q} b_j c_j^{k-1} = \frac{1}{k}, \qquad k = 1, 2, \ldots, q.$$

The final remaining step was to derive explicitly the transformation matrices required to implement these singly implicit formulae and this was done by Butcher (1979).

The importance of singly implicit Runge-Kutta formulae from an algorithmic point of view is that it is possible to derive a whole class of formulae of increasing order and that for each order q it is possible to derive an embedded formula of order $q + 1$ which is used for the purpose of local error estimation. This had led to the development of the program STRIDE due to Butcher, Burrage, and Chipman (1979a, b), which is based on the singly implicit formulae of Burrage. STRIDE is a variable step/variable order program which allows the use of formulae of orders 1, 2, . . . , 15 all of which are A(α)-stable with $\alpha > 83°$. The implementation details of these formulae are too numerous to describe here but are fully described in Butcher, Burrage, and Chipman (1979a). Although STRIDE still does not seem to be generally competitive with codes based on backward differentiation formulae (see also Gaffney 1982a for further comments), it should be remembered that linear multistep methods have benefited from a vast amount of testing and reprogramming. It is therefore quite possible that the gap between the performance of BDF and R-K methods will continue to decrease as more experience is gained with STRIDE and its strengths and weaknesses become better understood.

3.1.4 *Diagonally implicit and mono-implicit Runge-Kutta methods* (*Cash*)

Closely related to the singly implicit Runge-Kutta formulae implemented in STRIDE is the class of diagonally implicit Runge-Kutta formulae. This class of formulae was originally proposed by Nørsett (1974a) and was further developed by Alexander (1977) and Crouzeix (1975). A diagonally implicit Runge-Kutta formula is one whose defining matrix $\boldsymbol{A}$ is lower triangular. As a result, the use of a Newton iteration scheme to solve for the $\boldsymbol{k}_i$ associated with these methods calls for the solution of q sets of s algebraic equations. Normally these methods, called DIRK methods by Alexander and SDIRK by others, are constructed so that the diagonal elements are all equal, but formulae where the diagonal elements are "almost" equal also seem worthy of study.

Alexander proved that a q-stage, A-stable DIRK formula is strongly S-stable if $a_{qi} = b_i$, $1 \leqslant i \leqslant q$, and $c_q = 1$. Alexander (1977) derived strongly S-stable DIRK formulae of order two in two stages, order three in three stages and showed that it is impossible to derive a strongly S-stable DIRK formula of order four in four stages. Cash (1979) gave a strongly S-stable DIRK formula of order four in five stages and Cooper and Sayfry (1979) gave one of order five in six stages. In Alexander's paper some numerical results are given which indicate that DIRK formulae can in some cases be competitive with linear multistep methods, especially when low precision is requested. However, Alexander and others have pointed out that efficient error estimation and more particularly change of order are problems with DIRK formulae since it seems difficult to derive families of DIRK formulae suitable for inclusion in a VSVO algorithm.

There are several important theoretical results known concerning the maximum attainable order of q-stage DIRK formulae. Nørsett (1974) has shown that the maximum order attainable by a DIRK formula with q stages is generally $q + 1$ but if $q = 2\mu$, $2 \leqslant \mu \leqslant 10$, this maximum order is q. Nørsett and Wolfbrandt (1977) have given an important, and somewhat surprising, result concerning singly implicit Runge–Kutta formulae and DIRK formulae in cases where all the diagonal elements of the defining matrix A are not necessarily equal. They examine rational approximations of the form

$$\sum_{i=0}^{m} a_i z^i \Big/ \prod_{i=1}^{n} (1 + \gamma_i z) \text{ to } e^{-z} \text{ where } z \in \mathbb{C}, \gamma_i \in \mathbb{R}, \tag{3.21}$$

and show that the maximum attainable order is $m + 1$ with the least absolute value of the error constant being when $\gamma_1 = \gamma_2 = \ldots = \gamma_n$. This result immediately implies the result of Nørsett that the order of a q-stage semi-implicit Runge–Kutta formula cannot exceed $q + 1$. An additional, rather negative result from the point of view of DIRK formulae has been given by Hairer (1980a), who has shown that the order of an algebraically stable DIRK formula cannot exceed four.

The computational aspects of DIRK formulae have not received much attention so far. The step control procedure used by Alexander (1977) was based on Richardson extrapolation. However, in view of the widely used technique of embedding for explicit Runge–Kutta formulae (Fehlberg 1964), it was a natural progression for the error estimation in DIRK formulae to be carried out using pairs of embedded formulae with orders differing by one. Nørsett (1974a) derived some low-order embedded DIRK formulae and Cash (1979) extended this to derive a third-order DIRK formula in three stages and a fourth-order one in five stages, both of which are strongly S-stable and contain an embedded formula of one order lower. Present implementations of DIRK formulae often use fixed-order methods. There has not been any extensive testing of DIRK formulae to date although a research-level code due to Alexander has performed well in the integration of a certain class of oscillatory problems (Gaffney 1982a). Also a DIRK code SIRKUS has been given by Nørsett (1974a) and a semi-implicit Runge–Kutta code for large sparse systems has been developed by Houbak and Thomsen (1979).

The possibility of changing both order and step size efficiently with DIRK formulae came with the investigation of block DIRK formulae. Block implicit Runge–Kutta formulae have been around for some time (Shampine and Watts 1972, Watanabe 1978, Williams and de Hoog 1974), but generally these are based on fully implicit Runge–Kutta formulae and often they are only proposed as a starting procedure (Gear 1980d). Block DIRK formulae have been considered in Cash (1982) and such formulae offer important computational advantages over conventional DIRK formulae. For example, the second-order block formula given by Cash is

$$
\begin{array}{c|ccccc}
1 & 1 & & & & \\
2 & 1 & 1 & & & \qquad b = 0.256 \\
1 & \frac{1}{2} & -\frac{1}{2} & 1 & & \\
2 & 3 & -1 & -1 & 1 & \\
3 & -\frac{7}{6}-6b & 3b & \frac{11}{3}+6b & -\frac{1}{2}-3b & 1 \\
\hline
 & -\frac{7}{6}-6b & 3b & \frac{11}{3}+6b & -\frac{1}{2}-3b & 1 \quad \text{at } n+3 \\
 & 3 & -1 & -1 & 1 & \quad \text{at } n+2 \\
 & \frac{1}{2} & -\frac{1}{2} & 1 & & \quad \text{at } n+1\,.
\end{array}
\tag{3.22}
$$

This formula is strongly S-stable and gives second-order solutions at $n + 1$, $n + 2$, $n + 3$ as well as first-order solutions at $n + 1$ and $n + 2$. It can be seen that (3.22) gives second-order solutions at three step points using a total of five stages, so we describe this formula as being second-order and requiring "$1\frac{2}{3}$ stages per step". Also given by Cash is a third-order formula requiring $2\frac{1}{4}$ stages per step, a fourth-order one requiring $2\frac{4}{5}$ stages per step, and a fifth-order one requiring $3\frac{1}{6}$ stages per step. All these formulae are strongly S-stable, or very nearly so, and the stages required per step are considerably less than for conventional strongly S-stable DIRK formulae. Cash also describes procedures for varying both the order and step size of these formulae and some numerical results are given.

In Cash (1982) it is argued strongly that in the case of implicit Runge–Kutta formulae it may not be valid to compare the efficiency of two formulae by counting the number of stages. It is clear that we must also take account of the computational effort required to evaluate the k_i. This point is highlighted if we consider an extrapolation method. The trapezoidal rule using h, $\frac{1}{2}h$ Richardson extrapolation without smoothing can be written as

$$y_{n+1,h} = y_{n,h} + \frac{h}{2}\,[f(x_{n+1}, y_{n+1,h}) + f(x_n, y_{n,h})], \tag{3.23a}$$

$$y_{n,h} = y_{n,h/2} = y_n.$$

$$
\begin{aligned}
y_{n+i/2,h/2} = y_{n+(i-1)/2,h/2} &+ \\
&\frac{h}{4}[f(x_{n+i/2}, y_{n+i/2,h/2}) + f(x_{n+(i-1)/2}, y_{n+(i-1)/2,h/2})], \\
&i = 1, 2.
\end{aligned}
\tag{3.23b}
$$

The extrapolated solution y_{n+1} is given by

$$y_{n+1} = \tfrac{1}{3}\,(4y_{n+1,h/2} - y_{n+1,h}). \tag{3.23c}$$

Formally this process for computing y_{n+1} can be written as the diagonally implicit Runge–Kutta formula

$$
\begin{array}{c|cccc}
0 & 0 & & & \\
1 & \frac{1}{2} & \frac{1}{2} & & \\
\frac{1}{2} & \frac{1}{4} & 0 & \frac{1}{4} & \\
1 & \frac{1}{4} & 0 & \frac{1}{2} & \frac{1}{4} \\
\hline
 & \frac{1}{6} & -\frac{1}{6} & \frac{2}{3} & \frac{1}{3}
\end{array} . \tag{3.24}
$$

We see that this formula has four stages but this does not tell the whole story. This is because the stage k_4 is likely to be inexpensive to evaluate since, when computing the approximation $y_{n+1,h/2}$, we already have the approximations $y_{n+1,h}$ and $y'_{n+1,h}$ available. Many extrapolation methods can be regarded as high-order diagonally implicit Runge–Kutta methods which require more than the minimum possible number of stages to achieve a given order but which are such that some of the stages are inexpensive to evaluate. Note, however, that extrapolation methods are such that the diagonal elements for the corresponding DIRK method are not all equal.

We now consider the class of Mono-Implicit Runge–Kutta (MIRK) formulae introduced by Cash (1975a). The idea of introducing this class of formulae came from the observation that if we take the forward Euler rule

$$y_{n+1} - y_n = hf_n,$$

which has only a small region of absolute stability, and replace h by $-h$ we obtain the backward Euler rule

$$y_n - y_{n-1} = hf_n,$$

which is known to be L-stable. This idea can be extended by taking any explicit Runge–Kutta formula and replacing h by $-h$ to obtain what is called the "backward version" of the Runge–Kutta formula. If the explicit Runge–Kutta formula when applied with fixed h to the equation $y' = \lambda y$ gives

$$y_{n+1}/y_n = R(z), \qquad z = h\lambda,$$

and the backward version gives

$$y_{n+1}/y_n = \bar{R}(z),$$

it can easily be shown that

$$\bar{R}(z) = \frac{1}{R(-z)}. \tag{3.25}$$

This leads us to the "reflection principle": the backward formula's stability region is the complementary set of the image of the classical explicit formula's set reflected in the imaginary axis. For further extensions of this idea see Stetter (1973).

This class of backward formulae generally has very good stability properties and it can be shown to be a subset of the more general class of formulae

$$y_{n+1} - y_n = h \sum_{j=0}^{k} b_j f(x_{n+\alpha_j}, \bar{y}_{n+\alpha_j}), \tag{3.26}$$

$$\bar{y}_{n+\alpha_j} = \delta_j y_{n+1} + (1 - \delta_j) y_n + h \sum_{i=0}^{m} a_{ji} f(x_{n+\alpha_i}, \bar{y}_{n+\alpha_i}). \tag{3.27}$$

If in (3.27) we take $m = j - 1$, $\delta_j \equiv 0$, we obtain a classical explicit Runge-Kutta formula. If $m = j$, $\delta_j = 0$, $a_{jj} = \beta$ for all j, we obtain a diagonally implicit Runge-Kutta formula and if $m = k$, $\delta_j = 0$, we obtain a fully implicit Runge-Kutta formula. If, however, we take $m = j - 1$ and $\delta_j \neq 0$ for at least one j, we obtain a formula which is implicit in the single unknown y_{n+1}. In Cash (1975a) the general class of formulae proposed was of the form (3.26) and (3.27) with

$$\delta_j = 0, m = j - 1, 0 \leqslant j \leqslant r,$$

$$\delta_j = 1, m = j - 1, a_{ji} = 0, r + 1 \leqslant j \leqslant k, 0 \leqslant i \leqslant r,$$

where $r \leqslant [\tfrac{1}{2}k]$. Some numerical testing of these formulae was done by Birnbaum and Lapidus (1978) and they were found to perform well on a set of test problems from a chemical engineering environment. These formulae were further investigated by van Bokhoven (1980) who derived the general order relations to be satisfied by formulae of order $\leqslant 6$. As with fully implicit Runge-Kutta formulae, the way in which MIRK formulae are implemented is of great importance. The system of algebraic equations to be solved is of dimension s but the coefficient matrix of the modified Newton iteration scheme is of the form

$$\sum_{i=0}^{t} \alpha_i h^i J^i, \tag{3.28}$$

where J is an approximation to the Jacobian matrix. In Cash and Singhal (1982) a special class of MIRK formulae were investigated such that (3.28) can be approximated by

$$(I - \beta h J)^t,$$

since in this case no matrix-matrix products are called for. Some L-stable formulae of this class with order $\leqslant 4$ have been given in Cash and Singhal (1982) and some numerical results are presented.

3.1.5 *Defect correction* (*Cash*)

The technique of defect correction is a further attempt to overcome the difficulty of solving large systems of nonlinear algebraic equations when applying high-order implicit Runge-Kutta methods. The basic idea behind the Iterated Defect Correction (IDeC) technique is to obtain an approximate solution of the differential equation using a "cheap" low-order implicit R-K method and then to improve this approximate solution iteratively by means of a correction procedure. The technique of IDeC has been given by Frank and Ueberhuber (1977, 1978). They show that every IDeC method based on the backward Euler rule, using piecewise polynomial interpolation, has a fixed point which coincides with the solution obtained using an appropriate polynimial collocation scheme. However, in general the IDeC technique calls for not much more computational effort than is required by a linear multistep method to solve the algebraic equations arising from a modified Newton iteration. The technique of IDeC as described by Frank and Ueberhuber (1977) is as follows.

Consider the solution of the stiff system

$$y' = f(x, y), \; y(0) = y_0, \; 0 \leqslant x \leqslant H.$$

Define a grid,

$$\pi \equiv \{x_\nu : 1 \leqslant \nu \leqslant m\} \text{ where } x_\nu = H\zeta_\nu \text{ and } 0 < \zeta_1 < \ldots < \zeta_m = 1.$$

On this grid apply the backward Euler rule,

$$\eta_{\nu+1}^{(0)} = \eta_\nu^{(0)} + h_\nu f(x_{\nu+1}, \eta_{\nu+1}^{(0)}), \quad 0 \leqslant \nu \leqslant m-1, h_\nu = x_{\nu+1} - x_\nu. \tag{3.29}$$

This gives a solution $\eta^{(0)} = [\eta_0^{(0)}, \eta_1^{(0)}, \ldots, \eta_m^{(0)}]^{\mathrm{T}}$. Now construct an interpolating polynomial $P^0(x)$ of degree m which interpolates the point set $\{(x_\nu, \eta_\nu^{(0)}): 0 \leqslant \nu \leqslant m\}$, i.e.

$$P^0(x_\nu) = \eta_\nu^{(0)}, \qquad 0 \leqslant \nu \leqslant m. \tag{3.30}$$

Having constructed this polynomial we can define the defect, which is a continuous function of x, as

$$d^0(x) = \frac{\mathrm{d}}{\mathrm{d}x}[P^0(x)] - f[x, P^0(x)]. \tag{3.31}$$

This allows the following initial value problem to be constructed

$$\begin{aligned} y'(x) &= f(x, y) + d^0(x), \; y(0) = y_0 \\ &\equiv f(x, y) + \frac{\mathrm{d}}{\mathrm{d}x}[P^0(x)] - f[x, P^0(x)], \end{aligned} \tag{3.32}$$

which has the known solution $y(x) = P^0(x)$. We now solve the initial value problem (3.32) in exactly the same way, i.e. using the backward Euler rule on the same grid π, to obtain a numerical solution,

$$\tau^{(0)} \equiv [\tau_0^{(0)}, \tau_1^{(0)}, \ldots, \tau_m^{(0)}].$$

We now use the known discretization errors $\tau_\nu^{(0)} - P^0(x_\nu)$ as estimates of the unknown errors $\eta_\nu^{(0)} - y(x_\nu)$. This approach follows an original idea due to Zadunaisky (1976). We now replace the unknown errors in the identity

$$y(x_\nu) = \eta_\nu^{(0)} - (\eta_\nu^{(0)} - y(x_\nu)), \qquad 0 \leqslant \nu \leqslant m,$$

by the known error estimates $\tau_\nu^{(0)} - P^0(x_\nu)$. This leads to the following formula to improve the first approximate solution $\eta^{(0)}$:

$$\eta_\nu^{(1)} = \eta_\nu^{(0)} - [\tau_\nu^{(0)} - P^0(x_\nu)], \qquad 0 \leqslant \nu \leqslant m. \tag{3.33}$$

Following an idea due to Stetter (1974) this whole procedure can used iteratively as

$$\eta_\nu^{(j+1)} = \eta_\nu^{(0)} - [\tau_\nu^{(j)} - P^j(x_\nu)], \tag{3.34}$$

where $P^j(x)$ is the polynomial of degree m interpolating the point set $\{(x_\nu, \eta_\nu^{(j)}): 0 \leqslant \nu \leqslant m\}$. Subject to some general differentiability conditions on the function f, Frank and Ueberhuber (1977) have proved the following result for the IDeC method applied on an equi-distant grid for a fixed degree, m, of the polynomial and for $h = H/m$.

If an arbitrary Runge–Kutta scheme of order p ($\leqslant m$) is used and if f satisfies suitable differentiability conditions, then

$$\eta_\nu^{(j)} - y(x_\nu) = O[h^{\min\{p(j+1),m\}}].$$

Thus the technique of IDeC achieves the high order attainable by implicit Runge-Kutta formulae but at a significantly less computation cost. Much of the theorectical analysis concerning IDeC has now been completed (Stetter 1974, 1978, Hairer 1978, Frank and Ueberhuber 1978) and this approach does seem to be a very promising one.

Actual implementations of defect correction have been largely based on the backward Euler rule although Butcher (1979a) has considered the application of defect correction to a more general class of Runge–Kutta methods. Ueberhuber (1979) has considered many of the computational details associated with a particular implementation of IDeC but it is clear that much more experience is needed in this area before a really successful implementation can be contemplated.

3.1.6 *Runge–Kutta methods using an approximation to the Jacobian* (*Cash*)

For the numerical integration of the autonomous differential system

$$\frac{dy}{dx} = f(y), \qquad y(x_0) = y_0, \tag{3.35}$$

Rosenbrock (1963) proposed the class of q-stage integration methods defined by

$$y_{n+1} = y_n + h \sum_{r=1}^{q} b_r k_r, \tag{3.36a}$$

$$k_1 = f(y_n) + \alpha_1 h J(y_n) k_1,$$

$$k_r = f\left(y_n + h \sum_{s=1}^{r-1} a_{rs} k_s\right) + \alpha_r h J\left(y_n + h \sum_{s=1}^{r-1} \beta_{rs} k_s\right) k_r, \quad 2 \leqslant r \leqslant q, \tag{3.36b}$$

where $J(y_n) \equiv \partial f/\partial y|_{y=y_n}$. Such formulae have the major computational advantage that it is necessary to solve only linear systems of algebraic equations to find the k_r but have the disadvantage that the exact Jacobian is required at each step. In view of this, Rosenbrock methods are only likely to be competitive with other classes of methods on problems for which the Jacobian matrix is not too expensive to evaluate. An important class of problems of this type has been given by Edsberg (1974). Edsberg considered problems in which the elementary steps of some composite reaction taking place in a homogeneous solution according to the mass action law are known. In such cases the system of ODEs describing the kinetic behaviour of the system can be set up from the coefficients and the structure of the reactions. Edsberg showed that such problems can often be written in the form

$$y' = Ap, \qquad y(0) \text{ given}, \tag{3.37}$$

where A is an $M \times N$ matrix with integer entries and p is an N-vector with

$$p_j = k_j \prod_{i=1}^{M} y_i^{r_{ji}}. \tag{3.38}$$

Here the $r_{ji} \geqslant 0$ are integers describing the reactions and the $k_j > 0$ are the rate constants. For such problems we can obtain the Jacobian matrix f_y directly from the relation

$$\frac{\partial p_j}{\partial y_i} = r_{ji} p_j / y_i. \tag{3.39}$$

Most early studies of Rosenbrock methods appeared in the engineering literature. In particular we mention the work of Calahan (1968), Allen and Pottle (1966), Caillaud and Padmanabhan (1971), Lapidus and Seinfeld (1971), and Michelson (1976). In Cash (1976) the suggestion was made to look for Rosenbrock formulae with $\beta_{rs} = 0$ since these require only one Jacobian evaluation per step. Also given by Cash was a novel form of error estimation which is applicable to a relatively wide class of Runge–Kutta methods. To describe it we consider the second-order Rosenbrock method,

$$\begin{aligned} y_{n+1} - y_n &= h(b_1 k_1 + b_2 k_2), \\ k_1 &= [I - ahJ(y_n)]^{-1} f(y_n), \\ k_2 &= [I - ahJ(y_n)]^{-1} f(y_n + hc_1 k_1). \end{aligned} \tag{3.40}$$

The coefficients of this formula can be chosen so that it is L-stable (Cash 1976) and we call the resulting formula $R_2(x_n, \omega_1, \omega_2, a, c_1, h)$. The procedure used to obtain an estimate of the local truncation error is a modified form of Richardson extrapolation whereby two approximate solutions are found at each step point, one using two steps of size $h/2$ and the other using one step of length h. Thus, starting from the point x_{n-1} we use the integration formulae $R_2(x_{n-1}, \omega_1, \omega_2, a, c_1, h/2)$, $R_2(x_{n-\frac{1}{2}}, \omega_1, \omega_2, a, c_1, h/2)$ to compute an approximate solution y_n at x_n. A second approximate solution $\bar{y}_n$ is computed using the formula $R_2(x_{n-1}, \bar{\omega}_1, \bar{\omega}_2, a/2, c_1/2, h)$. The crucial point to note in using the formula R_2 with a step length h is that it uses exactly the same k_1 and k_2 as were used by $R_2(x_{n-1}, \omega_1, \omega_2, a, c_1, h/2)$. This means that in general a negligible extra computational effort is required to compute the error estimate. In Cash (1976) an L-stable formula of order two in two stages and an L-stable formula of order three in three stages using a Merson-type error estimate were derived. This technique has recently been used by Bui (1981) to obtain a fully embedded formula of order three.

A modification of Rosenbrock formulae was given by Wolfbrandt (1977). He introduced the class of ROW methods given by

$$[I - \gamma hJ(y_n)]\, k_i = hf\left(y_n + \sum_{j=1}^{i-1} \alpha_{ij} k_j\right) + hJ(y_n) \sum_{j=1}^{i-1} \gamma_{ij} k_j,$$

$$1 \leqslant i \leqslant q, \tag{3.41}$$

$$y_{n+1} = y_n + \sum_{i=1}^{q} b_i k_i.$$

Kaps and Wanner (1981) have shown that mathematically this formulation is equivalent to the computationally more efficient class of formulae where

$$[I - \bar{\gamma} hJ(y_n)]\, \bar{k}_i = hf\left(y_n + \sum_{j=1}^{i-1} \bar{\alpha}_{ij} \bar{k}_j\right) + h \sum_{j=1}^{i-1} \bar{\gamma}_{ij} k_j. \tag{3.42}$$

We shall refer to both these classes as ROW methods and we note that if $\gamma_{ij} = \bar{\gamma}_{ij} \equiv 0$, they reduce to Rosenbrock methods. If (3.41) is applied with a fixed step-size h to the scalar equation $y' = \lambda y$, we obtain

$$y_{n+1} = R(z) y_n, \qquad z = h\lambda,$$

where, if the method is of order q at least,

$$R(z) = \left[\sum_{j=0}^{q} z^j \sum_{i=0}^{j} \binom{q}{i} \frac{(-\gamma)^i}{(j-i)!}\right] \Big/ (1 - \gamma z)^q.$$

An investigation of how to choose γ to give optimal stability and accuracy has been carried out by Wanner (1980a). Higher-order formulae of the Rosenbrock, or ROW, class have been derived by several people and in particular we mention the work of Bui (1979), who derives an L-stable formula of order four in four stages; Kaps and Wanner (1981), who derive methods of order five and six requiring just one Jacobian evaluation per step (but the step control is by Richardson extrapolation); Kaps and Rentrop (1979), who derive a fourth-order method with an embedded third-order method for error estimation; Nørsett and Wolfbrandt (1979) who derive the order relations for Rosenbrock methods by extending the Butcher series approach; and the thesis of Wolfbrandt (1977) which investigates ROW methods. Finally, we mention a report by Shampine (1980b), who raises several interesting points concerning the implementation of Rosenbrock methods. Generally speaking, Runge-Kutta formulae evaluate the function $f(x, y)$ several times in the interval $[x_n, x_{n+1}]$. Shampine argues that it is desirable that these function evaluations should span $[x_n, x_{n+1}]$ so that the formula is able to spot any "quasi-discontinuities" in the solution. An example of a formula which does not satisfy this condition, the implicit midpoint rule, is discussed. By considering a differential equation of the form

$$y'(t) = d - by + aE(t), \tag{3.43}$$

where the forcing function $E(t)$ is a square wave, Shampine shows that the implicit midpoint rule can sometimes give a very poor solution to (3.43) due to its inability to spot a quasi-discontinuity.

Another important point raised by Shampine is that the linear system of algebraic equations defining $\boldsymbol{k}_i$ is almost bound to be ill-conditioned. This is also the case for BDF methods but for these formulae this is not serious since they solve for the change in the solution and normally it is only necessary to get the first few digits correct. However, for Rosenbrock methods this ill-conditioning can be serious and special care must be taken to monitor it. Shampine derives a fourth-order formula with an embedded formula of order three, both of which have rational coefficients and satisfy the spanning conditions, and discusses their implementation and practical performance in some detail.

A class of formulae similar to Wolfbrandt's methods was proposed by Cash (1980b). These take the form

$$y_{n+1} - y_n = \sum_{i=1}^{q} \omega_i \boldsymbol{k}_i, \tag{3.44a}$$

$$\left(\boldsymbol{I} - h\gamma \frac{\partial f}{\partial y}(y_n)\right)^2 \boldsymbol{k}_i = h\left(\boldsymbol{I} + ch\frac{\partial f}{\partial y}(y_n)\right) f\left(y_n + \sum_{j=1}^{i-1} b_{ij}\boldsymbol{k}_j\right). \tag{3.44b}$$

Each $\boldsymbol{k}_i$ calls for the solution of a linear system of equations of the form

$$\boldsymbol{A}^2\boldsymbol{k}_i = \boldsymbol{b}.$$

This can be done efficiently, i.e. without the need for matrix products, by LU decomposing $\boldsymbol{A}$ and solving the systems

$$\boldsymbol{LUz} = \boldsymbol{b},$$

$$\boldsymbol{LUk}_i = \boldsymbol{z}.$$

Cash gives an L-stable formula of order three, with an embedded L-stable formula of order two, of the form (3.44) and the results of some numerical computations are reported.

An obvious extension of Rosenbrock methods is to see what can be done with formulae of the general form (3.36) where $\boldsymbol{J}$ is no longer the exact Jacobian. An interesting contribution in this area is due to Steihaug and Wolfbrandt (1979) who consider methods of the form

$$(\boldsymbol{I} - hd_{ii}\boldsymbol{A})\boldsymbol{k}_i = f\left(y_n + h\sum_{j=1}^{i-1} a_{ij}\boldsymbol{k}_j\right) + h\boldsymbol{A}\sum_{j=1}^{i-1} d_{ij}\boldsymbol{k}_j, \qquad 1 \leqslant i \leqslant q. \tag{3.45}$$

Here $\boldsymbol{A}$ is a real square matrix and h is chosen so that $\boldsymbol{I} - hd_{ii}\boldsymbol{A}$ is nonsingular. Since $\boldsymbol{A}$ is no longer the exact Jacobian, there are considerably more order relations to be satisfied than for ordinary ROW methods. In particular Steihaug and Wolfbrandt show that to get order three it is necessary to satisfy eight order relations. They give a second-order formula with an embedded error estimate and present the results of some numerical computations.

An alternative approach to the problem of deriving Runge–Kutta formulae with coefficients depending on an inexact Jacobian is the one due to Verwer (1977). Verwer has investigated a class of formulae considered by van der Houwen (1972a, b) and defines a generalized m-point Runge-Kutta method to be

$$y_{n+1} = y_n + \sum_{j=0}^{m-1} \Lambda_{m,j}(h_n J_n) k_n^{(j)}, \tag{3.46}$$

$$k_n^{(j)} = h_n f\left(x_n + \mu_j h_n, y_n + \sum_{l=0}^{j-1} \Lambda_{j,l}(h_n J_n) k_n^{(l)}\right), \qquad h_n = x_{n+1} - x_n,$$

where

$$\mu_j = \sum_{l=0}^{j-1} \Lambda_{j,l}(0), \qquad j = 0, 1, \ldots, m-1$$

and $\Lambda_{j,l}$, $j = 1, 2, \ldots, m$, $l = 0, 1, \ldots, j-1$ are rational functions with real coefficients, Verwer derives a class of generalized Runge-Kutta formulae which have order independent of the choice of J_n. Associated with these formulae, Verwer defines the concept of internal S-stability as follows.

Define $y_{n+1}^{(j)}$ by the relation

$$k_n^{(j)} = h_n f(x_n + \mu_j h_n, y_{n+1}^{(j)}), \qquad j = 1, 2, \ldots, m.$$

The Runge-Kutta formula (3.46) can then be written as

$$y_{n+1}^{(0)} = y_n,$$

$$y_{n+1}^{(j)} = y_n + h_n \sum_{l=0}^{j-1} \Lambda_{j,l}(h_n J_n) f(x_n + \mu_l h_n, y_{n+1}^{(l)}), \qquad j = 1, 2, \ldots, m,$$

$$y_{n+1} = y_{n+1}^{(m)}. \tag{3.47}$$

The usual approach in the stability analysis of methods for stiff problems is to examine the stability of the solution only at the end point of the current step. However, for those Runge-Kutta schemes in which the solution is built up successively in several intermediate stages, it is also of interest to examine the stability properties of the intermediate solutions $y_{n+1}^{(j)}$. To allow for this, Verwer gives the following definition of internal S-stability.

Definition 3.8 The integration formula (3.47) is said to be internally S-stable if at each jth stage, $j = 1, 2, \ldots, m$, the corresponding scheme at stage j is S-stable.

Verwer considers the performance of three second-order formulae of the form (3.46) on a set of three test problems. The first formula is L-stable but not S-stable, the second is S-stable but not internally S-stable, and the third is internally S-stable. The numerical results presented suggest that the third formula is superior to the other two on the problems considered.

3.1.7 *Nonlinear methods* (*Cash*)

Finally, we mention briefly an interesting class of nonlinear Runge-Kutta methods. It seems likely that such methods will experience difficulties in solving some classes of stiff systems but numerical experience indicates that they may have real potential in solving the stiff equations arising from the methods of lines solution of parabolic PDEs. Their main feature is that, being nonlinear, they are able to achieve A-stability while still being explicit. Lambert (1974) seems to have been the first to consider nonlinear Runge-Kutta methods. One particular example he gives is

$$y_{n+1} - y_n = hf(x_{n+\frac{1}{2}}, y_n + \tfrac{1}{2}hy_n f_n/(y_n - \tfrac{1}{2}hf_n)), \tag{3.48}$$

in which the expression $\boldsymbol{ab}/\boldsymbol{d}$ must be interpreted for real or complex vectors as

$$\frac{\boldsymbol{ab}}{\boldsymbol{d}} = \frac{\boldsymbol{a}\,\mathrm{Re}(\boldsymbol{b}, \boldsymbol{d}) + \boldsymbol{b}\mathrm{Re}(\boldsymbol{d}, \boldsymbol{a}) - \boldsymbol{d}\mathrm{Re}(\boldsymbol{a}, \boldsymbol{b})}{(\boldsymbol{d}, \boldsymbol{d})},$$

where $(\boldsymbol{a}, \boldsymbol{b})$ denotes the scalar product of $\boldsymbol{a}$ and $\boldsymbol{b}$. This method has the advantage of being "component applicable" (see Lambert 1974, p. 176) but has the disadvantage that, although having order two in general, it only has order one if $y = 0$. Also, of course, care must be taken to ensure that $y_n - \frac{1}{2}hf_n$ does not vanish. Nonlinear Runge-Kutta methods were also examined by Wambecq (1978) and Hairer (1980b). The general class of formulae they consider is given by

$$y_1 = y_0 + \sum_{i=1}^{s} \sum_{j=1}^{i} \omega_{ij} \frac{g_i g_j}{\sum_{k=1}^{s} b_k g_k}, \tag{3.49}$$

$$g_i = hf\left(y_0 + \sum_{j=1}^{i-1} a_{ij} g_j\right).$$

These methods are more expensive to implement than those due to Lambert but do not suffer from the problem that the order changes if y passes through zero. However, there is still the problem that the denominator may become zero as is emphasized by the second-order equation

$$g_1 = hf(y_0),$$

$$g_2 = hf(y_0 + \tfrac{1}{2}g_1),$$

$$y_1 = y_0 + \frac{g_1 g_1}{2g_1 - g_2},$$

so they need to be implemented with particular care.

Hairer (1980b) gives a second-order A_0-stable method with an embedded error control. In numerical tests this method was found to perform poorly on some stiff problems arising in chemical kinetics, so they are probably ruled out for the integration of general stiff systems. However, good results were obtained for the integration of some parabolic PDEs using the method of lines. Because of these results and the very low storage requirement of the formulae (3.49) (they do not require the evaluation or storage of a Jacobian), they seem worthy of further attention as candidates for use with the method of lines solution of parabolic partial differential equations.

3.1.8 *Generalizations of Runge–Kutta methods: solvers of stiff problems* (*Burrage*)

At first sight, Runge–Kutta methods present an attractive alternative to linear multistep methods because of their excellent stability properties and high accuracy. In particular, the s-stage methods of order $2s$ are known to be not only A-stable but also algebraically stable, whereas an A-stable linear multistep method cannot have order greater than two. However, there are severe implementational problems when applying these high order Runge–Kutta methods to the solution of stiff problems.

Consider the solution of the initial value problem

$$y'(x) = f[x, y(x)], \qquad y(x_0) = y_0, \tag{3.50}$$

by the s-stage Runge–Kutta method

$$Y_i = y_{n-1} + h \sum_{j=1}^{s} a_{ij} f(x_{n-1} + hc_j, Y_j), \qquad i = 1, \ldots, s, \tag{3.51a}$$

$$y_n = y_{n-1} + h \sum_{j=1}^{s} b_j f(x_{n-1} + hc_j, Y_j). \tag{3.51b}$$

The implementational cost in carrying out the linear algebra computations required for solving for the Y_i ($i = 1, \ldots, s$), using some form of Newton iteration, is of the order of $(sN)^3/3$ multiplicative operations, as compared with $N^3/3$ operations for the implementation of linear multistep methods. However this problem can nevertheless be overcome by considering special families of Runge–Kutta methods. Here we will consider some of these methods and their generalizations to a wider class of methods.

One way of overcoming this implementational difficulty is to impose a special structure on the matrix $\boldsymbol{A}$. Nørsett (1974a) suggested letting $\boldsymbol{A}$ be lower triangular with all elements on the diagonal being equal, the DIRK method. As a result, the use of Newton iterations to solve for the $\boldsymbol{Y}_i$ involves only a single LU factorization of $\boldsymbol{I} - h\lambda\boldsymbol{A}$, λ being the repeated eigenvalue of $\boldsymbol{A}$, and the implementational cost is similar to that of an implicit linear multistep method.

A number of codes based on DIRK's have been developed including SIRKUS (Nørsett 1974a), DIRK (Alexander 1977) and a code for large sparse systems (Houbak and Thomsen 1979). Gaffney (1982a) has done some testing of DIRK (in conjunction with other codes) on oscillatory problems and it seems to perform well especially when low precision is required. On such problems DIRK-based codes seem well suited because they are at most of order four or five and are strongly S-stable. However, because of the extreme difficulty in deriving families of DIRK's suitable for error estimation and change of order, most codes are of fixed order and suffer from a lack of robustness.

Butcher (1976b) proposed a generalization of the DIRK idea in which A, instead of being triangular, merely has a one-point spectrum and is related by a similarity transformation to a matrix with λ down the diagonal and $-\lambda$ down the first subdiagonal. Transformations can then be made in the Newton iteration stage which reduces the implementational cost to essentially that of a linear multistep method. Such methods were called by Burrage (1978a) singly-implicit, and a class of s-stage collocation methods of order s was constructed in which the internal abscissae $c_1, \ldots, c_s$ are proportional to the zeros of the Laguerre polynomial of degree s. The algorithm STRIDE (Burrage, Butcher, and Chipman 1980, and Butcher, Burrage, and Chipman 1979) is based on these methods.

These methods possess three important properties crucial for any stiff algorithm. First, error estimates are provided by embedding an s-stage method of order s in an $(s + 1)$-stage method of order $s + 1$. Since both methods have the same eigenvalue, a new LU factorization is unnecessary, merely one extra f evaluation. Second, the methods in STRIDE form a sequence of methods of differing orders so that error estimates are available for all members in the family, particularly those adjacent to one another. With such estimates the appropriateness of choosing a method with a different order as well as with possibly varying step-size, can be automatically evaluated. Finally, for each order there is a range of possible values of λ which guarantees A(α)-stability with $\alpha > 83°$.

STRIDE also possesses a number of other features which should be applicable to many other algorithms for solving differential equations. These include the use of the Gill-Møller algorithm (Møller 1965) for reducing round-off error growth and the concept of the invariance of results under separate rescaling of each component of the solution vector and of the independent variable.

In numerical tests (Burrage 1978, Gaffney 1982a) STRIDE has been shown to have a number of good points but also some serious disadvantages. On the credit side, its good stability properties mean that it performs well on problems whose eigenvalues are near the imaginary axis and also for problems in which high accuracy is required. It has also been shown to be very robust and portable. On the debit side, testing by Gaffney indicates that it is not generally competitive with codes based on backward differentiation formulae. The reason for this is twofold. First, there is the lack of fine-tuning experience with Runge–Kutta codes as compared with the more mature backward differentiation codes. Second, and more serious is the fact that STRIDE appears to spend approximately half of its time transforming and untransforming the Y_i and $F(Y_i)$ in the Newton iteration stage. This is a serious drawback which can be ameliorated by performing fewer iterations, but is inherent to STRIDE and singly-implicit methods in general.

Thus although STRIDE and other DIRK-based codes have their strong points, they do not appear as generally competitive as the BDF codes. Can this situation be improved upon?

In the solution of (3.50) consider a step-by-step method in which, at the start of the nth integration step, r vectors $y_1^{(n-1)}, \ldots, y_r^{(n-1)}$ contain all the information about the solution that is to be used in carrying out the step. To compute the corresponding information at the end of a step, suppose that s internal stages are needed and denote the values computed in these stages as $Y_1, \ldots, Y_s$. The multivalue (or general linear) method can then be written as

$$Y_i = \sum_{j=1}^{r} a_{ij}^{(1)} y_j^{(n-1)} + h \sum_{j=1}^{s} b_{ij}^{(1)} f(x_{n-1} + hc_j, Y_j), \qquad i = 1, \ldots, s, \quad (3.52a)$$

$$y_i^{(n)} = \sum_{j=1}^{r} a_{ij}^{(2)} y_j^{(n-1)} + h \sum_{j=1}^{s} b_{ij}^{(2)} f(x_{n-1} + hc_j, Y_j), \qquad i = 1, \ldots, r. \quad (3.52b)$$

Such a method can be viewed as a multistep Runge–Kutta method and includes a wide variety of methods as particular cases. Before any computation can proceed r initial vectors $y_1^{(0)}, \ldots, y_r^{(0)}$ are needed and these can be computed by any suitable starting procedure.

Although multivalue methods have been previously studied, for example by Butcher (1966, 1973), Hairer and Wanner (1974), and by Cooper (1978), little attention has been paid to them. However, Burrage (1983) has recently shown how to construct families of high-order methods, and Butcher (1981b) has proposed a special family of multivalue methods based on the concept of singly-implicitness and with the $y_i^{(n)}$ representing $h^n y^{(i)}(x_n)$, the scaled derivatives.

It is clear that the concepts of single and diagonal implicitness can be extended to multivalue methods. Since most of the computational work is done in computing the Y_i ($i = 1, \ldots, s$), the extra work in producing the y_i ($i = 1, \ldots, r$) is negligible. The advantage is that there exist families of singly-implicit methods of order $s + r$ and families of diagonally-implicit methods of order s (if $s \approx r$). Furthermore the A-stability properties of such methods are superior to the corresponding Runge–Kutta methods (for example there exist A-stable singly-implicit multivalue methods of order ten).

The study of such methods is still in its infancy but they appear to hold great promise in combining the best features of Runge–Kutta and linear multistep methods. In particular, a variable-order code based on diagonally implicit multivalue methods seems very attractive.

As well as their implementational properties, multivalue methods have other attributes which Runge–Kutta and linear multistep methods do not possess. By choosing the $y_i^{(n)}$ to represent $y^{(i)}(x_n)$ it may be easier to detect discontinuities than with the more traditional methods. Furthermore, there is the possibility of being able to use algebraically stable methods. The concept of algebraic stability, which appears under various guises as G-stability, B-stability, and nonlinear stability, deals with the boundedness of numerical solutions in an appropriate norm when applied to a problem given by (3.50) which satisfies

$$\langle u, f(x, u) \rangle \leqslant 0,$$

where $\langle \ldots, \ldots \rangle$ is a semi inner product, and has been applied to linear multistep methods (Dahlquist 1978), Runge–Kutta methods (Burrage and Butcher 1979, Crouzeix 1979), and extended to multivalue methods (Burrage and Butcher 1980).

Unfortunately, there are some negative results concerning algebraic stability and the more traditional methods (for example, the maximum orders of an algebraically stable linear multistep method and algebraically stable DIRK are two and four, respectively), so that this concept has not proved very practical when developing stiff packages. However, from only a cursory study of the algebraic stability of multivalue methods (see Dekker 1981, for example), the above restrictions on the order of algebraically stable DIRK's do not appear so severe when applied to multivalue methods in general.

Thus although much of the work is still in its infancy, it seems likely that the various strong points of linear multistep and Runge-Kutta methods could be combined via the multivalue approach to produce efficient stiff packages.

Codes based on Runge-Kutta methods have been shown to be effective on certain classes of problems but not as generally competitive as BDF codes. However, by moving in the direction of multivalue methods, the author hopes that codes based on certain families of this class of methods could prove more competitive.

Nor is this the only direction. Frank and Ueberhuber (1977), for example, have proposed the technique of Iterated Defect Correction to obtain an approximate solution using a simple implicit Runge-Kutta method (such as the backward Euler) and to improve this solution iteratively by a correction procedure. This idea has been extended to a more general class of methods by Butcher (1981b) and Cooper and Butcher (1983). Although there is not yet a "production" code based on this technique, the approach seems promising.

Alternatively, a number of people including Kaps and Wanner (1981), Wolfbrandt (1977) and Hairer, Bader, and Lubich (1982) have studied the class of semi-implicit methods given by

$$(I - \lambda hA)k_i = hf(x_{n-1} + c_i h, u_i) + hA \sum_{j=1}^{i-1} d_{ij}k_j, \qquad i = 1, \ldots, s, \tag{3.53a}$$

$$u_i = y_{n-1} + \sum_{j=1}^{i-1} a_{ij}k_j, \qquad i = 1, \ldots, s, \tag{3.53b}$$

$$y_n = y_{n-1} + \sum_{j=1}^{s} b_j k_j, \tag{3.53c}$$

where A is an arbitrary matrix of the same dimension as the differential equation. If A is the Jacobian of the problem, (3.53) becomes a Rosenbrock method. There are packages already in existence (for example Kaps and Rentrop 1979) based on Rosenbrock methods which work well for low accuracy but as with DIRK's suffer from the disadvantage that it is extremely difficult to construct a variable-order package based on such methods. However, it may be possible to combine this approach with multivalue methods, for example. Thus it can be seen that there is still a great deal to be done in the study and selection of methods suitable for stiff packages.

3.2 Partitioning

3.2.1 *Introduction* (*Aiken*)

The practitioner may very well know which components are highly stable, that is stiff, and which are not. This information can be used to make the numerical solution more efficient for a variety of techniques. Such knowledge could also be used to make a modelling simplification to remove the highly stable component from the model, or to make a mathematical simplification to the problem such as the steady-state approximation (§7.2).

McRae, Goodin, and Seinfeld (1982) use the steady-state approximation once having identified stiff species in a large set of atmospheric kinetic equations by eigenvalue analysis. Still the size and stiffness of the full diffusion with reaction equations were too much for EPISODE and they used instead an asymptotic integration scheme. Soderlind (1979) discusses some stability properties of linear multistep compound methods on a system partitioned into stiff and nonstiff sections. Different techniques are used on each section. Soderlind and Dahlquist (1981) show through stability and error analysis why this is advantageous. The method chosen for the stiff section need be highly stable, while the method for the other section may focus on high accuracy. Thus an explicit technique can be used on the nonstiff set and this set tends to be much larger than the stiff set.

Palusinksi and Wait (1978) examine methods on stiff systems partitioned into one linear and one nonlinear system and into two nonlinear systems. Andrus (1979) also took this two-section, two-method route. See also Soderlind (1980, 1981c) and Dahlquist (1981). Enright and Kamel (1979), Carver and MacEwen (1982), and Dahlquist and Zu-Fan (1982) have done work on automatic partitioning (see also §2.10.2).

Most methods applied to a partitioned system use an implicit method on the stiff set and an explicit method on the nonstiff set. The sucess of this approach is dependent on the proper indentification of the partition and to the coupling between the two partitioned sets. This is dangerous since the best partition can change during the solution. Enright and Kamel (1980) apply the same implicit method to the entire equation set but explicit partitioning in the resulting iteration matrix. The iteration matrix is reduced using a series of similarity transformations, and a test is made to identify the stiff components. This leads to an iteration scheme that applies a modified Newton iteration to the transient components and repeated substitution to the smooth components. Eitelberg (1982) applies an implicit numerical method to sections of a large system divided into modules on physical grounds; the connections between the modules are considered in an explicit manner. This works only for weakly connected modules. Fagley and Carnahan (1982) discuss the equipment grouping strategy in a large interacting system that allows the GEAR method to be used on smaller systems at a time; some strategy for the iterations among the groupings is necessary for strongly coupled subsystems.

In addition to the obvious merits of smaller size, stiffness may find a better home in a "decoupled" or "semidecoupled" subsystem. The steady-state approximation is the best known way of reducing order on a set of ODEs, but the differential–algebraic set may offer its own problems. O'Malley and Anderson (1979) discuss how to find the small parameters automatically for a steady-state approximation; this is related to automatic partitioning.

Hofer (1976) advocates decoupling the stiff from the nonstiff in large systems with only a few stiff variables and using explicit techniques on the nonstiff part and implicit methods on the stiff part. Nandakumar and Andres (1978) explore a new class of algorithms that first heuristically decompose large systems into groups of smaller subsystems that share similar integration scales; they then solve individual subsystems and combine them iteratively. This is one of a very few studies in partitioning that recognizes the existence of a spectrum of large absolute-valued eigenvalues.

Refer also to the decomposition methods for stiff equations such as those of Clasen, Garfinkel, Shapiro, and Roman (1978), Burka (1982), and the problem-oriented studies of Anderson (1980) for control systems, and Humpage, Wong, and Lee (1974) for power systems (see §3.4.2).

3.2.2 *Notes on partitioning in the solution of stiff equations* (*Krogh*)†

The solution of stiff ODEs via the BDF involves at each step the solution of an equation of the form

$$f(t_{n+1}, y_{n+1}) - p'_{n+1} - \alpha(y_{n+1} - p_{n+1}) = 0. \tag{3.54}$$

where the subscript $n+1$ indicates the current step number, y_{n+1} is the numerical solution at the current time, t_{n+1}, to the differential equation

$$y' = f(t, y), \tag{3.55}$$

p_{n+1} and p'_{n+1} are initial estimates for y_{n+1} and $f(t_{n+1}, y_{n+1})$, and α is a parameter proportional to $1/h$, where h is the step size. The solution of (3.54) is usually obtained from a constant-slope Newton method of the form

$$\hat{G}\delta y_{n+1}^{(j)} = -r_{n+1}^{(j)} = -\,[f(t_{n+1}, y_{n+1}^{(j)}) - p'_{n+1} - \alpha(y_{n+1}^{(j)} - p_{n+1})], \tag{3.56}$$

where $y_{n+1}^{(0)} = p_{n+1}, y_{n+1}^{(j+1)} = y_{n+1}^{(j)} + \delta y_{n+1}^{(j)}$, $\hat{G}$ is an approximation to $G = J - \alpha I$, and $J = \partial f/\partial y$ evaluated at $(t_{n+1}, y_{n+1}^{(j)})$.

Iteration (3.56) has linear convergence with $\|\delta y_{n+1}^{(j)}\| = O(|\mu|^j)$. where

$$\mu = \text{largest eigenvalue of } \hat{G}^{-1}(\hat{G} - G). \tag{3.57}$$

In addition, one can use the value of μ as a guide to the number of iterations that are required to maintain stability of the numerical method, see Krogh and Stewart (1982a) and Klopfenstein (1971). The larger μ, the more corrector iterations required, until a point is reached that a new $\hat{G}$ is essential.

Errors in J and in α both contribute to μ. Errors in J outside of our direct control are assumed to be zero for the purposes of what we develop here. Of course a code must be prepared to cope with such errors.

At some point the code will obtain a factorization of

$$G_f = J - \hat{\alpha} I, \tag{3.58}$$

and we wish to use that factorization as long as possible. Initially we will have $\hat{\alpha} = \alpha$ (or perhaps $\hat{\alpha}$ = expected future α) and we wish to keep a useful $\hat{G}$ as α changes due to changes in the step size.

We propose here a factorization that is characterized by having low cost and

† The research described here was carried out at the Jet Propulsion Laboratory under contract with NASA. I would like to thank Mac Hyman for suggesting the use of $L(U - \beta I)$ for G, Linda Petzold for pointing out that all is not so simple if there are row permutations, Kris Stewart for getting me back in this arena, and finally my supervisor Charles Lawson for giving me a week to prepare this when there were things he would much rather have me doing.

good fill-in characteristics when being used on problems which have a sparse $\boldsymbol{J}$. A byproduct of the factorization is that equations which can be integrated with explicit methods are identified. What is given here is in a very unfinished form.

We first suggested using an automatic partitioning of stiff systems into equations integrated with Adams methods and equations integrated with BDF methods in Krogh (1968), and gave some results with such a code in Krogh (1973). This early work on automatic partitioning attempted to do this job without requiring a Jacobian matrix. We now believe it was a mistake to attempt so much with so little information. A test suggested by Shampine (1980c) does a good job in identifying when at least one equation in a system is stiff, and at such a point it seems to us (now) that there is little reason to avoid the calculation of the full Jacobian matrix.

Enright and Kamel (1979a) give an excellent summary of work that has been done on partitioning. Enright (1981) has done some work on investigating stability for algorithms which involve mixing different methods.

Row and column interchanges play an important role in the algorithm which we describe, but in order to bring out the essential ideas they are little more than alluded to in this section. An interchange of two rows is frequently accompanied by an interchange of the corresponding two columns thus mapping diagonals to diagonals. The factorization of $\boldsymbol{G}_{\mathrm{f}}$ has the form

$$\boldsymbol{G}_{\mathrm{f}} = \boldsymbol{LUR},$$

where $\boldsymbol{U}$ is upper triangular, and $\boldsymbol{L}$ and $\boldsymbol{R}$ are both unit lower triangular composed of a product of elementary matrix transformations. The total number of elementary matrix transformations in $\boldsymbol{L}$ and $\boldsymbol{R}$ is $\leqslant N$ (the dimension of $\boldsymbol{G}_{\mathrm{f}}$) and all are of a different size (i.e. the numer of rows or columns affected by the transformations are all different). Thus there is room in the space occupied by $\boldsymbol{G}_{\mathrm{f}}$ to store $\boldsymbol{L}$, $\boldsymbol{U}$, and $\boldsymbol{R}$. For example if $N = 7$, and the factorization consists of removing factors in the order R L L R L R, then the lower triangle has the appearance

$$\begin{matrix} l \\ l & l \\ l & l & l \\ l & l & l & r \\ l & l & r & r & r \\ r & r & r & r & r & r \end{matrix}$$

where l indicates an element of $\boldsymbol{L}$ and r an element of $\boldsymbol{R}$.

Let $\beta = \alpha - \hat{\alpha}$, i.e. (the current value of α) $-$ (the value of α used in forming $\boldsymbol{G}_{\mathrm{f}}$). We want to define a matrix $\hat{\boldsymbol{G}}$ which is a function of β such that for $\beta \neq 0$ the spectral radius of $\hat{\boldsymbol{G}}^{-1}(\boldsymbol{G} - \hat{\boldsymbol{G}})$ is less than the spectral radius of $\boldsymbol{G}_{\mathrm{f}}^{-1}(\boldsymbol{G} - \boldsymbol{G}_{\mathrm{f}})$. In addition the factorization of $\hat{\boldsymbol{G}}$ should be readily available from the factorization of $\boldsymbol{G}_{\mathrm{f}}$. In Krogh and Stewart (1982a) we examined $\boldsymbol{G} = c\,\boldsymbol{G}_{\mathrm{f}}$ where c is a parameter depending on β. That approach may give a fruitful mix with what we describe below, but such considerations would complicate considerably what we present. The factorization suggested by Enright (1978a, 1981) might also be useful in combination with what we suggest here.

If diagonals of G_f are mapped to diagonals of U, then we can write

$$G_f = LUR, \qquad G = LUR - \beta I, \qquad \hat{G} = L(U - D\beta)R$$
$$\hat{G}^{-1}(G - \hat{G}) = \beta G^{-1}(LDR - I),$$

where D is a diagonal matrix selected to minimize $\|LDR - I\|$.

Let $D = \text{diag}\,(d_k)$, $L = (l_{i,j})$, $R = (r_{i,j})$, where $l_{i,j} = r_{i,i} = 1$, and $l_{i,j} = r_{i,j} = 0$ for $j > i$. Because of the way L and R are constructed, if $j < i < k$ either $l_{i,j} = 0$ or $r_{k,i} = 0$. Thus LDR has the form $D + \bar{L}D + D\bar{R}$ where $\bar{L}$ and $\bar{R}$ are the parts of L and R below the diagonal. Thus the matrix norm is minimized if

$$d_k = 1/(1 + \|\bar{L}_k\|^2 + \|\bar{R}_k\|^2),$$

where $\bar{L}_k$ is the kth column of $\bar{L}$, $\bar{R}_k$ is the kth row of $\bar{R}$, and either $\bar{L}_k$ or $\bar{R}_k$ is zero. Observe that the smaller we can make $\bar{L}$ and $\bar{R}$ the smaller the norm of $G - \hat{G}$. Thus if a diagonal can be used as a pivot, we multiply on the left by L if $\|\bar{L}_k\| \leqslant \|\bar{R}_k\|$, and otherwise multiply on the right by R.

When diagonals in G_f do not map to diagonals in U, things become significantly more complicated. In this case D is permuted and some elements of βD fall in L, some in U, and some in R, The formula given above for d_k is still a good approximation for those elements of βD that fall in U; for those that fall in L or R we use

$$d_k = |u_{k,k}|/(|u_{k,k}|^2 + \|\bar{U}_k\|^2),$$

where $\bar{U}$ is the part of U above the diagonal, $\bar{U}_k$ is the kth row of $\bar{U}$, and $u_{k,k}$ is the kth diagonal element of U.

Although the algorithm appears fairly complicated, the actual arithmetic required should usually be less than that required for a simple LU factorization. (We sometimes set matrix elements to zero to save Gaussian elimination steps.) More details and the algorithm may be found in Krogh (1982).

3.2.3 *Riccati type transformations and decoupling of singularly perturbed ODE (Mattheij)*†

Consider the linear n-dimensional ODE

$$\epsilon \frac{\mathrm{d}x^1}{\mathrm{d}t} = A^{11}x^1 + A^{12}x^2 + r^1,$$
$$\frac{\mathrm{d}x^2}{\mathrm{d}t} = A^{21}x^1 + A^{22}x^2 + r^2, \qquad 0 \leqslant t \leqslant 1, \tag{3.59}$$

where A^{11}, A^{12}, A^{21} and A^{22} are matrix functions of t and r^1, r^2 are vector functions of t. We assume that A^{11} is of order k and is nonsingular for all t. In fact for our discussion only the homogeneous part of (3.59) is of importance, as we shall see. Problems with slow and fast time scales (as expressed by the small parameter ϵ) arise

† Supported in part by the Netherlands Organization for the advancement of Pure Research (Z.W.O) and the Niels Stensen Stichting.

both in initial value problems (IVP) and boundary value problems (BVP). We restrict ourselves to the latter class of problems. They differ from IVPs in that they usually deal with ODEs that have increasing and decreasing modes. Due to the instability with respect to initial values of the former, serious numerical difficulties may arise if one tries to compute them (as may be necessary to obtain a fundamental solution in solving the BVP). However, for the same reason that the increasing modes make forward integration an unstable affair, the decreasing modes render backward integration unstable. Therefore we look for methods that decouple the differential equation into slow and fast modes on the one hand, and in increasing and decreasing modes on the other hand. We shall do this by transforming the homogeneous part of (3.59) onto (block) uppertriangular form, using linear transformations $\boldsymbol{T}(t)$, i.e. we determine $\boldsymbol{U}^{11}$, $\boldsymbol{U}^{21}$, and $\boldsymbol{U}^{22}$, such that by setting

$$\begin{bmatrix} \boldsymbol{x}^1 \\ \boldsymbol{x}^2 \end{bmatrix} = \boldsymbol{T} \begin{bmatrix} \boldsymbol{y}^1 \\ \boldsymbol{y}^2 \end{bmatrix}, \tag{3.60}$$

we have

$$\epsilon \frac{\mathrm{d}\boldsymbol{y}^1}{\mathrm{d}t} = \boldsymbol{U}^{11}\boldsymbol{y}^1 + \boldsymbol{U}^{12}\boldsymbol{y}^2 + (\boldsymbol{T}^{-1}\boldsymbol{r})^1, \tag{3.61a}$$

$$\frac{\mathrm{d}\boldsymbol{y}^2}{\mathrm{d}t} = \boldsymbol{U}^{22}\boldsymbol{y}^2 + (\boldsymbol{T}^{-1}\boldsymbol{r})^2. \tag{3.61b}$$

The ODE (3.61b) has smooth solutions. Hence one may apply a multiple shooting technique for determining a fundamental solution and a particular solution of it, see Mattheij and Staaring (1982). Once the component $\boldsymbol{y}^2$ of a solution has been determined we are left with an (inhomogeneous) stiff ODE for $\boldsymbol{y}^1$. $\boldsymbol{U}^{11}$ now has an uppertriangular form,

$$\boldsymbol{U}^{11} = \begin{bmatrix} \boldsymbol{B} & \boldsymbol{C} \\ \boldsymbol{O} & \boldsymbol{E} \end{bmatrix}, \boldsymbol{B} \text{ of order } m, \tag{3.62}$$

such that $\boldsymbol{B}$ describes the system of increasing modes and $\boldsymbol{E}$ the system of decreasing modes. If we partition the vector $\boldsymbol{y}^1$ accordingly as

$$\begin{bmatrix} \boldsymbol{z}^1 \\ \boldsymbol{z}^2 \end{bmatrix},$$

then $\boldsymbol{z}^2$ should be computed in forward direction for which we may use a stiff integrator, and similarly $\boldsymbol{z}^1$ should be computed in backward direction using such a stiff integrator (note that $\boldsymbol{C}\boldsymbol{z}^1$ then appears as a forcing term). In a practical setting this should apply both to the computation of a fundamental solution and a particular solution, after which the desired $\boldsymbol{y}$ can be found from superposition. Finally we have to transform this solution back using (3.60). We shall describe first the decoupling of slow and fast time scales, then show how the fast system may be decoupled further. The latter decoupling is performed by an adapted predictor–corrector technique, the convergence and stability of which is shown here. Finally, we analyze the effects of perturbations in the (block) uppertriangular $\boldsymbol{U}$ on the computed solution.

For the matrix $\boldsymbol{T}$ we choose the (generalized) Riccati transformation

$$T(t) = \begin{bmatrix} M(t) & 0 \\ P(t)M(t) & I \end{bmatrix}, M(t) \text{ nonsingular.} \tag{3.63a}$$

As one may simply check, its inverse is given by

$$[T(t)]^{-1} = \begin{bmatrix} [M(t)]^{-1} & 0 \\ -P(t) & I \end{bmatrix}. \tag{3.63b}$$

Now write

$$A(t, \epsilon) \equiv \begin{bmatrix} \frac{A^{11}}{\epsilon} & \frac{A^{12}}{\epsilon} \\ A^{21} & A^{22} \end{bmatrix}. \tag{3.64}$$

Then we require the transformation $T(t)$ to be such that in the Liapunov equation

$$U = T^{-1}(AT - \dot{T}), \tag{3.65}$$

the system matrix U is block uppertriangular, with a kth order left upper block. Straightforward computation shows that, provided P satisfies the Riccati equation

$$\epsilon\dot{P} = \epsilon A^{21} + \epsilon A^{22}P - PA^{11} - PA^{12}P, \tag{3.66}$$

we obtain

$$\begin{bmatrix} U^{11} & U^{12} \\ 0 & U^{22} \end{bmatrix} = \begin{bmatrix} M^{-1}\left[\left(\frac{A^{11}}{\epsilon} + \frac{A^{12}}{\epsilon}P\right)M - \dot{M}\right] & M^{-1}\frac{A^{12}}{\epsilon} \\ 0 & -P\frac{A^{12}}{\epsilon} + A^{22} \end{bmatrix}. \tag{3.67}$$

It is of importance to realize that we have the freedom to choose the initial value $P(0)$, i.e. we may pick that integral curve that has favourable properties. In particular we try to choose $P(0)$ such that P is smooth. From power series expansion arguments (Wasow 1965), we therefore expect $P(0, 0) = 0$ to be a good choice. Rather than computing P by some stiff integrator, we suggest to compute a few terms of its power series expansion in ϵ, which has the form

$$P(t, \epsilon) = \sum_{j \geq 1} \epsilon^j P_j(t). \tag{3.68}$$

Substitution in (3.66) and comparing powers of ϵ, we obtain from the first-order term

$$0 = A^{21} - P_1 A^{11}, \tag{3.69}$$

whence

$$P_1 = A^{21}[A^{11}]^{-1}. \tag{3.70}$$

From the second-order term, we find

$$\dot{P}_1 = A^{22}P_1 - P_2A^{11} - P_1A^{12}P_1. \tag{3.71}$$

Since $\dot{P}_1$ can be derived from

$$\dot{P}_1 = \dot{A}^{21}[A^{11}]^{-1} - A^{21}[A^{11}]^{-1}\dot{A}^{11}[A^{11}]^{-1}, \tag{3.72}$$

we therefore can obtain P_2 in an analytical form from (3.71) and (3.72).

Altogether this gives us a decoupled ODE, as (3.61) and (3.67), of which the homogeneous form reads

$$\epsilon\frac{\mathrm{d}y^1}{\mathrm{d}t} = M^{-1}\left[\left(A^{11} + A^{12}\sum_{j\geq 0} \epsilon^j P_{j+1}\right)M - \epsilon\dot{M}\right]y^1 + M^{-1}A^{12}y^2, \tag{3.73a}$$

$$\frac{\mathrm{d}y^2}{\mathrm{d}t} = \left[A^{22} - \left(\sum_{j\geq 0} \epsilon^j P_{j+1}\right)A^{12}\right]y^2. \tag{3.73b}$$

The attractive feature of this approach is that it can be carried out analytically and therefore is very much suited to be combined with a numerical method for computing discrete solutions of the smooth part (3.73b), or for further decomposition of the singularly perturbed part (3.73a) and computing discrete solutions thereof. Note that it does not make much sense to go to very high order terms in the expansion as we are expecting to make even larger discretization errors anyway.

Thus far we didn't specify the matrix M. We now require U^{11} to have an uppertriangular form and therefore we have to determine M such that

$$\epsilon\dot{M} = FM - MU, \text{ where } F = A^{11} + A^{12}P,\ U = U^{11}. \tag{3.74}$$

If we assume that the directions of the solutions of the homogeneous part of (3.59) are only slowly varying, we may expect that (3.74) has a smooth solution M, such that $M(t)$ is orthogonal for all t. To understand this one should realize that the first column of M represents the direction of some solution of the ODE $\epsilon\dot{y}^1 = Fy^1$, and the first two columns span a two dimensional solution subspace. In order to find such M and U we propose a predictor-corrector technique, where the corrector utilizes a kind of QR algorithm that ultimately gives the orthogonal matrix M and the uppertriangular matrix U. We shall only sketch the method. Suppose we have a corrector

$$\sum_{j=0}^{n} \alpha_j z^j,\ \sum_{j=0}^{n} \beta_j z^j,\ \alpha_0 = 1,\ \beta_0 \neq 0, \tag{3.75}$$

then a typical corrector step has the form

$$M_i^j U_i^j = F_i M_i^{j-1} - \frac{\epsilon}{h\beta_0} M_i^{j-1} + G_{i-1}. \tag{3.76}$$

In (3.76) we have set $F_i = F(t_i)$, M_i^j = the jth correction at time t_i for $M(t_i)$ and similarly U_i^j; finally G_{i-1} contains the rest of the terms, depending only on previous time steps.

We next deal with the convergence of the iteration (3.76) and later with the accuracy of the approximation M_i^{∞}.

We write more conveniently the iteration (3.76) as

$$M^j U^j = HM^{j-1} + G, \tag{3.77}$$

i.e. we omit the indices and write

$$H = \left(F - \frac{\epsilon}{h\beta_0} I\right). \tag{3.78}$$

We shall restrict ourselves to matrices F having a complete system of eigenvectors $e_1, \ldots, e_n$, corresponding to the (ordered) eigenvalues $\lambda_1, \ldots, \lambda_n$ ($\lambda_1 \geqslant \lambda_2 \geqslant \ldots \geqslant \lambda_n$). Moreover we shall only investigate the convergence of the first column of M^j, m^j. Let $G(1)$ denote the first column of G, then there exist scalars $\gamma_1, \ldots, \gamma_n$ such that

$$G(1) = \sum_{j=1}^{n} \gamma_j e_j. \tag{3.79}$$

It is not restrictive to assume that all $\gamma_j \leqslant 0$. We now first investigate the fixed points (if any) of the iteration:

$$m^j \nu^j = Hm^{j-1} + G(1), \ \|m^j\|_2 = \|m^{j-1}\|_2 = 1. \tag{3.80}$$

Hence suppose for some vector m with $\|m\|_2 = 1$ and some scalar ν there holds

$$m\nu = Hm + G(1). \tag{3.81}$$

Note that (3.81) might be regarded as an inhomogeneous eigenvalue problem. Now write

$$m = \sum_{l=1}^{n} \mu_l e_l. \tag{3.82}$$

then we have

$$(\nu - \lambda_l)\mu_l = \gamma_l, \ l = 1, \ldots, n. \tag{3.83}$$

Hence if $\gamma_l \neq 0$ for any l, then $\mu_l \neq 0$. Now let k be such that $\gamma_k \neq 0$, then we obtain from (3.83)

$$\mu_l = \mu_k \frac{\gamma_l}{\gamma_k + (\lambda_k - \lambda_l)\mu_k}. \tag{3.84}$$

The actual values for μ_k (and hence for the other μ_l) follow from the requirement $\|m\|_2 = 1$. If we write for short

$$e_{ij} = (e_i, e_j) \tag{3.85}$$

(the natural inner product), then we must require that

$$f_k(\mu_k) = 1, \tag{3.86}$$

where

$$f_k(x) = x^2 \sum_{i,j=1}^{n} \frac{e_{ij}\gamma_i\gamma_j}{(\gamma_k + (\lambda_k - \lambda_j)x)\,(\gamma_k + (\lambda_k - \lambda_i)x)} \tag{3.87}$$

It is not restrictive to take k minimal. Since by definition f_k is positive definite ($x \neq 0$) we have a graph like in Figure 3.1.

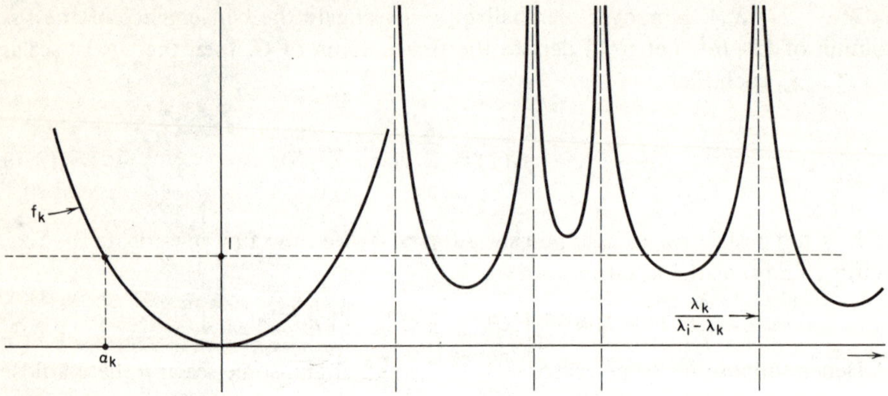

Fig. 3.1 The function $f_k(x)$, (3.87).

Hence there are at least two values of x such that $f_k(x) = 1$. Denote the negative root of $f_k(x) = 1$ by α_k. We have then a theorem.

THEOREM 3.01 The choice $\mu_k = \alpha_k (\leqslant 0)$ is the only stable one and gives a fixed point $\boldsymbol{m}$ of (3.88) with the largest possible $\nu = \lambda_k + \gamma_k/\mu_k$.

That we should choose $\mu_k < 0$ for obtaining the largest possible ν is trivial (n.b. $\gamma_k < 0!$). The stability follows from a perturbation argument.

Hence we may conclude that the sequence $[\boldsymbol{m}^j]$ converges to a unique vector $\boldsymbol{m}$ and the sequence $[\nu^j]$ converges at the same time to a scalar ν, which is the largest possible fixed point value. Now if $\boldsymbol{m}(t_{i-1})$ was the approximate direction of the most increasing solution at time $t = t_{i-1}$, then it can be expected that $\boldsymbol{m}(t_i)$ computed this way will approximately have the direction of this solution at $t = t_i$. Indeed, to see this let for simplicity $\boldsymbol{F}$ be a constant matrix, and $\boldsymbol{m}(t_i) = \boldsymbol{e}_1/\|\boldsymbol{e}_1\|_2$. Then apparently $\boldsymbol{G}_{i-1}(1)$ has the direction of $\boldsymbol{e}_1$, whereas $\gamma_1 \approx \epsilon/h\beta_0$ and $\gamma_2, \ldots, \gamma_n$ are small. From (3.86) with $k = 1$, we thus see

$$1 \approx \mu_1^2, \tag{3.88}$$

whence $\nu \approx \gamma_1 + \epsilon/(h\beta_0)$.

Since λ_1 is the largest eigenvalue of $\boldsymbol{H}$ $(= \boldsymbol{F} - [\epsilon/(h\beta_0)]\boldsymbol{I})$ we therefore see that ν is approximately the largest eigenvalue of $\boldsymbol{F}$!

By similar arguments as are used to show convergence of the QR algorithm one can expect that the general process (3.77) will converge in a stable way to some $\boldsymbol{M}$ and $\boldsymbol{U}$.

In order to find initial values for $\boldsymbol{M}$ and $\boldsymbol{U}$, one may use an appropriately ordered Schur normal form of $\boldsymbol{F}(0)$.

Since the above described method necessarily introduces discretization errors, the actually found uppertriangular matrices U_i will in general not correspond to an exactly transformed ODE. The following qualitative discussion is concerned with this problem.

Let $M(t)$ be a piecewise polynomial that is sufficiently differentiable (at least as many times as the order of the corrector) and moreover such that

$$M(t_i) \equiv M_i. \tag{3.89}$$

For this M, which is a certain transformation function now, define a matrix function $\hat{U}(t)$ by

$$\hat{U} = M^{-1}(AM - \epsilon\dot{M}). \tag{3.90}$$

If we (theoretically) discretize (3.90) we make a discretization error $h\Delta_i$. Apparently, using the notation $\hat{U}_i \equiv \hat{U}(t_i)$, we must then have

$$h \sum_{j=0}^{n} \beta_j M_{i-j}(U_{i-j} - \hat{U}_{i-j}) + h\Delta_i \tag{3.91}$$

(for simplicity we assume constant step sizes). Suppose we use BDF methods, then $\beta_j = 0$ for $j > 1$. This means that we have

$$\| U_i - \hat{U}_i \|_2 = O(\| \Delta_i \|_2) \tag{3.92}$$

(note that M_i is orthogonal and $\| \ldots \|_2$ is orthogonally invariant). In other words, the accuracy in U_i is not dependent on what happened at previous time steps! For more general methods we have to require that the polynomial

$$\sum_{j=0}^{n} \beta_j z^j$$

should have stable roots only. From a straightforward analysis it then follows that

$$\| U_i - \hat{U}_i \|_2 = O\left(\sum_{l=1}^{i} \| \Delta_l \| / h \right), \tag{3.93}$$

i.e. of $(p-1)$th order for a pth-order method.

Since we necessarily introduce discretization errors and truncation errors by the numerical scheme and the expansion respectively, we only have approximately correct block uppertriangular matrices U. We now briefly analyze their effects on the solution space. It is important to realize that any transformation $\tilde{T}$ corresponds to some exact $\tilde{U}$ by the relation:

$$\tilde{U} = \tilde{T}^{-1}(A\tilde{T} - \dot{\tilde{T}}). \tag{3.94}$$

Of course, in general $\tilde{U}$ will not be uppertriangular, but if $\tilde{T}$ is close to a triangularizing T, then we can expect $\tilde{U}$ to be close to an uppertriangular U defined by (3.65) and vice versa.

Now the truncated P and the specific M we have obtained give us a certain $\tilde{T}$ at some grid. By interpolation we may as well assume T to be continuous and sufficiently differentiable. Then by the transformation

$$x = \tilde{T}z, \tag{3.95}$$

each homogeneous solution of (3.59) uniquely corresponds to a solution of

$$\dot{z} = \tilde{U}z. \tag{3.96}$$

On the other hand, we actually integrate solutions of the system

$$\dot{y} = \hat{U}y, \tag{3.97}$$

where $\hat{U}$ is block uppertriangular and should be identified with an interpolated version of the system matrices we obtained above. So for the difference

$$e \equiv y - z, \tag{3.98}$$

we obtain the ODE

$$\dot{e} = \hat{U}e + (\hat{U} - \tilde{U})z. \tag{3.99}$$

From what we derived in the previous sections it follows that $\|\hat{U} - \tilde{U}\|$ is small. We then can give fairly sharp bounds for $y - z$ by using as boundary conditions

$$e^2(0) = 0, \; e^3(0) = 0 \text{ and } e^1(1) = 0. \tag{3.100}$$

Here the superscript ^{3} denotes the last $n - k$ coordinates. In fact the condition $e^3(0) = 0$ may be replaced by a condition at $t = 0$ and $t = 1$ if a more precise knowledge of the growth behavior of the slow modes is available. The actual estimates are obtained by integrating in the stable directions, utilizing the partitioned form of $\tilde{U}$ (Mattheij 1981a).

3.2.4 *Decoupling of boundary-value problems for two-time-scale systems (Mattheij and O'Malley)*

We wish to consider linear two-point boundary value problems of the form

$$\frac{\mathrm{d}x}{\mathrm{d}t} = a(t)x + f(t), \quad 0 \leqslant t \leqslant 1 \tag{3.101}$$

$$M_0 x(0) + M_1 x(1) = b, \tag{3.102}$$

where the homogeneous system involves coupled slow and fast dynamics. This contrasts sharply with most mathematical discussions of singularly perturbed systems where a small parameter multiplies some of the derivatives and slow and (potentially) fast components of the solution vector are immediately recognized. Special numerical methods are available for stiff initial-value problems with slow and fast components identified (Söderlind 1981c), and Riccati-like transformations are available to decouple such problems analytically and/or numerically (O'Malley 1969, Mattheij 1979, and O'Malley and Anderson 1982). In this paper, we propose a different type of transformation which will be useful in more general situations. We expect to implement numerically our algorithm and to pursue important generalizations in several directions. Relation to ongoing work of past and present colleagues and friends will be obvious to those familiar with earlier work (Ferguson 1975, Flaherty and O'Malley 1980, Mattheij and Staarink 1980, and Chow, Avramovic, Kokotovic, and Winkelman 1981). They are all thanked for their stimulation and encouragement.

Our basic assumption is that the homogeneous system has both smooth (or "slow") and rapidly varying (or "fast") solution modes. This requires the matrix $\boldsymbol{a}(t)$ to have large entries, so to make our hypotheses more explicit, we introduce a small positive scaling parameter ϵ and write

$$\boldsymbol{a}(t) = \frac{1}{\epsilon}\boldsymbol{A}(t, \epsilon) \sim \frac{1}{\epsilon}\sum_{j=0}^{\infty} \boldsymbol{A}_j(t)\epsilon^j.$$

We do not actually intend to use a full power series expansion here (or below), but would in practice use a truncated expansion (approximation) involving only a few terms. Thus, our n-vector system takes the form

$$\epsilon\frac{\mathrm{d}\boldsymbol{x}}{\mathrm{d}t} = \boldsymbol{A}(t, \epsilon)\boldsymbol{x} + \epsilon\boldsymbol{f}(t),$$

where the presence of slow modes implies that $\boldsymbol{A}(t, 0) = \boldsymbol{A}_0(t)$ is a singular matrix. Since the nonhomogeneity can be dealt with by variation of parameters, most of our attention will be restricted to the homogeneous problem.

We shall seek a change of variables

$$\boldsymbol{x} = \boldsymbol{T}(t, \epsilon)\boldsymbol{y},$$

where $\boldsymbol{T}$ and $\boldsymbol{T}^{-1}$ are both smooth (slowly varying) throughout the interval $0 \leqslant t \leqslant 1$. The transformed problem will take the form

$$\epsilon\frac{\mathrm{d}\boldsymbol{y}}{\mathrm{d}t} = \boldsymbol{U}(t, \epsilon)\boldsymbol{y} + \epsilon\boldsymbol{T}^{-1}\boldsymbol{f},$$

where

$$\boldsymbol{U} = \boldsymbol{T}^{-1}\left(\boldsymbol{A}\boldsymbol{T} - \epsilon\frac{\mathrm{d}\boldsymbol{T}}{\mathrm{d}t}\right), \tag{3.103}$$

We shall determine $\boldsymbol{T}$ and $\boldsymbol{U}$ simultaneously so that $\boldsymbol{U}$ has the block triangular form

$$\boldsymbol{U} = \begin{bmatrix} \boldsymbol{U}^{11} & \boldsymbol{U}^{12} & \boldsymbol{U}^{13} \\ \boldsymbol{0} & \boldsymbol{U}^{22} & \boldsymbol{U}^{23} \\ \boldsymbol{0} & \boldsymbol{0} & \epsilon\boldsymbol{U}^{33} \end{bmatrix}$$

where the eigenvalues of $\boldsymbol{U}^{11}$ and $-\boldsymbol{U}^{22}$ all have strictly positive real parts (at least for ϵ small). Thus, the rapidly varying solution modes of the transformed homogeneous problem will be either fast growing or fast decaying, but not (indefinitely) rapidly oscillatory. We shall assume that the partitioning holds throughout the interval $0 \leqslant t \leqslant 1$, thereby implicitly eliminating possible turning points within the interval (Wasow 1965). The matrix $\boldsymbol{T}$ then corresponds to a decoupling transformation which separates rapidly growing, rapidly decaying, and slow modes. The time scales for the rapid modes will be roughly $O(1/\epsilon)$ times those for the slow modes. The reason we don't allow purely imaginary eigenvalues of $U(t, 0)$ is that we wish to base our analysis on boundary layer methods (O'Malley 1974). We shall further presume that we can obtain smooth dependence of the U^{ij} on ϵ, although dependence on some root $\epsilon^{1/r}$, for some integer $r > 1$, would be necessary in some degenerate situations involving defective eigenvalues. These neglected possibilities should be considered in later work.

In order to proceed, let us fix the dimensions of U^{11} and U^{22} to be $k \times k$ and $l \times l$, respectively, and write T as

$$T(t, \epsilon) = T(t, 0)\,[I_n + \epsilon G(t, \epsilon)]\,,$$

where G will ultimately have the special, compatibly partitioned form

$$G(t, \epsilon) = \begin{bmatrix} 0 & G^{12} & 0 \\ G^{21} & 0 & 0 \\ G^{31} & G^{32} & 0 \end{bmatrix}$$

$$\sim \sum_{j=0}^{\infty} [G_j^{\alpha\beta}(t)]\,\epsilon^j.$$

At several stages, we make somewhat arbitrary choices for the form of our transformations. Though these cause us no complications, others might find different choices more appropriate. Our decoupling will be done in two steps. First, we shall separate fast and slow time scales. Then, we split fast-growing and fast-decaying modes to higher orders in ϵ. We shall let $m = n - k - l$ designate the number of slow modes with bounded derivatives as $\epsilon \to 0$.

At our first decoupling stage, it is temporarily convenient to combine notation. Splitting T after its first $k + l$ columns, we set

$$T(t, \epsilon) = [R(t, \epsilon) \quad Z(t, \epsilon)]\,.$$

Introducing

$$B(t, \epsilon) = \begin{bmatrix} U^{11} & U^{12} \\ 0 & U^{22} \end{bmatrix}$$

and

$$C(t, \epsilon) = \begin{bmatrix} U^{13} \\ U^{23} \end{bmatrix}$$

defining U implies that the fast decoupling transformation $R(t, \epsilon)$ must satisfy the linear matrix (Liapunov) equation

$$\epsilon\dot{R} = AR - RB.$$

Since A and B have power series expansions in ϵ, it is natural to seek a smooth matrix solution R in the form

$$R(t, \epsilon) \sim \sum_{j=0}^{\infty} R_j(t)\epsilon^j.$$

We shall simultaneously determine a formal series solution for this differential equation and for the block-triangular system matrix $B(t, \epsilon)$ for the transformed fast subsystem. Specifically, by equating coefficients we obtain

$$A_0 R_0 - R_0 B_0 = 0,$$

and, for each $j \geqslant 1$,

$$A_0 R_j - R_j B_0 = R_0 B_j + \alpha_{j-1}, \tag{3.104}$$

where the α_{j-1} are known in terms of preceding coefficients.

To solve our problem, we introduce the Schur canonical form

$$A_0 = Q_0 V_0 Q_0^{\mathsf{T}}$$

for the principal part $A_0(t)$ of our system matrix $\epsilon a(t)$ (here, $Q_0(t)$ is orthogonal and $V_0(t)$ is upper triangular). So

$$A_0 Q_0 = Q_0 V_0.$$

We observe that these matrices are convenient to obtain numerically through the QR algorithm (Golub and Wilkinson 1976). This provides a full triangularization for A_0, though a block-triangularization would suffice for much of the work described here. We shall set

$$V_0 = \begin{bmatrix} B_0 & C_0 \\ 0 & 0 \end{bmatrix}$$

and, without loss of generality, we suppose that the first k (and remaining l) eigenvalues of the $(k + l) \times (k + l)$ matrix B_0 will have strictly positive (negative) real parts throughout the interval $0 \leqslant t \leqslant 1$. This clearly, but definitely restricts the matrices $A_0(t)$ under consideration. The matrices $B_0(t)$ and $C_0(t)$ so specified determine our limiting transformed system matrix $U(t, 0)$. Moreover, with such an ordering of the eigenvalues of B_0, we separate the fast growing (first k) and fast decaying (next l) modes at this first (and critical) order (with respect to ϵ). If we further let R_0 coincide with the first $k + l$ columns of Q_0, we satisfy the desired linear system $A_0 R_0 - R_0 B_0 = 0$ and specify the fast part $R_0(t)$ of our limiting transformation matrix $T(t, 0)$. Its rank, $k + l$, coincides with that of $A_0(t)$.

Since

$$Q_0^{\mathsf{T}} R_0 = \begin{bmatrix} \mathrm{I}_{k+l} \\ 0 \end{bmatrix}$$

multiplying the linear system (3.104) by Q_0^{T} and setting

$$R_j = Q_0 \begin{bmatrix} \tilde{R}_j^1 \\ \tilde{R}_j^2 \end{bmatrix}, \qquad \text{for } j \geqslant 1,$$

implies the pair of equations

$$B_0 \tilde{R}_j^2 + C_0 \tilde{R}_j^2 - \tilde{R}_j^1 B_0 = B_j + \alpha_{j-1}^1,$$

and

$$-\tilde{R}_j^2 B_0 = \alpha_{j-1}^2,$$

where the α_{j-1}^i are known successively.

Since the triangular matrix B_0 is simply inverted by back subsitution, $\tilde{R}_j^2$ is uniquely determined and a linear system

$$B_0 \tilde{R}_j^1 - \tilde{R}_j^1 B_0 = B_j + \beta_{j-1}$$

with $\beta_{j-1} = \alpha_{j-1}^1 + C_0 \tilde{R}_j^2$ remains. We now seek to specify $\tilde{R}_j^1$ so that the resulting fast-growing and fast-decaying modes are separated (to the appropriate order) by making B_j upper triangular. Substituting for B_0 and B_j, we seek a solution of the form

$$\tilde{R}_j^1(t) = \begin{bmatrix} 0 & G_j^{12} \\ G_j^{21} & 0 \end{bmatrix}.$$

This provides the four linear matrix equations

$$\begin{aligned} U_0^{11} G_j^{12} - G_j^{12} U_0^{22} &= \beta_{j-1}^{12}, \\ U_0^{22} G_j^{21} - G_j^{21} U_0^{11} &= \beta_{j-1}^{21}, \\ U_0^{12} G_j^{21} &= U_j^{11} + \beta_{j-1}^{11}, \\ -G_j^{21} U_0^{12} &= U_j^{22} + \beta_{j-1}^{22}, \end{aligned}$$

presuming we can take $U_j^{12} = 0$ for $j > 0$. The first two systems are uniquely solvable for G_j^{12} and G_j^{21}, respectively, since U_0^{11} and U_0^{22} have no eigenvalues in common. Indeed, the triangular form of U_0^{11} and U_0^{22} allows a simple solution for these nonzero blocks of $\tilde{R}_j^1$ by forward and backward substitution. Then, the last two equations directly specify the block-diagonal matrix B_j. The procedure can be continued to any desired order.

To determine the full transformation $T(t, \epsilon)$, we must select a matrix $Z(t, \epsilon)$ whose span is complementary to that of $R(t, \epsilon)$. It is natural to attempt to take $T(t, 0) = Q_0(t)$. Then

$$\tilde{R}^1(t, \epsilon) = \begin{bmatrix} 0 & G^{12} \\ G^{21} & 0 \end{bmatrix} \text{ and } \tilde{R}^2(t, \epsilon) = (G^{31} G^{32}).$$

The T matrix so obtained is readily inverted since

$$T = Q_0 \begin{bmatrix} \mathrm{I}_{k+l} + \epsilon\tilde{R}^1 & 0 \\ \epsilon\tilde{R}^2 & \mathrm{I}_m \end{bmatrix}$$

implies that

$$T^{-1} = \begin{bmatrix} (I_{k+l} + \epsilon\tilde{R}^1)^{-1} & 0 \\ -\epsilon\tilde{R}^2 & I_m \end{bmatrix} Q_0^{\mathrm{T}},$$

and

$$(I_{k+l} + \epsilon\tilde{R}^1)^{-1} = \begin{bmatrix} I_{k+1} & -\epsilon G_{12} \\ -\epsilon G_{21} & I_l \end{bmatrix} \begin{bmatrix} (I_k - \epsilon^2 G_{12} G_{21})^{-1} & 0 \\ 0 & (I_l - \epsilon^2 G_{21} G_{12}) \end{bmatrix}$$

at least for sufficiently small ϵ.

The matrices T and U are, of course, determined simultaneously. We should check that our transformed state matrix U attains the desired block structure. It follows that Z, the last m columns of T, must satisfy

$$RC + \epsilon Z U^{33} = AZ - \epsilon \frac{\mathrm{d}Z}{\mathrm{d}t}.$$

Since

$$Z = Q_0 \begin{bmatrix} 0 \\ I_m \end{bmatrix},$$

multiplication by Q_0^{T} implies that

$$\epsilon Q_0^{\mathrm{T}} \frac{\mathrm{d}Q_0}{\mathrm{d}t} \begin{bmatrix} 0 \\ I_m \end{bmatrix} + \epsilon \begin{bmatrix} 0 \\ U^{33} \end{bmatrix} = Q_0^{\mathrm{T}} A Q_0 \begin{bmatrix} 0 \\ I_m \end{bmatrix} - Q_0^{\mathrm{T}} RC. \qquad (3.105)$$

Since

$$Q_0^{\mathrm{T}} A_0 Q_0 = \begin{bmatrix} B_0 & C_0 \\ 0 & 0 \end{bmatrix} \text{ and } Q_0^{\mathrm{T}} R_0 = \begin{bmatrix} I_{k+l} \\ 0 \end{bmatrix}, \text{ we have}$$

equality when $\epsilon = 0$. From the coefficient of each ϵ^j with $j > 0$, we directly obtain

$$C_j = \begin{bmatrix} U_j^{13} \\ U_j^{23} \end{bmatrix},$$

(in terms of preceding terms) from the first $k + l$ rows of (3.105) and

$$U_j^{33}$$

from the last m rows. This completes the specification of $T_j(t)$ and $U_j(t)$.

We should verify that the truncated series actually provides decoupling to a corresponding order and that analytic transformations exist with asymptotic expansions coinciding with the formal series generated. It should, however, be realized that our approach is not necessarily limited to asymptotic validity as $\epsilon \to 0$, but that the decoupling achieved should be numerically valid for any realistic small ϵ for which our algorithm is defined.

The principal value to our approach is the simplicity it provides for solving two-point boundary value problems. In $\boldsymbol{y}$-coordinates, our problem becomes

$$\begin{aligned} \epsilon \frac{d\boldsymbol{y}}{dt} &= \boldsymbol{U}(t, \epsilon)\boldsymbol{y} + \epsilon \boldsymbol{T}^{-1}(t, \epsilon)\boldsymbol{f}(t), \\ \boldsymbol{M}_0\boldsymbol{T}(0, \epsilon)\boldsymbol{y}(0) &+ \boldsymbol{M}_1\boldsymbol{T}(1, \epsilon)\boldsymbol{y}(1) = \boldsymbol{b}. \end{aligned} \tag{3.106}$$

If we first find a particular bounded solution of (5.106), using boundary conditions appropriate to the conditional stability of our fast subsystem, we then need only consider the homogeneous system.

All solutions of the homogeneous system have the form

$$\boldsymbol{y}(t) = \Omega(t, \epsilon)\boldsymbol{c}, \tag{3.107}$$

where Ω is any fundamental matrix for $\epsilon(d\Omega)/(dt) = \boldsymbol{U}(t, \epsilon)\Omega$. We shall use a fundamental matrix

$$\Omega(t, \epsilon) = \begin{bmatrix} \Omega^{11} & \Omega^{12} & \Omega^{13} \\ \boldsymbol{0} & \Omega^{22} & \Omega^{23} \\ \boldsymbol{0} & \boldsymbol{0} & \Omega^{33} \end{bmatrix},$$

subject to special boundary conditions at the endpoints $t = 0$ and $t = 1$ which will assure us boundedness of the $\Omega^{\alpha\beta}$. Specifically, we will define Ω on $0 \leqslant t \leqslant 1$ by asking that

$$\frac{d\Omega^{33}}{dt} = \boldsymbol{U}^{33}\Omega^{33}, \quad \Omega^{33}(0) = \boldsymbol{I}_m,$$

$$\epsilon \frac{d\Omega^{22}}{dt} = \boldsymbol{U}^{22}\Omega^{22}, \quad \Omega^{22}(0) = \boldsymbol{I}_l,$$

$$\epsilon \frac{d\Omega^{23}}{dt} = \boldsymbol{U}^{22}\Omega^{23} + \boldsymbol{U}^{23}\Omega^{33}, \quad \Omega^{23}(0) = \boldsymbol{0},$$

$$\epsilon \frac{d\Omega^{11}}{dt} = \boldsymbol{U}^{11}\Omega^{11}, \quad \Omega^{11}(1) = \boldsymbol{I}_k,$$

$$\epsilon \frac{d\Omega^{12}}{dt} = U^{11}\Omega^{12} + U^{12}\Omega^{22}, \ \Omega^{12}(1) = \mathbf{0},$$

$$\epsilon \frac{d\Omega^{13}}{dt} = U^{11}\Omega^{13} + U^{12}\Omega^{23} + U^{13}\Omega^{33}, \ \Omega^{13}(1) = \mathbf{0}.$$

As $\epsilon \to 0$, we note that Ω^{22} decays rapidly to zero for $t > 0$ (i.e. away from an initial boundary layer), while Ω^{12} rapdily decays to zero away from a terminal boundary layer near $t = 1$. Away from $t = 0$, Ω^{23} tends to an outer limit $(U^{22})^{-1} U^{23}\Omega^{33}$ as $\epsilon \to 0$. Ω_{12} and Ω_{13} involve boundary layer behavior at $t = 1$. Such asymptotic results readily follow from singular perturbations theory (O'Malley 1974).

Since any solution of our homogeneous problem (5.101-5.102) has the form (3.107), we obtain a unique solution if and only if the matrix

$$D \equiv M_0 T(0, \epsilon)\Omega(0, \epsilon) + M_1 T(1, \epsilon)\Omega(1, \epsilon)$$

is nonsingular. Moreover, the special structure of $T(t, \epsilon)$ and of $\Omega(0, \epsilon)$ and $\Omega(1, \epsilon)$ can be utilized to simplify the checking of this invertibility condition. If $D = 0(\epsilon^p)$ as $\epsilon \to 0$, we can anticipate having solutions which become algebraically unbounded like ϵ^{-p}. We can even identify the limiting behavior as $\epsilon \to 0$ within $(0, 1)$ with the solution of a reduced boundary value problem of order m (O'Malley 1969 and Harris 1973). Special numerical interest, however, concerns problems where ϵ is small but nonvanishing.

3.3 An overview of the highly oscillatory initial-value problem (Miranker)

3.3.1 *Introduction*

The highly oscillatory problem is particularly difficult since the rapidly changing behavior of the solution is an ever present feature and not a transient phase. Thus most approaches to the highly oscillatory problem abandon the customary procedure of approximating the solution pointwise (at least on a mesh). Instead some functional of the solution such as its smooth part or some running mean is accepted as a meaningful description and it is this functional which is approximated through computation. This overview will deal with several methods which take this approach. (For methods with the classical pointwise approach see Amudursky and Ziv 1977, or Snider and Fleming 1974).

We being this review with highly oscillatory recurrences since there are many applications for them, several of which are reviewed here. These recurrences are treated by the two-time technique. We then turn to the initial-value problem for differential equations. We do not explicitly cite applications for the latter, but we do illustrate computations on a model problem. Applications are to be found in circuit analysis and orbiting motions, for example. We demonstrate three approaches to the differential problem. The first, which is most common, is through singular perturbation methods and in particular, multitime techniques. The third is a method of averaging which is independent of singular perturbation methods. The second method, the extrapolation method, falls in between.

References are given to source material; also see Miranker (1981).

3.3.2 *Recurrences*

Consider first the model

$$x_{n+1} = (A + \epsilon B)x_n, \ x_0 \text{ given},$$

where $x \in \mathbb{R}^p$ and A and B are $p \times p$ constant matrices. To illustrate the behavior of this recurrence it is convenient to suppose for the time being that A and B commute. In this case

$$x_n = (A + \epsilon B)^n x_0 = A^n(I + \epsilon A^{-1}B)^n x_0.$$

Then

$$x_n = A^n \exp(A^{-1}B\epsilon n)x_0 \, [1 + n\epsilon O(\epsilon)] .$$

This demonstrates the composition of x_n as it develops on two scales: a rapidly developing part (A^n) and a slowly developing part ($\exp[A^{-1}B\epsilon n]$). In many cases the rapidly varying part is akin to noise from which the more meaningful part of the solution needs to be separated. How this separation may be performed in more general cases (i.e. noncommutative and nonlinear cases) has been demonstrated in Hoppensteadt and Miranker (1977), where a two-time formalism for recurrences is developed.

We continue by describing three applications, and we illustrate the separattion of the scales achieved by use of a two-time technique.

3.3.3 *Training algorithm*

The so-called training algorithm (Greenberg and Konheim 1964) arises in pattern recognition and corresponds to the mathematical problem of determining a separating hyperplane. Such a hyperplane is specified by a vector $w \in \mathbb{R}^{p+1}$ to which the following recurrence may be made to converge (in finitely many steps).

For fixed $\theta > 0$ and $w_0 \in \mathbb{R}^{p+1}$, define $w_1, w_2, \ldots$ as follows:

$$w_{n+1} = w_n + x_n S\left(\frac{w_n \cdot x_n}{\theta}; x_n\right),$$

where

$$S\left(\frac{w \cdot x}{\theta}; x\right) = \begin{cases} \left.\begin{matrix} 1, & w \cdot x \leqslant \theta \\ 0, & w \cdot x > \theta \end{matrix}\right\} \text{and } x \in A^*, \\ \left.\begin{matrix} -1, & w \cdot x \geqslant -\theta \\ 0, & w \cdot x < -\theta \end{matrix}\right\} \text{and } x \in B^*. \end{cases}$$

Here A^* and B^* are the finite point sets in $\mathbb{R}^{p+1}$ being separated. Setting $w_n = \theta z(n)$ and $\epsilon = \theta^{-1}$, the recurrence relation for w_n becomes

$$z(n + 1) = z(n) + \epsilon x_n S[z(n) \cdot x_n; x_n],$$

while in the definition of S, w is changed to z and θ is replaced by unity.

The two-time formalism specifies the slowly evolving aspect of the trajectory defined by this recurrence for $z(n)$ as a function $z_0(s)$ where moreover $z(n) = z_0(\epsilon n) + O(\epsilon)$. Indeed z_0 is the solution of the following nonstiff problem:

$$\frac{dz_0}{ds} = \overline{xS}(z_0),$$

where the average

$$\overline{xS}(z_0) = \lim_{n\to\infty} \frac{1}{n}\sum_{k=0}^{n-1} x_k S(z_0 \cdot x_n; x_k).$$

(In applications this average is finitely computable). Computation of z_0 is straightforward and leads to a limiting vector which is an appropriate w sought by the training algorithm.

3.3.4 *A population genetics model*

In a large population of diploid organisms having discrete generations, the genotypes determined by one locus having two alleles A and a, divide the population into three groups of types AA, Aa, and aa, respectively, Suppose that the gene pool carried by this popultion is in proportion p_n of type A in the nth generation. It follows (Crow and Kimura 1970) that

$$p_{n+1} = p_n + \frac{p_n(1-p_n)[(w_{11}-w_{12})p_n + (w_{21}-w_{12})(1-p_n)]}{w_1 p_n^2 + 2w_{12}p_n(1-p_n) + w_{22}(1-p_n)^2},$$

where w_{11}, w_{12}, and w_{22} are the relative fitnesses of the gentotypes AA, Aa, and aa, respectively. If the selective presses act slowly, i.e. if $w_{ii} = 1 + \epsilon a$, $w_{12} = 1$ and $w_{22} = 1 + \epsilon\beta$ then

$$p_{n+1} = p_n + \frac{\epsilon p_n(1-p_n)[(\alpha-\beta)p_n + \beta]}{1 + O(\epsilon)}.$$

The two-time formalism tells us that

$$p_n = P(\epsilon n) + O(\epsilon),$$

wherc

$$\frac{dP}{ds} = P(1-P)[(\alpha-\beta)P + \beta],$$

and this simple nonstiff differential equation is easily solved numerically.

3.3.5 *Regression analysis*

Let $g(w)$ be a function with a unique root $\hat{w}$ and such that $g(w)(w - \hat{w}) > 0$, $w \neq \hat{w}$. Let z_k, $k = 0, 1, \ldots$ be a sequence of identically distributed random variables with mean zero and unit variance. The Robbins-Munro algorithm for approximating $\hat{w}$ is specified by the following recurrence:

$$w(k+1) = w(k) - \epsilon\alpha_k\{g[w(k)] + \sigma z_k\}.$$

$g[w(k)] + \sigma z_k$ is a noisy measurement of $g[w(k)]$ so that the recurrence specifies a very chaotic behavior for $w(k)$. However, the two-time methodology asserts that

$$w(k) = W_0(\epsilon k)[1 + O(\epsilon)],$$

where

$$\frac{\mathrm{d}W_0}{\mathrm{d}s} = -\bar{\alpha} g(W_0),$$

and

$$\bar{\alpha} = \lim_{k \to \infty} \frac{1}{k} \sum_{j=0}^{k} \alpha_j.$$

The differential equation for W_0 is simple to solve, and for appropriate g, its solutions converge to the equilibrium value $\hat{w}$, the root which is sought.

3.3.6 *Two-time methods*

The model highly oscillatory problem is taken to be

$$\epsilon \frac{\mathrm{d}\boldsymbol{u}}{\mathrm{d}t} = (\boldsymbol{A} + \epsilon \boldsymbol{B})\boldsymbol{u}, \quad t \in (0, T],$$

$$\boldsymbol{u}(0) = \boldsymbol{u}_0,$$

which $\boldsymbol{u}$ is an n-vector and $\boldsymbol{A}$ and $\boldsymbol{B}$ are $n \times n$ matrices. The solution is

$$\boldsymbol{u} = \exp[(\boldsymbol{A} + \epsilon \boldsymbol{B})t/\epsilon]\, \boldsymbol{u}_0$$
$$= \exp(\boldsymbol{A}\tau + \boldsymbol{B}t)\boldsymbol{u}_0$$

where $\tau = t/\epsilon$. τ is called the fast time and t the slow time. If $\boldsymbol{A}$ and $\boldsymbol{B}$ commute the dependence of the solution on these two time scales separates,

$$\boldsymbol{u} = \exp(\boldsymbol{A}\tau)\exp(\boldsymbol{B}t)\boldsymbol{u}_0$$

and in principle, each of the factors here could be determined separately and without computational difficulty. When $\boldsymbol{A}$ and $\boldsymbol{B}$ do not commute this separation is no so readily available, and moreover, it is not necessarily the case that the development of the solution on the t-scale is even meaningful to approximate numerically. To treat this case we employ the method of two-times. We suppose that the initial condition has the form

$$\boldsymbol{u}(0) = \sum_{r=0}^{\infty} \boldsymbol{a}_r \epsilon^r,$$

and that the solution of the initial value problem has an approximation in the form of a general two-time expansion:

$$\boldsymbol{u} = \sum_{r=0}^{\infty} \boldsymbol{u}_r(t, \tau)\epsilon^r.$$

We take the leading term $\boldsymbol{u}_0(t, \tau)$ of the expansion as an approximation to the solution of the initial value problem. The method two times (Miranker and Hoppenstaedt 1979) specifies $\boldsymbol{u}_0(t, \tau)$ as follows

$$\boldsymbol{u}_0(t, \tau) = \Phi(\tau)\tilde{\boldsymbol{v}}_0(t),$$

where the fundamental matrix $\Phi(\tau)$ is given by

$$\Phi_\tau = \boldsymbol{A}\Phi, \quad \Phi(0) = \boldsymbol{I}. \tag{3.108a}$$

Further

$$\frac{\mathrm{d}\tilde{\boldsymbol{v}}_0}{\mathrm{d}t} = \bar{\boldsymbol{B}}\tilde{\boldsymbol{v}}_0, \quad \tilde{\boldsymbol{v}}_0(0) = \boldsymbol{a}_0, \tag{3.108b}$$

where

$$\bar{\boldsymbol{B}} = \lim_{T\to\infty} \frac{1}{T}\int_0^T \Phi^{-1}(\sigma)\boldsymbol{B}\Phi(\sigma)\mathrm{d}\sigma. \tag{3.108c}$$

(For the development of this two-time methodology in a more general nonlinear setting see Hoppenstaedt and Miranker 1976.) The specification of $\boldsymbol{u}_0(t, \tau)$ leads to the following numerical algorithm:

(i) Solve (3.108a) on a mesh of increment k in the τ-scale by some self-starting numerical method, obtaining the sequence $\Phi(jk), j = 0, \ldots, N$.

(ii) Using the values $\Phi(jk)$ obtained in (i), approximate $\bar{\boldsymbol{B}}$ by truncating the limit of τ integration and replacing the integral in (3.108c) by a quadrature formula,

$$\bar{\boldsymbol{B}} = \frac{1}{N}\sum_{j=0}^{N} c_k\Phi^{-1}(jk)\boldsymbol{B}\Phi(jk).$$

The integer N is determined by a numerical criterion which insures that the elements of the matrix $\bar{\boldsymbol{B}}$ are calculated to within some desired accuracy.

(iii) With $\bar{\boldsymbol{B}}$ (approximately) determined in (ii), solve (3.108b) for $\tilde{\boldsymbol{v}}_0(h)$ by some self-starting numerical method.

(iv) Compute $\boldsymbol{u}_0(h, Nk) = \Phi(Nk)\tilde{\boldsymbol{u}}_0(h)$, and take this as the approximation for $\boldsymbol{u}(h)$.

The approximation $\boldsymbol{u}_0(h, Nk)$ produced by this algorithm is not an approximation to $\boldsymbol{u}(h, Nk)$ (i.e. is not a pointwise approximation in the conventional sense of numerical analysis). The approach of this algorithm takes pointwise approximations as ill-conditioned (and meaningless). Indeed approximation on the fast time scale is abandoned. In fact, $\boldsymbol{u}_0(h, Nk)$ is an approximation to $\boldsymbol{u}(t, Nk)$ for t near h and not necessarily for $t = h$.

In Figure 3.02 we schematize the computation. Of course, in practice ϵ will be extremely small so that unlike the schematic an enormous number of oscillations of Φ will occur in the t interval $[0, h]$. Notice how far the computed answer $\phi(Nk)\tilde{\boldsymbol{v}}_0(h)$ may be from the customary approximation to the solution, $\boldsymbol{u}_0(h, h/\epsilon)$.

The fundamental matrix $\Phi(\tau)$ is composed of modes corresponding to the eigenvalues of $\boldsymbol{A}$. Since the eigenvalues of $\boldsymbol{A}$ lie in the closed left half plane, the profile for (a component of) Φ will, after some moderate number of cycles, settle down to an (almost) periodic function. Thus the set of mesh points $\{jk : j = 0, \ldots, N\}$ may be expected to extend over just these cycles (approximately).

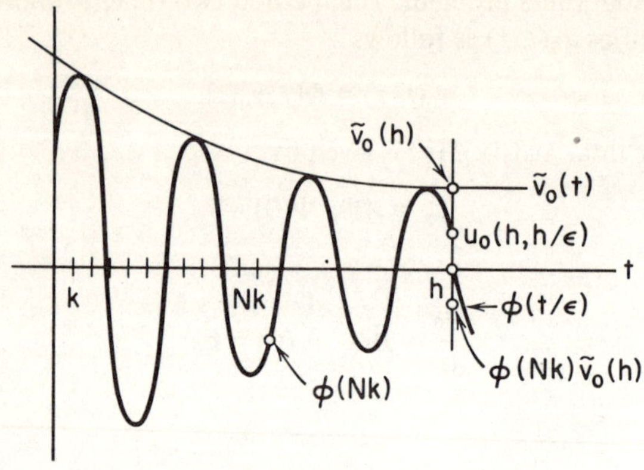

Fig. 3.2 Illustrations of numerical algorithm

3.3.7 *The extrapolation method*

The exploitation of singular perturbation theory for the development of numerical techniques for stiff ODEs usually proceeds with the numerical determination of values of one or more terms in the asymptotic expansion supplied by that theory. The extrapolation method (Hoppensteadt and Miranker 1982) which we will now review is a way to break through this limitation of approach.

Consider the following nonlinear analogue of the model problem which we have been discussing

$$\frac{d\boldsymbol{x}}{dt} = f(t/\epsilon, \boldsymbol{x}), \quad \boldsymbol{x}(0) = \xi.$$

where $\boldsymbol{x}, f, \xi \in \mathbb{R}^n$ and where $f(\tau, \ldots)$ is an almost periodic function of τ. Multi-time perturbation methods lead to the approximation

$$\boldsymbol{x}(t, \epsilon) = \boldsymbol{x}_0(t) + \epsilon \boldsymbol{x}_1(t, t/\epsilon) + O(\epsilon^2),$$

where $\boldsymbol{x}_0$ is determined from the initial value problem

$$\frac{d\boldsymbol{x}_0}{dt} = \bar{f}(\boldsymbol{x}_0), \quad \boldsymbol{x}_0(0) = \xi.$$

$\bar{f}$ is the average of f, defined by

$$\bar{f}(\boldsymbol{x}_0) = \lim_{T\to\infty} \frac{1}{T}\int_0^T f(\tau, \boldsymbol{x}_0)\, d\tau.$$

The coefficient $\boldsymbol{x}_1$ is determined from the formula

$$\boldsymbol{x}_1(t, t/\epsilon) = \tilde{\boldsymbol{x}}_1(t) + \int_0^{t/\epsilon} [f(\tau, \boldsymbol{x}_0) - \bar{f}(\boldsymbol{x}_0)]\, d\tau.$$

In this formula, $\tilde{\boldsymbol{x}}_1$ is determined at a later step in the perturbation scheme. Since it will not be needed here, it is not discussed further. Thus

$$x(t,\epsilon)=x_0(t)+\epsilon\left(\tilde{x}_1(t)+\int_0^{t/\epsilon}[f(\tau,x_0)-\bar{f}(x_0)]\,d\tau\right)+O(\epsilon^2).$$

We seek to determine certain larger values ϵ' of ϵ. The initial-value problem is solved with these larger values of ϵ and an appropriate extrapolation determines an approximation to $x(t,\epsilon)$ itself.

First the average $\bar{f}$ must be determined. A straightforward numerical approximation of $\bar{f}$ can be costly, but there is the possibility of accelerating this computation by a second difference method which we now describe.

In most applications, the integral of the almost periodic function f has the form

$$V(T,x)=\bar{f}(x)T+p(T,x).$$

where p is an almost periodic function of its first argument and which has mean zero. Thus, given a tolerance δ, there is a δ-translation number $\mathcal{T}(\delta,x)$ such that

$$\|p[T+\mathcal{T}(\delta,x),x]-p(T,x)\|<\delta$$

for all $T\geqslant 0$; in particular, since $p(0,x)=0$, then $|p[\mathcal{T}(\delta,x]|<\delta$. To find candidates for $\mathcal{T}$ note that

$$V(2T,x)-2V(T,x)=p(2T,x)-2p(T,x).$$

In particular for $T=\mathcal{T}(\delta,x)$, we have that

$$\begin{aligned}V(2\mathcal{T},x)-2V(\mathcal{T},x)&=p(2\mathcal{T},x)-p(\mathcal{T},x)-p(\mathcal{T},x)+p(0,x)\\&=O(\delta).\end{aligned}$$

Thus, any δ-translation number of p makes this second difference of order δ. Unfortunately, the converse does not hold; in particular, $\|V(2T,x)-2V(T,x)\|$ may be small while $\|p(T,x)\|$ is not small.

Still, by tabulating

$$V(2T,x)-2V(T,x),$$

candidates for $\mathcal{T}(\delta,x)$ can be found and tested by comparing the values of $V(T,x)/T$ for several of them, since these should all approximate $\bar{f}(x)$. In practice, this method is no worse than the direct calculation of $\bar{f}(x)$, and in periodic cases, it reliably gives $\bar{f}(x)$ after calculation over one period.

Thus we use

$$\frac{1}{(\delta,x)}\int_0^{\mathcal{T}(\delta,x)}f(\tau,x)d\tau$$

as an approximation to $\bar{f}$.

The extrapolation method proceeds by choosing an appropriate value T which represents a time at which rapid motions can be ignored. We pick T to be a δ-translation number of $p(\tau,x)$. Then in particular,

$$\frac{p(2T,x)-2p(T,x)}{T}=\frac{1}{T}\int_T^{2T}[f(\tau,x)-\bar{f}(x)]\,d\tau=O(h^p),$$

for $x=\xi+O(h)$. The existence of such a value of T follows from viewing this equation as the statement that T is an approximation to a δ-translation number with $\delta = O(h^p)$. Such a T value must be found, perhaps using the second difference method just described or additional knowledge about a specific problem being studied.

Once a T value is found we define

$$\epsilon' = h/T,$$

and then we calculate $\boldsymbol{x}(h, \epsilon'/2)$ and $\boldsymbol{x}(h, \epsilon')$ from the initial-value problem by a pth-order numerical method. It follows from the form of the asymptotic expansion for $\boldsymbol{x}(t, \epsilon)$ above that

$$2\boldsymbol{x}(h, \epsilon'/2) - \boldsymbol{x}(h, \epsilon) = \boldsymbol{x}_0(h) + \epsilon' \int_T^{2T} \{\boldsymbol{f}[\tau, \boldsymbol{x}_0(h)] - \bar{\boldsymbol{f}}[\boldsymbol{x}_0(h)]\}\mathrm{d}\tau + O[(\epsilon')^2]$$

$$= \boldsymbol{x}_0(h) + O(h^{p+1}) + O\left[\left(\frac{h}{T}\right)^2\right].$$

On the other hand,

$$\boldsymbol{x}(h, \epsilon) = \boldsymbol{x}_0(h) + \epsilon\bar{\boldsymbol{x}}_1 + \epsilon \int_0^{h/\epsilon} \{\boldsymbol{f}[\tau, \boldsymbol{x}_0(h)] - \bar{\boldsymbol{f}}[\boldsymbol{x}_0(h)]\}\mathrm{d}\tau + O(\epsilon^2)$$

$$= \boldsymbol{x}_0(h) + O(\epsilon),$$

since

$$\int_0^{h/\epsilon} \{\boldsymbol{f}[\tau, \boldsymbol{x}_0(h)] - \bar{\boldsymbol{f}}[\boldsymbol{x}_0(h)]\}\mathrm{d}\tau = O(1).$$

Therefore

$$\boldsymbol{x}(h, \epsilon) = 2\boldsymbol{x}(h, \epsilon'/2) - \boldsymbol{x}(h, \epsilon') + O(\epsilon) + O(h^{p+1}) + O\left[\left(\frac{h}{T}\right)^2\right].$$

This formula gives the extrapolation metod for calculating $\boldsymbol{x}(h, \epsilon)$.

3.3.8 *A method of averaging*

We now describe an example of a numerical method which is independent of singular perturbation methodology (Miranker and Wahba 1976). For other examples of methods which are likewise independent see Amdursky and Ziv (1977) and Crow and Kimura (1970), the latter employing aliasing.

Consider the following model problem

$$\ddot{x} + \lambda^2 x = f(t), \quad t \in (0, T).$$

When $f(t) = \lambda^2 \sin t$, this problem has the following family of solutions

$$x(t) = a \sin \lambda t + \frac{\sin t}{1 - 1/\lambda^2}.$$

For λ large, this solution family consists of a high frequency carrier wave, $a \sin \lambda t$, modulated by a slow wave, $(\sin t)/(1 - 1/\lambda^2)$.

The linear multistep class of methods is highly desirable for numerical analysis since these methods are easy to use and easy to analyze. However these methods consist of a linear combination of unstable functionals; namely, solution values and values of solution derivatives at points. The method of averaging replaces these unstable functionals by stable ones, thereby producing a class of linear multistep methods suitable for the highly oscillatory problem. We suppose that the stable functionals provide information about the solution being sought, and (subject to a process like mesh refinement) that the stable functionals furnish as adequate a description of the solution as is needed.

Let r, s, and N be positive integers, let $h = T/N$ and let $t_i = ih$, $i = 0, \pm 1, \ldots$ be the points of a mesh. Let $z(t)$ be a functional of x which can be calculated at each mesh point. Then we seek to determine $y_n = y(t_n)$, in terms of y_{n-i}, $i = 1, \ldots, r$ and $z_{n-i} = z(t_{n-i})$, $i = 0, 1, \ldots, s$ by means of the linear multistep formula

$$\sum_{i=0}^{r} a_i y_{n-i} + \sum_{i=0}^{s} b_i z_{n-i} = 0, \; n = 0, 1, \ldots, N.$$

We choose $y(t)$ to be

$$y(t) = \int_{-\infty}^{\infty} k(t-s)x(s)\mathrm{d}s,$$

where

$$k(z) = \begin{cases} \dfrac{1}{\Delta} & (-\Delta < z < 0), \\ 0 & \text{otherwise.} \end{cases}$$

Thus $y(t)$ represents the average of $x(t)$ over the interval $[t-\Delta, t]$.

The functional $z(t)$ is chosen to be $(\mathrm{d}^2/\mathrm{d}t^2 + \lambda^2)x(t)$, i.e. $f(t)$, which we suppose can be stably calculated at each mesh point. Thus with a change in normalization, the linear multistep formula may be written as

$$y_n = \sum_{i=1}^{r} c_i y_{n-i} + h^2 \sum_{i=0}^{s} d_i f_{n-i}.$$

Notions of local truncation error and of order of accuracy of such generalized linear multistep method (for functionals) are introduced. They are analogous to and generalizations of the traditional ones. For example, let

$$s_0 = 1 \text{ and } s_j = -c_j, \; j = 1, \ldots, r,$$

and let

$$L = \Delta/h.$$

Then

$$m_l = \frac{1}{(l+1)!} \sum_{k=1}^{l+1} \binom{l+1}{k} L^{k-1} \sum_{j=0}^{r} j^{1+l-k} s_j$$

$$-\frac{h^2\lambda^2}{l!}\sum_{j=0}^{s} j^l d_j - \frac{1}{(l-2)!}\sum_{j=0}^{s} j^{l-2} d_j,\ l=0,1,\ldots,$$

are so-called generalized moments of the method. The equations $m_l = 0$, $l = 0, 1, \ldots$ are generalized moment conditions.

View the equations $m_l = 0$, $l = 0, \ldots, r-1$ as r equations for the r unknowns s_j, $j = 1, \ldots, r$. The lth row of the resulting coefficient matrix which has as its jth term

$$\frac{1}{(l+1)!}\sum_{k=1}^{l}\binom{l+1}{k} L^{k-1} j^{1+l-k},$$

is a linear combination of the first l rows of the Vandermonde matrix. Thus the system of r equations has a solution in this case. Indeed by choosing the d_j, $j = 0, \ldots, s$ to be proportional to λ^{-2}, we obtain a solution for the s_j, $j = 1, \ldots, r$ which is $O(1)$.

If the coefficients s_j, $j = 1, \ldots, r$ and d_j, $j = 0, \ldots, s$ are chosen as solutions of the generalized moment equations $m_l = 0$, $l = 0, \ldots, p$, we may obtain an estimate of the local truncation error which is $O(h^{p+1})$.

In addition to notions of local accuracy a notion of stability for generalized linear multistep methods may be introduced. One such condition is quite analogous to the classical one; namely, that the polynomial

$$S(z) = \sum_{j=0}^{r} s_j z^{r-j}$$

obeys the root condition. Then combining notions of local accuracy and of stability, a global error estimate may be obtained.

We now consider some examples of these methods in which the coefficients are determined by the generalized moment conditions. In particular, we have for $l = 0, 1$ and 2, respectively:

$$m_0 \equiv \sum_{j=0}^{r} s_j - h^2\lambda^2 \sum_{j=0}^{s} d_j,$$

$$m_1 \equiv \sum_{j=0}^{r} j s_j + \frac{L}{2}\sum_{j=0}^{r} s_j - h^2\lambda^2 \sum_{j=0}^{s} j d_j,$$

$$m_2 \equiv \tfrac{1}{2}\sum_{j=0}^{r} j^2 s_j + \frac{L}{2}\sum_{j=0}^{r} j s_j + \frac{L^2}{6}\sum_{j=0}^{r} s_j - \frac{h^2\lambda^2}{2}\sum_{j=0}^{s} j^2 d_j - \sum_{j=0}^{s} d_j.$$

Consider the following case where the first two generalized moment conditions are satisfied.

A. $m_0 = m_1 = 0$.

For $r = s = 1$, we get

$$c_1 = 1 - \frac{2}{L} + \frac{2}{L} h^2 \lambda^2 d_0,$$

$$d_1 = \frac{2}{h^2 \lambda^2 L} - \left(\frac{2}{L} + 1\right) d_0.$$

In the special case $d_0 = 0$, this becomes

$$c_1 = 1 - \frac{2}{L},$$

$$d_1 = \frac{1}{h^2 \lambda^2 L}. \tag{I}$$

These coefficients (i.e. c_1) obey the root condition if and only if $L \geqslant 1$. In the special case $d_0 = d_1$, we obtain

$$c_1 = 1 - \frac{2}{L+1},$$

$$d_0 = d_1 = \frac{1}{h^2 \lambda^2} \frac{1}{L+1}. \tag{II}$$

Under the restriction $L \geqslant 0$, the root condition is equivalent to $L \geqslant 0$ for these coefficients.

For $r = s = 2$,

$$c_1 = 1 - \frac{2}{L} - \left(1 + \frac{2}{L}\right) c_2 + \frac{2}{L} \lambda^2 h^2 (d_0 - d_1),$$

$$d_1 = \frac{2}{\lambda^2 h^2 L}(1 + c_2) - \left(1 + \frac{2}{L}\right) d_0 - \left(1 - \frac{2}{L}\right) d_2.$$

In the special case $d_0 = 0, c_1 = c_2, d_1 = d_2$, this becomes

$$c_1 = c_2 = \frac{L-3}{2L},$$

$$d_1 = d_2 = \frac{3}{2\lambda^2 h^2 L}. \tag{III}$$

In this case,

$$S(z) = z^2 - \frac{L-3}{2L} z - \frac{L-3}{2L},$$

and this polynomial $S(z)$ obeys the root condition for a set of values of L which includes all $L \geqslant 1$.

In the special case $c_1 = c_2$, $d_1 = d_2 = 0$, c_1 and d_1 become

$$c_1 = c_2 = \tfrac{1}{2} \frac{L}{3+L}, \tag{IV}$$

$$d_0 = \frac{1}{\lambda^2 h^2} \frac{3}{3+L}. \tag{IV}$$

Here

$$S(z) = z^2 - \tfrac{1}{2}\frac{L}{3+L}z - \tfrac{1}{2}\frac{L}{3+L}.$$

This polynomial obeys the root condition for a set of values of L which includes all $L > 0$.

In the special case $c_1 = c_2$, $d_0 = d_1 = d_2$, we obtain

$$c_1 = c_2 = \tfrac{1}{2}\frac{L-1}{L+1},$$

$$d_0 = d_1 = d_2 = \frac{2}{3\lambda^2 h^2}\frac{1}{1+L}. \tag{V}$$

In this case, the root condition is obeyed for $L > 0$.

Now we consider a case corresponding to three generalized moment conditions.

B. $m_0 = m_1 = m_2 = 0$.

For $r = s = 1$, we get

$$c_1 = 1 - L\left(\frac{L^2}{3} + \frac{L}{2} - \frac{2}{h^2\lambda^2}\right)^{-1},$$

$$d_0 = \frac{1}{\lambda^2 h^2}\left[1 - L + L^2\left(\tfrac{2}{3}L^2 + L - \frac{4}{h^2\lambda^2}\right)^{-1}\right]. \tag{VI}$$

$$d_1 = \frac{1}{\lambda^2 h^2}\left[-1 + L + (2L - L^2)\left(\tfrac{2}{3}L^2 + L - \frac{4}{h^2\lambda^2}\right)^{-1}\right].$$

Notice that the root condition is obeyed for L large and positive but is violated for $h\lambda$ small compared to L.

In all of these examples as in the general case, we see that the coefficients obtained as solutions of the moment conditions depend on λ^2. At first sight this seems to be more restrictive than the case of the classical linear multistep formulae where the coefficients of the formula do not depend on the coefficients of the differential equation. In fact there is no such distinction. In the classical case, the coefficients of the differential equation enter into the method when it is used to approximate the differential equation, e.g. when $\dot{y}_{n-i}$ is replaced by $f(t_{n-i}, y_{n-i})$. It is essential after all that the numerical method at some point be dependent on the equation to be solved. In the present development, this dependence occurs at the outset in the determination of coefficients and in the error analysis. In the classical case it enters in the error analysis and in the use of the methods.

3.3.9 *Computational experiments*

We now apply the six sets of methods labelled I, II, . . . , VI to the model problem:

$$\ddot{x} + \lambda^2 x = \lambda^2 \sin t,$$

$$x(0) = 0, \ \dot{x}(0) = \frac{\lambda}{2} + \frac{1}{1 - 1/\lambda^2}.$$

Computations are made over the interval $[0, T] = [0, \pi]$. In Table 3.1 we display $h^{1/2}$ times the l^2-norm of the global error:

$$\|e\|_{l^2} \equiv \left(h \sum_{n=0}^{[\pi/h]} e_n^2 \right)^{1/2},$$

for a set of various combinations of $h = 0.1, 0.01, \lambda = 10, 10^3, 10^5$ and $L = 1, 2, 3$ and for each of the six methods cited. Here $[\pi/h]$ denotes the integer part of π/h.

To illustrate both the favorable and unfavorable effects, the table contains cases for which the methods are designed to operate well along with cases to which correspond poor or nonsensical results.

For example, although the cases corresponding to $\lambda = 10$ give fair results, these cases are not highly oscillatory, and we should not expect good results. When h is decreased, improvement should occur but only for the highly oscillatory cases. Method VI is used in some unstable cases. The highly oscillatory cases for moderate L give extremely good results as we expect.

In the systems case, the model problem is replaced by the second order system

$$\ddot{\boldsymbol{x}} + \Lambda^2 \boldsymbol{x} = \boldsymbol{f}(t, \boldsymbol{x}).$$

Here $\boldsymbol{x}$ and $\boldsymbol{f}$ are q-vectors and Λ is a $q \times q$ matrix. The coefficients c_j (and s_j) and d_j of the numerical method are replaced by $q \times q$ matrices (denoted by the same symbols). Many such formal replacements of the scalar development follow. For example, the first two generalized moments becomes

$$\boldsymbol{m}_0 = \left(\sum_{j=0}^{r} s_j - h^2 \Lambda^2 \sum_{j=0}^{s} d_j \right) \zeta_q,$$

$$\boldsymbol{m}_1 = \left(\sum_{j=0}^{r} j s_j + \frac{L}{2} \sum_{j=0}^{r} s_j - \Lambda^2 \sum_{j=0}^{s} j d_j \right) \zeta_q,$$

where ζ_q is the q-vector all of whose components are unity.

Referring to the remark above and noting the dependence of the coefficients of the numerical method on the coefficients of the differential equation, we see from $\boldsymbol{m}_0$ and $\boldsymbol{m}_1$ here the way in which the dependence appears in terms of the matrix Λ^2, for the coefficients determined by the generalized moment conditions. It is important to take note that the coefficients depend on the matrix Λ^2 and not explicitly on eigenvalues of Λ^2. Thus, if we know that a system has highly oscillatory components, we may use the methods described here without having to calculate the eigenvalues of Λ^2.

TABLE 3.1
$h^{1/2}\|e\|_{l^2}$ *for six averaging methods*

		$h = 0.1$			$h = 0.01$		
		$L\ (=\Delta/h)$					
Method	λ	1	2	3	1	2	3
I	10	.273	.108	.112	.133	.126	.126
	10^3	.113	.00217	.0611	.0283	.00683	.0083
	10^5	.112	.00209	.0611	.0111	.000106	.00627
II	10	.122	.133	.155	.126	.127	.128
	10^3	.00125	.0622	.177	.0241	.00926	.0136
	10^5	.00104	.0621	.177	.000118	.00627	.0125
III	10	.242	.111	.0872	.136	.126	.126
	10^3	.0032	.00422	.00317	.0294	.00684	.00546
	10^5	.0034	.00419	.00313	.00023	.00112	.89 E –6
IV	10	.123	.111	.0938	.126	.126	.126
	10^3	.00627	.0144	.0244	.0241	.00684	.00546
	10^5	.00623	.0144	.0244	.000133	.000179	.000264
V	10	.144	.152	.156	.127	.127	.128
	10^3	.0657	.094	.119	.0249	.0116	.0136
	10^5	.0657	.0939	.119	.0063	.00942	.0125
VI	10	.758 E 4	.66 E 11	.124	.195 E 1	.471 E 1	.11 E 2
	10^3	.0447	.0639	.244	.0246	.00901	.0253
	10^5	.0447	.0639	.244	.00421	.00629	.0251

3.4 Solution of stiff equations resulting from partial differential equations (Finlayson)

Engineers often need to solve mathematical models that are expressed as PDEs. Several case studies are described from the author's experience, emphasizing the methods chosen to integrate the equations in the time domain, with special regard for the constraints prescribed by the fact that the time-dependent ODEs are derived from PDEs in space and time.

There are four themes illustrated in the case studies:

(i) When a PDE is discretized in the spatial domain by using a finite-difference, finite-element, or global-collocation method, the resulting equations in time are stiff. The degree of stiffness can often be estimated a priori.

(ii) Integration schemes that work well for large error tolerances are valuable. First, the spatial truncation error may be large and it is wasteful to make the temporal truncation error very small. Second, the solution may have the same properties throughout the time interval considered; that is, the type of solution which causes the problem to be very stiff in the first place may move as a wave through space, so that as time proceeds the only thing that changes is the location of the stiffness-producing feature, not the stiffness itself. In that circumstamce, methods which rely on taking large time steps during part of the time interval to achieve their success may be less useful.

(iii) Schemes and packages must be applicable to combined ODEs and algebraic equations, since these often occur in practice. The time derivative must be allowed to enter the problem as a matrix multiplied by the vector of time derivatives, since that form arises in most Galerkin finite element methods.

(iv) For large time-dependent problems in two and three spatial dimensions, extensive codes have already been developed using schemes that may not be the best for stiff problems. The techniques developed for stiff problems, though, must be done in a way that the schemes can be easily fit into those codes.

The themes are illustrated in case studies. The first study is for chemical reactors in the form of packed beds. While kinetic behavior has traditionally been stiff, the packed-bed reactor can become more stiff due to sharp changes in temperature. Such reactors provide a useful and demanding testing ground for stiff integrators. The second case study is for flow through porous media. Here the problem is extremely stiff and good numerical solutions are difficult to obtain under certain conditions. The spatial variation of the solution plays a role both in defining the stiffness and in limiting certain advantages of stiff integration codes. The final example is the convective diffusion equation. This equation has similar difficulties to the equation for flow through porous media but is linear. Even so, in two-dimensional time-dependent simulations the most sophisticated techniques are required.

3.4.1 *Chemical reactors*

A prototype model of a packed-bed reactor is given by the following equations:

$$\frac{\partial c}{\partial r} = \frac{\alpha}{r}\frac{\partial}{\partial r}\left(r\frac{\partial c}{\partial r}\right) + \beta R(c, T), \tag{3.109}$$

$$\frac{\partial T}{\partial t} = \frac{\alpha'}{r}\frac{\partial}{\partial r}\left(r\frac{\partial T}{\partial r}\right) + \beta' R(c, T), \tag{3.110}$$

$$\frac{\partial c}{\partial r} = \frac{\partial T}{\partial r} = 0 \quad \text{at } r = 0, \tag{3.111}$$

$$c = c_0, \quad T = T_0 \quad \text{at } t = 0,$$

$$\frac{\partial c}{\partial r} = 0, \quad -\frac{\partial T}{\partial r} = \mathrm{Bi_w}(T - T_w) \quad \text{at } r = 1. \tag{3.112}$$

The physical meaning of the dimensionless coefficients is given elsewhere (Finlayson 1980). One method for solving these equations is to use the finite difference method to represent the spatial variation of the solution and then use some other method of solution in time, such as Runge–Kutta, trapezoid rule (giving the Crank–Nicolson method), or a GEAR package. If we divide the distance $0 \leqslant r \leqslant 1$ by a uniformly spaced set of grid points, with spacing h, we can write

$$T_j(t) = T(r_j, t); \; r_j = h(j-1). \tag{3.113}$$

The differential equation (3.110) then becomes (for an internal point not on the boundary):

$$\frac{\mathrm{d}T_j}{\mathrm{d}t} = \frac{\alpha'}{h^2}(T_{j+1} - 2T_j + T_{j-1}) + \frac{\alpha'}{r_j 2h}(T_{j+1} - T_{j-1}) + \beta' R(c_j, T_j). \tag{3.114}$$

This equation can then be written as

$$\frac{\mathrm{d}T_j}{\mathrm{d}t} = \alpha' \Sigma B_{ji} T_i + \beta' R(c_j, T_j), \tag{3.115}$$

where the matrix $[B_{ij}]$ is derived from (3.114). The task is then to integrate the set of ODEs as an initial-value problem.

Another method of solution is to represent the spatial variation of the solution with a polynomial defined over the entire domain, $0 \leqslant r \leqslant 1$. Collocation is applied at the Gaussian quadrature points, giving the orthogonal collocation method (Finlayson 1972, 1980). The result of applying this method to (3.110) has the same form as (3.115) except that the matrix $[B_{ij}]$ is different. In particular, it is dense (every entry is nonzero) instead of banded.

If a Galerkin finite element method is applied to (3.110), the result is slightly different:

$$\Sigma C_{ji} \frac{\mathrm{d}T_i}{\mathrm{d}t} = \alpha' \Sigma B_{ji} T_i + 2\beta' \int_0^1 N_j R(c_j, T_j) r \mathrm{d}r. \tag{3.116}$$

Notice here that the left-hand side involves a matrix multiplication. This means that explicit schemes for integrating (3.116) are ruled out, or must at least involve one LU decomposition of the matrix $[C_{ij}]$. Furthermore some packages for integrating ODEs do not allow equations in such a form, or require extra manipulation. For example, the code LSODI, developed by Hindmarsh at the Lawrence Livermore Laboratory (see §4.3), can be applied to equations in this form, whereas the GEAR or GEARB packages developed by Hindmarsh must be applied to the revised equation,

$$\frac{\mathrm{d}T_i}{\mathrm{d}t} = \Sigma (C^{-1})_{ij} (\alpha' \Sigma B_{jk} T_k + 2\beta' + \ldots),$$

where the notation C^{-1} is meant to denote an LU decomposition rather than an inversion, to preserve the sparse nature of the matrices $[B_{ij}]$, and $[C_{ij}]$.

All methods lead to sets of ODEs that must be integrated in time. The various packages RKF45, GEAR, GEARB, LSODE, LSODI can all be applied to some subset of the problems. The finite difference method and the finite element method lead to problems with banded matrices, whereas the orthogonal collocation method leads to problems with dense matrices. Despite this difference it turns out that all methods lead to stiff equations if very many grid points are used. This point is illustrated with the diffusion equation.

For the diffusion problem

$$\frac{\partial T}{\partial t} = \frac{\partial^2 T}{\partial x^2}, \tag{3.117}$$

We can apply finite-difference, finite-element or othogonal collocation methods to obtain a discretized set of equations:

$$\Sigma C_{ji} \frac{dT_i}{dt} = \Sigma B_{ji} T_i. \tag{3.118}$$

The difficulty of integration of these equations depends on the eigenvalues of the matrix, in particular:

$$\det\{[C_{ji}]\lambda - [B_{ji}]\} = 0. \tag{3.119}$$

For collocation and finite difference methods the matrix $[C_{ij}]$ is the identity matrix. If we apply separation of variables to (3.09) we are led to the eigenvalue problem:

$$\frac{d^2 X}{dx^2} + \lambda X = 0. \tag{3.120}$$

If we apply collocation, finite difference, or finite element methods to this eigenvalue problem we get (3.119). Thus the first eigenvalue from (3.119) represents an approximation (and usually a quite good approximation) to the first eigenvalue of the physical problem (3.120). The largest eigenvalue, however, is found to be represented by the following equation,

$$\lambda_{\max} = \text{LB}/h^2, \tag{3.121}$$

where the value of LB is listed in Table 3.2 for a variety of methods (Finalyson 1980). What this means is that the stiffness ratio for a linear diffusion problem is given by

$$\text{SR} = \text{LB}/\lambda_1 h^2, \tag{3.122}$$

and that as h becomes smaller, when we add more grid points, the problem becomes stiffer. These features are magnified for the nonlinear problems, such as (3.109, 3.110).

The implication of this information is that if the problem demands a large number of finite difference grid points or finite elements to represent the spatial changes in the solution, then the temporal problems are guaranteed to be stiff, and integration methods must be used that are applicable to stiff equations.

The solution methods described above are embodied in two computer codes, REACOL for orthogonal collocation, and REACFD for finite difference. The temporal integration is done by one of three methods: a fixed-time-step second-order Runge–Kutta method, a variable-time-step fourth-order Runge–Kutta method, and the GEARB package (Hindmarsh version; GEARB was used for orthogonal collocation because it was available, even though the appropriate package is GEAR). The program is arranged so that the user can easily change the reaction rate expression and the number of components to fit the situation. The advantage to using such a code in an academic setting is that every student can obtain results for his or her reactor model even though the problem may be very difficult. It is very discouraging (and not too conducive to learning) to try to solve a difficult problem,

TABLE 3.2
Values of LB in (3.121)

Method	LB
Finite difference	4
Collocation, FEM	
cubic functions	36
quartic functions	98
quintic functions	222
Collocation, Hermite	
cubic functions	36
Galerkin, FEM	
linear	12
quadratic	60
Galerkin, FEM, lumped	
linear	4
quadratic	24

get unstable numerical results, and not know whether the difficulty lies in the coding or in the problem. Use of GEARB obviates that uncertainty; if GEARB has difficulty the student may still want to look at the coding for possible errors, but half the battle is won.

As a simple example take (3.109, 3.110) with

$$R = c\exp(\gamma - \gamma/T),$$

and the parameters $\alpha = \alpha' = 1$, $\beta = 0.3$, $\beta' = 0.2$, $\gamma = 20$, $\mathrm{Bi}_\mathrm{w} = 20$, $T_\mathrm{w} = 1$, $c_0 = T_0 = 1$. Solutions obtained with the orthogonal collocation method are shown in Table 3.3. Note particularly that if only two collocation points are used in the radial direction (a number which suffices for some problems) all the temporal methods take about the same computation time and have comparable errors. The error is almost entirely due to the spatial approximation, as can be seen by comparing the results for $N = 2$ and $N = 5$. For $N = 5$, however, the problem is stiffer, and the methods designed for stiff equations are faster, although not more accurate. When the finite difference method is employed with a fixed time-step and a second-order Runge-Kutta method, the solution varies with grid spacing and time step:

$$\langle c \rangle = a + b\Delta r^2 - c\Delta t^2. \tag{3.123}$$

There is no need to take t very small to have an accurate answer in time when there is still some truncation error in space. In fact because of the cancellation of errors in (3.123) this method may be more accurate than a solution derived using a variable time-step Runge-Kutta method or GEAR method.

A phthalic anhydride reactor provides a real case with interesting mathematical results (Finlayson 1982b, Froment 1967). This time the packed-bed reactor is operated under conditions in which the temperature of the packing need not be

TABLE 3.3
Solutions to reaction problem giving values of average concentration at one t = 0.6.

N in orthogonal collocation	c	Method in time	CPU time sec
2	0.8068	RK-2, $t = 0.005$	0.97
	0.8031	BKINIT, eps = 10^{-5}	0.97
	0.8088	GEARB, eps = 10^{-5}	1.00
5	0.9197	RK-2, $t = 0.001$	14.1
	0.9198	RKINIT, eps = 10^{-5}	7.9
	0.9206	GEARB, eps = 10^{-5}	3.1
exact	0.91926		

the same as the temperature of the fluid flowing past the packing. This temperature difference occurs because reaction occurs on the catalytic packing and energy is given off, raising its temperature. The energy must be transferred to the fluid, but encounters a heat transfer resistance between the packing and the fluid. The resulting equations are similar to (3.109-3.112), except that there are several species to consider:

$$\frac{\partial c_i}{\partial t} = \frac{\alpha}{r}\left(r\frac{\partial c_i}{\partial r}\right) + \mathrm{Da_I} R_i\,[(c_i), T_\mathrm{s}], \tag{3.124}$$

$$\frac{\partial T}{\partial t} = \frac{\alpha'}{r}\frac{\partial}{\partial r}\left(r\frac{\partial T}{\partial r}\right) + \gamma(T_\mathrm{s} - T), \tag{3.125}$$

$$\gamma(T_\mathrm{s} - T) = \mathrm{Da_{III}} R\,[(c_i), T_\mathrm{s}]. \tag{3.126}$$

R is a nonlinear function of concentration and packing temperature, T_s. Thus (3.126) represents an algebraic equation to be solved at each position r and t. When orthogonal collocation is applied to (3.124, 3.125) one obtains a set of ODEs as in (3.114), coupled with a set of algebraic equations. This type of problem can be solved using REACOL and GEARB provided the reaction rate expression solves the algebraic equation for T_s, given the fluid temperature and all the concentrations. This was done using Newton–Raphson to solve the nonlinear equations.

This reactor is parametrically sensitive. Figure 3.3 shows the solution when the inlet and wall cooling temperatures are 620 K and 630 K. In the first case, the conversion of ortho-xylene to phthalic anhydride is well-behaved and the temperature of the reactor remains within bounds. If the temperature is raised just 10 K, the temperature rises very sharply and all the phthalic anhydride is further reacted to form carbon dioxide. The GEARB package is capable of performing the integration up until the point where the reaction temperature rises precipitously; there the integration stops because it cannot proceed with a very small required step size. This is a good example of the power of having a robust stiff integrator available. When a robust integrator is used and the reactor blows up we know that

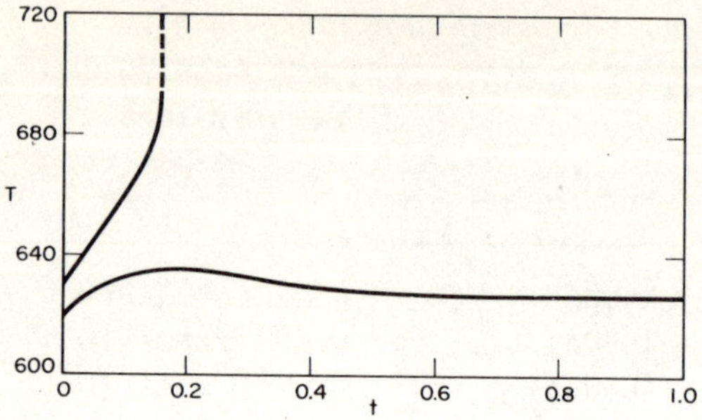

Fig. 3.3 Average temperature in phthalic anhydride reactor

something dramatic is happening in the physics and chemistry, because otherwise the integrator would handle the problem easily. The same feature of reliability was especially valuable in a pioneering study of the kraft pulping process (Gustafson, Sleicher, McKean, and Finlayson 1983). In this case reaction and diffusion occured in a wood chip, and the equations are similar to (3.109-3.112). The reaction rate expressions are unusual, however, and lead to some difficult integration problems. The REACOL program with orthogonal collocation and GEARB performed admirably and gave valuable results that otherwise might have been clouded by numerical problems and uncertainties.

3.4.2 *Flow through porous media*

The next case study concerns the movement of water through very dry soil. The equation is

$$-\frac{dS}{dp_c}\frac{\partial p}{\partial t}=\frac{\partial}{\partial x}\left(k_r(p_c)\frac{\partial p}{\partial x}\right), \quad p_c=-p, \tag{3.127}$$

with boundary conditions appropriate to injecting water at one end by keeping the pressure or saturation at a fixed value;

$$p(0,t)=\mathrm{BP1}, \quad \partial p/\partial x=0 \ \text{at}\ x=1.$$

The initial conditions are

$$p(x,0)=\mathrm{BP0}.$$

When the soil is very dry, the relative permeability, k_r, can vary over many orders of magnitude; factors of 10^6 are not uncommon. The derivative of a saturation with capillary pressure can also make large variations. For the cases considered here we take these functions as

$$k_r=1/(1+p_cL/146)^{6.65}, \quad S=0.32+0.68/(1+p_cL/231)^{3.65}.$$

Typical solutions are shown in Figure 3.4 as a function of the initial dryness; as can be seen, if the soil is initially very dry the profile is very steep and the water front moves through the porous media as a wave.

It is clear that to interpolate such a function it is necessary to have a very fine mesh near the front. Unfortunately, the front moves in time, so that the fine

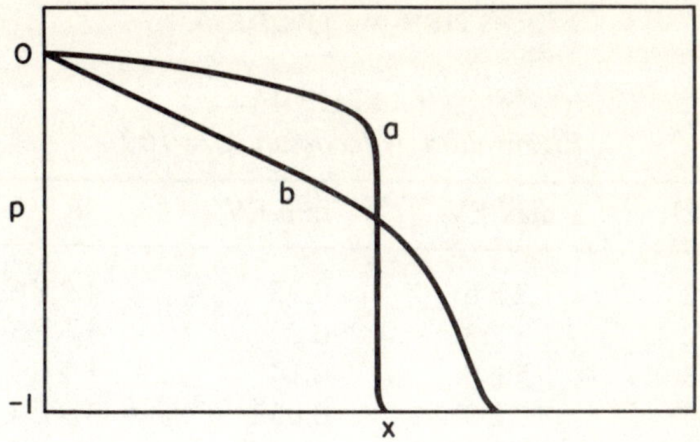

Fig. 3.4 Pressure profiles for flow through porous media. a, $L = 1000$ cm; b, $L = 300$ cm.

mesh must either be everywhere or else it must somehow be moved to occur in the right location at the right time. A straightforward application of the finite difference method to (3.127) gives

$$-\left.\frac{\mathrm{d}S}{\mathrm{d}p_c}\right|_i \frac{\mathrm{d}p_i}{\mathrm{d}t} = \frac{1}{\Delta x^2}\,[k_{i+1/2}(p_{i+1}-p_i) - k_{i-1/2}(p_i - p_{i-1})]\,. \qquad (3.128)$$

One method of evaluating the permeability is to take an average value:

$$k_{i-1/2} = \tfrac{1}{2}(k_i + k_{i-1}).$$

These equations can be integrated using a package for stiff equations, such as GEARB. At any time one can stop the integration and determine the eigenvalues of the Jacobian of (3.128). Typical values are listed in Table 3.4. Notice that some of the stiffness ratios are quite large, e.g. 10^9.

We can also define a coefficient ratio as follows:

$$\mathrm{CR} = \max_i \left|\frac{1}{k_r}\frac{\mathrm{d}S}{\mathrm{d}p_c}\right|_i \Bigg/ \min_i \left|\frac{1}{k_r}\frac{\mathrm{d}S}{\mathrm{d}p_c}\right|_i .$$

This ratio can be calculated a priori since it depends only on the pressure, and the range of pressure values is known at the outset. Fortunately, it correlates very well with the calculated stiffness ratio, as indicated in Figure 3.5. This coefficient ratio then provides a guide for other problems to help identify which problems are going to be especially stiff. It is useful for choosing a useable time-step for a fixed-time-step method. Jensen and Finlayson (1980) show how optimal discretization can reduce the stiffness ratio from 10^{11} to 10^6 for one of these problems.

One of the difficulties with this problem is that the pressure can be positive if the soil is saturated. Under those conditions $\mathrm{d}S/\mathrm{d}p_c = 0$. If the soil is initially dry, and then one end is put in contact with water (say a layer of water is placed on top of it) then that end will be saturated. The saturated region will move in time. The region of space which is saturated then is governed by an equation like (3.127) but with no time derivative, i.e. an algebraic constraint. Thus to solve this problem it is necessary to have a stiff integration package which can handle these algebraic constraints, and the number of algebraic constraints depends on the solution and may change in time. A modification of the original GEAR program (Finlayson and Nelson, 1977) was made in order to do this.

TABLE 3.4
Eigenvalues of Jacobian, L = *100*

BPO	BP1	max EV	min EV	SR	CR
0.0	−2	1.2 E 6	0.25	4.9 E 6	1.4 E 5
−0.5	−2	1.3 E 3	0.22	5.6 E 3	1.4 E 2
0.05	−3	1.5 E 8	0.14	1.1 E 9	1.3 E 8
−0.5	−3	1.2 E 3	0.032	3.9 E 4	1.2 E 3
0.0	−5	7.5 E 3	0.0053	1.4 E 7	3.0 E 5

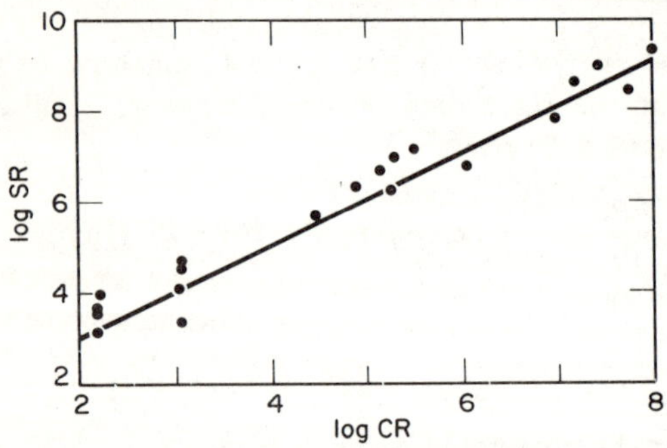

Fig. 3.5 Coefficient ratio for several solutions to (3.128)

Supposing the time-dependent nature of the solution is properly handled, what type of behavior usually results? One solution is shown in Figure 3.6; it was generated using the Galerkin method with some upstream dispersion and a trapezoid rule in time with a fixed time-step. The solution is clearly in error. Unfortunately, the errors are not caused by the time step used; solving the problem more accurately in time leads to no better solution. The difficulty lies with the spatial approximation. In this case only twenty elements were used and it is clear that the type of solution expected cannot be interpolated with twenty elements. For another problem discussed below it is possible to show how many elements are needed for a given solution. Here we cannot do that, but other calculations indicate something like 200 elements are needed if a cubic trial function is used, giving 601 total points. Sometimes it is not possible to compute with this many points for economic reasons. In that case one might try to use only twenty or forty elements. If one does so with a stiff approximation, it will be carefully followed by the integration package and large time steps will never be achieved. In such cases, then, it does not make sense to use a small error-tolerance since the spatial contribution to the error is large anyway, except that the integration packages may not work other than with small values. The author has found that values of eps below 10^{-3} are generally necessary for the GEARB package developed by Hind-Marsh to work efficiently for these problems.

If one is dissatisfied with a solution containing oscillations, since the oscillations are known to be due to numerical errors, and the calculations must be performed

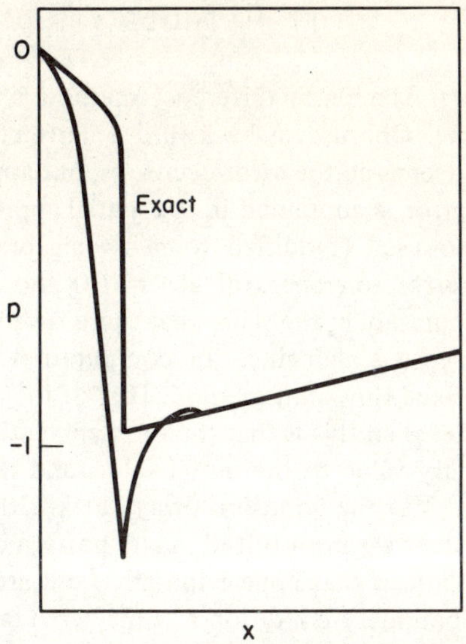

Fig. 3.6 Pressure profile by Galerkin method.

on a fixed budget, it is necessary to resort to introducing numerical dispersion. In this problem the simplest way to do so is to evaluate the permeability by using the value of permeability at the nearest node upstream (with upstream defined as the direction with the largest pressure):

$$k_{i-1/2} = k_{i-1}.$$

There are other more sophisticated and more accurate methods as well. If one does this then the oscillations are damped; the front is smooth but may be much smoother than the true solution. When this approach is taken it clearly does not make sense to solve the equations in time very accurately; the spatial contribution to the error is large. Again stiff integration packages that work well with large error tolerances are important.

Figure 3.7 shows the error in the pressure at a particular position and time (chosen so that the sharp front is nearby and any errors in the position of the

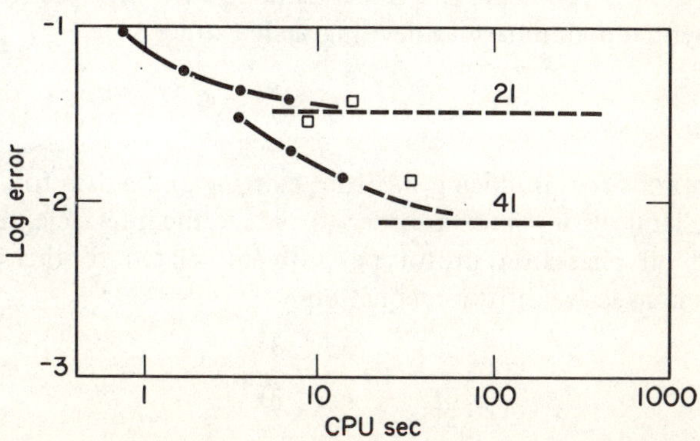

Fig. 3.7 Error in pressure solution. Finite difference; GEAR.

front cause a large error). The finite difference method is used in space and two methods are used in time. One uses a backward Euler method with a fixed time step. As the time step is decreased the error decreases, but approaches an asymptote since eventually all the error is contained in the spatial approximation. A modified version of GEAR is also used (modified to allow algebraic equations) and the problem is solved with error tolerances of eps = 1.0 and 0.1. With these values the solution is solved accurately in the sense that there is very little time truncation error, but the economic cost is high since the computation time is more than that obtained with a simple fixed-time-step method. Part of the reason the stiff integration package did no better than this is that the problem is stiff because the pressure changes from the boundary value to the initial value, and this leads to large variations in the coefficients. Yet the solution always makes this variation, is always this stiff, and large time steps are not utilized in any portion of the time integration. Furthermore, small oscillations that appear initially (and are inevitable when there is a step change in the boundary condition at time zero) are carefully tracked by the stiff integration package. This example is not meant to be presented as a failure of the stiff integration package—indeed the package made possible many solutions that were otherwise inaccessible—but to indicate that the dramatic results achieved in other contexts do not always obtain for solving the ODEs derived from PDEs.

If one wants to solve this problem really well, it is necessary to either use a very fine mesh or a moving mesh. This was done by one of the author's students (Jensen and Finlayson, 1980). The problem was transformed to a moving coordinate system that moved with the velocity of the front. Then the front stayed in the same position for all time, and the location of the boundaries moved in time. The solution required defining a set of nodes and solving the problem on a subset of nodes, and the subset kept changing in time. Rather than trying to force such a problem into the structure of the existing stiff integration packages (such as GEARB) it was easier to use the trapezoid rule, with a fixed time-step. Since the solution hardly changed in time, large time steps could be used. indeed the major time-saving resulting from the technique was the large time steps possible; in the moving coordinate system the solution appeared as if it were in steady state. While a stiff integration package was not used in the solution, the experience of using them on similar problems and understanding why they did not always work well was an essential preliminary to devising such a strategy.

3.4.3 *Chemical flooding*

One of the methods for producing oil from existing oil fields is to inject a solution of surfactant chemicals in water in order to reduce the interfacial tension between the water and oil phases. A prototype problem related to this problem is the solution of the convective diffusion equation:

$$\frac{\partial c}{\partial t} + \text{Pe}\frac{\partial c}{\partial x} = \frac{\partial^2 c}{\partial x^2},$$

$$c(x, 0) = 0,$$

$$c(0, t) = 1,$$

$$c(1, t) = 0.$$

A numerical solution is shown in Figure 3.8. This solution was obtained using the finite difference equation in the form

$$\frac{dc_i}{dt} + \text{Pe}\,\frac{c_{i+1} - c_{i-1}}{2\Delta x} = \frac{1}{\Delta x^2}(c_{i+1} - 2c_i + c_{i-1}). \tag{3.129}$$

The ODEs were then solved using GEARB. The error criterion was set low enough that the solutions are essentially the exact solutions of the difference equations (3.129). If only fifty grid points are used, oscillations develop, as shown. These oscillations are inherent in the spatial approximation—no better solution of the ODEs is possible. The GEARB routine faithfully follows the development and propagation of each oscillation. If 500 grid points are used, then the numerical solution is essentially exact, as shown. Here, too, the time-dependent equations are solved essentially exactly.

Now consider the problem when the parameter Pe, the Peclet number, increases by 1000 times. Then we need 500 000 points for a good solution, and such a solution is very expensive to obtain. Yet if fewer points are used, oscillations develop due to the spatial approximation, and very accurate time-dependent methods are not helpful. In this case we can introduce upstream dispersion by changing the convection term to the following form:

$$\left.\frac{\partial c}{\partial x}\right|_i = \frac{c_i - c_{i-1}}{\Delta x}.$$

Figure 3.8 shows the solution using this form when only fifty grid points are used. The solution is smooth. The solution to the time-dependent equations is found very accurately by the GEARB routine, but it is clear that the error obtained when comparing the numerical solution to the exact solution is large. Again it makes little sense to obtain very accurate answers to a problem when upstream

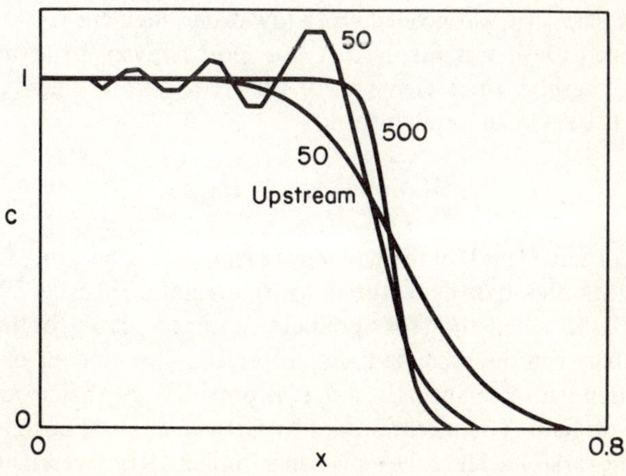

Fig. 3.8 Solution of convective diffusion equation. Finite difference, centered, n = 50 500. Finite difference, upstream, n = 50.

dispersion has been introduced; a stiff integrator is needed that works well for large values of the error-control parameter.

This problem, too, can be solved by a moving-coordinate system, and this has been done by Jensen and Finlayson (1980). Very small meshes are used near the front, and the front is stationary in the moving-coordinate system, but the location of the boundaries changes in time. In order to achieve economical results it is necessary to use a method of integration which is at least A-stable, and the trapezoid rule was used. Large time steps were then possible. Because some of the nodes are in the solution domain and others are not at any particular time the stiff integration packages were not used.

For the chemical flooding application it is necessary to solve equations like the convective-diffusion equation in a two-dimensional region:

$$\frac{\partial c}{\partial t} + u\frac{\partial c}{\partial x} + v\frac{\partial c}{\partial y} = D\left(\frac{\partial^2 c}{\partial x^2} + \frac{\partial^2 c}{\partial y^2}\right). \tag{3.130}$$

Typically fluid is injected in one corner of a square and removed in the corner diagonally opposite. The Peclet number is very large, requiring a small grid spacing to prevent oscillations. The small grid spacing makes the calculations very expensive, and numerical dispersion cannot be used because it changes the peak concentration of the surfactant in an unrealistic way, which changes the phenomenon. One solution to this problem is to solve the equations on a moving-coordinate system, this time one that moves from the injection well to the production well. Small finite elements are placed near the front, and the front always stays in the region where the elements are small. To efficiently handle the time-dependent problem, though, it is necessary to use an integration method that is A-stable, so that large time steps can be taken, but also one that is efficient in solving the linear algebra problem which must be solved each time step. Solving these equations with a banded solver is much less efficient than solving with a profile or frontal routine. Thus it was necessary to develop a method which was compatible with a frontal routine, was A-stable, and had an automatic step-control feature. In essence, what was needed was a low-order, high-eps version of GEARB.

Since a graded mesh was involved it was most convenient to use the Galerkin finite element method since elements of different sizes are easily handled. This means that (3.130) is in an implicit form:

$$M_{ij}\frac{\mathrm{d}c_i}{\mathrm{d}t} + A_{ij}c_j = DB_{ij}c_j.$$

Gresho, Lee, and Sani (1980) of the Lawrence Livermore Laboratory have provided an integration routine that has the features of stiff integration packages. It uses a second-order Adams–Bashforth method for a predictor, a trapezoid rule for the corrector, and the difference between the predicted and corrected value to control the step size to meet the specified error tolerance. Gresho, et al. provided both a first- and second-order method, as we do here. The method used by Gresho, et al., though is based on time derivatives of the variables. Since the equation is implicit, it is necessary to solve this implicit equation first even to get started. Rather than doing that, we devised a method based upon extrapolation. The derivation is given here for the first-order method.

First we write an equation for an extrapolation of the polynomial going through the solution at t_{n-1} and t_n (see Figure 3.9). The method is illustrated for $dy/dt = f(y)$:

$$y^{\mathrm{p}}_{n+1} = y_n + \frac{h_n}{h_{n-1}}(y_n - y_{n-1}). \tag{3.131}$$

We next write a Taylor series for the exact solution, assuming the exact solution is known at the beginning of a time step, $y(t_n) = y_n$ and $y(t_{n-1}) = y_{n-1}$:

$$y(t_{n+1}) = y(t_n) + h_n y'(t_n) + h_n^2 y''(t_n)/2 + \dots.$$

Subtracting this from (3.131) gives

$$y^{\mathrm{p}}_{n+1} - y(t_{n+1}) = \frac{h_n}{h_{n-1}}(y_n - y_{n-1}) - h_n y'(t_n) - h_n^2 y''(t_n)/2.$$

Using a Taylor series for y_{n-1},

$$y_{n-1} = y_n - h_{n-1} y'_n + h_{n-1}^2 y''_n/2$$

gives

$$y^{\mathrm{p}}_{n+1} - y(t_{n+1}) = -h_n(h_n + h_{n-1})y''_n/2. \tag{3.132}$$

The corrector is given by the backward Euler method:

$$y^{\mathrm{c}}_{n+1} = y_n + h_n f(y^{\mathrm{c}}_{n+1}). \tag{3.133}$$

The truncation error of the backward Euler method is

$$y^{\mathrm{c}}_{n+1} - y(t_{n+1}) = h_n^2 y''_n/2. \tag{3.134}$$

The values of y^{p}_{n+1} and y^{c}_{n+1} are available from (3.131-3.133). We solve (3.132 and 3.134) for the unknown $y(t_{n+1})$ and $y''_n(\zeta)$. The result is

$$y^{\mathrm{p}}_{n+1} - y^{\mathrm{c}}_{n+1} = -(2h_n^2 + h_n h_{n-1})y''_n/2.$$

The truncation error is then estimated using

$$d_{n+1} = y^{\mathrm{c}}_{n+1} - y(t_{n+1}) = \frac{y^{\mathrm{c}}_{n+1} - y^{\mathrm{p}}_{n+1}}{2 + h_{n-1}/h_n}. \tag{3.135}$$

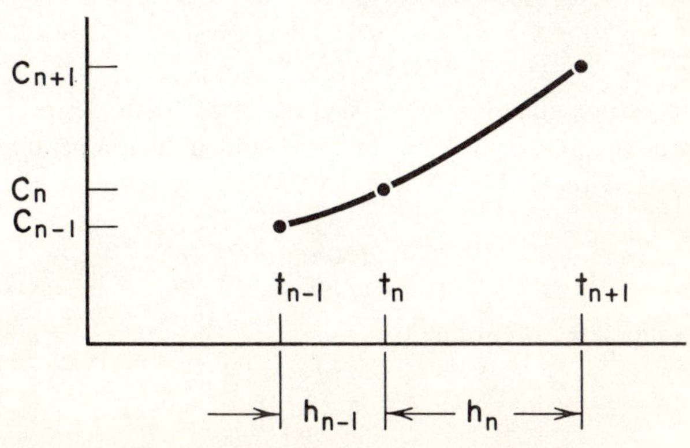

Fig. 3.9 Diagram for time-step control.

To summarize we use (3.131) to predict y_{n+1}, (3.133) to correct, and (3.135) to estimate the truncation error.

The estimate of the truncation error is used to pick the next time-step:

$$h_{n+1} = h_n(\epsilon/\|d_{n+1}\|),$$

where ϵ is the user-specified accuracy. In Jensen (1980) this time step was used for the next calculation. Jensen also used the second-order version for the convective–diffusion equation.

In practice, of course, the estimate of truncation error may not be a good estimate, and after picking a time step and doing the calculations the estimate of the truncation error in the new step may be larger than ϵ. Consequently, Jensen (1980) devised the following strategy to allow recalculation when necessary. Define the error for sets of equations as

$$d = \left(\frac{1}{N} \sum_{i=1}^{N} d_{n+1,i}^2 \right)^{1/2},$$

or

$$d = \max_i |d_{n+1,i}|.$$

If

$$\beta\epsilon < d,$$

the time step is not accepted since the truncation error estimate acutally achieved is greater than expected from the previous steps. The step is repeated with a step size half as big. The value of beta is empirically chosen in the range 1 to 1.5. If

$$d \leqslant \beta\epsilon,$$

then the step is accepted. However, if

$$\epsilon < d \leqslant \beta\epsilon,$$

the next time step is decreased according to

$$h_{n+1} = h_n(\epsilon/d)^{1/2}.$$

This allows the estimated error to be slightly larger than desired, as large as $\beta\epsilon$. If $d > \epsilon/\alpha$, the next time-step is kept the same, and alpha is a parameter empirically chosen in the range 1 to 10. Finally if

$$d \leqslant \frac{\epsilon}{\alpha},$$

the time step is increased according to

$$h_{n+1} = h_n \left(\frac{\epsilon}{\alpha d} \right)^{1/2},$$

with a maximum ratio of 3 to prevent introducing excessive error. The corresponding formulae based on interpolation using second-order expressions are given in Jensen (1980).

The time-step scheme described above is based on the same ideas contained in the GEARB package developed by Hindmarsh, but has the advantage that it is easily implemented in a code that already uses the backward Euler method. The second-order scheme can likewise be used in a code that now uses the trapezoid rule, or Crank–Nicolson. The second-order scheme was applied to solve for the movement of chemical in the two-dimensional flow pattern described above. The predicted step size was always accepted even if the truncation error turned out to be too large. When the solution reached a point in which smaller time steps had to be taken, it sometimes broke down. Later Jensen devised the strategy given above for accepting a step size and applied this to a large code for steam injection of an oil field, where it worked with good results (Jensen 1980). Approximately 30 to 50 per cent of the computation time was saved compared to the previous step-size control strategy, and cases converged that had not converged with the previous strategy. The previous strategy was one in which the change in saturation or pressure from one time step to another was kept below a prescribed value. Obviously this limits the truncation error only in some vague, general sense. The schemes discussed here, however, have a theoretical justification and work even better.

3.4.4 *Conclusion*

The concepts of stiff integration codes can sometimes be applied directly to solve the ODEs arising from PDEs. In some cases, however, the problems are so difficult or so large that a straightforward application of the stiff integrators would be prohibitive in cost. In such cases the principles of the stiff integrators can sometimes be incorporated with other solution techniques and tricks, leading to a viable solution method. These principles have been illustrated in three case studies.

3.5 Some characteristics of ODE problems generated by the method of lines (Schiesser)†

3.5.1 *Introduction*

The major advances in stiff ODE integrators of the past ten years have had a major impact on the numerical integration of PDEs primarily from the use of the Numerical Method of Lines (NMOL), which can generally be applied to the following initial-value PDE system:

$$\boldsymbol{u}_t = \boldsymbol{f}(\boldsymbol{x}, t, \boldsymbol{u}, \boldsymbol{u}_{\boldsymbol{x}}, \boldsymbol{u}_{\boldsymbol{xx}}, \boldsymbol{u}_{\boldsymbol{xxx}}, \ldots), \tag{3.136}$$

$$\boldsymbol{g}(\boldsymbol{x}_\mathrm{b}, t, \boldsymbol{u}, \boldsymbol{u}_{\boldsymbol{x}}, \boldsymbol{u}_{\boldsymbol{xx}}, \ldots) = \boldsymbol{0}, \tag{3.137}$$

$$\boldsymbol{u}(\boldsymbol{x}, t_0) = \boldsymbol{u}_0(\boldsymbol{x}). \tag{3.138}$$

Subscripts t and $\boldsymbol{x}$ denote partial derivatives with respect to the initial-value and boundary-value independent variables, t and $\boldsymbol{x}$, respectively, and:

$$\boldsymbol{u} = \text{vector of dependent variables,}$$

† The author is indebted to Dr S. S. Hu of Beijing and Zhejiang Universities, PRC, for the development of the adaptive grid applications.

f = vector of derivative functions,

g = vector of boundary condition functions,

x_b = vector of boundary values of x,

t_0 = initial value of t,

u_0 = vector of initial conditions.

Basically, the NMOL involves replacing the spatial (boundary-value) partial derivatives of the PDEs, e.g. u_x, with algebraic approximations. If the PDEs have only one initial-value variable, typically time t, a set of ODEs in that initial-value variable results. Although this idea is straightforward conceptually, it has surprising generality. In fact, the NMOL may become a standard approach to PDEs, largely replacing the more conventional algorithms for the numerical integration of PDEs.

3.5.2 *Geometric classification of PDEs*

The NMOL can, in principle, be applied to all three classes of PDEs, parabolic, elliptic, and hyperbolic, in one, two, and three dimensions (the vector x can be three-dimensional). The application of the NMOL to a particular PDE problem is characterized by the form of the algebraic approximation of the spatial derivatives. Generally, these approximations can be classified as direct and indirect. The direct methods involve direct replacment of the PDE spatial derivatives with algebraic approximations, e.g. finite differences, while the indirect methods involve some operations performed on the PDE residuals, i.e. the method of weighted residuals such as Galerkin and collocation on finite elements. The first approach leads to systems of explicit ODEs of the form:

$$\mathrm{d}y/\mathrm{d}t = f(y, t), \tag{3.139}$$

while the second approach leads to systems of implicit (implicitly-coupled) ODEs of the form:

$$A\,\mathrm{d}y/\mathrm{d}t = f(y, t), \tag{3.140}$$

Where A is a coupling matrix that may be a function of y. ODE integrators are now available as library routines for both types of ODE integrators (Hindmarsh 1983).

With regard to the geometric classification, a significant volume of research has been performed on the particular approximations of the spatial derivatives which have proven effective. For example, in the case of direct methods, centered approximations are recommended for parabolic PDEs, while noncentered (biased upwind) approximations are recommended for hyperbolic PDEs which take into account the directional nature of the solution (typically resulting from fluid flow). Boundary-value ODEs and elliptic PDEs can be formulated in terms of associated parabolic problems that are then integrated to equilibrium ($u_t = 0$ in (3.136)).

For both direct and indirect methods, the variation of the solution of (3.136–3.138), $u(x, t)$, with respect to the boundary-value variable, x, is in most cases based on polynomials. For example, for indirect methods, linear or quadratic finite elements or B-splines are typically used as the basis functions during application of the method of weighted residuals. Clearly, many combinations of methods are possible, e.g. Galerkin on finite elements, collocation on orthogonal polynomials, etc., but all of these seemingly different methods have the same conceptual framework. In other words, the NMOL is a loosely defined approach to PDEs with the one basic feature that a set of approximating ODEs is eventually defined by the analyst. This flexible structure is reflected in the extensive body of literature pertaining to the NMOL developed by mathematicians, scientists, and engineers reporting many variations on the basic approach. This situation is also reflected in the spectrum of computer codes now available which have been developed with the objective of facilitating the computer solution of PDE problems without requiring the user to become enmeshed in the details of approximating the spatial derivatives, defining the corresponding approximating ODEs, etc.

3.5.3 *Dimensionality*

The grid on which the approximation of the PDE spatial derivatives is based is an important consideration, particularly as the dimensionality of the problem system increases. For example, if M PDEs are to be approximated in D-dimensional space with N grid points in each direction, the number of ODEs is approximately MN^D, which clearly can become a very large number (consider $M = 2, D = 3, N = 20$ for which the number of ODEs is 16 000, and this is a rather modest problem).

Obviously, every effort must be made to accommodate PDE problems by the NMOL as efficiently as possible, particularly for two- and three-dimensional problems involving sets of PDEs. Generally, this means taking into account the structure of the approximating ODE Jacobian matrix if implicit integration in t is used.

3.5.4 *Explicit and implicit integration*

Since the system of ODEs approximating the original PDE system must be numerically integrated with respect to t, the choice of an initial-value integrator is required. Explicit integration, of course, has a stability limit which may require excessively small integration step-sizes. However, our experience has indicated that this limitation should not be assumed to preclude the use of explicit integration. In fact, we routinely try explicit integration of the ODE system first. It is easy to program and is surprisingly effective in many NMOL applications. Only if the computer run times are excessive, or more specifically, the integration time-steps are small, or the eigenvalues of the ODE system Jacobian matrix are widely separated, should an implicit integrator be used. In other words, using a stiff integrator for a nonstiff problem can be wasteful in computer time because of the linear algebra of implicit integration which may be performed unnecessarily.

3.5.5 *The ODE-system Jacobian matrix*

The approximating ODE system typically has a block-banded structure. To demonstrate this characteristic, the Jacobian matrix of a nine-PDE, five-ODE problem is shown in Figure 3.10. The physical system that generated the ODE–PDE problem is the simple four-pass heat exchanger in Figure 3.11. This exchanger is modelled by first-order hyperbolic PDEs reflecting the flow through the passes and surrounding shell. Five-point, biased upwind approximations (Carver and Hinds 1977; Carver and Schiesser 1980) of the first-order spatial derivatives in the PDEs, applied over an eleven-point spatial grid, produce 100 ODEs (note that the main diagonal of the Jacobian matrix mapped in Figure 3.10 has three diagonals on one side and one

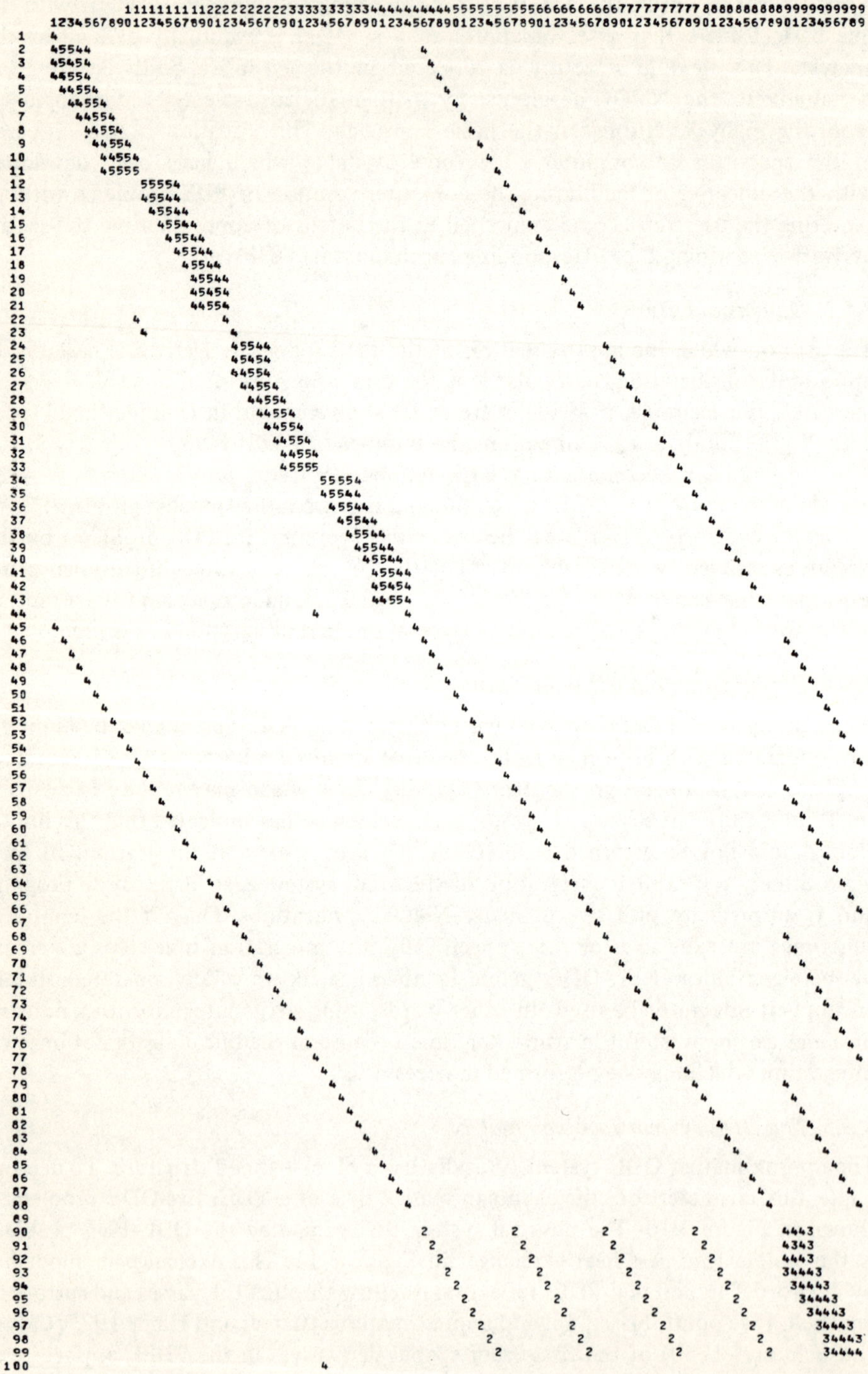

Fig. 3.10 Jacobian matrix map of the nine-PDE/five-ODE heat exchanger model.

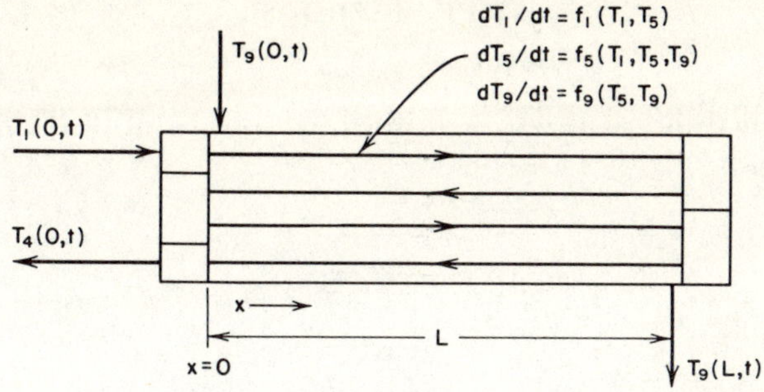

Nine PDEs

Four tube fluid temperatures: $T_1(x, t)$, $T_2(x, t)$, $T_3(x, t)$, $T_4(x, t)$.

Four metal temperatures: $T_5(x, t)$, $T_6(x, t)$, $T_7(x, t)$, $T_8(x, t)$.

Shell fluid temperature: $T_9(x, t)$.

11-point grid in x.

Number of ODEs $= 9 \times 11 + 1 = 100$.

Fig. 3.11 Four-pass heat exchanger modelled by nine PDEs and five ODEs.

diagonal on the other side resulting from the five-point biased upwind approximations). The outlying bands result from the coupling between the PDEs. Four of the PDEs do not have a spatial derivative (the metal tubes do not move) so that just the main diagonal appears for these PDEs. This Jacobian matrix is rather typical in the sense of the requirements for selecting an implicit integrator. The system is clearly not banded in the sense that a banded integrator can be used effectively (the full bandwidth is large).

The Jacobian-matrix map in Figure 3.10 suggests that efforts to reduce the overall bandwidth may be worthwhile, particularly as the size of the ODE system increases (two- and three-dimensional PDE problems). We have explored this idea to some extent, and illustrate what can be done with an intuitive approach based on physical reasoning.

The differentiator computes the first-order spatial derivatives over the eleven-point grid by operating on vectors of the dependent variables, in this case $T1$, $T2$, . . . , $T9$, defined along the one-dimensional system. However, any given derivative of the 100-ODE system is more likely to depend on the dependent variables in the neighborhood of its associated grid point than on the full set of dependent variables stored in a vector along the system. Thus, the dependent variables are reordered to reflect this local dependency (physically, rates of heat transfer are more directly determined by local temperatures than by temperatures along the full length of the exchanger).

After the 100 derivatives have been computed exactly, the derivatives are reordered (in the reverse fashion). The resulting vector of 100 derivatives in array F is then set in the initial-value ODE integrator.

The effect of this reordering on the Jacobian matrix is seen in Figure 3.12. Clearly, significant badnwidth reduction has been achieved yet the two numerical solutions are identical to within the prescribed accuracy of the integration in t, and the eigenvalues are also essentially identical. Thus, even in the case of this

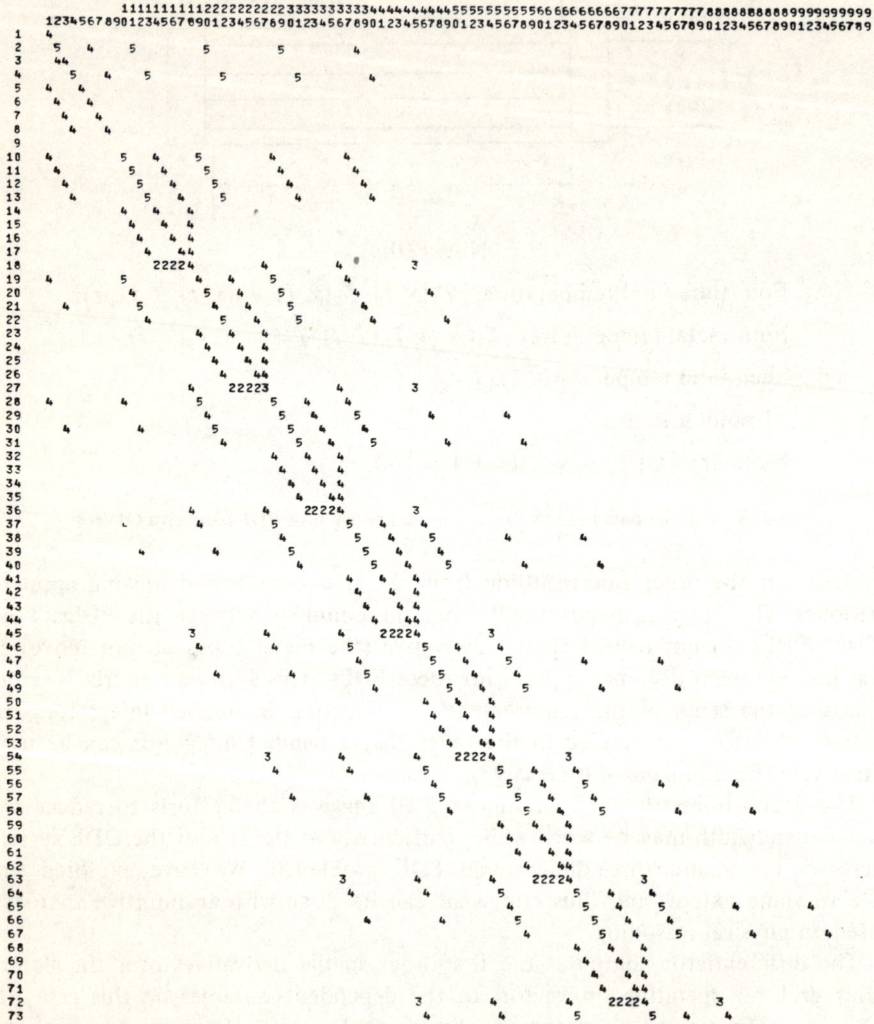

Fig. 3.12 Bandwidth-reduced Jacobian matrix for the reordered 100-ODE system.

small problem, significant bandwidth reduction is possible which facilitates the integration of the ODE system, since a banded integrator with a smaller declared bandwidth can be used.

The preceding approach to bandwidth reduction is based on physical reasoning. Automatic bandwidth-reduction, as implemented in finite-element analysis (Lewis 1982; Collins 1973), has also been applied to the NMOL solution of PDE problems.

In summary, the mapping of the ODE system Jacobian matrix is recommended to indicate:

(i) the ODE system overall structure (bandedness, sparseness, etc.);
(ii) order of magnitude of the matrix elements (the numbers in the map of Fig. 3.10 are order-of-magnitude indicators).
(iii) the detailed relationships between dependent variables and their derivatives;
(iv) the degree of nonlinearity as reflected by changes in the matrix;
(v) the time scale, stiffness and stability of the NMOL approximation.

3.5.6 *The steady-state approximation*

Scientific and engineering applications frequently lead to ODE/PDE models for which the individual equations represent phenomena of widely differing rates of change in time. For example, in modelling a tubular catalytic reactor, the thermal mass (capacitance) of the catalyst may be so large that the temperature response defined by the energy balance PDE is orders of magnitude slower than the composition responses modelled by component balance PDEs. A commonly accepted method of analyzing such a system is to set the partial derivatives in time of the fast PDEs to zero; if the reactor model is one-dimensional, a mixed system of boundary-value ODEs (the fast component balances) and a slow PDE (the energy balance) results. When the NMOL is applied to this mixed system a set of ODEs in time results which is significantly smaller than if the NMOL were applied to the original set of PDEs without making the ssa. For example, if the reactor model has two component balances and one energy balance, the number of approximating ODEs in time will be reduced by two-thirds. In other words, by using the ssa, the size of the ODE problem is significantly reduced. However, the reduced ODE problem generally will not have the block-banded structure of Figure 3.10, and cannot be bandwidth reduced, as in Figure 3.11. Rather, the Jacobian matrix of the reduced ODE system will be much fuller, typically lower-triangular full. The question then naturally arises whether the reduction in the number of ODEs is sufficient to give overall increased computational efficiency with the fuller Jacobian matrix. Of course, this question presupposes that the ssa was used in the first place so as not to significantly affect the numerical solution.

Although the ssa has been widely used, and generally is considered to be valid if used carefully, some recent research has indicated that it does not lead to any significant computational efficiencies if a quality banded implicit ODE integrator is used; in other words, the original system of PDEs can be integrated numerically by the NMOL just as efficiently as the reduced system of ODEs resulting from the application of the ssa. This conclusion has been reached in recent studies for both stiff ODEs and PDE systems. Consequently, the question of when to use the ssa can possibly be avoided by using implicit integration directly on the original problem rather than on the reduced problem.

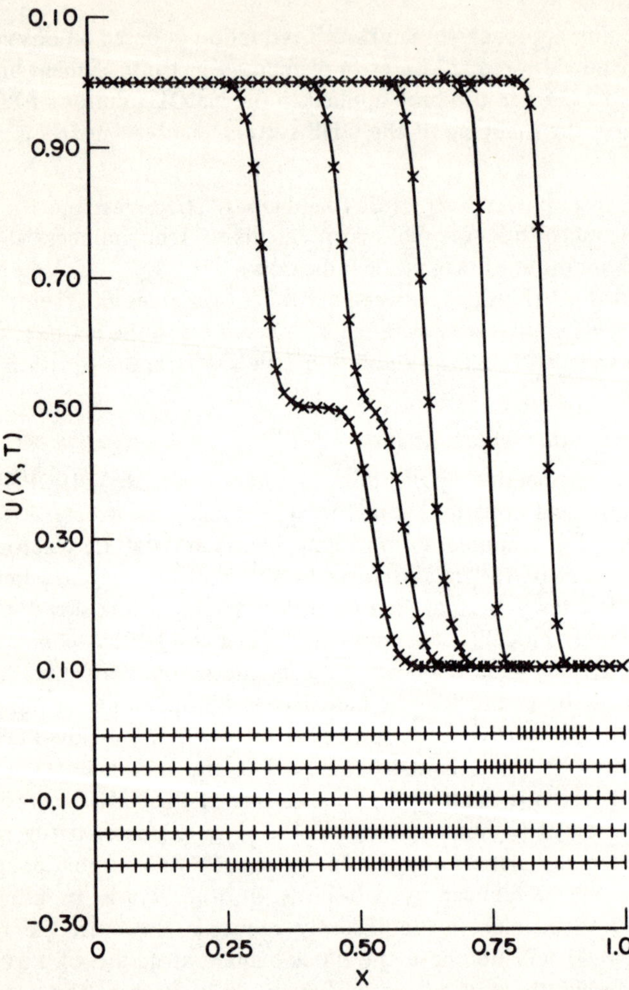

Fig. 3.13 Adaptive grid solution of Burger's equation.

3.5.7 *Adaptive grids*

Recently, attention has been given to the problem of computing accurate numerical solutions to PDE systems which exhibit steep moving fronts. Typically, such solutions are generated by systems of first-order hyperbolic PDEs. Any methods for approximating the spatial derivatives of such PDE systems on a fixed spatial grid are severely tested. The solution to this problem, generally, has been to implement moving grids which automatically adapt to the local needs of the problem, e.g. concentrate spatial grid points where the solution is changing rapdily in space.

If such adaptive grids are used within the NMOL, they impose special requirements on the ODE integrator, particularly if the total number of spatial grid points changes as the solution evolves (essentially, this means that the number of approximating ODEs also changes during the solution).

To illustrate the effectiveness of an adaptive grid, we present the NMOL solution to Burger's equation:

$$u_t = Du_{xx} - uu_x, \tag{3.141}$$

$$u(x, 0) = f(x, 0), \tag{3.142}$$

$$u(0, t) = f(0, t), \tag{3.143}$$

$$u(1, t) = f(1, t), \tag{3.144}$$

$$f(x, t) = f_1/f_2 \quad \text{(exact solution)}, \tag{3.145}$$

$$f_1 = 0.1\mathrm{e}^{-A} + 0.5\mathrm{e}^{-B} + \mathrm{e}^{-C},$$

$$f_2 = \mathrm{e}^{-A} + \mathrm{e}^{-B} + \mathrm{e}^{-C},$$

$$A = (0.05/D)\,(x - 0.5 + 4.95t),$$

$$B = (0.25/D)\,(x - 0.5 + 0.75t),$$

$$C = (0.5/D)\,(x - 0.375).$$

Burger's equation is an ideal PDE test problem since it exhibits steep moving fronts, is nonlinear and has an exact solution. In Figure 3.13, the points are the computed NMOL solution, and the solid curves are the exact (analytical) solution. The behavior of the adaptive grid is demonstrated at different values of t at the base of the figure. The change in the number of ODEs (spatial grid points) during the solution of Burger's equation is indicated in Figure 3.14. In order to produce this solution, the mechanics of the adaptive grid had to be developed, and an integrator was required which could change the number of ODEs as the solution evolved (Hu and Schiesser 1981).

3.5.8 *Software*

With the increasing interest in the NMOL, software packages have been developed and are generally available. The sub-class of problems within the general class defined by (3.136–3.138) accommodated by each code should be appreciated by the prospective user; also, the particular combination of approximations used varies widely from one code to another, so that at this point in time, it appears a collection (library) of software is the best approach for taking advantage of the substantial programming efforts that have already been made in implementing NMOL solutions of ODE–PDE problems. The more widely distributed codes include PDECOL, DPDE, DSS/2 (Rice 1983), DYLA (Gelinas, Doss, Vajk, Djomehr, and Miller 1982) and FORSIM.

3.5.9 *Summary*

This has been a brief introduction to the NMOL integration of ODE–PDE systems. A major body of literature now exists for this technique which could not be discussed within the space limitations. Clearly though, the NMOL approach to ODE–PDE problems produces sets of approximating ODEs that require quality ODE integrators; typically, the ODE systems are large, stiff, nonlinear, block-banded, or nearly full, and change in number and character as the solution evolves. Additional requirements for ODE integrators may include implicit coupling, accommodation of discontinuities, and automatic partitioning of the problem system (perhaps into stiff and nonstiff components) to take advantage of parallel computation and other features of the next generation of large-scale scientific computers.

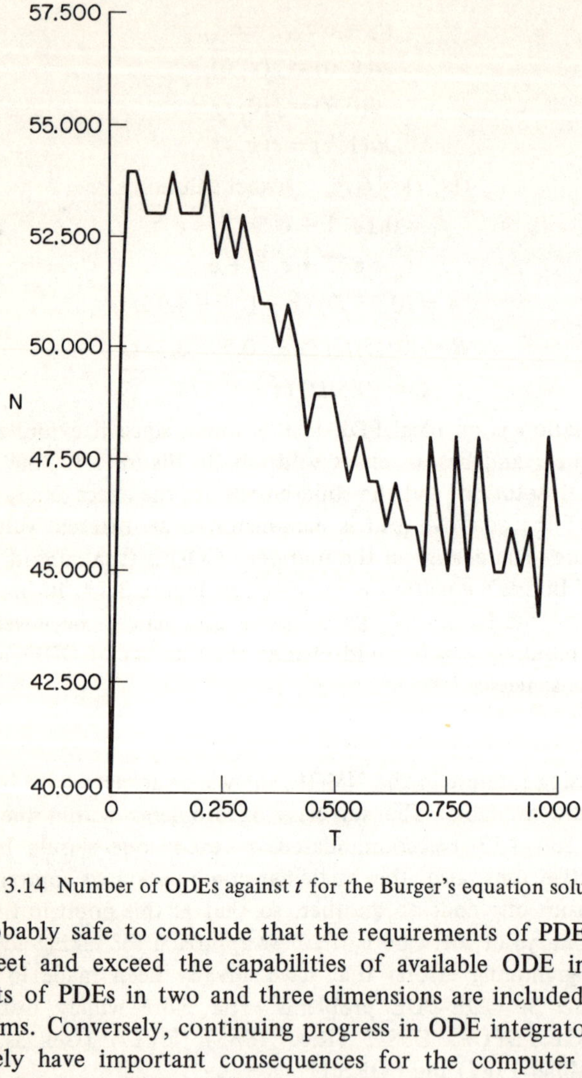

Fig. 3.14 Number of ODEs against t for the Burger's equation solution.

Thus it is probably safe to conclude that the requirements of PDE problems will expand to meet and exceed the capabilities of available ODE integrators, particularly if sets of PDEs in two and three dimensions are included as anticipated problem systems. Conversely, continuing progress in ODE integrator development will most likely have important consequences for the computer integration of PDE systems; the large PDE problems will become more amenable to solution as progress in ODE integration continues.

4

STIFF PACKAGES

4.1 Review (Aiken)

4.1.1 *Introduction*

The user would like to obtain the solution to an ODE set quickly (in both user time and in the computer's time) and in a convenient form, as plots or a tabulation with, say, a constant whole-number increment between the independent variable.

The vast majority of stiff users at the beginning of the 1980s used early versions of GEAR, commonly available through users' computer centers. By numbers of users, the most common problem is probably small (less than ten or twenty equations), moderately stiff, and nonlinear. It needs only to be solved to accuracy of about one per cent for qualitative understanding of the dynamic relationships among the variables.

Necessary information from the user obviously includes functional relationships and initial conditions; other important but possibly expendable instructions would be accuracy requirements and integration interval (see Klinzing (1980) for a scheme to determine this interval automatically). Additional information often proves to increase efficiency of the solution, but would probably not be welcome to the average user. This includes making a decision on whether or not the system is stiff.

Formation of an ODE package as automatic as possible (but with options should the user wish to specify; see §8.2), winning approval with numerical analysts and big users, and then the public, is difficult and related to the problems associated with making comparisons among packages. See §4.5 for a description of the evolution of an ODE package in an industrial environment.

Education of the user to the point where the true power of an approach can be utilized is another problem area; quite often a good numerical methods course, including solution of ODEs, is not in an engineer's curriculum. See "Getting the Power to the People" by Hindmarsh (1980b), "What Everyone Solving Differential Equations Numerically Should Know," by Shampine (1980d), and Curtis (1980).

Future directions to go in the development of general software is discussed in §4.2 and in Chapter 8. They include ultimate automation such as automatic method selection for special problems like stiff and oscillatory systems, differential-algebraic sets, differential-difference sets, systems with discontinuities, sparsity, and such. Better interaction between experimentation, differential equation-model identification, and parameter estimation with numerical solution requirements also needs to be developed.

The following subsections contain a review of some features involved in the construction of automatic packages for solving stiff ODEs.

4.1.2 *Start-up*

Gear (1980a) discusses method and initial step-size selection in multistep ODE solvers. Early automatic codes like Gear's original DIFSUB required the user specify the method (stiff or nonstiff; analytic or numerically determined Jacobian) and the initial step size. Usually a user does not know to the appropriate order of magnitude what size the first step should be and probably does not know the best method to use. LSODE, the latest successor to the GEAR package, does not require an initial step size from the user. The easier question of what method to use is the more important, but a badly guessed initial step can waste a significant amount of time or cause one to miss some interesting behavior. Gear (1980) examines an approach that uses the initial local eigenvalues to make the (nonstiff) method selection, and high accuracy estimates at nearby points to select a good initial step sequence and allow a high order to be used from the beginning (where it can be particularly valuable for stiff equations and problems with initial zeros).

Shampine (1978a) studied the effect that inherent machine limitations on the minimum step size have on solutions containing zeros, discontinuities, and stiff transients. He finds only in the third category should there be limitations in practice; suggestions for algorithmic ways out are given. Thus beginning with extremely small steps can be feasible as long as protective measures are present to avoid over-zealous increases from there on, skipping interesting phenomena. Hindmarsh (1977) gives reasons why an ODE solver may take steps smaller than the machine precision of representing the independent variable.

Gear (1980d) presents Runge–Kutta-like formulae which enable a high-order multistep method to be used from the start. Only one Runge–Kutta step is needed to do this and efficiency can be nicely increased automatically. The Runge–Kutta method can be viewed as an extrapolation technique.

4.1.3 *Step-size and order control*

This idea enabled Gear's backward-difference algorithm to take a "step jump" in advancing automatic solution of stiff (and nonstiff) ODEs. Gear, Tu, and Watanabe (1974) and Gear and Tu (1974) examined the effect of a changing mesh size on the stability of multistep methods. Skelboe (1977) studied control of order and step size for multistep methods when one of the eigenvalues is close to the imaginary axis; an instability test is provided to determine automatically when this is the case. Lindberg (1977) characterized the optimal step-size sequence for stiff methods. Dwyer, Kee, and Sanders (1979) and Rai and Anderson (1980) constructed adaptive mesh generators to control the spacing of zones and increase the resolution in regions of strong gradients.

4.1.4 *Global error*

Users often do not realize that their integration package uses their requested accuracy requirement to match against an estimate of the local error, not the actual error in the solution. Lindberg (1977) showed, for stiff problems, the advantage of keeping the global error at the maximum allowable level during long intervals. Dahlquist (1981a) reported work in progress to extend Lindberg's ideas to automatically control step size on the basis of global error; application is made to a system in partitioned form. See Shampine and Watts (1976) for software which generates a global-error estimate; a problem perhaps greater than expense is reliability.

Dew and West (1978) considered estimating and controlling global error in Gear's method. Stetter (1974) considered global error estimation for nonstiff problems; Stetter (1979) global estimation in Adams predictor-corrector codes. See Thompson and Tuttle (1982) for a discussion and example of the disastrous effects of not knowing global error along a solution trajectory.

4.1.5 *Stiffness detection*

The automatic detection of stiffness is related to the question of automatic method selection, i.e. the choice of a stiff or nonstiff method. Gear (1980d) recommended that initially a nonstiff method always be used. Petzold (1980, 1982b) discussed a technique that uses information available at the end of each step to make a decision between continuing with a stiff or a nonstiff method. Shampine and Hiebert (1977) reported a simple but effective method for following occurrence of stiffness by use of the Runge-Kutta-Fehlberg (4,5) formulae; also see Shampine 1977b).

Kennealy and Moore (1977) showed a heuristic method for detecting stiffness in mass-action kinetics. Braekhus and Aasen (1981) explored use of various explict methods for detecting stiffness in problems of structural mechanics; see too Gladwell (1980).

Gordon and Shampine (1977) mentioned a code at Sandia, called DE, that solves nonstiff ODEs but keeps a "computer-eye" out for the occurrence of stiffness. These authors also identified two other very important ways of detecting stiffness nonautomatically: based on physical reasoning and based on computational experience with similar equations. There is another nonautomatic way: through proper scaling of all variables, as would be done for a singular perturbation analysis or in preparation for putting the equations on an analog computer. See Flaherty and O'Malley (1979) for thoughts on automatic scaling; also Dahlquist and Soderlind (1982).

Shampine (1980a) reported a special definition of stiffness appropriate for implicit A-stable formulae; this definition can be recognized using information already available during the integration. See also Sacks-Davis and Shampine (1981). Shampine (1981) discussed a code DEASY that automatically recognizes stiffness and selects a suitable method.

4.1.6 *Algebraic equations*

There has been little comparative advancement during recent years in solving linear or nonlinear equations relevant to ODE solvers, except for large sparse systems. However, see Hindmarsh, Sloan, and Dubois (1978) for algorithmic advancements for dense linear LU decomposition. Johnson and Barney (1976) tested five conventional methods for solving linear algebraic equations (MINV, SOLVE, DECOMP-SOLVE, JACOBI, and GAUSS-SEIDEL). Byrne and Hindmarsh (1976, 1977) consider solution of linear block tridiagonal forms arising from PDE descretization. Sherman and Hindmarsh (1980) consider solving the linear equations from Newton iteration on a nonlinear sparse set by the Yale Sparse Matrix Package (YSMP).

Shampine (1979a) stated that the solution of the algebraic equations from implicit ODE formulae is special. He found that the residual was the appropriate measure for acceptance of an approximate solution; a way to do this and the advantages are detailed.

Hindmarsh (1977c) considered the idea of rank-one updates for the inverse of the Newton iteration matrix in the context of solving stiff ODEs, but the results are disappointing. Enright (1978a) gave us a more efficient method for matrix factorization after a change in step size or order, particularly good for large dense systems. See Shampine (1981b) for a pertinent discussion on Jacobians and stiff methods; also Eitelberg (1979).

Hiebert (1982) of Sandia evaluated software that solves systems of nonlinear equations and found none to be "best." Poor variable scaling was noted as an area particularly important to address in the future. Also, readily available software that incorporates constraints or non-negativity is needed.

4.1.7 *Comparison of stiff methods*

Enright (1982) explained why it is not possible in general to fairly compare completely different methods or given packages containing different methods in order to arrive at a "best" method. Such comparisons, however, can point to weaknesses in a method or code (Enright, Hull, and Lindberg 1975; Hull 1980; Enright and Hull 1976).

As Enright stated, it makes much more sense to compare different implementations of a given method, or to compare two related methods or packages. Bryne, Hindmarsh, Jackson, and Brown (1977) compared GEAR and EPISODE with respect to appearance to the user, features of software engineering, and the basic algorithms. EPISODE performs better than GEAR for problems involving waves or reoccurring stiffness, but GEAR is better for simple decaying problems.

Enright, Hull, and Lindberg (1975) offered the program package STIFF–DETEST for comparison of stiff ODE solvers; this was also used by Brown (1978). See Weimar and Clough (1979) for a critical evaluation of the semi-implicit Runge-Kutta methods for stiff systems. Thompson and Tuttle (1980) and Thompson (1977) and Bushard (1976) also have performed extensive comparisons of popular stiff methods on a variety of test problems. See also Scherer (1976), Carver, Lloyd, Sprung, and Pitts (1979), and Chan, Birnbaum, and Lapidus (1978).

4.1.8 *Test problems*

Packaging considerations make it hard enough to compare methods, but the bottom line in evaluating software is how well it works. Within selected stiff application areas, there may be typical nonlinear structure, degree of stiffness, and size, but in general there is not. One reasonable approach is to examine one application area only, or to examine many different test problems, representative of a spectrum of applications.

Whichever approach is taken, there is virtue in consistency of choice. By far the most commonly used stiff ODE test problem is the kinetics rate equation set of Robertson (1975; this cited the original 1966 article). We have counted 23 uses of this equation set for this purpose in the literature in the period January 1974 through March 1982, and there are undoubtably many more. Other favorites are given in Robertson (1975).

Enright, Hull, and Lindberg (1975) listed five classes of stiff problems with a number of examples of each class: linear with real eigenvalues, linear with nonreal eigenvalues, nonlinear coupling (smooth to transient and transient to smooth), nonlinear with real eigenvalues (most mass action kinetics problems are here), and nonlinear with nonreal eigenvalues; eigenvalue ranges are given for most of these. However, Shampine (1977b) and Shampine and Hiebert (1977) found several of the examples did not qualify, by their definition, as being stiff—although some have other types of anomalous behavior. Shampine (1981a) pointed out a number of shortcomings with this test set and enumerated ways of improving it.

Enright and Hull (1976) gave ten test problems involving chemical kinetics, in batch and more complex reactors. Johnson and Barney (1976) documented eleven problems they used for testing. Hindmarsh and Byrne (1976) gave a diurnal kinetics problem with the reoccurring stiffness feature characteristic of atmospheric reactions; they also gave a simple diffusion-convection problem for use with the method of lines. A variety of test systems were offered by Michelsen (1976), including a large one and a differential-algebraic one. Chan, Birnbaum, and Lapidus (1978) listed eight stiff problems.

There are also available several testing equations of a rather special nature. Skelboe (1977) suggested stiff problems of a highly oscillatory nature, as did Gaffney (1982a). Fatunla (1980) listed six examples, some of which are stiff and highly oscillatory. Dahlquist, Edsberg, Sollermo, and Soderlind (1980) gave a simple stiff equation with a turning point, a stiff nonlinear oscillator, and a combustion example. Kreiss and Kreiss (1981) consider an example of a stiff two-point boundary value problem. Carver (1980) gave two simple hyperbolic equations for testing: Burger's equation and a model for a counter-current heat exchanger.

4.1.9 *Packages*

Most recent stiff packages generally available are mentioned in various subsections of this review. Table 4.1 presents a summary. Lawrence Livermore National Laboratory has led the way in developing general and special purpose stiff packages. Outside of LLNL, LSODE can be obtained from the National Energy Software Center (NESC), Argonne National Laboratory, 9700 South Cass Avenue, Argonne, Illinois 60439; identify LSODE as NESC No. 592. EPISODE is available as NESC No. 675 and EPISODEIB as NESC No. 705.

4.2 Stiff software: what do we have and what do we need? (Gear)[†]

4.2.1 *Introduction*

In addition to the large amount of fine theoretical work that has been done on stiff equations in the last decade, there have been a great number of new methods developed for solving stiff equations in the last fifteen years, and a great many improvements to existing methods. This much is obvious from the literature. In preparing to write this, I surveyed a few people in major laboratories and looked over some of the literature concerned with libraries and codes, and was surprised to find that in spite of this large amount of work, relatively little of it

[†] Supported in part by the U.S. Department of Energy, Grant DOE DEACO276ER02383.

TABLE 4.1
Stiff software

Name	Comment	Reference
General		
ACSL	Language; GEAR-type	Mitchell (1982)
BLEND	Adams and BDF	Skeel and Kong (1977)
DEASY	Recognize stiffness	Shampine (1981)
DGRUNG	Semi-implicit R–K	Boness (1979)
DIRK	Diagonally implicit R–K	Alexander (1977)
EPISODE	Reoccurring stiffness	Byrne (1981)
FACSIMILE	Parameter estimation	Curtis (1980)
GEAR	From DIFSUB (obsolete)	Hindmarsh (1974)
GRK4T	Rosenbrock methods	Kaps and Rentrop (1979)
IMP	Implicit midpoint	Lindberg (1974)
METAN1	Semi-implicit midpoint	Bader and Deuflhard (1981)
PSCSP	Discont; new language	Halen and Benz (1982)
QUADPACK	Jump discontinuities	De Doncker (1978)
SDBASIC	Second-derivative	Enright (1975)
STIFF 3	Semi-implicit R–K	Michelsen (1976)
STIFF–DETEST	Compare stiff methods	Enright et al. (1975)
STINT	Cyclic composite	Tendler et al (1978)
STRIDE	Implicit R–K	Butcher, et al. (1979)
STRUT	Stiffness detect	Krogh and Stewart (1983)
TRAPEX	Extrapolation	Enright et al. (1975)
Direct Variants of GEAR		
DSTPGT		Thompson and Tuttle (1982)
EPISODEB	Banded Jacobian	Byrne and Hindmarsh (1976)
GEARS	Sparsity (obsolete)	Hindmarsh (1979)
GEARBI	2–3 dimensional PDE	Hindmarsh (1979)
GEARV/GEARST	For parallel or vector processors	Morris et al. (1979)
LSODA	Automatic method select	Hindmarsh (1982)
LSODE	More ease, flexibility	Hindmarsh (1980)
LSODAR	Rootfinding	Hindmarsh (1982)
LSODES	Sparsity	Hindmarsh (1982)
LSODEI	Linearly implicit	Painter (1981)
LSODIS	Sparse implicit	Seager and Balsdon (1982)
Special Applications		
CAKE	Kinetics; sparsity	Ridler (1977)
BELLCHEM	Kinetics	Edelson (1976)
FRST	Nuclear power plants	Earles et al. (1982)
KINRATE	Kinetics	Edsberg (1974)
KISS	Kinetics	Gottwald (1981)
LARKIN	Large kinetics	Deuflhard et al. (1981)
MACKSIM	Kinetics; sparsity	Boyd et. al. (1980)
SETKIN	Kinetics preprocessor	Dickinson and Gelinas (1976)
Differential–Algebraic		
DASP3	Also partitioning	Soderlind (1980)
DASSL	Uses BDF	Petzold (1982)
EPISODEIB	Banded Jacobian	Hindmarsh (1979)
FAST	Translator	Stutzman et al. (1976)
GEARIB	Banded Jacobian (obsolete)	Hindmarsh (1979)
GEMS	Extension of IMP	Babcock et al. (1981)
LSODI	Linearly implicit	Painter (1981)
Partial Differential		
CSDT	Adjusts mesh	Janac (1978)
DISPL2	Galerkin, B-splines	Byrne (1981)
FORSIMVI	Nuclear; sparsity	Carver (1979)

Name	Comment	Reference
PDECOL	Collocation, B-splines	Madsen and Sincovec (1976)
MOL1D	Automatic discretization	Hyman (1976)
Process Simulation		
CSMP		Hylton (1982)
CSMP		Watson and Gourlay (1976)
DPS	Stiff option	Sebastian et al. (1981)
DYFLO2	Stiff option	Franks (1982)
DYNSYL	Stiff option	Patterson and Rozsa (1978)

has found its way into the hands of the users. In fact, except for those fortunate enough to have direct access to some of the people active in ODE code development, virtually all users work with codes based on BDF methods from the NAG or IMSL libraries, from Hindmarsh's LSODE set, or derivatives of earlier codes on which these were based. This is probably no surprise to users on the other side of the fence, as they have long said that until we numerical analysts take time to write good software and get it out to the users, our ideas will not be put into action. However, here in 1982 many of "we numerical analysts" feel that we have seen the light and now think of ourselves as "numericians"—people who are concerned about the real numerical aspects of problems and produce mathematical software to present our results. Yet, in spite of this, very few of the new methods are actually in use and most of the highly-used software is based on early methods.

Does this mean that the user has needs that are not being met by existing codes? I think so. The main development that has affected the average user over the last decade is that earlier software has been greatly improved in two ways: fine tuning and user interfaces. Although some codes were developed in that period which embody new methods, they have not undergone the careful modification which would tune them for the user, such as Hindmarsh has applied in developing the LSODE set of codes from earlier codes. There are a number of user needs that are not met by the currently highly-used codes. For some of these needs there are methods, but the software is in various states of development from experimental (a euphemism for badly written) to what we might call preproduction versions. For other needs there currently does not exist software, or in some cases even reasonable methods.

In the following subsections, I will examine what methods and software are available for solving the problem areas that are part of or impact on the solution of stiff equations. I shall also point out where some of the currently recognized difficulties lie.

4.2.2 *Classification of problems*

The phenomenon of stiffness that first caused numerical difficulty was due to eigenvalues with large negative real parts. The methods developed to solve those are generally effective for problems which are such that all eigenvalues, $\lambda(J)$, of J satisfy $\arg[-\lambda(J)] < 60^\circ$. For example, the BDF formulae are A(73.23°)-stable for orders up to four, while the fifth-order formula is A(51.83°)-stable (Gaffney 1982a). We will refer to this angle as the argument of stability. A sequence of codes has been written based on these methods, and the current versions of these codes are the ones in common use. I suspect the reason that newer methods are having very little impact in this area is that the underlying method in these older codes is within a small factor of being as good as one can hope to develop, and the codes based on these methods have evolved over a decade to be so finely tuned that recent codes based on slightly better methods show very little improvement. Furthermore, the fact that these older codes have been developed in a "user-active" environment has caused them to be well-tailored to user needs as far as interfaces go.

However, there are other types of problems that are frequently classified as "stiff" and additional features of stiff problems that cause difficulties for many existing codes. There are also a number of features that are needed in codes but are not yet commonly available. Among these are the abilities to cope with equations of the following types.

(i) Equations with lightly-damped oscillatory features.
(ii) Linear constant-coefficient equations (efficiently).
(iii) Large systems of equations.
(iv) Implicit differential equations.
(v) Differential–algebraic equations.
(vi) Differential equations with discontinuities.
(vii) Equations involving problems with high-frequency components.
(viii) Equations partitioned into "fast" and "slow" sets (efficiently).
(ix) Equations with delayed terms present.

Also needed are codes with the following features.

(x) Global error estimation and control.
(xi) Atuomatic switching between appropriate methods.
(xii) Output solutions which are smooth functions of input parameters.

In the following subsections some of these problem areas are discussed as well as the state of software development for their solution.

4.2.3 *Current software for the standard stiff problem*

In addition to the codes in the IMSL and NAG libraries and those of Hindmarsh (Hindmarsh 1979 b, c, 1980a) all of which are widely available, there are a number of newer codes in various stages of development which may have an impact in the near future. I have discussed some of these because they appear to be particularly suitable for some of the special problems. Among the other codes are two recent entries to the market based on techniques not used in common stiff codes. Bader and Deuflhard (1981) have developed an extrapolation code based on a semi-implicit midpoint rule. This midpoint rule has the advantage that it does not require iteration of nonlinear equations involving the Jacobian and yet has a quadratic error expansion so that each extrapolation step adds two orders of accuracy. Their program, METAN1, is still experimental but is a promising candidate for future use. A second program, DSTIFF, is based on multistep formulae chosen to provide a least-squares fit to a number of accuracy requirements as well as stability requirements. The argument of stability reported by Gupta (1981) exceeds 62° all the way to order ten, so these methods may also be useful for the problems described in the next subsection.

Mention should also be made of many codes designed primarily for initial value problems in PDEs and of codes designed for particular user groups such as people solving chemical kinetic problems, network simulation problems, etc. PDE codes often make use of ODE codes via the method of lines. In some cases they use standard ODE codes. (See, for example, Hyman 1979; Madsen 1979; Melgaard and Sincovec 1981a; Melgaard and Sincovec 1981b; and Sincovec and Madsen 1975. For a survey, see Machura and Sweet 1980.) Other codes are applicable to PDEs or ODEs, or combinations of the two (for example, the FORSIM package, Carver 1973). There are also many other codes which have been tailored for specific users, e.g. FACSIMILE (Chance, Curtis, Jones, and Kirby 1977), and there are undoubtedly many other experimental codes of which I am unaware.

4.2.4 *Lightly damped oscillatory problems*

When the Jacobian has large eigenvalues near to but to the left of the imaginary axis, the solution to the problem does not exhibit oscillation (except possibly in a transient region) provided the driving terms are not adding energy at the corresponding frequency; but many numerical methods will have difficulty because they will fail to damp the oscillation, or they will even amplify it. The BDF methods are particularly bad because they are not A-stable above second order and are strongly amplifying for eigenvalues close to the imaginary axis at moderate distances from the origin. These problems are very similar to stiff problems in that common numerical methods tend to amplify components that are not present in the true solution. Miranker (1981) has also pointed out that they are similar in that both can be viewed from the singular perturbation point of view. However, they are not similar when one considers that the classical stiff problem is "superstable" and any perturbation is rapidly damped in the actual system, whereas the lightly damped oscillating problem is not superstable and a perturbation can cause an effect in the solution for some time. If the eigenvalue is very close to the imaginary axis, an oscillatory component could take a long time to damp out, and until that happens, almost all integrators would be forced to take a small step to follow the high-frequency component. (It is, of course, possible to use methods which are exact for certain oscillatory functions. This is a form of exponential fitting and requires knowledge of the eigenvalues of the Jacobian. Such methods are not usually too useful for general-purpose software.) It is tempting to consider methods that damp the oscillations rapidly so that they do not hinder the numerical method. This is a satisfactory approach in those problems in which the oscillating term does not have a strongly nonlinear effect, so that damping, which is equivalent to a time averaging, does not change the slow components. A problem such as

$$y' = \mathrm{i}wy, \quad w \geqslant 0,$$

$$z' = y^2 + y'^2,$$

cannot be damped because the value of $z(t)$ is strongly dependent on the amplitude of the oscillation. Such problems will be discussed later. We examine here those problems in which the oscillation is either absent or can be damped without harm. For these problems we need methods that are "more stable" than the underlying problem near the imaginary axis to damp the unwanted components.

Gaffney has examined several codes for these types of problems, having found codes based on the BDF formulae to be unsatisfactory. Four codes are compared in Gaffney (1982a). They are STRIDE (Butcher, Burrage, and Chipman 1979), STINT (Tendler, Bickart, and Picel 1978), DIRK (Alexander 1977), and BLEND (Skeel and Kong 1977). Of these, BLEND and DIRK are experimental codes which in no sense are in a form for the general user, having been developed as a test bed for the underlying ideas in the papers cited. STRIDE (available as a Fortran or Algol code) uses singly-implicit R–K methods (Burrage 1978) which give stability out to an argument of at least 83°. STRIDE can also be used for nonstiff equations by using fixed-point iteration instead of Newton's methods to solve the nonlinear equations at each step, but it appears that its strength is for stiff equations with eigenvalues near the imaginary axis. STINT is an implementation of cyclic composite methods. These achieve larger regions of absolute stability than is possible with simple linear multistep methods by applying a sequence of different multistep methods. If its order is limited to four, it is stable out to arguments of almost 81°, far superior to the BDF methods. BLEND uses a combination of Adams and BDF methods in a way that can give the stability of second-derivative methods. The basic idea is that if the Adams method is written as $\boldsymbol{A} = \boldsymbol{0}$, the BDF method as $\boldsymbol{B} = \boldsymbol{0}$, and $\boldsymbol{J}$ is the Jacobian, the linear combination

$$\boldsymbol{A} + \boldsymbol{J} \cdot \boldsymbol{B} = \boldsymbol{0}$$

becomes the Adams method when $\boldsymbol{J}$ is small, and the BDF method when $\boldsymbol{J}$ is large. However, it also involves a term of the form $\boldsymbol{J}\boldsymbol{y}'$ in the $\boldsymbol{J} \cdot \boldsymbol{B}$ component. For the linear ODE $\boldsymbol{y}' = \boldsymbol{J}\boldsymbol{y}$ we see that $\boldsymbol{y}'' = \boldsymbol{J}\boldsymbol{y}'$, so this term is equivalent to the second derivative. Consequently, it can achieve the stability of second derivative methods —for example, it can be A-stable up to the fourth order. Furthermore, its accuracy is obtained from the accuracy of the underlying Adams and BDF methods, so it does not depend on the accuracy of the Jacobian for integration accuracy; an error in the Jacobian only affects the region of stability. Consequently, it is not necessary to re-evaluate the Jacobian frequently, as would be required in second-derivative methods to be discussed later. Gaffney (1982a) compared the maximum argument of stability for STINT, BLEND, and the BDF methods. Even through order seven, BLEND is competitive with the 83° of the STRIDE implementation. Unfortunately, BLEND has never been developed into a production code. (We do have a version embedded in LSODE in addition to the original modification of DIFSUB, but it is inadequately documented.) DIRK is an experimental implementation of diagonally implicit Runge–Kutta methods with a fixed order. The diagonal coefficients are chosen to be the same so that although at each step of an R-stage method it is necessary to solve R systems of nonlinear equations versus the one system of multistep methods, the same LU factorization can be used for all equations. Gaffney presented a table of recommendations based on his tests on page 80 of his 1982(a) paper. Since the method of choice varies with accuracy requirements, argument of stability needs, and user demands on the interface to the code, I refer the reader who needs to solve such problems to Gaffney (1982a) for further details. Second-derivative methods may also be competitive for these problems because they have a large argument of stability, although they tend to be inefficient for large problems with a nonconstant Jacobian.

4.2.5 *Linear constant-coefficient problems*

Although there are probably not many problems in this category, considerable savings are possible when they occur. Clearly, it is not necessary to re-evaluate a Jacobian matrix, and techniques to avoid refactoring can be exploited. To my knowledge, there has been very little code development in this area except for the work by Enright (1977). He modified two codes—one of Hindmarsh's GEAR codes and one of his second derivative codes, SDBASIC (Enright 1974 and Addison 1979). A typical stiff code has to solve a nonlinear equation at each step and this is done with a Newton iteration which requires solving a linear system involving the matrix $\boldsymbol{I} - h\beta\boldsymbol{J}$ where $\boldsymbol{J}$ is the Jacobian. Refactorization is necessary for two reasons: the Jacobian changes or the step size h and/or the coefficient β changes. If the system is linear, the Newton iteration converges in one step. Furthermore, the Jacobian does not change, so it is necessary to refactor only when h or β change. This is only a shift to the matrix, but unfortunately there are no known fast techniques for updating the LU decomposition after a shift of a matrix—indeed, it appears that it cannot be done any faster than the original decomposition in general, so Enright uses a different decomposition, $\boldsymbol{LHL}^{-1}$ where $\boldsymbol{L}$ is unit lower triangular and $\boldsymbol{H}$ is upper Hessenberg. A shift of this is obtained by a shift of the $\boldsymbol{H}$, while $\boldsymbol{H}$ can be factored in $O(n^2)$ steps. A further advantage of a constant Jacobian accrues in the case of second-derivative methods, namely that the calculation of the second derivative, done by forming $\boldsymbol{J}\boldsymbol{y}'$, does not require re-evaluation of the Jacobian at each step. It is that re-evaluation that makes the second-derivative methods less competitive than they would otherwise be when compared to methods that only lose stability but not accuracy when the Jacobian is somewhat inaccurate.

4.2.6 *Large systems of equations*

The major difficulty with large problems is matrix manipulation due to the need to solve linear equations a little more than once per time step. As the size of the problem increases, the percentage of time in this part of the program increases and it becomes important to look for fast methods. When the problem gets very large, the storage for the Jacobian and its decomposition may also be a problem. Fortunately, the Jacobian for large problems is almost always sparse because at least part of the problem has arisen from the partial discretization of a PDE. The use of direct sparse techniques has received a fair amount of attention and one version of LSODE, namely LSODES which uses the Yale sparse matrix package, is available. Less attention has been paid to iterative methods which will be needed for the really large problems, although much of the work done on iterative methods for symmetric problems is applicable when the system of ODEs arises from a method-of-lines approach to many PDEs. Miranker (1981) discusses the use of conjugate-gradient methods for symmetric Jacobians. An earlier code of Hindmarsh, GEARBI (Hindmarsh 1976c), uses a block SOR method for problems arising from the method of lines applied to two space-dimension PDEs. Hindmarsh has indicated that this will be upgraded to an LSODE code. However, in problems in which the Jacobian is nonsymmetric and even has an irregular nonzero structure, most of the currently-used methods for solving equations iteratively are not too satisfactory. There is some work continuing on this problem—for example, Gear and Saad (1981), and this has been put in a version of LSODE, but it has not been distributed because it still represents an initial attempt.

If the problem is not so large that direct methods spill out of primary storage, they can be used but can be very expensive. In that case, there is a lot of interest in techniques which minimize the number of times the Jacobian must be recomputed and/or the matrix refactored. There are few codes generally available based on these techniques, but some of them are likely to appear in future codes. For example, Krogh and Stewart are working on a new code STRUT, based on BDFs for stiff equations, which makes considerable effort to detect whether the Jacobian need be re-evaluated or the matrix refactored (Krogh and Stewart 1981a, b). The "constant leading coefficient" technique (Jackson and Sacks-Davis 1980) has been used to reduce matrix refactorization by holding the β coefficient constant. This changes as the step-sizes change in variable-step methods. By sacrificing a small amount of accuracy it can be held constant.

4.2.7 *Implicit differential equations*

An implicit differential equation is written in the form

$$\boldsymbol{F}(\boldsymbol{y}, \boldsymbol{y}', t) = \boldsymbol{O}$$

and can be handled directly by a number of methods including BDF and certain implicit Runge–Kutta methods. A few codes have been written using the BDF methods but until recently they have not been widely distributed. Some of them have been embedded in proprietary network-analysis software (Hachtel, Brayton, and Gustavson 1971), so for a long time the only code available was a modified version of DIFSUB which was "experimental" in the worst meaning of the word (Gear 1971c). Recently, two better codes have appeared, LSODAI (Painter 1981) and DASSL (Petzold 1982a). Both of these are based on the BDF methods which replace the derivative in the implicit equation with a difference approximation. For example, the first-order method is the Backward Euler method and is obtained by substituting $(\boldsymbol{y}_n - \boldsymbol{y}_{n-1})/h$ for $\boldsymbol{y}_n'$ to get the nonlinear equation

$$\boldsymbol{F}[\boldsymbol{y}_n, (\boldsymbol{y}_n - \boldsymbol{y}_{n-1})/h, t_n] = \boldsymbol{O}$$

to solve for $\boldsymbol{y}_n$ at each step. To my knowledge, no codes are available based on implicit Runge–Kutta methods, although such methods would work provided that the final stage is evaluated at the end point of the step and is used to provide the output value.

4.2.8 *Differential–algebraic equations*

These problems arise when an implicit system represents fewer differential equations than there are variables and equations because the Jacobian $\partial \boldsymbol{F}/\partial \boldsymbol{y}'$ is singular. For some time it has been known that difficulties can be encountered with some such systems, but only recently has the extent of some of these difficulties been comprehended. The current situation is that we do not know how to solve the general problem, but we do know that all conventional ODE methods break down. Fortunately, the difficulties do not arise very frequently, so large classes of problems can be solved by the software mentioned previously; however, it is possible to construct examples which are physically reasonable but impossible to handle, and there are cases in which examples have arisen in practical simulation studies.

The nature of the difficulty can be seen in some very simple model problems. Consider the system of three equations

$$x = g(t),$$
$$y = x',$$
$$z = y'.$$

These are clearly not differential equations: their solution is uniquely defined independently of initial values. If a fixed-step integrator is applied to these it is equivalent to applying numerical differentiation to $g(t)$. For example, if the Backward Euler method is applied, we calculate

$$x_1 = g_1,$$
$$y_1 = (x_1 - x_0)/h,$$
$$z_1 = (y_1 - y_0)/h.$$

This numerical calculation uses initial values for x and y, which are probably incorrect, so the values of y_1 and z_1 will be incorrect. In fact, they will diverge as h tends to zero. However, it can be shown that, for fixed step-size methods such as for BDF, after a number of steps the answers have the required degree of accuracy. However, if the step varies or the problem has time-dependent coefficients, we may run into difficulty. Precisely, we must talk about the index of nilpotency of the system. It is the maximum size of a subsystem of the sort above. In the example, this size is three. If this index exceeds one, the time-dependent problem cannot be solved by known methods. If the index exceeds two, variable step-size methods also fail for the constant coefficient case. Details can be found in Petzold (1981b) and Gear, Hsu, and Petzold (1981). A further report is also in preparation by the same authors.

These problems do arise occasionally. For example, Painter (1981) stated that the Navier–Stokes equations have an index of two. They were integrated successfully by an implicit ODE code, perhaps because the critical parts of that problem are linear. It would be highly desirable for a code to detect the existence of a high index, but at the present we do not know how to do that. The symptom of such a problem in an automatic code is that the step-control mechanism gets into trouble and reduces the step endlessly without reducing the error—possibly even increasing it, so that eventually the code reports that the error tolerance cannot be achieved, leaving the user to guess the reason. Furthermore, since it happens only when the Jacobian with respect to y' is singular, there are conventional stiff equations arbitrarily close to these very hard problems. If such systems are integrated with existing integrators it is possible that they will fail also.

4.2.9 *Differential equations with discontinuities*

Discontinuities in the derivatives of solutions cause difficulties in codes unless they are handled carefully. The difficulty arises because the code is proceeding under the assumption that the solution is well approximated by a smooth form such as a polynomial. This is clearly not true when there is a discontinuity in a derivative of an order lower than the degree of the polynomial. If a one-step method is being used, the problem can be overcome by placing a mesh point at the discontinuity so that separate smooth approximations are used on each side of the discontinuity. Even then, the code has to be organized carefully. If, for example, a trapezoidal rule is being used to integrate the problem $\boldsymbol{y}' = f_i(t)$ where $i = 0$ for $t < t_c$ and $i = 1$ for $t \geqslant t_c$, the step ending at t_c must be integrated using f_0, even when $\boldsymbol{y}'$ is evaluated at t_c in that step. After that step has been completed, the switch to $\boldsymbol{f}_1$ should be made. This means that the code to evaluate the derivatives must contain a switch and be told by the integrator when the discontinuity has been passed. If a multistep method is being used, the order of the multistep method must not exceed the order of the discontinuity in any steps which use information from both sides of the discontinuity, meaning that we also need to know the order of the discontinuity. Alternatively, we can simply restart a multistep method to avoid this problem.

From this it is apparent that we need to know the location of the discontinuity and possibly its order. If the discontinuities occur at known values of the independent variable t the problem is fairly straightforward and is simply a matter of code organization. If the discontinuities occur at known values of the dependent variables, or given functions of them, it is necessary first to locate a discontinuity by monitoring the values of these functions, which we call switching functions. Carver's FORSIM code provides such options for locating discontinuities. For a discussion, see Carver (1977). The location of a particular point by monitoring functions of dependent variables goes back, I think, to an early Illiac I R-K code that was probably written by Nordsiek (I do not have any records on that; I just remember using it around 1958). It was also put into one of Fred Krogh's pioneering variable-order, variable-step codes under the name of the GSTP feature.

In the two cases above, the detection of a discontinuity is possible because the user provides switching functions to indicate their presence. In the worst scenario, we do not even have that information (perhaps because the derivatives are being evaluated by a large section of coder that cannot be analyzed reasonably or depend on tabular data). Then we must also attempt to detect the presence of a discontinuity. Thus, we see that there are three problems to handle with discontinuities:

(i) detection,
(ii) location,
(iii) restarting.

Detection has been investigated for nonstiff problems in Gear and Osterby (1981). It is not clear that the technique used can be extended to stiff problems. Fortunately, we usually have switching functions to take care of this. The second step is equivalent to that of root finding for the equation $\boldsymbol{g}(\boldsymbol{y}, t) = \boldsymbol{0}$, where $\boldsymbol{y}(t)$ is given by the numerical integrator. Hindmarsh will have a new version of LSODE—called LSODAR—for this problem, but it is not yet ready for distribution.

Restarting is important in multistep codes because they are usually less efficient in the first few steps, which are taken at lower order and small step size. R-K codes can commence with a normal-sized step immediately. The problem of starting at a modest order (four) with a reasonable step-size is examined in Gear (1980a) where a special form of an R-K step is used to generate enough information for a four-step multistep method to continue. A code for the nonstiff method is available in a related report. There is a version of the technique for stiff problems, but as a general rule a problem will not be stiff after a discontinuity because transient components requiring a small step-size for accuracy will have been stimulated. However, this is not always true. I have encountered one case in which the problem remained stiff although there were discontinuities fairly frequently. A one-dimensional PDE describing a heat-conducting material which was melting as heat propagated had been converted into a system of ODEs by the method of lines. The underlying heat equation kept the problem stiff at all times. As the solid-liquid interface passed over a line, there was a discontinuity because of the latent heat effects and the jump change in the heat conductivity. This problem was causing difficulty for any of the integrators the user was trying. (As is common, I do not know if it finally was solved. The user did not return after the third visit, either because my suggestion of that meeting worked or he decided that I was wasting his time!)

4.2.10 *Problems with high-frequency components*

These problems are ones in which the high-frequency components are an integral part of the problem and must not be damped because such damping would change the solution, or cannot be damped because they arise from nonlinear oscillators. Petzold (1978) developed a preliminary code as a modification of DIFSUB. The underlying method is called a multirevolutionary method because it attempts to step over many cycles (or revolutions, in the case of satellite orbits where the method was first investigated). The basic idea is that a single cycle is investigated numerically by integrating the original differential equation with a conventional method. The results from this are used to predict the rate of change of the behavior of the starting point of each cycle. If the behavior of the starting points is "smooth" it is possible to extrapolate forward over many cycles without the need to integrate over each one. This extrapolation looks like an integration method with modifications proportional to the ratio of the period to the step size. We are using generalizations of the Adams and BDF methods for this extrapolation. Complications arise because the period may be changing so must be recalculated from the solution, and because the problem may be very stable, creating the same effect as stiffness in the multirevolutionary method. Gallivan and I have continued that work at Illinois and are writing a code embodying a number of new techniques for detecting the oscillatory behavior, deciding when to use the multirevolutionary method because it might be more efficient than a conventional method, and when to use the generalized BDF rather than the generalized Adams methods (Gear and Gallivan 1981).

Problems with more than one frequency probably cannot be handled by this method, so these remain something of an open problem. Methods employing Fourier expansions are one possibility, but I know of no general software embodying these.

4.2.11 *Partitioned systems*

It may be profitable to split a large system of equations into one or more subsystems in order that each subsystem can be handled by different methods and/or different step sizes. Different methods could be profitably used if the stiffness is confined to a subsystem. Different step sizes could be used if the bandwidth of one subsystem is significantly less than that of another. Even if the same method and step size are used in all subsystems, gains are still possible if the matrix arithmetic associated with stiffness is avoided in the nonstiff subsystems. There are relatively few codes available embodying these techniques. Soderlind (1980) describes a code, DASP3, which uses a conventional four-stage R-K explicit method for one subsystem, and a three-step BDF method for the other. It is necessary for the user to partition the system into the form

$$y' = f(y, z, t),$$
$$\epsilon z' = g(y, z, t),$$

a priori. The code does not make this determination automatically. Various approaches to using different step sizes are discussed in Gear (1980), and a code which will allow a system to be divided into a number of subsystems—each to be handled by different, automatically chosen step sizes and orders—is being written by D. Wells at the University of Illinois as part of his Ph.D thesis. This code also requires the user to do the partitioning. Of course, one approach to automatic partitioning is to treat each equation as a separate subsystem and leave it free to select its own step size, order, and even method (e.g. stiff versus nonstiff). Unfortunately, it appears that the overhead of such an approach will be so large than any savings in the actual integration will be negated.

One difficulty with partitioning is that it is based on the hope that certain properties are confined to subsystems. In general, this is not true. For example, in a linear system in which the Jacobian is not reducible, the dominant eigenvectors (those corresponding to the large eigenvalues) will have components in all dependent variables, so that stiffness cannot be confined to a subsystem. If transformation to a basis of eigenvectors were to be made, we could confine stiffness to a subsystem, but this is not usually practical. However, it is not necessary to have a basis of eigenvectors; it would suffice to have a basis, part of which spanned the dominant eigenspace and the other part of which spanned the remainder. This is really what the iterative methods are attempting to do in Gear and Saad (1981). An approach to this is the technique in Enright and Kamel (1979) in which the iteration matrix is decomposed into an $\boldsymbol{LHL}^{-1}$ form where $\boldsymbol{H}$ is upper Hessenberg and $\boldsymbol{L}$ is a combination of a lower triangular matrix and permutations chosen to put the large components into the first part of $\boldsymbol{H}$. In effect, the lower part of $\boldsymbol{H}$ is ignored, corresponding to ignoring components due to the nondominant eigenspace. In practice, the reduction is terminated after the large components have been handled. Since this matrix is used to solve the linear equations arising in the Newton iteration, the effect of discarding the small part of $\boldsymbol{H}$ is to handle the stiff components by a Newton step and the nonstiff components by a functional iteration step. In Gear and Saad (1981) it is shown how this can be used to treat the stiff components with BDF and the nonstiff components with Adams methods. However, to my knowledge there are no distributed codes based on these techniques.

4.2.12 *Delayed equations*

Delayed, or retarded, differential equations can also exhibit stiffness. There has been some work recently on codes for these problems. Among these are codes based on STFDIF, a composite, block, one-step method by the group at Syracuse, and a code based on DIFSUB by Roth and Watanabe at Illinois (Roth 1980; Watanabe and Roth 1982).

4.2.13 *Global error estimation and control*

Most codes attempt to control no more than an estimate of the local error, i.e. the error introduced in a single step. If "error-per-unit-step" control is used, the step size is adjusted so that the error permitted is believed to be less than a tolerance, TOL, scaled by the step size used. Then, if there is no error growth due either to the problem or the method, the error after an interval of length L should be bounded by L*TOL. However, the actual error depends on the amplification, or otherwise, of the errors by the differential equations and method. Some codes have been written for the nonstiff problem to estimate global errors, for example, GERK (Shampine and Watts 1978). When the problem is stiff, errors introduced in the early stages into the stiff components decay due to the superstability of the problem (provided the method also exhibits the same stability). In that case, the best step size strategy is not an error-per-unit step strategy, but something closer to an error-per-step strategy in which the local error estimate is controlled to be less than TOL. Dahlquist (1981) is investigating the control of global error in stiff equations. It is an important problem because the relationship beteeen the user-supplied error-control parameter TOL and the global error is even less understood by the typical user than in nonstiff problems.

4.2.14 *Automatic method switching*

Until recently, all codes either required the user to specify whether a stiff or non-stiff method should be used, or provided no choice. Since most problems are not stiff everywhere, as the behavior in an initial transient is best handled by a non-stiff method, it is highly desirable to switch between nonstiff and stiff methods. A number of Shampine's codes for nonstiff problems have detection mechanisms that warn users of potential problems such as stiffness. Stewart has investigated using these to switch automatically from Adams methods, as implemented in Shampine's code STEP, to BDF methods for stiff equations. The code STRUT being written by Stewart and Krogh is expected to include this mechanism. Petzold (1982b) has also investigated techniques for this, and has modified a version of LSODE with Hindmarsh. Although not distributed yet, it is receiving experimental use at Livermore under the name LSODA.

At Illinois we are using similar techniques to switch to and from stiff methods in the code being written for oscillatory methods. The basic idea is to estimate the step sizes that can be used in the stiff method and in the nonstiff method. In the case of nonstiff methods, the step size must be limited to keep an estimate of the largest eigenvalues inside the region of absolute stability. These are then divided into estimates of the cost per step of each method (stiff methods are much more expensive per step because of the matrix operations), and the smallest resulting value is used to select the method. In our codes this is done for each of several orders in Adams and BDF methods so that order selection is handled at the same time. One difficulty that has been encountered (and requires careful tuning of the code) arises if the step size in the nonstiff integrator is not controlled strongly enough to keep the method absolutely stable. In that case the numerical solution exhibits small oscillations at the error tolerance level. The integrator calculates step size estimates by estimating derivatives of the solution using numerical differentiation. The oscillating errors give rise to artificially large derivative estimates which cause the step size estimates to be too small, so that it appears that stability is not a problem. Since the BDF methods have larger error coefficients than Adams methods, a switch is never made to the BDF methods. It has proved necessary to ensure that the step size is kept small enough in Adams methods to be certain that the method is well inside the region of absolute stability.

4.2.15 *Smooth solutions*

The output of most integrators is far from being a smooth function of its inputs because if a change is made to one of the inputs, say an initial value, the step size and order sequence may change drastically, causing a significant change in the output. In applications in which the user is only concerned with the difference between the numerical and true solutions, this is not a problem. However, there are many situations in which the output of the integrator is the input to another program such as an optimizer. An example of this occurs in the estimation of reaction rates in chemical kinetic equations. These are often solved by performing a least-squares fit to observed concentrations by reintegrating the equations with different values of the reaction rates. We have been particularly concerned with this problem because we use an integrator to determine the behavior of one cycle for use in the multirevolutionary methods for oscillating problems.

We would like to see methods developed which obey certain smoothness properties. In particular, we would like methods for which we can get not only bounds on the global error of the solution, but also bounds on the difference between the numerical derivative of the solution with respect to any input parameter and the corresponding derivative of the true solution. A student at Illinois, Vu Thu, is working on this problem for ODEs at the time of this writing. The critical difficulty is that a code has to avoid making any "absolute" decisions such as step rejection, because these invariably lead to discontinuous behavior. The code cannot include any "if" statements, MAX and MIN functions, or similar functions with discontinuous properties.

4.2.16 *Conclusions*

Although there are a number of good codes for standard stiff problems, there are many areas for which code development is still needed, and other areas in which we do not yet have adequate methods. The experience with the early codes suggests that more effort in code development after the first "production" version has been written is needed if a code is to become widely used.

4.3 The ODEPACK solvers (Hindmarsh)

The DIFSUB, GEAR, and EPISODE packages were added to a list of available general purpose initial-value solvers that was growing quite sizable by 1975. The length and diversity of this list caused some concern to users and software developers alike. There was much duplication of capabilities offered, but at the same time there was very little in common among the solvers in terms of either their external appearance or their internal structure. This situation was in sharp contrast to that in other areas in which "systematized collections" of Fortran routines were being developed. The earliest examples were EISPACK (Smith, Boyle, Garbow, Ikebe, Klema, and Moler 1976) for computing matrix eigensystems, LINPACK (Dongarra, Bunch, Moler, and Stewart 1979) for solving linear systems, and FUNPACK, for certain special functions.

4.3.1 *The ODEPACK concept*

The idea of a systematized collection of initial-value ODE solvers, tentatively called ODEPACK, was discussed informally as early as 1974, in workshops attended by people from all over the world (Byrne 1976). However, it was quickly realized that the task was much larger in the ODE case than in other areas, partly because of the complexity of the subject, and partly because of widely divergent views of what ODEPACK should look like. Starting in 1976, attempts were made to reduce the problem by involving only people at U.S. Department of Energy laboratories, and LLNL received funding to study the feasibility of ODEPACK from the Applied Mathematical Sciences Research Program under the Office of Basic Energy Sciences in DOE.

The natural first step, and a necessary preliminary to any actual development of an ODEPACK, is the setting of standards for the interface between the user and the ODE solvers. The user interface to a solver consists mainly of the call sequence of the routine the user must call, together with definitions of the one or more user-supplied routines called by the solver. To the extent that solvers for various problem types using various methods must all communicate certain specific things to and from the user, it is possible to formulate a loose set of standards for the user interface. An early proposal is given in Hindmarsh and Byrne (1976b). A sequence of workshops and discussions on user interface standards for ODE solvers succeeded in producing a reasonable consensus in 1978 (Hindmarsh 1978, 1979a). This tentative interface standard was achieved only through considerable compromise by the various participants, which included ODE software authors and users at various DOE laboratories.

At that time, it was agreed that several of the more popular ODE solvers, including GEAR, GEARB, DE/STEP (Shampine and Gordon 1975) and RKF45 (Forsythe, Malcom, and Moler 1977), would be rewritten to conform with the tentative standard interface, resulting in a small collection that was at least systematized in its external appearance. The first result of that agreement was a package based on the GEAR and GEARB packages, called LSODE (Livermore Solver for ODEs) (Hindmarsh 1980). The LSODE solver and variants of it written subsequently (all in accordance with the tentative standard interface with minor modifications) are briefly described in the following subsections. In the meantime, unfortunately, the other software authors involved withdrew from the agreement, and so this collection does not yet have analogous rewritten versions of their codes.

4.3.2 *LSODE*

LSODE combines the capabilities of GEAR and GEARB. Thus it solves explicitly given stiff and nonstiff systems $\dot{y} = f(t, y)$, and in the stiff case it treats the Jacobian as either full or banded, and as either user-supplied or internally approximated by difference quotients (see §4.6 for an interesting application). By comparison with GEAR and GEARB, LSODE offers a number of new features that make it more convenient, more flexible, more portable and easier to install in software libraries. Some of these are the following:

(i) Through the redesigned user interface, many new options and capabilities are available, and others are much more convenient than before. Some examples are: more flexible error tolerance parameters, independent flags for starting and stopping options, internally computed initial step size, two work arrays in the call sequence for all internal dynamic work space, user names for f and J in the call sequence, easy changing of input parameters in mid-problem, convenient optional inputs (such as maximum method order), convenient optional outputs (such as step and function evaluation counts), optional provision of derivatives of the solution (of various orders) at any point, and real and integer user data space (of dynamic length) available in the f and J routines (with no extra burden on the casual user).

(ii) The user documentation, which is contained in the initial comment cards of the source, is given in a two-level form. A short and simple set of instructions, with a short example program, is given first, for the casual user. Then detailed instructions are given for users with special problem features or a desire for nonstandard options. The latter is also organized so as to allow selective reading by a user who wants only a fraction of the nonstandard capabilities.

(iii) When stiff options are selected, linear systems are solved with routines from LINPACK (Dongarra, Bunch, Moler, and Stewart 1979), which is becoming a widely accepted standard collection of linear system solvers.

(iv) Some retuning of various heuristics was done so that performance is more reliable than for GEAR/GEARB.

(v) The core routine which takes a single step, called STODE, is independent of the way in which the Jacobian matrix (if used) is treated. Thus as variant versions of LSODE are written for other matrix structures (such as LSODES), these will share the same subroutine STODE.

(vi) The writing of all error messages is done in a small isolated general-purpose message-handler called XERRWV. Two other small subroutines are user-callable for optional changing of the output unit number and optional suppression of messages. This trio of routines is compatible with a much larger error package (the SLATEC Error Handling Package) written elsewhere (Jones 1978).

(vii) LSODE easily allows a user to interrupt a problem and restart it later (e.g. in switching between two or more ODE problems). Also, using LSODE in overlay mode is very easy, with no loss of needed local variables.

(viii) The various lists of constants needed for the integration, formerly appearing in a subroutine called COSET, are now computed (once per problem). This adds to the portability of LSODE.

4.3.3 *LSODI*

The LSODI solver (Hindmarsh 1980), written jointly with J. F. Painter (LLNL), treats systems in the linearly implicit form $\boldsymbol{A}(t, \boldsymbol{y})\dot{\boldsymbol{y}} = \boldsymbol{g}(t, \boldsymbol{y})$, where $\boldsymbol{A}$ is a square matrix. Many problems, including PDEs treated by finite elements and the like, result in such systems, and it is almost always more economical to treat the system in the given form than to convert it to an explicit form $\dot{\boldsymbol{y}} = \boldsymbol{f}$. LSODI allows $\boldsymbol{A}$ to be singular, but the user must then input consistent initial values of both $\boldsymbol{y}$ and $\dot{\boldsymbol{y}}$. In the singular case, the system is a differential–algebraic system, and then the user must be much more cautious about formulating a well-posed problem, as well as in using LSODI, which was not designed to be robust in this case. LSODI is based on (and supersedes) the GEARIB package, but corrects a number of deficiencies, as follows:

(i) The matrices involved can be treated as either full or banded, by use of the method flag.

(ii) The dependence of $\boldsymbol{A}$ on $\boldsymbol{y}$ is automatically and inexpensively accounted for whether partial derivatives are supplied by the user or computed internally by difference quotients.

(iii) When $\boldsymbol{A}$ is singular, the user needs to supply only the initial value of $\mathrm{d}\boldsymbol{y}/\mathrm{d}t$, and this array (along with the initial $\boldsymbol{y}$) is passed through the call sequence, rather than computed in a user-replaceable package routine. (Admittedly, correct initial data can be difficult to obtain for some types of problems.) When the initial $\mathrm{d}\boldsymbol{y}/\mathrm{d}t$ is not being supplied, an input flag instructs LSODE to compute it on the assumption that $\boldsymbol{A}$ is initially nonsingular. Thereafter, no such assumption is made, but ill-conditioning can be a problem when $\boldsymbol{A}$ is singular.

(iv) The user-supplied residual routine includes a flag which allows the user to signal either an error condition or an interrupt condition.

(v) To the maximum extent possible, LSODI shares the same user interface as LSODE and so reflects all the advantages over GEARIB that LSODE has over GEAR and GEARB.

The differences between the LSODI and LSODE user interfaces occur primarily in the user-supplied subroutines. With LSODI, one must supply a routine to compute the residual function $\boldsymbol{r} = \boldsymbol{g}(t, \boldsymbol{y}) - \boldsymbol{A}(t, \boldsymbol{y})\boldsymbol{s}$ for a given t, $\boldsymbol{y}$, and $\boldsymbol{s}$, and another routine to add the matrix $\boldsymbol{A}$ to a given array. Optionally, the user can supply a routine to compute the Jacobian matrix $\partial\boldsymbol{r}/\partial\boldsymbol{y}$.

By virtue of the modular and systematized organization of LSODE and LSODI, the two packages share most of their routines with each other.

Some examples of the use of LSODE and LSODI on systems arising from PDE problems can be found in Hindmarsh (1981) and Painter (1981). In the latter, experiments on incompressible Navier–Stokes problems shed some light on the difficulties involved with differential–algebraic systems.

4.3.4 *LSODES*

The LSODES package solves explicit systems $\dot{y} = f$, but treats the Jacobian matrix J as a general sparse matrix in the stiff case. LSODES was written jointly with Sherman (Exxon Production Research Co.), and supersedes a sparse variant of GEAR called GEARS (Spellmann and Hindmarsh 1975; Sherman and Hindmarsh 1980). In LSODES, linear systems are solved using parts of the Yale Sparse Matrix Package (YSMP) (Eisenstat, Gursky, Schultz, and Sherman 1977a, b). This involves several phases:

(i) Determination of sparsity structure. This is either inferred from calls to the f routine, inferred from calls to a J routine (if one is supplied), or supplied directly by the user. A user input flag determines which is done.

(ii) Determination of pivot order. Diagonal pivot locations are chosen, and the choice is based on maximizing sparsity. This is done by YSMP.

(iii) Symbolic LU factorization of the matrix $\boldsymbol{P}$. This is based only on sparsity and the pivot order, and uses the module in YSMP designed for nonsymmetric matrices with compressed pointer storage.

(iv) Construction of $\boldsymbol{J}$. This can be done internally by difference quotients, or with a user-supplied routine. In the difference quotient case, the number of f evaluations needed is kept to minimum by a column grouping technique due to Curtis et al. (1974). In the other case, the user-supplied routine provides one column of $\boldsymbol{J}$ at a time, in the form of a vector of length N (although only non-zero elements need be computed and stored), so that users need never deal with the internal data structure for $\boldsymbol{J}$ and $\boldsymbol{P}$. In any case, $\boldsymbol{J}$ is stored internally in an appropriate packed form. Evaluations of $\boldsymbol{J}$ are done only occasionally, as explained below.

(v) Construction of $\boldsymbol{P} = \boldsymbol{I} - h\beta_0\boldsymbol{J}$. In contrast to GEARS, LSODES does not force a re-evaluation of $\boldsymbol{J}$ whenever the existing $\boldsymbol{P}$ is deemed unsuitable for the corrector iterations. Instead, when the value of $\boldsymbol{J}$ contained in the stored value of $\boldsymbol{P}$ is likely to be usable (and $\boldsymbol{P}$ is not, only because $h\beta_0$ has changed significantly), then a new matrix $\boldsymbol{P}$ is constructed from the old one, with careful attention to round-off error. This cuts down greatly on the number of $\boldsymbol{J}$ evaluations necessary.

(vi) Numerical LU factorization of $\boldsymbol{P}$. This is done by YSMP in sparse form, and the array containing $\boldsymbol{P}$ is saved in the process (this allows for updating $\boldsymbol{P}$ as described above). Because of the absence of partial pivoting for numerical stability, this operation can conceivably fail. However, this has only rarely been observed in practice, and if it does occur (with a current value of $\boldsymbol{J}$), the step size h gets reduced and the problem disappears.

(vii) Solution of $\boldsymbol{Px} = \boldsymbol{r}$. This is done by YSMP using the sparse factorization of $\boldsymbol{P}$. Because a modified Newton iteration is used, many values of $\boldsymbol{r}$ (i.e. many linear systems) can arise for the same $\boldsymbol{P}$, and the separation of the various phases takes advantage of that fact.

The first three phases, and part of the fourth (column grouping for difference quotients), are normally done only at the start of the problem. However, the user can specify that the sparsity structure is to be redetermined in the middle of the problem, and then these operations are repeated.

Actually, the matrix operated on by YSMP is $\boldsymbol{A} = \boldsymbol{P}^{\mathrm{T}}$, not $\boldsymbol{P}$, because $\boldsymbol{P}$ is generated in column order while YSMP requires the matrix to be described and stored in row order. This causes no difficulty, however because YSMP includes a routine for solving the transpose problem $\boldsymbol{x}^{\mathrm{T}}\boldsymbol{A} = \boldsymbol{r}^{\mathrm{T}}$ (which is equivalent to $\boldsymbol{P}\boldsymbol{x} = \boldsymbol{r}$) as well as for the direct problem $\boldsymbol{A}\boldsymbol{x} = \boldsymbol{b}$.

A package called LSODIS, similar to LSODI (for the $\boldsymbol{A}\dot{\boldsymbol{y}} = \boldsymbol{g}$ problem) but using YSMP for general sparse treatment of matrices as in LSODES, is in the process of being written.

4.3.5 *LSODA*

LSODA is a variant of LSODE of yet another kind. It was written jointly with Petzold (Sandia-Livermore), and switches automatically between nonstiff (Adams) and stiff (BDF) methods, by an algorithm developed by Petzold (1980). (The suffix *A* is for Automatic.) Thus it is more convenient than LSODE for users who would rather not be bothered with the issue of stiffness. Also, it is potentially more efficient than LSODE (when used with a fixed method option), when the nature of the problem changes between stiff and nonstiff in the course of the solution. In place of the method flag parameter of LSODE, the user of LSODA supplies only a Jacobian-type flag. The storage space supplied to the solver can be either static (and thus allow for either the stiff or nonstiff method), or dynamic (and altered each time there is a method switch, to an amount specified by the solver).

4.3.6 *LSODAR*

LSODAR combines the capabilities of LSODA with a rootfinder. It allows one to find the roots of a set of functions $g_i(t, \boldsymbol{y})$ of the independent and dependent variables in the ODE system. Thus, for example, it could be used in a particle-tracking problem to determine when a particle path reaches any of the walls of a container. LSODAR was also written jointly with Petzold based on an algorithm developed by Hiebert and Shampine (1980) of Sandia-Albuquerque. The user must supply, in addition to the LSODA inputs, a subroutine that computes a vector-valued function $\boldsymbol{g}(t, \boldsymbol{y}) = (g_i, i = 1, 2, \ldots, NG)$ such that a root of any of the NG functions g_i is desired. Of course there may be several such roots in a given interval, and LSODAR returns them one at a time, in the order in which they occur along the solution, with an integer array to tell the user which g_i (if any) were found to have a root on a given return.

4.3.7 *An atmospheric model for package comparison*

In order to illustrate the various solvers (new and old) described above, and to demonstrate their relative merits on a realistic problem, we consider here an example. The problem is a simple atmospheric model (Sherman and Hindmarsh 1980) with two chemical species undergoing diurnal kinetics and transport in two space dimensions. The independent variables in the PDE system are horizontal position x, altitude z (both in kilometers), and time t (in seconds), with

$$0 \leqslant x \leqslant 20, \; 30 \leqslant z \leqslant 50, \; 0 \leqslant t \leqslant 86400.$$

The dependent variables are $c^1(x, z, t)$ = the concentration of the oxygen singlet [O], and $c^2(x, z, t)$ = that of ozone $[O_3]$ (both in moles/cm^3). The concentration of molecular oxygen $[O_2]$ is assumed constant. The equations of the model are:

$$c_t^i = (K_v(z)c_z^i)_z + K_h c_{xx}^i + R^i(c^1, c^2, t) \quad (i = 1, 2), \tag{4.1}$$

where R^1 and R^2 represent the chemistry and are given by

$$R^1(c^1, c^2, t) = -(k_1 + k_2c^2)c^1 + k_3(t)c^2 + k_4(t) \times 7.4 \times 10^{16},$$
$$R^2(c^1, c^2, t) = (k_1 - k_2c^2)c^1 - k_3(t)c^2.$$

Subscripts t, z, and x denote partial derivatives. The various coefficients are as follows:

$$K_v(z) = 10^{-8}\exp(z/5),$$
$$K_h = 4 \times 10^{-6}, \quad k_1 = 6.03, \quad k_2 = 4.66 \times 10^{-16},$$
$$k_3(t) = \begin{cases} \exp[-7.601/\sin(\pi t/43200)] & \text{for } t < 43\,200 \\ 0 & \text{for } t \geqslant 43\,200 \end{cases},$$
$$k_4(t) = \begin{cases} \exp[-22.62/\sin(\pi t/43200)] & \text{for } t < 43\,200 \\ 0 & \text{for } t \geqslant 43\,200 \end{cases}.$$

The initial conditions are

$$c^i(x, z, 0) = 10^{6i}\left[1 - \left(\frac{x-10}{10}\right)^2 + \tfrac{1}{2}\left(\frac{x-10}{10}\right)^4\right] \times$$
$$\times \left[1 - \left(\frac{z-40}{10}\right)^2 + \tfrac{1}{2}\left(\frac{z-40}{10}\right)^4\right]$$
$$(i = 1, 2),$$

and both c^1 and c^2 are required to satisfy homogeneous Neumann boundary conditions along all the x and z boundaries.

To solve the system (4.1) numerically, we apply the method of lines using a regular rectangular mesh with constant mesh spacings

$$\Delta x = 20/(M_x - 1), \quad \Delta z = 20/(M_z - 1).$$

The spatial derivatives are approximated by standard five-point central differences in each direction. The boundary conditions are similarly replaced by difference relations. The resulting ODE system $\dot{y} = f(t, y)$ has size $N = 2M_xM_z$. The initial value vector y_0 is taken from the initial condition functions given above. The system Jacobian J is sparse, with roughly $12M_xM_z = 6N$ nonzero elements. As a banded matrix, with component ordering first by species, then by x, and finally by z, it has a half-bandwidth of $2M_x$, and thus a full bandwidth of $4M_x + 1$.

As a nominal case, consider the choice $M_x = M_z = 10$. As to accuracy, a crude model of this type calls for no more than a few significant figures. To be conservative in recognizing that tolerance parameters are applied to local errors, which can accumulate into global errors, we might impose a local relative tolerance of 10^{-4}. We must also specify a positive absolute tolerance on the values of c^1 because it decays to negligible values at night. A reasonable absolute tolerance is 10^{-2}. With the ODEPACK solvers, specifying such a mixed relative/absolute error control is trivial, but with the GEAR and EPISODE families, a slight modification to the driver is necessary.

Of the various solvers mentioned, five are suitable for this particular problem—LSODE, LSODA, LSODES, EPISODEB, and GEARBI. Recall that LSODES uses a general sparse treatment of the Jacobian matrix, GEARBI uses block-SOR, and the others (in this case) treat the Jacobian as banded. The problem was set up for each of these five solvers and run on a CDC-7600 computer. For all but GEARBI, both the user-supplied Jacobian option and the internal difference quotient Jacobian option were tested. (For GEARBI, there is no difference quotient option.) The results of the various runs are given in Table 4.2. The tabulated quantities are:

R.T. = CPU run time in sec
NST = number of steps
NFE = number of f evaluations
NJE = number of J evaluations
NLU = number of LU decompositions
W.S. = total size of work space arrays

In the table, the notation USJ denotes the user-supplied Jacobian option, and DQJ denotes the internal difference quotient Jacobian option.

For the sake of illustration, the GEARBI test was repeated on a 38 × 38 grid, and the results given in the last line of the table. This is the largest square grid that could be accommodated with that solver on the CDC-7600 within its Small Core Memory (about 57,000 words). As noted, the Large Core Memory (about 400,000 words) was used for the larger actual atmospheric models. The Cray-1 computer will accept even larger problem sizes.

Several points of interest can be noted in the table, for the 10 × 10 problem. First, the number of steps does not vary greatly from solver to solver because that is determined almost entirely by the accuracy requirement, and the accuracy is much the same for all these runs. The relative merits of the solvers must be judged from other statistics.

The performance characteristics of LSODE, LSODA, and EPISODEB are similar, as expected, since they all use a banded Jacobian. The variations are partly attributable to differences in the fine tuning, but the fact that EPISODEB has the smallest NST and (with the USJ option) the shortest run time can be attributed to its use of variable-step formulae. The use of a difference-quotient Jacobian is invariably more expensive here, owing to its cost of 41 evaluations of f for each evaluation of J.

The LSODES results show that a general sparse matrix treatment gives a significant speedup over the band treatment. This results partly from the matrix software itself, but mostly from the algorithm of effectively saving old values of J for greater reuse. Note that each computed value of J is used for 36 to 46 steps, as opposed to only three to six steps with the solvers using a banded Jacobian. Also, the cost penalty for a difference quotient Jacobian is much smaller with LSODES, because each J evaluation here costs only eight f evaluations. The storage requirement is only slightly smaller, reflecting the need for sparsity information arrays

TABLE 4.2
Results of kinetics–transport test problem

Solver	R.T.	NST	NFE	NJE	NLU	W.S.
LSODE (USJ)	23.2	344	519	68	68	14 242
LSODE (DQJ)	28.4	337	3338	69	69	14 242
LSODA (USJ)	21.3	339	584	55	55	14 242
LSODA (DQJ)	24.6	339	2795	55	55	14 242
LSODES (USJ)	13.1	364	529	10	70	12 455
LSODES (DQJ)	13.5	369	602	8	72	12 664
EPISODEB (USJ)	18.6	264	461	81	81	14 400
EPISODEB (DQJ)	25.1	264	3782	81	81	14 400
GEARBI (10 × 10)	6.3	316	526	50	50	3 004
GEARBI (38 × 38)	199	393	698	96	96	43 324

and the fact the Newton matrix $\boldsymbol{P}$ is not overwritten with its LU decomposition, as it is in the banded case.

The best performance on this problem, however, is that of GEARBI. This should not be a surprise, since the Jacobian has a very regular block structure of which the block-SOR method in GEARBI is taking full advantage, both in storage and computation. The LU decompositions here are only those of the block diagonal part of the Newton matrix (with 2×2 blocks). The total number of block-SOR iterations for the 10×10 grid was 607, or an average of less than two per step. For the 38×38 grid this cost rose to 2122 iterations, or an average of 5.4 per step. The latter run also shows that spatial discretization errors in the 10×10 grid answers are as large as two percent.

For large stiff systems with wide-bandwidth coupling, of which this is an example, the single most important criterion for selecting a method or solver usually turns out to be the storage requirement. For this problem, on an $m \times m$ grid, the storage for a band-oriented solver is roughly $22m^2 + 12m^3$, while that for GEARBI is $30m^2$. The storage for LSODES is harder to predict, but behaves very roughly like $150m^2$ in the range $m = 15$ to 25. Thus LSODES has an increasing storage advantage over LSODE etc. for $m \geqslant 10$, but GEARBI has a lower storage requirement than any of them. However, if the problem fails to have the regularity needed for such an approach, or if block-SOR is not appropriate for numerical reasons, then the choices seem to be reduced to:

(i) solvers using a general sparse direct linear system solver, such as LSODES;
(ii) solvers using more suitable (possibly *ad hoc*) iterative methods (and which economize on matrix storage in some way); and
(iii) radically different *ad hoc* treatments such as operator splitting, or methods that combine full implicitness and splitting ideas.

4.4 The comparison of numerical methods for stiff ODEs (Enright)

4.4.1 *Introduction*

During the past decade several new approaches for the numerical solution of stiff equations have been proposed, and many of these have been implemented as software packages intended for general use. In this investigation we will identify some of the pitfalls inherent in attempts to evaluate the performance of stiff methods, and we will review some of the progress that has been made. We will distinguish between the comparison of different methods and the assessment of an individual method. One of our recommendations will be that while the latter is feasible and can be very worthwhile, the former is controversial and subject to many difficulties.

It must be emphasized that comparing or assessing stiff methods should not be thought of as a static activity. While it is true that attempts to compare stiff methods have not led to many definitive statements about the relative merits of various approaches, they have led to a better understanding of numerical methods and have resulted in improvements and modifications which otherwise may not have been possible.

Before discussing the evaluation and comparison of methods in detail, we will review what a stiff problem is, what comprises a stiff method, and what a stiff method is attempting to do. In doing this we will identify those features which cause the most difficulty when one attempts to compare stiff methods.

We will introduce the various components that comprise a stiff method and discuss how each can contribute to the performance of a method. Indeed, one must appreciate that when assessing a stiff method, one is not only assessing the underlying formula (or class of formulae) but also the numerous other strategies which collectively determine the method's performance.

Next we will discuss the objectives of contemporary stiff methods and how these objectives are achieved, in particular how the accuracy of a numerical method is controlled, and suggest an approach which could lead to a more meaningful and uniform interpretation of accuracy for stiff solvers. Other special features of the problems or methods which complicate the evaluation of a method will be identified. Finally, we will review some of the progress that has been made and indicate where future advances are most likely.

4.4.2 *Properties of a stiff problem*

It must be understood that stiffness is a characteristic that depends on the differential equation, the length of the integration, the initial conditions and the accuracy requirement, and that, in addition, a problem can be stiff for parts of the integration and nonstiff for other parts. We will consider a problem to be stiff if during its solution by classical methods (such as an explicit Runge-Kutta method or an Adams multistep method) the step size is severely constrained by the numerical stability of negligible transient components rather than the accuracy of the significant components. In particular, we are assuming that some component of the solution of a stiff problem must become negligible (relative to the accuracy requirement). These components generally will be significant only in the initial or transient region. Problems with high frequency undamped components are a special case and can only be considered stiff if the initial conditions are such that the amplitudes of the high frequency components are small relative to the accuracy requirement. For this class of problems the inherent cost of the integration is extremely sensitive to the accuracy requirement and the specified initial conditions—a situation that is definitely not typical of stiff problems. (For this reason some people would not consider this type of problem to be stiff.)

Stiff problems often involve large systems of equations and when they do, it is essential that any special structure be exploited. For example, the problem could have a sparse Jacobian, it could possess only a few transient components, or it could be linear. In each of these cases, methods can exploit the corresponding structure, although the extent to which it can be exploited is very method-dependent as is the ability to detect automatically the existence of such structure.

4.4.3 *Components of a stiff method*

It has long been recognized that a method for initial-value problems in ODEs consists of much more than a formula (or a class of formulae) for advancing the solution in a step-by-step fashion. It must also include a local error estimator and a step-size choosing strategy designed to control the accuracy of the integration. In methods designed for stiff systems, these latter two strategies can have a significant impact on the overall efficiency and reliability of a method, as can the iteration control strategy and the linear equation solver that is used. Since the time required to solve a stiff problem is often dominated by the time required to solve the systems of equations that arise on each time step, strategies involved in this iteration can also be critical. While most methods solve these systems of equations using a modified Newton iteration, they differ in their respective stopping criterion.

Methods which are not based on A-stable formulae often encounter difficulties when solving problems having eigenvalues of the Jacobian close to the imaginary axis. This difficulty is often compounded when the step-choosing strategy is based only on asymptotic ($h \to 0$) properties of the local error estimator. Methods that include an heuristic for detecting when numerical stability is influencing the error estimates, and which choose the step-size (and possibly the order) accordingly, can deal with this difficulty in a more robust way.

4.4.4 *Objectives of a stiff method*

The methods that have been developed for stiff problems over the past decade have tended to have similar calling sequences, but the interpretation of some of the parameters can be quite different for different methods. This is particularly true of the error-control strategy and the way in which the error is measured.

Methods inevitably attempt to control some measure of the global error (relative to the requested accuracy, TOL) by monitoring and controlling some measure of the local error on each step of the integration. While the relationship between local and global errors for stiff systems is not well understood, most methods, primarily on the basis of numerical evidence, attempt to ensure that the magnitude of the local error per step is bounded by TOL on each step. When this type of error control is used, the relationship between the global error and TOL is inherently method-dependent as the bound on the global error depends on the step sizes used. Contrary to popular belief, it is not true that control of the magnitude of the local error per step alone will guarantee control of the global error. A method which chooses a sufficiently small constant step size and generates the sequence of approximation (x_i, y_i) i = 1, 2, . . . , N) with $x_i = x_0 + ih, y_i \equiv y_0$ when applied to the stiff problem will satisfy the local error per step criterion but will produce a numerical solution of dubious value. When interpreting the accuracy of a solution produced by a method which claims only to control the magnitude of the local error on each step, a user must know more information (such as the average step size) before he is justified in having any confidence in the accuracy of the solution. Since most methods do use this type of error control, they produce solutions whose global accuracy is related to the specified tolerance in a very method-dependent fashion. If one is only interested in endpoint accuracy, this difficulty is not usually as acute, but if one wishes accuracy throughout the interval, this difficulty can make comparison especially awkward.

One way to overcome this difficulty would be to ensure that the next generation of codes is based on a compatible interpretation of the parameter TOL. Direct control of the magnitude of the local error per unit step would result in such a method-independent relationship between a global error bound and TOL, but this is known to be inefficient for stiff problems. An alternative approach would be to control the magnitude of the local error per step but to combine this with local extrapolation. Since extrapolation would alter the stability properties of the method, one would have to develop new formulae or use a lower-order predictor in the local error estimator. A more promising approach which could also lead to a method-independent interpretation of tolerance is based on a variable local error-per-unit step designed to control the magnitude of the defect of the numerical solution. If a method, when applied to the stiff problem, produces the discrete numerical solution (x_i, y_i) $(i = 1, 2, \ldots, N)$ as an approximation to $x_i, y(x_i)$ $(i = 1, 2, \ldots, N)$ and there exists a continuous piece-wise differentiable function $z(x)$ defined on $[x_0, x_N]$ interpolating (x_i, y_i) $(i = 1, 2, \ldots, N)$ such that $z' = f(x, z) + \delta(x)$, then $\delta(x)$ is the defect associated with the approximate solution $z(x)$. This point of view has been applied successfully in the analysis of nonstiff methods (see, for example, Hanson and Enright 1981), and it is hoped that it will prove fruitful in the analysis of stiff methods as well. This analysis will be complicated by the fact that to account for the very stable global error control observed in practice, one will have to consider the direction of the local error (and hence the defect) as well as its magnitude. We will return to this point in §4.3.5.

Most codes provide options which dictate how the error should be measured, and this complicates the comparison of methods as the options provided by different methods often are not compatible. Pure absolute error control is the simplest, but it can be inappropriate for stiff problems as they often have solutions with components of widely different magnitude. Component-wise relative error has been suggested for stiff problems, but control of local error in a relative sense does not imply that the relative global error is being controlled, and indeed, it is very difficult to say what effect this has on the global error.

On the other hand, if one interprets global accuracy from the defect point of view, then component-wise control of the local error relative to the magnitude of the derivative would lead to a defect that is small relative to $\|f(x, z)\|$. Shampine has recently argued (Shampine 1981a) that for most problems a component-wise weighted absolute error control is appropriate with the weights depending on the solution. As he points out, this makes it awkward for users to choose the appropriate weights without a few preliminary integrations, but for test problems this is probably the best error measure to use.

4.4.5 *Other complications*

In addition to the difficulties discussed above that arise because of inherent properties of stiff problems, stiff methods, or the error-control strategy, there are other difficulties that can arise which complicate any attempt to compare stiff methods. Some stiff problems have an unstable manifold of solutions close to the solution that is being sought. For example, it is well known that in chemical kinetics the problems are stable and well conditioned, provided the concentrations remain nonnegative; but a problem can become mathematicaly unstable if a negative concentration is introduced.

During a numerical solution it is possible that a small negative concentration (relative to the accuracy requirement or the round-off level) will be introduced and this can lead to an unstable trajectory being followed (Edsberg (1976) discusses this difficulty in more detail). Although some methods may be more prone to this difficulty than others, it is not clear that it should be considered a fault of the method as long as the specified error control strategy is being respected. Certainly one instance of this difficulty can distort a comparison of methods.

Recently some new techniques have been proposed for which even linear analysis is difficult. The familiar stability definitions such as A-stability and A(α)-stability are based on the assumption that analyzing the scalar equation $y' = \lambda y$ for each eigenvalue λ of the matrix $\boldsymbol{A}$ will allow one to characterize the stability and propagation of local errors for the stiff system $\boldsymbol{y}' = \boldsymbol{A}\boldsymbol{y}$. This assumption is valid for most stiff methods since the nonsingular similarity transformation which reduces $\boldsymbol{A}$ to its Jordan canonical form will also "uncouple" most methods and reduce the original problem to an equivalent sequence of scalar problems (of course the equivalent error measures will change, but for a given sequence of step sizes and a fixed formula the solutions will be equivalent). Some recent approaches such as those based on nonlinear combinations of the derivative (see, for example, Lambert 1974) or those that use different formulae for different components of the system (see, for example, Hofer 1976 or Soderlind 1980) do not satisfy this assumption.

When a method does satisfy this assumption, then it is clear that when one is in the transient region $|h\lambda_i| \ll 1$ for eigenvalues λ_i corresponding to the smooth components, there will be virtually no error (except for roundoff) in the direction of these components, with the local error almost entirely in the subspace spanned by the transient components. Similarly, after the transient region is passed and the method is in the smooth region, we should have $|h\lambda_j| > 1$ for eigenvalues λ_j corresponding to the negligible transients of the numerical solution. For these components stability is ensured and their magnitude will remain negligible relative to the accuracy requirement, and hence the local error must be mostly in the direction of the smooth components. It is not obvious that this will also be the case for methods which do not satisfy this assumption.

Some methods require an analytic Jacobian while others can use either an analytic Jacobian or an approximate Jacobian based on numerical differences. Even when an approximate Jacobian is allowed, the performance of a method can be very sensitive to the step size used in determining the approximate Jacobian (Gaffney 1982a), and this can be very disconcerting.

4.4.6 *Progress that has been made*

Any comparison of stiff methods will be based on monitoring the performance of methods as they solve a representative collection of stiff problems over a range of tolerances. It is clear that one should include measures of reliability as well as efficiency when describing the performance of a method. It is also clear that there should be several measures of efficiency available. The relative importance of these efficiency measurements will depend, among other things, on the size and complexity of the differential equation. From our discussion of the "objective" of a stiff method, it should be apparent that the choice of an appropriate reliability measure is a controversial issue and while a few possibilities exist, none is entirely satisfactory.

In our early comparisons of stiff methods (Enright, Hull, and Lindberg 1975), we gathered a collection of test problems, proposed measures of reliability and efficiency, and reported a comparison of several modified versions of existing methods. The modification to existing methods was made necessary because of the variable local error per unit step control we assumed in order to justify our choice of reliability measure. It was clear to us that while this type of testing could be used to identify weaknesses of a particular method or to compare modified versions of the same code, it was not possible to identify the "best" method in any sense. Our primary goal in that study was to establish a testing methodology which could be used to compare the performance of methods on any class of stiff problems. It is rather unfortunate that the main impact of that investigation seems to have been that the test problems we used have been used as a representative set of stiff problems in several subsequent investigations. This was not our intention as the problem set was proposed more to illustrate and justify the methodology we were introducing. Shampine (1981a) has made several suggestions and corrections which certainly would be necessary if these test problems were to be used as a standard test set.

After our initial testing it became clear that while comparisons of methods was very difficult and controversial (since it involved modifying methods), the assessment of an individual method was feasible and much less controversial. In Enright (1980) we reported on the use of testing tools we had developed to assess automatically an unmodified stiff method on any class of stiff problems. The measurements of reliability were method-dependent, and various levels of detailed statistics could be provided. This package of testing tools has been widely distributed, and results obtained using this package for a selection of contemporary stiff methods are summarized in Addison (1982). As Addison notes, there are still few methods which operate efficiently at stringent tolerances (essentially only those based on variable-order multistep formulae), while at larger tolerances several good methods are available. It is also clear from Addison's result that a robust implementation of the BDF formulae such as the well known code GEAR (Hindmarsh 1974a) is still the benchmark against which new codes should be compared. Its main known weakness is on problems with eigenvalues of the Jacobian near the imaginary axis, and for this class of problems several acceptable alternatives are available.

For the future, we feel there are three important applications where the detailed evaluation of a stiff method's performance will prove invaluable. The first is when one has a class of problems of a special type and wishes to choose an appropriate method for dealing with them. An interesting example of such a study is that of Gaffney (1982a), in which he reports on the performance of a selection of methods on a particular stiff oscillatory problem.

The second application involves the comparison of different versions of the same basic code or methods with very similar strategies and objectives on general stiff problems. The former situation arises frequently when one is developing a method, and an example of the latter is the comparison of GEAR and EPISODE (Bryne, Hindmarsh, Jackson, and Brown 1977). The third application is in connection with quantifying the costs involved and the potential advantages in exploiting structure when solving problems of a special type. As an example of such an application we have used (Kamel and Enright 1980) the testing tools described above to quantify the advantages possible when one automatically partitions a problem with only a few transient components and subsequently exploits this partitioning in the linear algebra modules.

In summary, we feel that while a comparison of stiff methods for general stiff problems is not feasible, one can compare methods for a special type of stiff problem or versions of similar methods for general stiff problems with far fewer difficulties. Automatic testing tools and the collection of standard test problems can make this a straightforward, painless process which could prove essential in many applications.

4.5 The evolution of an ODE solver in an industrial environment: the solution of several representative stiff problems (Thompson and Tuttle)

The purpose of this section is to describe the initial selection and subseqeunt modifications of software for the automatic solution of ODEs in an industrial environment. The selection and modifications were in direct response to the types of problems to be solved, the staff, the hardware, and the interface requirements. Although one particular environment is at issue, these considerations are applicable to other industrial installations.

4.5.1 *Description of environment*

The NPGD Computer Center includes interconnected CDC CYBER 76, CYBER 750, and CYBER 720 computers. The CYBER 720 and CYBER 750 process all of the interactive work and act as a system interface for all of the batch processing. Most of the scientific processing is done on the CYBER 76 and the CYBER 750. The CYBER 76 has a 27.5 nanosecond clock period with 64 000 sixty-bit words of small core memory, having a 220 nanosecond access time per word. It also has available 256 000 sixty-bit words of large core memory, having a 687 nanosecond access time per octet of words. The CYBER 750 has a 25 nanosecond clock period with 198 000 sixty-bit words of memory, having a 575 nanosecond access time per word. Clearly, these systems are relatively large and fast. These facts should be kept in mind during the discussion of the difficulties that arise in terms of required storage and overall execution time.

4.5.2 *Description of problems*

The problems to be solved are primarily related to the design and analysis of nuclear reactors. The physical model is usually developed by the engineers, although it is becoming more frequent for this to be done in conjunction with the mathematicians. It is not uncommon to encounter problems involving time simulations ranging from a millisecond (§4.5.5) to hundreds of seconds (§4.5.6), or with solution components ranging in magnitude from 10^{-10} to 10^{10} in the same problem. Furthermore, it is common to encounter problems which, in spite of the simplifications in the physical model, remain extremely difficult from a numerical point of view. Indeed the simplifications involved in a crude model often render the simplified model much more difficult numerically. The software has been used to solve a wide variety of such problems. Table 4.3 contains a summary of several representative problems. These problems are relatively straightforward in the sense that they represent initial-value problems in the standard form

$$y' = f(t, y)$$
$$y(t_0) = y_0,$$

and do not possess any noteworthy special features (e.g. severe discontinuities in f). They are, therefore, amenable to solution by many of the available high-quality solvers (Thompson 1977). In fact, several such solvers are routinely used to solve the problems in this first category, both on a production and nonproduction basis.

A second category includes those in nonstandard form which possess unusual requirements and which are, therefore, not amenable to solution by previously available standard software. The most frequently encountered are: the equations of fluid flow, heat diffusion in a cylinder, and neutron kinetics equations. In each case, one-dimensionality and other simplifying assumptions are frequently introduced. Although the individual problem types are well known, several difficulties arise when they are solved simultaneously. The computational difficulties associated with the solution of these coupled systems are primarily related to the detail and number of equations, the complexity of the auxiliary computations involved in the calculation of derivatives, and the occurrence of special events which radically alter the character of the equation set (e.g. changing from stiff to nonstiff, changing the number of equations, or introducing derivative discontinuities).

It is this second category of problems which directly influenced the software evolution. Table 4.4 contains descriptions of four representative problems from this category. The descriptions include the special problem requirements which dictated the software development and other pertinent information for each problem. Each of these problems actually represents the basis for a corresponding large-scale software project. Therefore, the times during which the corresponding

TABLE 4.3
Summary of representative standard problems

Description of physical problem	Mathematical problem requirements
Coupled heat exchange performance	Ability to switch from a system of nonstiff equations to a different system of stiff equations.
Interacting species—point kinectics equations	Integration of small systems of extremely stiff equations.
Structural mechanics modelling of reactor performance during earthquake conditions.	Integration of rapid oscillatory transient followed by a long steady-state integration.
Finite-element models arising in structural mechanics problems.	Integration of large banded systems of equation (100–2000 equations).
Quench-front type problems.	Integration of large systems of non-stiff equations with mild derivative discontinuities (100–500 equations).
Analytic continuation. Dependence of the solution of systems of nonlinear equations depending on a parameter.	Ability to track solutions of a system of equations containing an imbedded parameter. Simple root-finding requirements.
Analysis of reactor vessel internal loads.	Integration of an extremely large (500–9000 equations) system of nonstiff equations.
Analysis of steam generator performance.	Integration of large banded extremely stiff equations (100–1000 equations). Root-finding to locate phase change boundaries.

project developments took place are also included. This places in perspective the specific ODE software development. Each of these projects represents an ongoing effort. The entries in the table correspond to the times when the bulk of the work was performed. Also included are indications of the complexity of the computer coding associated with the calculation of system derivatives. It should be noted that the number of lines of coding does not account for the additional use of a large mathematical software library of auxiliary subroutines. Use of these routines can effectively add anywhere from a few hundred to several thousand lines of additional coding.

4.5.3 *Core reflood analysis*

The performance of the reactor coolant system during the refill and reflood phases of a postulated loss-of-coolant accident is to be simulated. At the end of blowdown (i.e. the rupturing of a coolant pipe), the water level in the reactor vessel is usually calculated to have dropped below the elevation of the bottom of the core, leaving the core immersed in saturated or superheated vapor. As the emergency core coolant is injected into the system, the water level in the vessel lower plenum rises and eventually the bottom of the core is recovered. A set of one-dimensional fluid flow equations coupled to a simplified heat equation is used to model this process, thus resulting in a system of 100–150 problem-dependent ODEs. During the initial phase of the transient (before the water level rises to the bottom of the core), the problem is both nonstiff and relatively easy to solve. When the water initially comes in contact with the bottom of the core, additional equations are activated, derivative discontinuities occur in several components, and an extremely rapid transient ensues with the accuracy restrictions dominating the stability restrictions. As the core recovery proceeds, a steady-state condition is approached. Finally, the core flood tanks that injected the emergency core coolant empty and must be turned off.

The first difficulty that arises for the automatic software is in precisely locating the time at which the water strikes the bottom of the core. This point must be properly located in order to generate an accurate solution and also to avoid chattering in the integrator (Carver 1977; Shampine 1979c). A second difficulty is associated with the efficient switching from nonstiff to stiff solution techniques as core recovery proceeds. When the switch is made, it is also desirable that the integration methods have enlarged stability regions in order to handle the problem. Finally, the time corresponding to the emptying of the core flood tanks must be properly accounted for to prevent the errors in auxiliary computations from swamping the integrator's error control, which would result in excessive execution times.

4.5.4 *Reactor pressurizer*

Pressurized water reactor systems (Ransom 1976) normally use a steam pressurized surge tank to maintain and control primary system pressure. This surge tank is commonly referred to as a pressurizer. During normal operation, the upper portion of the vessel contains pressurized steam and the lower portion contains liquid. A decrease in the primary system's coolant pressure causes an outsurge of water from the pressurizer to the primary system, which compensates for any cooling or the removal of the primary system coolant. An increase in the primary system's coolant pressure causes an insurge of primary coolant into the pressurizer, which compensates for any heating or the injection of primary coolant. The pressurizer is maintained at a reasonably constant pressure through the use of heaters to raise the pressure, sprays to lower the pressure by condensing steam, and relief valves to rapidly lower the pressure. The fluid model consists of three regions which normally include: superheated steam, saturated mixture, and subcooled liquid. Various combinations of these regions are in existence at any given time, and a given region can be created or deleted in the course of a transient. The model consists of ODEs derived from the conservation of mass equations for the liquid, equations defining the rise of the bubbles, and equations describing the behavior

TABLE 4.4

Summary of representative nonstandard problems

Description of physical problem	Description of special problem requirements	Problem Characteristics			Chronological time in which the problem was solved
		Lines of active coding in DFN	Avg. time req'd to evaluate DFN (sec)	Typical execution time (sec)	
Modelling of reactor coolant system during refill and reflood stages of a postulated loss-of-coolant accident.	Coupled fluid and heat conduction equations. Initially nonstiff. Extremely rapid transient following a severe discontinuity. Mildly stiff approach to steady-state	2500	0.010	200	1977–1978

Modelling of reactor pressurizer system.	Coupled systems of differential-algebraic equations. Occurrence of special events.	2300	0.040	100	1979
Modelling of the growth of vapor bubbles in a time-dependent pressure field.	Coupled systems of ODEs and PDEs. Nonstandard coupling at a moving boundary. Rapid transient followed by gradual approach to steady state.	1300	0.005	50	1979-1980
Modelling of anticipated reactor response during a system transient resulting from a postulated upset condition in the primary system.	Coupled fluid, heat conduction, and kinetics equations. Steady-state initialization. Severe derivative discontinuities. Special trips & switches.	3000	0.050	75	1978-1981

of the heaters. This forms a system of fourteen ODEs. There are also many auxiliary computations related to the interface conditions between the regions and to condensation.

The problem is small, nonstiff and for the most part, relatively easy to solve. It is included here for several reasons. The first is that without an extremely accurate and consistent set of water property tables it is impractical to solve this problem purely as a system of ODEs. Rather it is desirable to recast the system into one consisting of a coupled system of ODEs and nonlinear algebraic equations. The next important feature is that the sprays, heaters, and valves form a set of nineteen special events that must be correctly accounted for in order to generate an accurate solution (each of these events is a function of the solution). Therefore, there are more special events than ODEs! Finally, it is necessary to restart the integration following most of these special events, in such a way as to avoid missing one of the other events.

4.5.5 *Bubble growth in a time-dependent pressure field*

Since it is conceivable that a reator coolant undergoing rapid decompression during a loss-of-coolant accident could become highly superheated before flashing, several well-known blowdown experiments have been performed relating to this question. In these experiments, the process is modelled by a system of coupled PDEs and ODEs consisting of the heat conduction equation (for the thermal layer enveloping a vapor bubble), the equation of motion (to describe the growth of a vapor bubble), and a description of the state of the surrounding bulk liquid (Aguilar and Thompson 1982). Since the subsystems are coupled at the bubble surface, a moving boundary problem must be solved. Furthermore, the coupling is algebraic in some cases. This model has been used to correctly analyze well-known bubble growth measurements in a decreasing pressure field, bubble growth in constant-pressure water, and novel chamber-and-piston experiments. Depending on the problem to be analyzed, the resulting system of ODEs contains 20–300 equations. Initially the problem is nonstiff. It is necessary to then determine precisely the time at which the bubble forms and begins to grow—which is again a function of the solution. At this point, the microscopic flashing model is activated—resulting in a very rapid transient, followed by a gradual approach to steady-state as pressure recovery proceeds.

The first difficulty that arises for the automatic software is the determination of the time corresponding to the onset of bulk nucleation. The next difficulty is related to the algebraic coupling at the bubble surface which results when the microscopic flashing model is activated. A third difficulty is related to the necessity to switch to stiff solution techniques to accommodate the mild stiffness associated with the gradual pressure recovery. The problem is solved by a method-of-lines approach in which any one of several transformations is used both to immobilize the moving boundary and to scale the variables. Depending on the various transformations used, root-finding involving the old independent variable and the new dependent variable is required for locating the output points (Shampine and Gordon 1975).

4.5.6 *Simulation of upset transients*

This problem simulates the performance of the reactor coolant system during transients caused by minor upset conditions (e.g. loss of off-site power, or loss of main heat-sink). Since all of the transients are relatively slow, a simplified mathematical model is used which consists of the following coupled subsystems:

(i) the one-dimensional cylindrical heat equation,
(ii) a simplified set of one-dimensional flow equations,
(iii) the time-dependent point kinetics equations, and
(iv) several auxiliary subsystems (as in §4.5.5).

The total system contains approximately 150 ODEs and is usually considered stiff due to the presence of subsystem (iii). The system is initialized by holding all boundary conditions constant and integrating until steady-state is achieved. Depending on the transient to be simulated, a time-dependent forcing function is then activated. One common simulation involves the step insertion of reactivity into the point kinetics equations. Owing to the system coupling, this effectively introduces a severe jump discontinuity in the time derivative. This discontinuity is quickly transmitted throughout the rest of the system, resulting in a radical change in the nature of the various subsystems (e.g. changes in the mode of heat transfer from the fuel pin to the reactor coolant). It is also necessary to incorporate approximately a dozen reactor trip functions in the model (e.g. system pressure too high or too low, outlet temperature too high, or outlet flow rate too low). Prior to the trip, the problem is stiff, but relatively easy to solve. After trips occur, portions of the system remain stiff and other portions undergo rapid transitions. Again it is imperative to precisely locate the times of the trips in order to generate an accurate solution. Finally as the transient proceeds, it again approaches steady state.

4.5.7 *Description of software*

In Table 4.5 we have indicated the general chronology for our software development. In addition to these items, we also wish to discuss several other issues:

(i) the initial assessment and selection of an integrator,
(ii) the addition of generalized Adams-type methods,
(iii) the addition of the capability to solve systems of coupled differential and algebraic equations,
(iv) the addition of root-finding capabilities,
(v) the addition of diagnostic tools, and
(vi) user support.

4.5.8 *Initial selection*

The specific software which was evaluated consisted of Runge–Kutta software, nonstiff Adams software, stiff collocation software, and several variants of GEAR and EPISODE. Several other commercially available solvers were also evaluated, but were excluded since they were not competitive. Fortunately, several previous evaluations had been performed elsewhere, notably Shampine, Watts, and Davenport (1975) and Enright, Hull, and Lindberg (1975). These previous evaluations significantly influenced our assessment. Subsequently, several of the above solvers

TABLE 4.5
Chronology of the software development

Activity	Time
Initial assessment and selection	1975–1976
Assessment and testing of GEARS.	1976
Dynamic dimensioning, subroutine standardization, and other related modifications.	1976–1977
Incorporation of extrapolatory root-finding.	1977–1978
Incorporation of Adams-type methods.	1978
Incorporation of ability to solve systems of coupled differential–algebraic systems.	1978–1979
Incorporation of extensive diagnostics and user-oriented debugging devices (e.g. auxiliary drivers).	1979–1980
Incorporation of interpolatory root-finding.	1980–1981
Incorporation of refinements in the root-finding capabilities.	1981

were selected for inclusion in our Standard Mathematical Subroutine Library (Barinka 1977). As indicated by Table 4.3, these solvers are still used in applications at B & W. Furthermore, the capabilities of DVDQ (Krogh 1976) have influenced the portion of the development of our software related to the addition of root-finding. However, it was decided that in our environment GEAR and EPISODE were the most appropraite candidates for a general-purpose solver.

Due to the sizes of the problems listed in Table 4.4, we found it difficult to use GEAR and EPISODE (recall that a large amount of available computer memory is usually required to evaluate system derivatives for our problems). It was also usually not feasible to use their banded counterparts GEARB and EPISODEB since the coupling of fluid (in a loop), heat, and kinetics equations results in a Jacobian that does not have a banded structure, or even a nearly banded structure in which elements outside a given band may be neglected. Therefore, it was decided to use the sparse solver GEARS as the base for our general-purpose solver. The present software evolved specifically from Spellman and Hindmarsh (1975).

4.5.9 *Adams-type methods*

It was noted in Example §4.5.3 that the problem becomes mildly stiff as the transient proceeds. It was found that the third- and fourth-order stiffly stable Adams-type methods in Rodabaugh and Thompson (1977), when used with a Newton iteration, handle the problem more efficiently than do the standard Adams methods or BDF methods. Because they were constructed as linear combinations of standard Adams methods, they were also easily incorporated into GEARS.

4.5.10 *Coupled differential–algebraic systems*

Systems of the form

$$y' = f(t, y, w), \quad y(t_0) = y_0 \tag{4.2}$$

$$0 = r(t, y, w), \quad r(t_0, y_0, w_0) = 0, \quad w(t_0) = w_0 \tag{4.3}$$

arise naturally in many applications. One accepted approach is to differentiate (4.3) with respect to t to obtain an initial value problem in standard form. As noted in §4.5.4 and §4.5.6, this may not be feasible since an explicit expression for (4.3) may not be available and the use of finite differences may be too inaccurate. Another approach is to solve (4.3) as part of the function (4.1). This frequently proves impractical for several reasons. The first is caused by the interaction of the convergence error in the iteration with error control in the integrator. If the convergence tolerance is too small, the program's execution time is high (due to the iteration); whereas if the tolerance is too large, the integrator error test fails repeatedly, leading to prohibitively small step-sizes and resulting in high execution times as well as possibly erroneous results. The second drawback is that the evaluation of r may require that most of the computations required for f be completed. This again leads to extremely high execution times. A third drawback is that questions related to problem stability can be masked by indirect solution techniques (§4.5.4). In order to circumvent these difficulties, the software was modified to directly solve these types of systems. This is a relatively straightforward modification since a Newton iteration is already available for the solution of the corrector equations. All that is required is to modify the usual iteration matrix

$$P = I - h\beta \frac{\partial f}{\partial y}$$

to

$$P' = \begin{bmatrix} I - h\beta \dfrac{\partial f}{\partial y} & \dfrac{\partial f}{\partial w} \\[2ex] \dfrac{\partial r}{\partial y} & \dfrac{\partial r}{\partial w} \end{bmatrix}$$

The predictor step is performed in the usual fashion for y while an algebraic extrapolation of the same order and based on back-points is used for w.

It is recognized that while there does exist some theory pertaining to the existence and uniqueness of solutions to (4.2) and (4.3), there does not at this time exist adequate justification for the numerics. However, the actual performance has been very satisfactory. The procedure has been used to solve problems accurately that otherwise could not have been solved. Furthermore, the integrator has failed with an error message each time we have attempted to solve a problem for which a solution is known not to exist. Consequently, we have found this capability to be a very useful one.

4.5.11 *Root-finding*

Variants of this capability previously have been used in other software, notably Carver (1977) and Gladwell (1979). Our approach was most influenced by Krogh (1970) where the procedure is referred to as g-stops. This terminology will also be used throughout this section.

The overall idea is to construct an auxiliary function whose zeros correspond to the occurrence of the special event. For instance, consider the problem introduced in §4.5.3. When the water level in the lower plenum reaches the bottom of the core, drastic changes in the nature of the system occur. It is necessary at this point to take special action, such as restarting the integration. Thus, the auxiliary function would be the height of the bottom of the core minus the height of the water. Hence, the occurrence of this special event corresponds to a zero of the auxiliary function. When the integrator locates this zero, the integration is then restarted with an appropriately small step size to accommodate the ensuing rapid transient. This same approach can be used in §4.5.4 for turning heaters on and off, or for opening and closing relief valves.

In general, we wish to locate the zeros of the equations

$$0 = g_i(t, y, y'), \quad i = 1, \ldots, k$$

simultaneously with the integration of the ODE. In each case, once an interval is located in which one of the components changes sign, a root-finder is used to locate the corresponding zero. A modification of Brent's one-dimensional algorithm (Brent 1971) is used to perform the root-finding in the present software.

The root-finding is of two types, interpolatory and extrapolatory. Interpolatory g-stops are used when it is possible to integrate past the time corresponding to the special event. For example, this would be the case if the event corresponded to a simple switch with an associated time delay to obtain output, or when it is used to locate times at which a given variable attains a particular value.

After the completion of each successful step (following the corrector iteration), the Nordsieck array is extrapolated backward over the interval corresponding to this step to examine for sign changes and to locate roots, if necessary. In this case, it is not necessary to restart the integrator unless, of course, the root corresponds to a severe discontinuity or requires that other special action be taken.

Extrapolatory g-stops are primarily intended for a totally different purpose. Generally, they are used to cope with severe derivative-discontinuities associated with special events—which frequently occur in the problems we must solve—for example, when the water reaches the bottom of the reactor core. Experience with these problems indicates that several commonly recommended techniques for dealing with such discontinuities do not always work well. For example, simply ignoring the discontinuity is usually disastrous. If it is severe enough, the presence of the discontinuity will at best result in excessive integrator-chattering, leading to extremely high execution times. Furthermore, the special event often necessitates the addition or deletion of portions of the model (e.g. the creation or disappearance of a steam region in §4.5.4), making it impossible to ignore the event. Another technique which is sometimes recommended is to complete the current step and then allow the changes corresponding to the special event to be made. Depending on the nature of the problem prior to the occurrence of the special event, the step size beng used is often much too large, resulting in either a program abort or inaccurate results.

Extrapolatory g-stops correspond to using the predictor to detect sign changes over the next step. In the present software, this is accomplished by extrapolating forward before each integration step is performed (i.e. before the corrector iteration is initiated). Since only the predictor is used, the discontinuity can be located without requiring any derivative evaluations for points past the discontinuity. The integration is then restarted with the appropriate changes in order and step size necessary to "transcend" the discontinuity.

Extrapolatory g-stops are generally less accurate than interpolatory g-stops since they are located using only the predictor. In fact, it is possible to miss roots of the auxiliary function if only extrapolatory g-stops are used. This can occur, for example, if the auxiliary function does not change sign following the prediction, but does change sign following a correction. In problems such as in §4.5.6, it is imperative that the times corresponding to all trips and switches be correctly located since the system response can be drastically altered otherwise. Therefore, in our implementation, extrapolatory root-finding is automatically backed up by interpolatory root-finding. While the possibility stills exists that a g-stop might be missed (the corrector polynomial simply may not reflect the sign change in the auxiliary function), we have not encountered a situation where a missed extrapolatory g-stop is not picked up by the back-up interpolatory calculations. Furthermore, the increased reliability of the software more than adequately compensates for the slight increase in overhead.

Scaling the auxiliary function is also related to the use of g-stops. It is usually necessary that the function be scaled appropriately in order that the root-finder's error-control be compatible with that of the integrator. While it is possible to leave the responsibility for such matters to the user, we consider it preferable to automatically accommodate this matter in the software.

4.5.12 *Diagnostics*

Since the evaluation of the derivatives may involve several thousand lines of executable code (e.g. the entire program for the relatively small problem in §4.5.5 comprises approximately 17 500 lines of coding), it is imperative that effective diagnostic tools be available. The tools described here evolved as the problems in Tables 4.3 and 4.4 were implemented and debugged. They represent devices which were found to be convenient and, in some cases, indispensible. While they probably do not encompass all devices that might be useful, they do give an indication of what is useful. The following is a brief description of each of several capabilities which are included as debug options in the software.

(i) Identification of troublesome components. Frequently, only a few components are causing the error test or the corrector iteration to fail repeatedly. An option is included to automatically identify these components by using a max norm and printing the number of each equation that caused failure.

(ii) Identification of oscillatory components. Model deficiencies or programming errors are often manifested as oscillatory components. Therefore, an option is included to detect a sign change in the derivative between successive steps and to print the components for which this occurs.

(iii) Checking the derivative function evaluation. Given the complexity of the problem, evaluation of some derivatives may not be defined, particularly during initial debugging. To detect this, the y' vector is preset to negative indefinite before the evaluation is called. After the function evaluation, y' is checked; if any components have not been defined, the component is identified and the integration is halted. A related problem is that from the use of fixed-step, first-order techniques, users have grown accustomed to naively resetting the y-vector if it falls outside of some desired physical range (e.g. resetting of negative flow rates). Needless to say, this usually has a detrimental effect on the integrator. In order to detect this, integer sums of the y-vector are formed before and after the function evaluation and on exits from and entries to the integrator. If the sums do not agree, an error message is issued and the integration is halted. This procedure requires only four additional storage locations and easily detects an error that otherwise is extremely difficult to locate.

(iv) Halting divergent corrector iterations. It has been observed that when the Newton iteration diverges, it does so quite rapidly. This can lead to abnormal terminations in the function evaluation (e.g. a pressure out of range in water property tables). To avoid this, a divergent corrector iteration is immediately halted.

(v) Printing the Jacobian. Some programming errors, such as the use of variables from a previous time-step, are easily detected if a "picture" of the Jacobian is printed. An auxiliary subroutine is provided which prints 1 for each nonzero value of the Jacobian and zero for the zero values.

(vi) Statistical information related to integrator performance. A good deal of other information regarding integrator performance is also collected, although it is only indirectly available to the user. This information is often of substantial interest to the numerical analyst, however. For example, counts of the number of order increases and decreases can give an indication of order cycling, which is often associated with model errors or instabilities. Other useful information includes the number of steps taken at each order, the number of integrator crashes, the number of error test failures, and the number of corrector convergence failures.

While most of the above devices are simple and relatively straightforward, they often mean the difference between making or missing a delivery date (and hence, whether or not the software will continue to be used by users who are already suspicious of a black box). They also relieve the numerical analyst of some of the burden of debugging a complicated program by providing meaningful diagnostics to the programmer and the engineer. As a result, these aids increase the user's confidence in the software by overcoming to some extent his fear of its black box nature.

4.5.13 *User support*

The typical scientific programmer or engineer is not usually equipped to utilize effectively a piece of numerical software as sophisticated and complex as an automatic integration package. To ease this difficulty, simplified driver subroutines that solve initial value problems in standard form have been written. These in turn preset many of the parameters required and then call the standard integrator. This allows the user to learn and grow as the complexity of his problem increases. At any given time, he needs only to access the portions of the integrator that are required by his problem. To increase the level of mathematical sophistication of the typical user, internal training courses in numerical analysis are also offered. Finally, the diagnostic capabilities of the integrator have been expanded to provide meaningful information to the user. In spite of the fact that this does not eliminate the need for the in-depth involvement of an experienced numerical analyst in any large development effort, it does contribute to the usefulness of the software by relieving the numerical analyst of many of the routine tasks that can be performed by engineers and trained programmers.

4.6 Some experiences in the selection of integrators for large-scale systems (Chen and Schiesser)

4.6.1 *Introduction*

There is increasing interest in applying first-principle models to chemical plants, i.e. models based on fundamental conservation principles, thermodynamics, kinetics, and transport processes, which offer the possibility of a fundamental understanding of the proposed units. Generally, these models are expressed as systems of differential equations, particularly if unsteady-state operation is to be considered. Dynamic simulation is indispensable for the understanding of the stability and controllability of proposed units and plants.

When a first-principle model is formulated, it typically involves a large (greater than 100) system of nonlinear equations. Such large-scale ODE problems are demanding in terms of the selection and use of a numerical integrator. In addition to dealing with the stability problems of stiff ODEs, these integrators generally achieve high accuracy by variable-order, variable-step procedures which are automatically invoked in accordance with the requirements of the problem system equations.

Here we relate some of our experiences in integrating a system of 128 nonlinear ODEs which are formulated for the dynamic simulation of a hydrogen plant in a high-Btu coal gasification plant. The individual units in the plant model and the number of equations for each unit are listed in Table 4.6. The interconnection of the processing units is shown in Figure 4.1.

To the best of our knowledge, this is the first reported attempt to simulate dynamically an entire plant with interconnected units modelled as both discrete (perfectly mixed) systems and distributed systems. The latter lead to PDEs which are approximated by the numerical method of lines (NMOL) thereby giving a final set of ODEs for the plant model. The origin of large sets of ODEs therefore is due to: (1) the PDEs approximated by the NMOL; and (2) relatively complex units, such as the stripper–absorber columns which require sets of ODEs for a reasonably comprehensive mathematical description.

TABLE 4.6
Components of simulated plant

Unit	No. of PDEs	No. of ODEs
Reformer	1	21-point NMOL
Shift reactor no. 1	1	21-point NMOL
Shift reactor no. 2	1	21-point NMOL
CO_2 Absorber		34
CO_2 Stripper		10
Clean-up methanator	1	21-point NMOL
Total	4	128

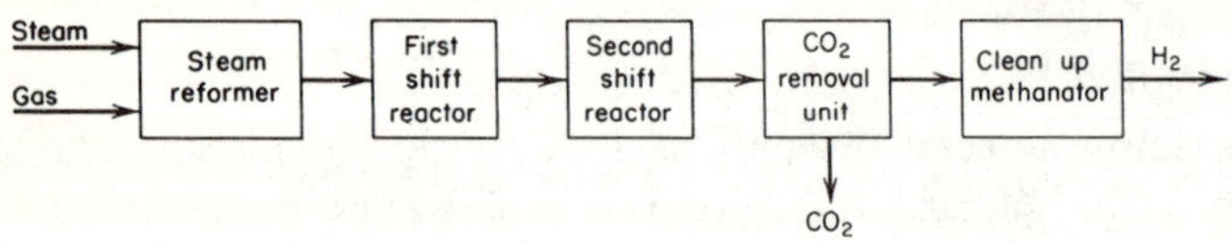

Fig. 4.1 The major units of the H_2 plant with the associated number of ODEs and PDEs.

4.6.2 *The ODE system*

The models for the distributed units of the hydrogen plant, i.e. the units with a significant spatial dimension, the reformer, shift reactors, and cleanup methanator, originally consisted of PDEs reflecting both the material and energy balances for these units. However, we soon observed that the material balance PDEs were dynamically much faster than the energy balances; this was due to the large thermal masses of these units and the rapid flow of material. Thus, the original system of PDEs for each distributed unit was replaced by a single PDE energy balance and one or more ODEs for the material balances; the ODEs resulted from dropping the partial time derivatives from the original PDE material balances so that only one independent variable remained, axial position along the unit. The mathematical model for each of the distributed units therefore consisted of a mixed ODE–PDE system.

This particular form of the distributed unit models imposed a special structure on the approximating ODEs developed by the NMOL. This structure is demonstrated in Figure 4.2, where a map of the ODE Jacobian matrix is printed. Generally, any horizontal position in this map corresponds to one of the 128 dependent variables of the 128-ODE system; any vertical position corresponds to one of the temporal derivatives of the 128-ODE system. For example, the first row of the map with elements 757 reflects the dependence of the first temporal derivative, dy_1/dt on the first three dependent variables, y_1, y_2, and y_3, since the latter appear in the first three columns of the map. In other words, the first ODE of the 128-ODE system is of the form:

$$dy_1/dt = f_1(y_1, y_2, y_3).$$

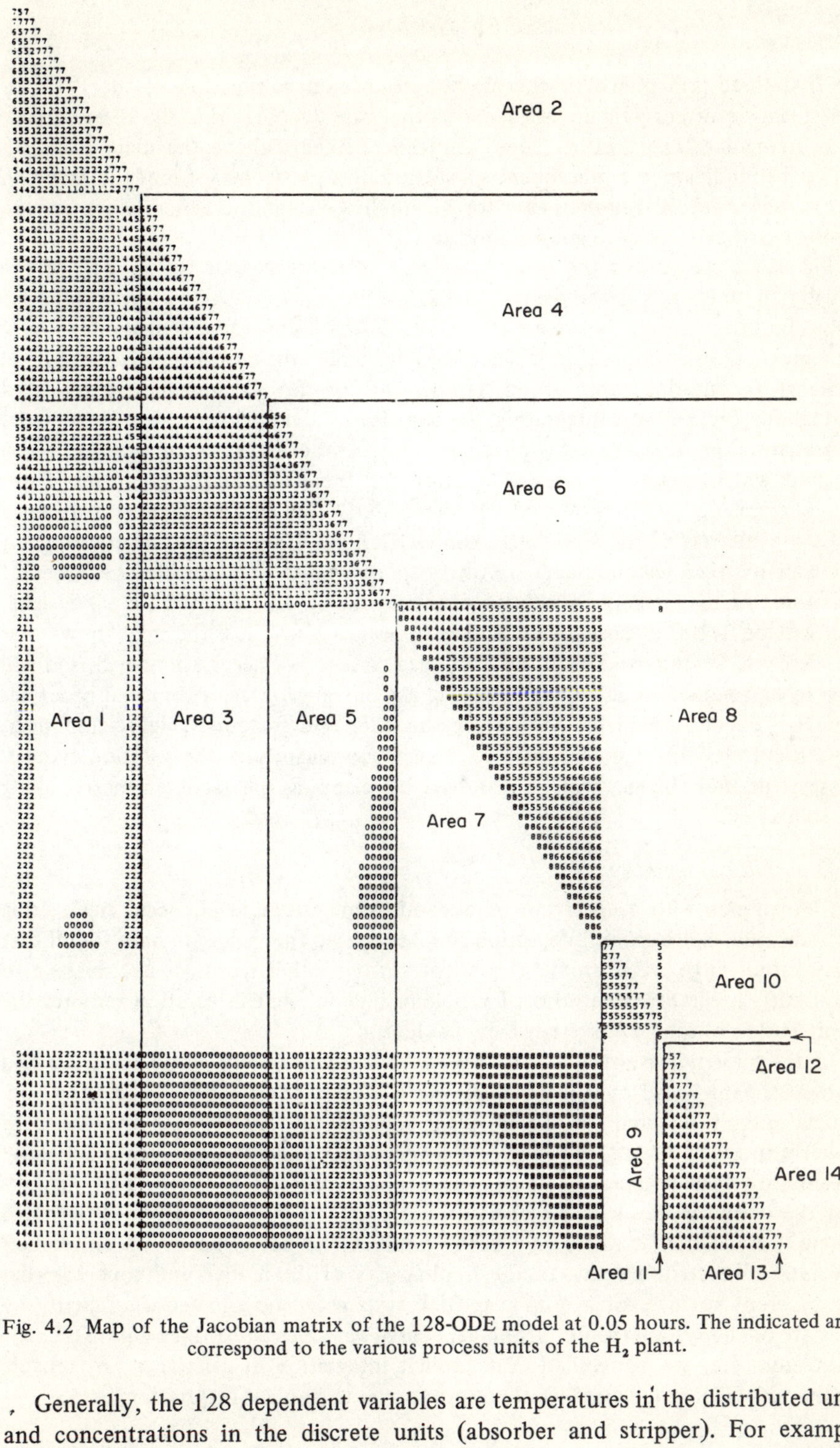

Fig. 4.2 Map of the Jacobian matrix of the 128-ODE model at 0.05 hours. The indicated areas correspond to the various process units of the H_2 plant.

Generally, the 128 dependent variables are temperatures in the distributed units and concentrations in the discrete units (absorber and stripper). For example, y_1, y_2, and y_3 in the preceding differential equation are the three temperatures for the first three grid points of the method-of-lines approximation of the PDE for the steam reformer (Figure 4.1). Note that the Jacobian matrix is essentially lower-triangular filled, with a few significant elements above the main diagonal. As we shall indicate in a subsequent subsection, this particular form of the Jacobian matrix poses special requirements for an implicit integrator which processes the Jacobian matrix to achieve good stability.

The basic reason for the lower-triangular structure is that the models of the distributed units mix steady-state ODEs for the material balance with dynamic PDEs for the energy balance. The ODEs essentially respond instantaneously throughout the distributed units since they have no time derivatives. For example, a change in entering composition, temperature or flow is immediately reflected throughout the unit by integrating the associated ODEs with respect to the axial coordinate along the unit. This produces immediate changes in composition along the unit which affect the rates of reaction and therefore the temperatures. Thus, the temperature ODEs produced by the NMOL applied to the energy balance PDEs are all related to the "upstream" ODEs resulting in the lower-triangular Jacobian matrix. The numbers in the map of Figure 4.2 indicate the order of magnitude of the elements of the Jacobian matrix, e.g. an element with a 4 designation will be 1/100 of an element with a 6 designation. Thus, the map shows that the elements far removed from the main diagonal are not necessarily small, and will have to be retained in general in applying any integration algorithm that processes the ODE Jacobian matrix. Also, since the 128-ODE system is highly nonlinear, the structure of the Jacobian matrix changes significantly as the solution evolves. These structural changes can be observed by mapping the Jacobian matrix along the solution.

4.6.3 *Explicit integration*

Our experience with a spectrum of dynamic simulations in chemical engineering has indicated that explicit integration should be tried first since we have found that many rather large ODE problems are not stiff. In other words, it is a mistake to use a stiff integrator on a nonstiff problem since implicit integration requires the manipulation of the ODE system Jacobian matrix.

In the present case, we found that the 128-ODE system could be integrated with reasonable cost by fourth-order explicit integration for the operating conditions initially selected for the H_2 plant. This would suggest that the 128-ODE model is in fact not stiff. However, we soon found that for a different plant inlet stream (lower methane flow rate), the explicit integration failed. This suggests that the operating level of the plant variables (or the absolute values of the ODE system dependent variables) significantly affect the performance of an explicit integrator. We attribute this to the nonlinearity of the model, and conclude that the apparent stiffness of a nonlinear ODE system can be affected signficantly by the state of the system (values of the dependent variables along the solution).

In any case, we concluded that explicit integration in general is not reliable for the 128-ODE model, and we therefore turned to implicit integration.

4.6.4 *Implicit integration*

In order to develop a solution procedure that would perform well for all anticipated operating conditions of the hydrogen plant, we next turned to implicit integration and in particular LSODE.

The problem of selecting the full or banded option to use is clear from the Jacobian matrix map. The full matrix is 128 × 128 = 16 384 elements which is not particularly demanding in terms of memory for present-day computers. However, if the model were significantly larger, the storage and processing of the Jacobian matrix could become a problem. On the other hand, because the Jacobian matrix is lower triangular, the lower bandwidth for LSODE integrator was set to 127. For the initial runs, the upper bandwidth was also set to 127.

The case of the higher methane entering flow rate, which was successfully integrated with the explicit integrator as mentioned previously, was repeated with LSODE (full Jacobian matrix evaluation). The numerical solutions with the explicit and implicit (LSODE) integrators were essentially identical. The computational statistics for the LSODE run, summarized in Table 4.7 show that the solution was generated with reasonable computational effort. Two points can be observed from these statistics:

(i) The 128-ODE system has an initial fast transient approximately in the interval of problem time from 0 to 0.10 hour. This is suggested by the relatively large computational effort at the beginning of the numerical solution. This also suggests that the 128-ODE is at least moderately stiff since the initial time step is constrained by the requirement to compute the fast transients accurately. Once these transients have decayed to insignificance, the implicit integrator accelerates as indicated by the larger step size and smaller number of steps, derivative evaluations, and Jacobian matrix evaluations. This acceleration is the hallmark of an implicit integrator. An explicit integrator could not accelerate in this way for a stiff problem; rather, the time step would be constrained to small values throughout the entire solution due to the stability constraint of the integrator.

(ii) To further substantiate this conclusion concerning the relative efficiency of the implicit integrator, the temporal eigenvalues of the 128-ODE system were computed. The largest eigenvalue (in the absolute sense) was 4000. Therefore, the largest time step that an explicit integrator could use and still maintain stability is approximately $h = 2.7/4000 = 0.000675$. The implicit integrator reported a maximum step-size of 0.02912 (from Table 4.7) so the ratio of the implicit to the explicit step-size is $0.02912/0.000675 = 43$. To look at this result in another way, the explicit integrator, to reach the final problem time of 0.5 hour, would have to take $0.5/0.000675 = 740$ steps, while the implicit integrator required 136 steps (the sum of the third column in Table 4.7). Thus, the ratio of the numbers of steps for the explicit/implicit integrators is $740/136 = 5.44$ (the explicit integrator requires 5.44 times as many steps as the implicit integrator).

The price paid for this relative computational efficiency of the implicit integrator is the requirement to perform linear algebra for the iterative solution of the nonlinear corrector equations. This in turn requires working with the ODE system Jacobian matrix. However, the last column in Table 4.7 indicates that LSODE handles the problem of updating the Jacobian matrix very effectively; the Jacobian matrix is updated only when required, and rather infrequently. In fact, during two

TABLE 4.7

Statistics for the LSODE solution—high methane feed rate

System time (hr)	Step size (last used)	No. of steps	*f*-evaluations	*J*-evaluations
0.05	0.00220	62	76	12
0.10	0.00298	26	42	4
0.15	0.00612	15	19	4
0.20	0.00869	7	10	1
0.25	0.01154	5	6	1
0.30	0.00828	8	10	3
0.35	0.01750	7	1	4
0.40	0.02750	2	4	0
0.45	0.02380	3	4	1
0.50	0.02912	1	1	0

intervals along the solution, the Jacobian matrix was not updated, and in three other intervals, it was updated only once. This is an important consideration since, in general updating the ODE Jacobian-matrix is a major part of the total computational effort in implicit integration.

The preceding analysis of the relative efficiency of the implicit integrator is not the whole story, however. LSODE was also applied to the 128-ODE model for the case of low methane feed rate. Whereas the explicit integrator completely failed for this case in the sense that a solution could not be generated with any reasonable computing effort, LSODE generated a solution with no more effort than in the preceding case of high methane feed rate. The computational statistics are summarized in Table 4.8.

TABLE 4.8

Statistics for the LSODE solution—low methane feed rate

System time (hr)	Step size (last used)	No. of steps	*f*-evaluations	*J*-evaluations
0.05	0.00225	62	76	12
0.10	0.00448	16	31	3
0.15	0.00241	16	37	5
0.20	0.00798	8	11	2
0.25	0.00807	6	9	1
0.30	0.01234	4	5	0
0.35	0.01520	8	13	3
0.40	0.02430	3	4	1
0.45	0.02430	2	2	0
0.50	0.03395	2	3	1

For this case of low methane feed rate, the total number of steps is 127 (the sum of the third column in Table 4.8), while the corresponding figure is 136 for the high methane feed (from Table 4.7). Thus in this sense LSODE generated a solution for the low methane feed case with slightly less effort than for the high methane case; yet, as mentioned previously, the fourth-order explicit integrator could not generate a solution for the low methane case with any reasonable computing effort. The point we are making is that implicit integration appears to be effective, relative to explicit integration, even for moderately stiff but highly nonlinear problems. In other words, nonlinear systems apparently can cause explicit integration to fail while implicit integration will generate a solution with reasonable computational effort. We do not suggest that we have a fundamental understanding of this phenomenon; rather we can only point to several nonlinear problems that were successfully integrated by implicit methods for which explicit integration failed even though the eigenvalues for the linearized systems were not widely separated. For example, we have found in simulating separation systems, e.g., distillation columns, that as the equilibrium relationships become severely nonlinear, or the separations become very "tight" in the sense that the components to be separated have very similar physical properties, the integration of the model ODEs can be accomplished only with implicit integration, even when the physical problem does not appear to be particularly stiff; that is, there are no obviously "fast" and "slow" physical components, and the eigenvalues of the linearized system are not widely separated.

One indication that nonlinearity has a significant influence on integrator performance can be inferred from the Jacobian matrix maps of the 128-ODE system. Referring again to Figure 4.2, we see that (1) the 128-ODE system has a lower-triangular structure, and (2) this structure changes significantly as the solution evolves as indicated in Figure 4.3, which is a map of the Jacobian matrix at the end of the solution (essentially infinite time). The changes in the shading of the areas in the figures demonstrate the effect of the nonlinearities in the ODE system. This effect is further amplified in Figure 4.4 which contains a portion of the maps from Figures 4.2 and 4.3. Note that the numbers at 0.50 hour (steady state) are generally larger than for 0.05 hour. Since these numbers indicate the orders of magnitude of the individual elements in the Jacobian matrix, they reflect large changes in the properties of the Jacobian matrix as the solution evolves. These changes appear to have a major effect on integrator performance as discussed previously.

The maps in Figures 4.2 and 4.3 also suggest that some reduction in the required memory and the computational effort to integrate the 128-ODE system might be achieved by using less than the full Jacobian matrix. For example, the upper bandwidth of the Jacobian matrix specified for LSODE could be reduced from 127 to a value which would still encompass all of the nonzero elements above the main diagonal. In fact, this was done; the integration of the 128-ODE system was repeated with an upper bandwidth of 44. However, the expected saving in memory and computation were not realized. The memory actually increased due to the increased size of the work arrays required to process a banded version of the Jacobian matrix (the lower and upper bandwidths were set to 127 and 44 respectively). This increase in memory is not due to the actual storage of the Jacobian matrix, which is significantly less than for the full 128 $\times$ 128 matrix

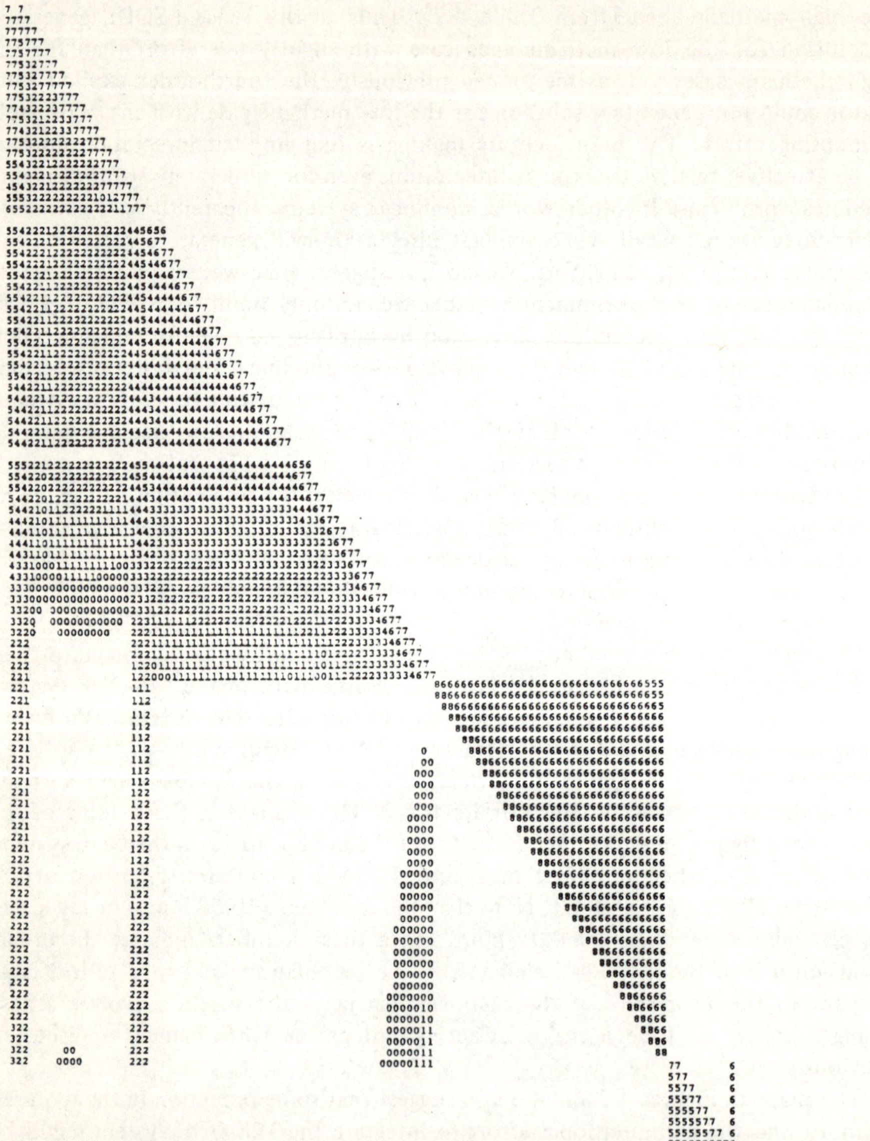

Fig. 4.3 Map of the Jacobian matrix of the 128-ODE model at 0.5 hour (steady-state)

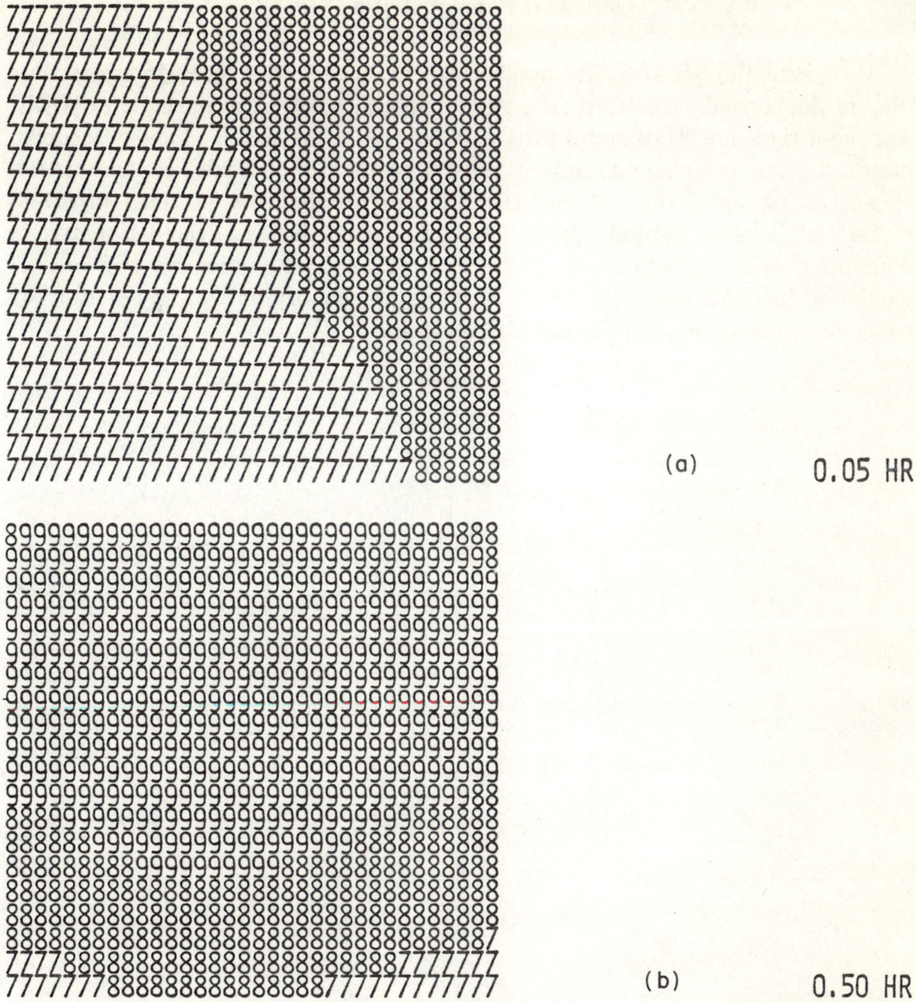

Fig. 4.4 Section of the Jacobian matrix map of the 128-ODE model at 0.05 and 0.5 hour

(44 × 128 = 5632 versus 128 × 128 = 16384 or 65.6 percent reduction in the storage of the Jacobian matrix); rather, the increased memory is required for the pointer (location) information to store the nonzero elements of the banded version of the Jacobian matrix. This point can be demonstrated by considering the formulae given in the LSODE manual for the storage requirements of the full and banded options:

Full matrix storage: 22 + 9*NEQ + NEQ**2

22 + 9*128 + 128**2 = 17558

Banded matrix storage: 22 + 9*NEQ + (2*ML + MU + 1)*NEQ

22 + 9*128 + (2*127 + 44 + 1)*128 = 39546

Clearly, the banded storage mode is advantageous only for ML and MU, which are small in comparison to NEQ.

Also, even though a smaller number of elements of the Jacobian matrix was used in the banded version, the overall computational effort to integrate the ODEs was about the same. Thus in this particular application the use of a banded Jacobian matrix did not offer any advantages relative to the use of a full Jacobian matrix. However, this application is somewhat unusual in that the Jacobian matrix is rather full; a more typical application with a truly banded Jacobian matrix (a concentration of diagonals around the main diagonal with zeros everywhere else) would be handled more efficiently in terms of memory and computer time by using the banded option of the LSODE integrator.

5

SPECIAL NUMERICAL APPROACHES TO SPECIFIC APPLICATIONS

5.1 Chemical kinetics

5.1.1 *The development of software* (*Edsberg*)

Since the beginning of the use of digital computers for numerical simulation of dynamic systems, the problem class of chemical kinetics has attracted much interest both among people developing numerical software and the users. Several program packages have been implemented during the last two decades. Apart from the practical importance and interest of the problem itself, one reason for this development is probably the appealing structure of the problem of chemical kinetics which, among other things, makes it possible to program a translator which transforms the chemical network into the set of ODEs. Another reason perhaps is that numerical analysts of this problem class have found an appropriate test-batch for stiff problems, making it possible to try an ODE method on problems of varying size and stiffness, and even instability phenomena.

Here we give a choice of software that has been published: BELLCHEM by Edelson (1976b), HAVCHM by Stabler and Chesick (1978), KINRATE by Edsberg and Uhlen (1979), FACSIMILE by Curtis (1979b), MACKSIM by Carver and Boyd (1979), KISS by Gottwald (1981) and LARKIN by Dueflhard, Bader, and Nowak (1981). The common feature about chemical kinetics programs is that the user can work interactively with the computer using an alpha-numerical or graphical terminal and/or plotter. The purpose for writing such programs also is that the user need not be acquainted with the numerical methods but should be able to concentrate upon the chemistry and have visualized how the kinetics of a certain reaction system behave for different values of initial concentrations $\boldsymbol{c}(0)$, rate constants $\boldsymbol{K}$, or the reaction structure itself, i.e. $\boldsymbol{Q}$ and $\boldsymbol{P}$ (see §2.10.1 for nomenclature).

When implemented on a computer, it is natural to split a chemical kinetics program into a number of segments; i.e. the program package consists of several parts which communicate through data files. This is the case for the programs KINRATE, FACSIMILE, and LARKIN, where you can distinguish between the following parts:

(i) Translator, which transforms the reactions given as input by the user into either code for the right-hand sides $\boldsymbol{f}(\boldsymbol{c})$ in (2.13) and the corresponding Jacobian $\partial \boldsymbol{f}/\partial \boldsymbol{c}$ or some sparse representation of the matrices $\boldsymbol{Q}$ and $\boldsymbol{P}$ (or $\boldsymbol{S}$ and $\boldsymbol{P}$), from which it is possible in a later phase of the program to compute $\boldsymbol{f}(\boldsymbol{c})$ and the Jacobian when they are needed.

(ii) Parameter-assigner, which obtains as input the rate constants $c_i(0)$. Here the user may also specify certain numerical parameters, such as the initial and end points of integration, the tolerance for the ODE-solver, scaling factors for the components, possibly also initial step size and the largest allowed step size in the integration interval.

(iii) Simulator, which performs the numerical integration of (2.13). The user may here prescribe the times t_i where the solution $\boldsymbol{c}(t_i)$ should be printed out. In this part it may also be possible for the user to decide how to act if the simulation fails.

(iv) Postprocessor, with the help of which the user may examine the solution at the terminal or plot the results from the simulation parts in certain interesting intervals or examine more closely just a few of the components c_i of the reaction system.

Most programs for chemical kinetics are implemented in FORTRAN, but the program package KISS by Gottwald is written in SIMULA, the better structuring possibilities of which have made it possible to write the whole program in one module.

We have already mentioned the possibility of translating a chemical network into the right-hand sides $\boldsymbol{f}(\boldsymbol{c})$ of (2.13) with the help of a program. Since the components of $\boldsymbol{f}(\boldsymbol{c})$ are built up by polynomial expressions, the Jacobian can be formed analytically also with the help of a program. The formal expression for the Jacobian is

$$\partial \boldsymbol{f}/\partial \boldsymbol{c} = \boldsymbol{SK}\,\mathrm{diag}(g_i)\boldsymbol{P}^{\mathrm{T}}\,\mathrm{diag}(1/c_i),$$

if $c_i \neq 0$. The Jacobian is then used in the Newton-iterations when solving numerically the set of nonlinear equations

$$\boldsymbol{c} - h\beta \boldsymbol{f}(\boldsymbol{c}) + \boldsymbol{d} = 0$$

in each time step. Here $\boldsymbol{c}$ is the approximate solution at time $t + h$, t the former time at which the solution was computed; β is a method-dependent constant, and $\boldsymbol{d}$ contains information from the solution at time t. Hence a set of linear equations

$$\boldsymbol{Ax} = (\boldsymbol{I} - h\beta\partial \boldsymbol{f}/\partial \boldsymbol{c})\boldsymbol{x} = \boldsymbol{b},$$

has to be solved for each Newton-iteration. This is done by using the LU-decomposition of $\boldsymbol{A}$, $\boldsymbol{L}$ and $\boldsymbol{U}$ being retained as long as convergence properties are satisfactory. Each time $h\beta$ is changed or some elements of the Jacobian have changed so much that convergence is too slow, the LU-decomposition is updated by first recomputing the Jacobian, then computing the new LU-decomposition of $\boldsymbol{A}$. Comments on the calculation of the Jacobian numerically or analytically are found in Curtis (1978a).

Sparse techniques are used in, e.g. FACSIMILE and LARKIN, while the full matrices are treated in, e.g. KINRATE and KISS. This sparse technique is a general-purpose method that can be used for any sparse system; thus no extra information based on the special structure of chemical kinetics problems is used.

So far, no attempt has been made to compare different programs for chemical kinetics. However, comparisons of stiff ODE solvers applied to chemical problems have been made, and results can be found in Enright and Hull (1976) and Gottwald (1982).

5.2 Combustion

5.2.1 *Introduction (Aiken)*[†]

There is a fundamental difference in the computational problems for combustion kinetics compared to the typical stiff mass-action chemical application. This difference is associated with the branching nature of the combustion reaction system that has the fast free radicals dictating the dynamics of the entire reaction set over some small interval, the flame zone. Also unlike the more typical situation in stiff kinetics, we are often primarily interested in this narrow interval of radical dominance, analogous to the initial boundary layer in stiff noncombustion kinetics. If only this interval were to be computed, the problem would not be difficult, since the time scale of interest would correspond to a fast-acting radical. It is the situation where a solution is to be found over a time interval (or space) much greater than the expolosion—but including it—that we address here. There are three main time frames involved, roughly corresponding to initiation, propagation and branching, and termination. Within the propagation and branching zone, there can be several subdomains where various reactions predominate.

Another interesting aspect of the combustion kinetic system is the existence of positive and negative eigenvalues. Thus integration in forward or reverse time is inherently unstable. This also arises, for example, in interactive counter-diffusion (Aiken 1981a). A method must be used that has a small or zero stability region in the right-half plane. Very little guidance is offered in the literature for such systems. Because the positive eigenvalues contain the information of interest during an explosion, focusing on them can provide a unique numerical approach (Hoppensteadt, Alfeld, and Aiken 1981).

An important yet simple example for illustrative purposes is the combustion of hydrogen. The reaction scheme considered will be:

$$\left.H_2 + O_2 \xrightarrow{k_0} H\cdot + HO_2 \right\}\ \text{initiation} \tag{5.1a}$$

$$\left.\begin{array}{l} H_2 + OH\cdot \xrightarrow{k_1} H\cdot + H_2O \\ \left.\begin{array}{l} O_2 + H\cdot \xrightarrow{k_2} OH\cdot + O\cdot \\ H_2 + O\cdot \xrightarrow{k_3} H\cdot + OH\cdot \end{array}\right\}\ \text{branching} \end{array}\right\}\ \text{propagation} \qquad \begin{array}{r} (5.1b) \\ (5.1c) \\ (5.1d) \end{array}$$

$$\left.\begin{array}{l} OH\cdot \xrightarrow{a_1} \\ H\cdot \xrightarrow{a_2} \\ O\cdot \xrightarrow{a_3} \end{array}\right\}\ \text{termination} \qquad \begin{array}{r} (5.1e) \\ (5.1f) \\ (5.1g) \end{array}$$

with rate constants (Semenov 1959):

† Aiken (1982b)

$$\begin{aligned}
k_0 &= 60\ \mathrm{cm^3\,mol^{-1}\,s^{-1}} \\
k_1 &= 2.3\,\mathrm{E}\,11\ \mathrm{cm^3\,mol^{-1}\,s^{-1}} \\
k_2 &= 4.02\,\mathrm{E}\,9\ \mathrm{cm^3\,mol^{-1}\,s^{-1}} \\
k_3 &= 2.82\,\mathrm{E}\,12\ \mathrm{cm^3\,mol^{-1}\,s^{-1}} \\
a_1 &= 92\ \mathrm{s^{-1}} \\
a_2 &= 8\ \mathrm{s^{-1}} \\
a_3 &= 92\ \mathrm{s^{-1}}
\end{aligned}$$

A typical initial condition, say in an experimental bomb, might be

$$\begin{aligned}
[\mathrm{H_2}] &= 1\,\mathrm{E}{-7}\ \mathrm{mol\,cm^{-3}} \\
[\mathrm{O_2}] &= 0.5\,\mathrm{E}{-7}\ \mathrm{mol\,cm^{-3}} \\
[\mathrm{H}\cdot] &= 0 \\
[\mathrm{OH}\cdot] &= 0 \\
[\mathrm{O}\cdot] &= 0
\end{aligned}$$

Note that the initiation reaction, (5.1a) is very slow compared with the propagation reactions, (5.1b-d). The three propagation reactions include two reactions, (5.1c) and (5.1d), in which two radicals are produced for every one consumed; this autocatalytic behavior is termed *branching*. The character of this branching can be seen from the following sequence of reactions:

$$\begin{array}{lllll}
 & & & \mathrm{H}\cdot & \\
 & & & + & \\
\mathrm{H_2 + O_2 \to HO_2} + & \mathrm{H}\cdot & \mathrm{O}\cdot + \mathrm{H_2} \to & \mathrm{OH}\cdot + \mathrm{H_2} \to & \mathrm{H_2O + H}\cdot. \\
 & + & + & & \\
 & \mathrm{O_2} \to & \mathrm{OH}\cdot + \mathrm{H_2} \to & \mathrm{H_2O} & \\
 & & & + & \\
 & & & \mathrm{H}\cdot &
\end{array}$$

After the initial production of a H· through (5.1a), three more H· are produced in this sequence which can be repeated at a sharply increasing rate, until fuel is consumed or the termination reactions become dominant. This single cycle can be written as

$$\begin{aligned}
&\mathrm{H}\cdot + \mathrm{O_2} \xrightarrow{k_2} \mathrm{OH}\cdot + \mathrm{O}\cdot \\
&\mathrm{O}\cdot + \mathrm{H_2} \xrightarrow{k_3} \mathrm{OH}\cdot + \mathrm{H}\cdot \\
&2(\mathrm{OH}\cdot + \mathrm{H_2} \xrightarrow{k_1} \mathrm{H_2O} + \mathrm{H}\cdot) \\
&\overline{\phantom{\mathrm{H}\cdot + \mathrm{O_2} + 3\mathrm{H_2} \to 3\mathrm{H}\cdot + 2\mathrm{H_2O}}}
\end{aligned}$$

or

$$\mathrm{H}\cdot + \mathrm{O_2} + 3\mathrm{H_2} \to 3\mathrm{H}\cdot + 2\mathrm{H_2O} \tag{5.4}$$

as the overall reaction of this cycle. Thus stoichiometrically, three hydrogen radicals are produced for every one consumed and three hydrogen fuel molecules are used for every oxygen molecule consumed.

The termination reactions given as (5.1e-g) are heterogeneous wall-capture-type adsorption reactions: the rates for such reactions can vary significantly for various reactor constructions. We shall consider here only termination by wall capture and thus only the first explosion limit.

The elementary reactions of (5.1) can be modelled by the law of mass action as:

$$\frac{d[H_2]}{dt} = -\{k_1[OH\cdot] + k_3[O\cdot]\}[H_2] - k_0[H_2][O_2], \tag{5.5a}$$

$$\frac{d[O_2]}{dt} = -k_2[H\cdot][O_2] - k_0[H_2][O_2], \tag{5.5b}$$

$$\frac{d[H\cdot]}{dt} = -\{k_2[O_2] + a_2\}[H\cdot] + k_1[H_2][OH\cdot] + k_3[H_2][O\cdot] + k_0[H_2][O_2], \tag{5.5c}$$

$$\frac{d[OH\cdot]}{dt} = k_2[O_2][H\cdot] - \{k_1[H_2] + a_1\}[OH\cdot] + k_3[H_2][O\cdot], \tag{5.5d}$$

$$\frac{d[O\cdot]}{dt} = k_2[O_2][H\cdot] - \{k_3[H_2] + a_3\}[O\cdot]. \tag{5.5e}$$

These equations, with the parameters of (5.2) and the initial conditions of (5.3), produce the numerical solution shown in Fig. 5.1. We define three regions where different chemical mechanisms are dominant: Region I, the initial region where initiation of the hydrogen radical is dominant; Region II, where the branching cycle is most important; and Region III, in which the termination reactions become important. We shall consider each of these regions in the next subsection.

5.2.2 *Steady-state approximation* (*Aiken*)

The usual implementation technique for the steady-state approximation would set the derivatives of the fast free radicals ($H\cdot$, $OH\cdot$, and $O\cdot$) to zero and solve the resulting two differential and three algebraic equations. It should be obvious from (5.3), that if (5.5c–e) are set to zero and solved for $[H\cdot]$, $[OH\cdot]$, and $[O\cdot]$, respectively, the system may change with time only from the very slow initiation reaction. Thus for the parameters of (5.2) over 0.1 second, O_2 and H_2 do not change from their initial conditions while $H\cdot$, $OH\cdot$, and $O\cdot$ instantly attain and hold constant at 0.37 E –13, 0.71 E –24, and 0.26 E –16 mol cm^{-3}, respectively.

Region I. Without any hydrogen radicals present, the branching cycle cannot go; there is in fact a critical minimum concentration below which the initiation reaction

$$H_2 + O_2 \xrightarrow{k_0} H\cdot + HO_2,$$

with $H\cdot$ production rate

$$\frac{d[H\cdot]}{dt} = k_0[H_2][O_2], \tag{5.6}$$

is faster than the rate at which $H\cdot$ is being used in the cycle, the maximum rate which it can be used being

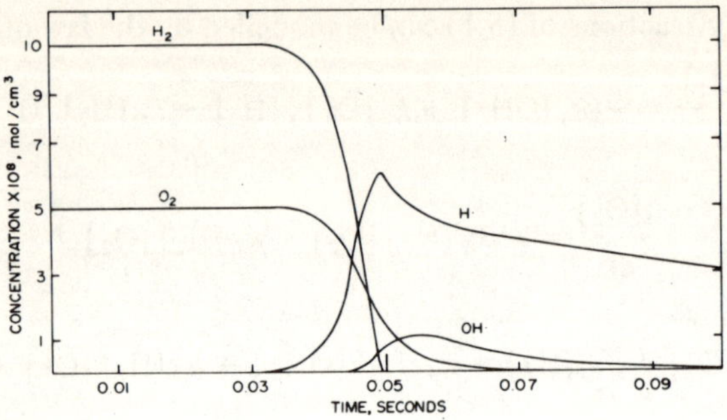

Fig. 5.1 A portion of the solution to hydrogen combustion in pure oxygen.

$$\frac{d[H\cdot]}{dt} = -k_2[H\cdot][O_2]. \tag{5.7}$$

For example, if $[H_2] = 1.0\,E{-7}$, $[O\cdot] = 0.5\,E{-7}$, $k_0 = 60$, (5.6) gives

$$\frac{d[H\cdot]}{dt} = 3\,E{-13}, \tag{5.8}$$

which, to match (5.7), would require $[H\cdot] = 1.5\,E{-15}$ ($k_2 = 4\,E9$, $[O_2] = 0.5\,E{-7}$).

Let us form a general set of scaled equations by use of

$$u = [H_2]/u_0, \quad v = [O_2]/v_0, \quad x = [H\cdot]/x_0,$$
$$y = [OH\cdot]/y_0, \quad z = [O\cdot]/z_0, \quad \tau = t/t_0,$$

where u_0, v_0, z_0, and t_0 are magnitude scaling factors, and choose scales as dictated by the particular chemical knowledge in each region. (5.5) becomes

$$\frac{u_0}{t_0}\dot{u} = -(k_1 y_0 y + k_3 z_0 z)u_0 u - k_0 u_0 v_0 uv, \tag{5.9a}$$

$$\frac{v_0}{t_0}\dot{v} = -(k_2 x_0 x)v_0 v - k_0 u_0 v_0 uv, \tag{5.9b}$$

$$\frac{x_0}{t_0}\dot{x} = -(k_2 v_0 v + a_2)x_0 x + (k_1 u_0 u)y_0 y + (k_3 u_0 u)z_0 z + k_0 u_0 v_0 uv, \tag{5.9c}$$

$$\frac{y_0}{t_0}\dot{y} = (k_2 v_0 v)x_0 x - (k_1 u u_0 + a_1)y_0 y + (k_3 u u_0)z_0 z, \tag{5.9d}$$

$$\frac{z_0}{t_0}\dot{z} = (k_2 v_0 v)x_0 x - (k_3 u_0 u + a_3)z_0 z. \tag{5.9e}$$

A pertinent scale for x in this initial region would be obtained by setting rates (5.6) and (5.7) equal:

$$k_0 u_0 v_0 = k_2 v_0 x_0,$$

or

$$x_0 = \frac{k_0 u_0}{k_2} \ll 1. \tag{5.10}$$

Thus x starts from zero in this region and reaches a maximum on the order of $k_0 u_0 / k_2$. Let us choose the time scale

$$t_0 = 1/(k_2 v_0) \ll 1,$$

so that we are stretching time and looking close to the beginning of the reactions. We expect u and v to change very little over this region so that $u_0 = v_0 = [H_2(0)]$.

Finally, since H• has not reached a concentration to allow reaction (5.1c) to really get started, certainly reactions (5.1b) and (5.1d), whose rates depend on the products from (5.1c) will be negligible, and will be omitted. Equation (5.9a) thus becomes

$$\dot{u} = \frac{k_0}{k_2} uv, \tag{5.11a}$$

$$\dot{v} = \frac{k_0}{k_2} (-vx - uv), \tag{5.11b}$$

$$\dot{x} = -ux + uv. \tag{5.11c}$$

This makes obvious the small changes in $[O_2]$ and $[H_2]$ over this region for a significant increase in the concentration of H•, as the right-hand sides of (5.11a) and (5.11b) are multiplied by small numbers.

Region II. Now allow x to attain a value on the order of u and v as the cycle (5.4) becomes the dominant mode for chemical change and significant amounts of u and v are used to produce x. Note also from (5.4) that no net y or z radicals are produced so that their characteristic scales (y_0 and z_0, respectively) should be smaller than for x. For extensive branching, however, at any instant there could be quite a significant number of these short-lived radicals.

Let us again take

$$t_0 = 1/(k_2 v_0),$$

to stretch time for this fast reaction sequence; also

$$u_0 = v_0 = x_0 = [H_2(0)],$$

and

$$y_0 = \frac{k_2 [H_2(0)]}{k_1}, \qquad \frac{k_2}{k_1} \ll 1,$$

$$z_0 = \frac{k_2}{k_3} [H_2(0)], \qquad \frac{k_2}{k_3} \ll 1.$$

These choices lead to

$$\dot{u} = -uy - uz, \tag{5.12a}$$

$$\dot{v} = -vx, \tag{5.12b}$$

$$\dot{x} = -vx + uy + uz, \tag{5.12c}$$

$$\frac{k_2}{k_1}\dot{y} = vx - uy + uz, \tag{5.12d}$$

$$\frac{k_2}{k_3}\dot{z} = vx - uz, \tag{5.12e}$$

where terms for initiation and termination have been omitted because very small numbers multiply them. For example, we have dropped from the right-hand side of (5.12c) $a_2x/\{k_2[\mathrm{H_2}(0)]\}$, since $a_2/\{k_2[\mathrm{H_2}(0)]\} \ll 1$. Thus we have only the cycle (5.4) in the model for this region. Since we have just those three reactions, two of the five equations in (5.12) must be redundant. In fact,

$$\dot{u} + \dot{x} = \dot{v}, \tag{5.13a}$$

$$\dot{u} = \frac{k_2}{k_1}\dot{y} + 2\frac{k_2}{k_3}\dot{z} + 3\dot{v}, \tag{5.13b}$$

so that (for example)

$$x = v - u + [u(0) - v(0)] + x(0),$$

$$y = \frac{k_1}{k_2}[u - u(0)] - 2\frac{k_1}{k_3}[z - z(0)] - 3\frac{k_1}{k_2}[v - v(0)] + y(0),$$

and (5.12) reduce to

$$\dot{u} = -u\frac{k_1}{k_2}[u - u(0)] - 2\frac{k_1}{k_3}[z - z(0)] - 3\frac{k_1}{k_2}[v - v(0)] + y(0) - uz, \tag{5.14a}$$

$$\dot{v} = -v\{v - u + [u(0) - v(0)] + x(0)\}, \tag{5.14b}$$

$$\frac{k_2}{k_3}\dot{z} = v\{v - u + [u(0) - v(0)] + x(0)\} - uz. \tag{5.14c}$$

Since $k_2/k_3 \ll 1$, a ssa for (5.19c) would reduce the system to:

$$z = \frac{v}{u}\{v - u + [u(0) - v(0)] + x(0)\}, \tag{5.15a}$$

$$\dot{u} = -u\frac{k_1}{k_2}[u - u(0)] - 2\frac{k_1}{k_3}[z - z(0)]$$

$$-\frac{3k_1}{k_2}[v - v(0)] + y(0) - uz, \tag{5.15b}$$

$$\dot{v} = -uz. \tag{5.15c}$$

It may in fact be true that the ssa is adequate to apply to (5.12d) directly to give (5.12e)

$$\begin{aligned} \dot{u} &= -3vx, \\ \dot{v} &= -vx, \\ \dot{x} &= 2vx, \end{aligned} \tag{5.16}$$

or since here $\dot{x} = -\frac{2}{3}\dot{u}, \dot{v} = \frac{1}{3}\dot{u}$,

$$\dot{v} = -v[1 - 2v - x(0)] \tag{5.17}$$

as the single equation! This can be solved analytically from t_0 to t:

$$v = \frac{1 - x(0)}{2 - \phi}, \tag{5.18}$$

where

$$\phi = \exp\left([x(0) - 1)]\ (t - t_0) - \ln \frac{[x(0) - 1] + 2v(0)}{v(0)} \right).$$

This is not likely to be as accurate as (5.15) since we use here both $k_2/k_1 \ll 1$ and $k_2/k_3 \ll 1$ while (5.15) uses only $k_2/k_3 \ll 1$; also $k_2/k_3 < k_2/k_1$ by about an order of magnitude. However, neither (5.18) nor (5.15) will strictly apply near to the point of maximum x ($\dot{x} = 0$). This can be seen by examining (5.13b) which, upon elimination of $\dot{v}$, gives

$$\dot{u} = -\frac{k_2}{2k_1}\dot{y} - \frac{k_2}{k_1}\dot{z} - \tfrac{3}{2}\dot{x}. \tag{5.19}$$

Now, near to the maximum in x (where $\dot{x} = 0$), the rate of using x (reaction 5.1c) is the same as the rate of its creation (reactions 5.1b and 5.1d). At this point, $\dot{x}$ is zero, but H_2 is still being used (although now in small concentrations) and thus $\dot{y}$ and $\dot{z}$ may not be zero. The degree of error incurred from setting them to zero is parameter dependent but definitely less error would result from setting only one of the two derivatives to zero, as is done for the scheme resulting in (5.6). Away from the maximum, however, when $(3/2)\dot{x} \gg k_2 y/(2k_1)$, $k_2\dot{z}/k_1$, the approximations $k_2 y/(2k_1) = k_2\dot{z}/k_1 = 0$ can be used. Therefore, in Region II, the ssa is not recommended in general on $\dot{x}$ at all and can be used for $\dot{z}$ and $\dot{y}$ everywhere except near the maximum in $\dot{x}$. Let us call the approach using (5.15) the *partial steady-state* and that for (5.23) the *complete steady-state* approach.

Once past the maximum in H•, fuel has been depleted and thus reactions (5.1b) and (5.1d) are no longer important; reaction (5.1c), in which O_2 and H• produce OH• and O•, is now of major importance. As such, these radicals reach substantial concentrations. We must therefore re-scale concentrations. Let:

$$v_0 = x_0 = y_0 = z_0 = [H_2(0)]; \quad u_0 = \frac{k_2[H_2(0)]}{k_3},$$

while keeping the cycle time scale $t_0 = 1/\{k_2[H_2(0)]\}$. This gives:

$$\frac{k_2}{k_3}\dot{u} = -\frac{k_1}{k_3}uy - uz, \tag{5.20a}$$

$$\dot{x} = -xv + \frac{k_1}{k_3}uy + uz - \frac{a_2 x}{k_2[H_2(0)]}, \tag{5.20b}$$

$$\dot{y} = xv - \frac{k_1}{k_3}uy + uz - \frac{a_1 y}{k_2[H_2(0)]}, \tag{5.20c}$$

$$\dot{z} = xv - uz - \frac{a_3 z}{k_2[H_2(0)]}, \tag{5.20d}$$

where initiation terms have again been omitted but termination terms have now been allowed in since they will come into importance sometime after the [H•] maximum, when depending on the values of the a_i. For the values specified in (5.2), they in fact do not immediately come into importance here and some additional saving in computation could be gained by their omission until later.

Now let us apply the ssa to (5.20):

$$u\left(\frac{k_1}{k_3}y + z\right) = 0.$$

Since neither y nor z may be zero, we take $u = 0$ as the root. Thus we may simplify:

$$\dot{v} = -xv, \tag{5.21a}$$

$$\dot{x} = -xv - \frac{a_2 x}{k_2[H_2(0)]}, \tag{5.21b}$$

$$\dot{y} = xv - \frac{a_1 y}{k_2[H_2(0)]}, \tag{5.21c}$$

$$\dot{z} = xv - uz - \frac{a_3 z}{k_2[H_2(0)]}. \tag{5.21d}$$

The advantage of omitting the termination terms until v has decayed sufficiently to make their presence felt is more obvious here as an analytical formulation would be possible. If this is performed, a periodic test for the importance of the termination term could be made and it could be included when important.

Region III. Now let us slow down our time scale to $t_0 = 1/a_3$ (assuming relatively slow termination) and examine the situation where H_2 and O_2 are low in concentration, having gone into the production of radicals whose concentrations are of the order of $[H_2(0)]$. Thus take

$$u_0 = \frac{a_3}{k_3},\ v_0 = \frac{a_3}{k_2},$$

$$x_0 = y_0 = z_0 = [H_2(0)];$$

then

$$\frac{a_3}{k_3[H_2(0)]}\dot{u} = -uy - uz, \tag{5.22a}$$

$$\frac{a_3}{k_3[H_2(0)]}\dot{v} = -vx, \tag{5.22b}$$

$$\dot{x} = -vx - \frac{a_2}{a_3}x + \frac{k_1}{k_3}uy + uz, \tag{5.22c}$$

$$\dot{y} = vx - \frac{k_1}{k_3}uy - y + uz, \tag{5.22d}$$

$$\dot{z} = vx - uz - z, \tag{5.22e}$$

where

$$\frac{a_3}{k_2[H_2(0)]} \ll 1, \quad \frac{a_3}{k_3[H_2(0)]} \ll 1, \quad \frac{a_2}{a_3} < 1, \quad \frac{k_1}{k_3} < 1.$$

The ssa may apply to (5.22a) and (5.22b) and, if so, would give $y = z$, $v = v = 0$ leading to

$$\dot{z} = -z, \tag{5.23a}$$

$$\dot{y} = -y, \tag{5.23b}$$

$$\dot{x} = -\frac{a_2}{a_3}x, \tag{5.23c}$$

which is obviously the correct limit of simple decay for each species.

Let us now use the simple perturbation analysis of the last section to construct a computational scheme that efficiently solves the H_2-O_2 reaction set of (5.5). We shall again consider each region in turn and the parameters and initial conditions of (5.2) and (5.3).

Region I. This region ends with [H•] achieving a magnitude of about 1.5 E−15 mol cm^{-3}, as given by the scale of (5.10). This occurs after about 0.002 second into the reaction sequence. The full equation set is very stiff, as three time scales are represented: the very slow initiation time scale, the faster termination rate, and the much faster time scale of the cycle reactions (maximum [H•] changes from these three causes are on the order 10^{-13}, 10^{-7}, 10^{-4} mol cm^{-3} s^{-1}). This stiffness is eliminated, however, through use of (5.11); order reductions as well as a simplification of the functions are achieved. Equation (5.11) may thus be used initially until [H•] reaches about 1.0 E−15 mol cm^{-3}, at which point we switch to the solution scheme for Region II. As this region lasts for only a very short time, use of (5.11) brings minimal returns. There is no reason why this region should not be omitted entirely, and a starting concentration of [H•] = about 1.5 E−15 mol cm^{-3} be used immediately in Region II. This can be estimated *a priori* as above, and the inaccuracy in time is quite small.

Region II. We have two possible equation sets to use initially in this region from our perturbation analysis that resulted from (5.12): the partial steady-state approach, (5.15), that resulted from first, an elimination of two dependent equations, then an application of the ssa to one equation; and the complete steady-state approach, (5.17), that resulted from use of the ssa twice, then elimination of two dependent equations. Each of these solutions eliminates the (moderate) stiffness of the cycle equations, reduces the order of the set, and simplifies the functional evaluation. The results of a numerical solution of each approach are shown in Figure 5.2. Although the partial steady-state method is clearly superior over all of Region II, either scheme could be considered adequate for most practical purposes, until the maximum in [H•] is approached. As was predicted in the perturbation analysis, the time derivatives of [OH•] and [O•] generally may not both be taken as zero when the time derivative of [H•] is also zero.

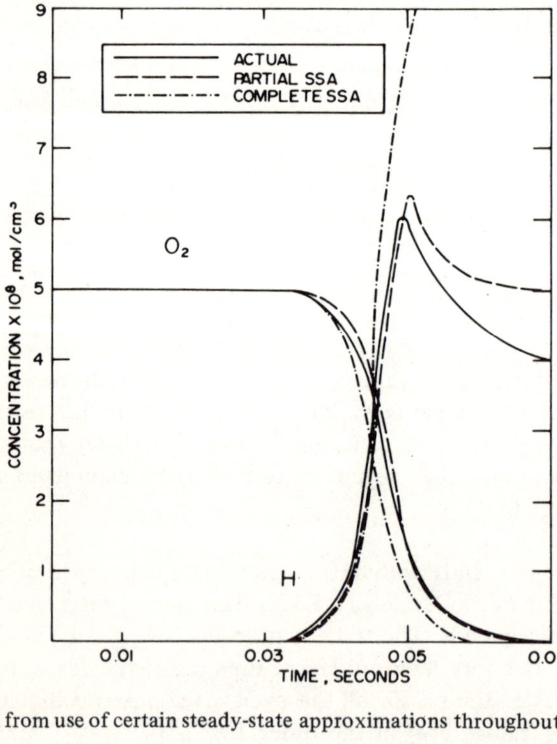

Fig. 5.2 Solution from use of certain steady-state approximations throughout explosion interval.

A time must be determined, near to the point of maximum [H•], after which a steady-state method may not be applied. An adequate measure is provided by the inflection point in the solution for [H•]. This can be monitored at no additional expense during the solution. Chemically, this criterion states that the [H•] atoms are being significantly depeleted by the cycle reactions (relative to the rate of generation). This switch takes place at about 0.048 second for the partial steady-state approach and at about 0.045 second for the complete steady-state approach.

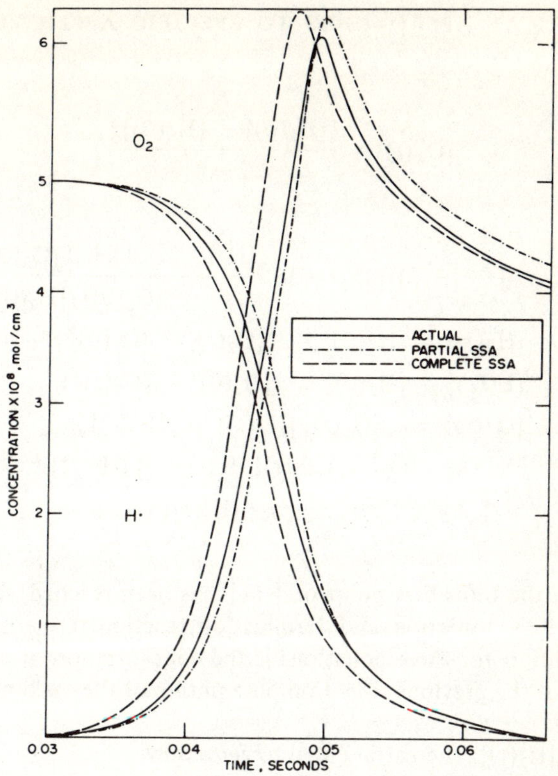

Fig. 5.3 Solution from use of steady-state approximation with switches to full solution around maximum in [H •], according to inflection criteria.

The solution thereafter by use of the equation set (5.12) is shown in Figure 5.3 for the partial and complete steady-state approaches. The larger equation set does indeed improve the solution over the [H •] turning point and after. We seek the most efficient solution, however, so use (5.21) as soon as possible.

We must face the problem, however, of when to make the switch to (5.21). The scales for (5.20) suggest this to occur when [H•] is quite small. The maximum in [H•] in fact happens at about that point, so we should be able safely to use (5.21) any time after the maximum. This has been verified computationally. For example, application after $t = 0.052$ second gives the exact solution.

Region III. After about 0.08 second, $[O_2]$ has a value less than 10^{-10} mol cm^{-3} and (5.28) may then safely be used without any loss in accuracy.

If the complete steady-state approach gives accuracy adequate for some predictive or design purpose, the following summary in dimensional form may be used:

Step 1. Determine the approximate concentration of H• where the depletion of H_2 by the cycle is comparable to the homogeneous generation; i.e. $k_0[H_2][O_2] \approx k_2[H\cdot][O_2]$. For the parameters of (5.7), this gives $[H\cdot] \approx 1.5\,\mathrm{E}{-5}$ mol cm^{-3}. We shall start the Region II equations with this value; denoting it by $[H\cdot(0)]$.

Step 2

$$[O_2(t)] = \frac{\{[H_2(0)] - [H\cdot(0)]\}}{2 - \phi},$$

where

$$\phi = \exp\left(\{[H\cdot(0)] - [H_2(0)]\}k_2 t - \ln \frac{\{[H\cdot(0)] - 1\} + 2\,[O_2(0)]/[H_2(0)]}{[O_2(0)]/[H_2(0)]}\right),$$

$$[H\cdot(t)] = [H\cdot(0)\} - 2\{[O_2(t)] - [O_2(0)]\},$$

$$[H_2(t)] = [H_2(0)] - 3[O_2(0)] - [O_2(t)]\},$$

$$[O\cdot(t)] = k_2[O_2(t)]\,[H\cdot(t)]/\{k_3[H_2(t)]\},$$

$$[OH\cdot(t)] = \frac{k_2}{k_1}\;\frac{[O_2(t)]\,[H\cdot(t)] + [O\cdot(t)]\,[H\cdot(t)]\,k_3/k_2}{[H_2(t)]}.$$

Step 3. When the inflection point of $[H\cdot]$ has been reached, switch to the full equation set (5.12). Initiation and termination reaction terms may be omitted. The initial condition for these equations is the concentration of all species at the inflection point of the previous step. Compute until past the maximum in $[H\cdot]$.

Step 4. Take $[H_2(t)]$, thereafter equal to zero. Solve

$$\frac{d[O_2]}{dt} = -k_2[H\cdot]\,[O_2],$$

$$\frac{d[H\cdot]}{dt} = -\{k_2[O_2] + a_2\}[H\cdot] + k_1[H\cdot]\,[OH\cdot] + k_3[H_2]\,[O\cdot],$$

$$\frac{d[OH\cdot]}{dt} = k_2[O_2]\,[H\cdot] - \{k_1[H_2] + a_1\}[OH\cdot] + k_3[H_2]\,[O\cdot],$$

$$\frac{d[O\cdot]}{dt} = k_2[O\cdot]\,[H\cdot] - \{k_3[H\cdot] + a_3\}[O\cdot],$$

until $[O_2] < k_2[H_2(0)]\,k_3$, with those conditions.

Step 5. Use

$$\frac{d[OH\cdot]}{dt} = -a_2[H\cdot],$$

$$\frac{d[H\cdot]}{dt} = -a_2[H\cdot],$$

$$\frac{d[O\cdot]}{dt} = -a_3[O\cdot],$$

until the desired time interval is completed. Use of such a five-step scheme results in a very substantial saving in computation time with a small loss in accuracy.

5.2.3 *Fast algorithms for combustion kinetics calculations* (*Pratt*)†

The problem of economical integration of the coupled nonlinear ODEs describing exothermic, homogenous-gas-phase combustion-reaction kinetics has not yet been adequately solved. Multistep methods, based on traditional explicit numerical integration schemes such as the fourth-order Runge–Kutta method, dominated the thinking of early workers performing single-point calculations modelling one-dimensional combustion processes in shock tubes and rocket nozzles. In the 1960s, considerable progress was made with respect to reliability and efficiency of single-point calculations, due to the introduction of implicit methods by Tyson (1964), and Treanor's introduction of locally exact solutions (exponential functions) to extend the convergence radius of explicit Runge–Kutta methods (Treanor 1966). Lomax and Bailey (1967) explored the combined use of explicit and implicit methods based on low-order polynomial approximations as an alternative to Treanor's method. In the 1970s, Treanor's algorithm was utilized by Dimitrov (1975) to model hydrogen-air combustion, and Tyson's and Lomax and Bailey's ideas were incorporated into production codes (Bittker and Scullin 1972; Nickerson, Frey, and Coats 1971).

Attention was focused next on multi-point calculations, required for modelling of multidimensional reactive flows (Young and Boris 1977; Pratt 1976; Pratt 1979). This increasing interest in modelling of furnaces and combustors resulted in a major reassessment of available algorithms. It was recognized that all proposed techniques for modelling of complex chemically reacting flows—whether Eulerian space discretization with operator-splitting for treating coupled phenomena, or Lagrangian mass or vorticity discretization with coupled phenomena treated by the method of fractional steps—*all* approaches require a very fast, very stable but only moderately accurate homogeneous batch chemistry integrator (Pratt 1979a). Such an algorithm must be:

(i) self-starting (automatic selection of step size),
(ii) single-step (to minimize storage), and
(iii) very stable (reliable and efficient over a wide range of initial conditions, fuel-air equivalence ratio, and kinetic mechanisms and data.)

At the time of this writing, the automatic stiff integrator LSODE (see §4.3.2), although state-of-the-art, is generally recognized by combustion-device modellers to be not fast enough to enable economical solution of multi-dimensional chemically reacting flowfields.

Is it possible to find a faster algorithm for application to the present problem? While LSODE is universally regarded as the best available packaged code for solving any arbitrarily given set of ODEs, it is always possible that there exists a superior method for solving any uniquely specified system. A clue to such a better method may be found in a statement of J. D. Lambert (Lambert 1973):

> For stiff equations, for which a typical solution is a rapidly decaying exponential, the error in interpolating such functions by a polynomial of high degree is very large, and correspondingly we find that linear multistep methods of high stepnumber cannot be used successfully for such problems.

† This work supported by NASA–Lewis Research Center.

The present paper describes the development of an algorithm which has been specifically designed to solve only the problem under consideration, and which is significantly faster than LSODE.

The equations for adiabatic, homogeneous-gas-phase chemical-reaction kinetics at constant pressure are given by

$$\frac{d\sigma}{dt} = f(\sigma, T), \tag{5.24}$$

where

$$f = [f_i, \ldots, f_N]^{\mathsf{T}}, \quad \sigma = [\sigma_i, \ldots, \sigma_N]^{\mathsf{T}},$$

$$f_i = -\rho^{-1} \sum_{j=1}^{J} (\alpha'_{ij} - \alpha''_{ij})(R_j - R_{-j}), \tag{5.25}$$

$$R_j = A_j T^{B_j} \exp(-E_j/RT) \prod_{k=1}^{N} (\rho\sigma_k)^{\alpha'_{kj}}, \tag{5.26a}$$

and

$$R_{-j} = A_{-j} T^{B_{-j}} \exp(-E_{-j}/RT) \prod_{k=1}^{N} (\rho\sigma_k)^{\alpha''_{kj}}. \tag{5.26b}$$

In (5.24–5.26), σ_i are the mole numbers of the ith species ($i = 1, \ldots, N$); T is the temperature; ρ is the mass density; α'_{ij} and α''_{ij} are the stoichiometric coefficients of species i ($i = 1, \ldots, N$) in reaction j ($J = 1, \ldots, J$) as a reactant and as a product species, respectively; A_j, A_{-j}, B_j, B_{-j}, E_j, and E_{-j} are constants in the modified Arrhenius rate expressions for R_j and R_{-j}, which in turn are the forward and reverse rates of the jth reaction ($j = 1, \ldots, J$), respectively.

For adiabatic reaction, an enthalpy conservation equation constitutes an algebraic constraint

$$\sum_{i=1}^{N} h_i\sigma_i = H_0 = \text{constant}, \tag{5.27}$$

where h_i is the sensible-plus-chemical enthalpy of the ith species, and where H_0 is the mass-specific enthalpy of the mixture.

The mass density ρ in (5.25) and (5.26) is determined by the equation of state for an ideal gas,

$$\rho = P/(RT\sigma_m), \tag{5.28}$$

where P is the absolute pressure, R is the universal gas constant, and

$$\sigma_m = \sum_{i=1}^{N} \sigma_i \tag{5.29}$$

is the reciprocal mean molecular weight of the mixture.

For purpose of subsequent exposition, (5.24-5.29) may be expressed in two different locally linearized formulations, which will be referred to as the "L-formulation" and the "Z-formulation" of the rate equations.

Equation (5.25) may be expressed as a difference between two positive terms, as follows:

$$f_i = Q_i - D_i, \tag{5.30}$$

where

$$Q_i \equiv \rho^{-1} \sum_{j=1}^{J} (\alpha'_{ij} R_{-j} + \alpha''_{ij} R_j), \tag{5.31}$$

and

$$D_i \equiv \rho^{-1} \sum_{j=1}^{J} (\alpha'_{ij} R_j + \alpha''_{ij} R_{-j}). \tag{5.32}$$

The terms Q_i and D_i represent the gross rates of production and consumption of species-i, respectively, due to the contributions of all J forward and reverse actions.

The objective of this decomposition is to enable factorization of the mole number from the destruction term:

$$D_i = L_i \sigma_i, \tag{5.33}$$

where L_i is simply

$$L_i = (\rho \sigma_i)^{-1} \sum_{j=i}^{J} (\alpha'_{ij} R_j + \alpha''_{ij} R_{-j}). \tag{5.34}$$

For example, in one of the j reactions,

$$H_2 + O_2 \underset{b}{\overset{f}{\rightleftharpoons}} OH + OH,$$

with $R_f = k_f \sigma_{H_2} \sigma_{O_2}$ and $R_b = k_b \sigma_{OH}^2$, H_2 and O_2 are consumed by the forward reaction, and are produced by the reverse reaction.

With this notation, (5.24) may be rewritten as

$$\frac{d\sigma_i}{dt} = Q_i - L_i \sigma_i. \tag{5.35}$$

This equation is seen to be formally linear in σ_i, and is therefore readily integrated to give the "locally exact" solution

$$\sigma_i^{n+1} = \sigma_i^n + h f_i^n \left(\frac{1 - \exp(-L_i^n h)}{L_i^n h} \right), \tag{5.36}$$

where the superscripts $(n, n+1)$ denote the time level, and where h is the time step-size $(t^{n+1} - t^n)$.

This may also be rewritten in the form

$$\sigma_i^{n+1} = (Q_i^n/L_i^n) + (\sigma_i^n - Q_i^n/L_i^n)\exp(-L_i^n h). \tag{5.37}$$

Since the loss coefficients L_i ($i = 1, \ldots, N$) are all positive numbers, the first terms in brackets may be regarded as the *local asymptotic* solution which represents the large time-step value of the locally exact solution.

Inspection reveals that the coefficients L_i ($i = 1, \ldots, N$) represent characteristic time constants for the L-formulated rate equations; that is, an accurate approximate solution would have to be resolved on a step size of order $(L_i)^{-1}_{\max}$. As may also be seen from inspection, as the temperature T increases (due to adiabatic exothermic chemical reaction), the L_i all increase exponentially; therefore the required step size, $h \approx (L_i)^{-1}_{\max}$, becomes unacceptably small. Traditionally, the steady-state approximation has been utilized to overcome this difficulty. However, the practice recommended here is to use the L-formulated equations *only* early in the induction phase, when some of the initial mole numbers σ_i^0 may be ill-posed or zero. Under these conditions, a large difference between the Q_i and D_i terms will result, so that the norm of the corresponding state variable differentials, $|Z_i|_{\max}$, is very much greater than $(L_i)_{\max}$. Use of the L-formulation therefore admits a larger step size, scaled on $(L_i)^{-1}_{\max}$ rather than on $|Z_i|^{-1}_{\max}$.

While the focus of the present paper is on solution of the gas-phase kinetic rate equations (5.24–5.29) it is emphasized that the "Z-formulation" exposition to follow is perfectly general to (5.24) and is not specific to the particular structure of the kinetic rate equations, as is the case for the "L-formulation." The ideas here are due to Brandon (1979).

The right-hand side of (5.24) may be expanded in a first-order-truncated Taylor's series about the current approximate solution at $\{(\sigma_k, t^n) : k = 1, \ldots, N\}$, as follows:

$$f_i = f_i^n + \left(\frac{\mathrm{d}f_i}{\mathrm{d}\sigma_i}\right)^n (\sigma_i - \sigma_i^n). \tag{5.38}$$

Here $(\mathrm{d}f_i/\mathrm{d}\sigma_i)$ is the *total* derivative of f_i with respect to σ_i, given by

$$\begin{aligned}\frac{\mathrm{d}f_i}{\mathrm{d}\sigma_i} &= \sum_{k=1}^{N} \frac{\partial f_i}{\partial \sigma_k}\frac{(\mathrm{d}\sigma_k/\mathrm{d}t)}{(\mathrm{d}\sigma_i/\mathrm{d}t)} + \frac{\partial f_i}{\partial T}\frac{(\mathrm{d}T/\mathrm{d}t)}{(\mathrm{d}\sigma_i/\mathrm{d}t)} \\ &= f_i^{-1}\left(\sum_{k=1}^{N} \frac{\partial f_i}{\partial \sigma_k} f_k + \frac{\partial f_i}{\partial T}\frac{\mathrm{d}T}{\mathrm{d}t}\right).\end{aligned} \tag{5.39}$$

To avoid confusion with the *partial* derivative of f_i with respect to σ_i, $(\partial f_i/\partial \sigma_i)$, a special name and notation will be employed:

$$Z_i \equiv \frac{\mathrm{d}f_i}{\mathrm{d}\sigma_i}. \tag{5.40}$$

The variable Z_i corresponds to Brandon's "diagonal transition matrix" or "state variable differential." It is worthy of note that Z_i may also be recognized as the ratio of second- to first-time derivatives of the mole number σ_i ($i = 1, \ldots, N$):

$$Z_i = \frac{d^2\sigma_i/dt^2}{d\sigma_i/dt} = \frac{df_i/dt}{f_i} = d\ln f_i/dt. \quad (5.41)$$

Therefore, Z_i is a measure of the *local curvature* of the graph of σ_i with time.

To evaluate Z_i analytically, the partial derivatives must be determined. They may easily be found by inspection of (5.39) and (5.60–5.62). The total temperature derivative in (5.39) is obtained from taking time derivatives of (5.27),

$$\frac{dT}{dt} = -\sum_{i=1}^{N} h_i f_i \Big/ \sum_{i=1}^{N} \sigma_i C_{pi}, \quad (5.42)$$

where $C_{pi}(i = 1, \ldots, N)$ are the constant pressure specific heat capacities.

With this new notation,

$$f_i = f_i^n + Z_i^n(\sigma_i - \sigma_i^n), \quad (5.43)$$

or, regrouping terms,

$$f_i = (f_i^n - Z_i^n\sigma_i^n) + Z_i^n\sigma_i. \quad (5.44)$$

This locally linearized equation is readily integrated to give the locally exact solutions,

$$\sigma_i^{n+1} = (\sigma_i^n - f_i^n/Z_i^n) + (f_i^n/Z_i^n)\exp(Z_i^n h), \quad (5.45)$$

where $(n, n + 1)$ denotes the time level, and where H is the time step-size $(t^{n+1} - t)$.

Equation (5.50) can be rewritten as

$$\sigma_i^{n+1} = \sigma_i^n + hf_i^n\left(\frac{\exp(Z_i^n h) - 1}{Z_i^n h}\right). \quad (5.46)$$

For graphs of σ_i versus time which have negative curvature, $Z_i < 1$, the first form (5.45) may be thought of as an asymptotic-plus-transient solution, the second (5.46) as a "filtered Euler" approximation (Miranker 1981).

Finally, an important relationship which follows from the locally exact solution may be obtained by substitution of (5.46) into (5.44):

$$f_i^{n+1} = f_i^n \exp(Z_i^n h). \quad (5.47)$$

This emphasizes the role of the state variable differential Z_i as a "diagonal transition matrix." Thus Z_i may be thought of as a finite-difference analog of, or approximation to, the eigenvalues of the Jacobian matrix of (5.24).

The "Z-formulated" equations (5.46) are formally identical to the "L-formulated" (5.36) if $-Z_i^n$ is substituted for L_i^n. By analogy, the state variable differential Z_i may be seen to be a negative or positive time constant, so that approximate solutions to the Z-formulated equations must be resolved on a time scale of order $|Z_i|_{max}^{-1}$. Fortunately, as can be seen by inspection of the curvature of the σ_i versus time traces in Figure 5.3, $|Z_i|_{max}$ decreases asymptotically as the reactions go to chemical equilibrium (at large t), so that the Z-formulated equations admit increased step sizes, whereas use of the L-formulated equations requires a steady-state approximation, asymptotic or other *ad hoc* assumptions to overcome the small step size scaled on $(L_i)_{max}^{-1}$ at large t, due to the exponentially increasing temperature. Therefore, the Z-formulation is used exclusively in all post-induction regimes.

The problem of calculating transient batch reaction in homogeneous gas-phase systems is analogous to the calculation of laminar flame propagation, except for the absence of diffusive phenomena associated with a flame front. However, both scenarios share the computational difficulties associated with three distinctively different regimes of time (batch reaction) or space (flame reaction). These three regimes, illustrated in Figure 5.1 are: induction regime (preflame zone); heat-release regime (flame zone), and equilibration regime (post-flame zone). The distinguishing physical characteristics of each regime will be summarized in turn, and the corresponding computational implications will be discussed. For reasons of simplicity and clarity, attention will be focused exclusively on the batch-reaction problem.

The induction period immediately follows some form of bulk ignition, for example, following the passage of a shock wave through a homogeneous near-stoichiometric mixture of fuel and air. Computationally, induction will occur whenever the properties of computational fluid elements from adjacent grid nodes are averaged to simulate mixing; for example, if a thermally and chemically "hot" fluid element is mixed with a "cold" one (Pratt 1976, 1979b). During the induction period, concentrations of reactant precursors and intermediate chain-carriers (e.g. O, OH, H, CH_x) must increase from their poorly characterized or zero initial values to a value sufficiently great to initiate the endothermic reactions which lead to oxidative and thermal pyrolysis of the hydrocarbon fuel to produce the pyrolyzates CO and H_2. Throughout the induction regime, the coupling with the energy equation (an algebraic constraint) is weak, so that essentially isothermal conditions obtain. The induction period ends when observable ignition occurs, identified by a rapid rate of temperature increase ($dT/dt \approx 10^7 \text{ K s}^{-1}$) and an accompanying rapid depletion of reactants and rapid increase of intermediates and products.

The principal computational difficulties in modelling the induction regime are two-fold. First, all species are initially in a chemically non-equilibrated state in which all of the chemical reactions are proceeding at bulk-average temperature rates, with concentrations of key species so low (they may not be measurable) that the corresponding reaction rates are not physically meaningful. These ill-posed initial conditions cause the governing set of ODEs to be extraodinarily stiff (Young and Boris 1977; Moore and Eccles 1974; McRae, Goodin, and Seinfeld 1982). Second, since the required time to achieve ignition is not known a priori, and the Jacobian matrix of the governing equations is dominated by positive real eigenvalues, it is necessary to take small time steps throughout the induction regime.

The induction regime ends and the heat release regime beings when a sharply defined increasing rate of change of temperature and mole numbers occurs. In this regime, the full chemical mechanism is active with very strong temperature coupling through the energy conservation equation (5.32). This regime is characterized by an exponentially increasing temperature, which in turn dominates the reaction rates through the exponential dependence on temperature. The heat release period ends after concentrations of the chain-carrying intermediate (O, OH, H) have passed their peak values, when all species have begun to approach their chemically equilibrated concentrations. During this regime, the eigenvalues of the Jacobian matrix change from predominately positively to predominately negative real values.

Following heat release, all species mole numbers are in monotonic, asymptotic approach toward their respective chemical equilibrium values. The describing ODEs are once again stiff, but now in a different sense than during the induction period: in the equilibration regime, many reactions achieve a condition in which forward and reverse rates are large in value but have vanishingly small differences. The Jacobian eigenvalues all have negative real values, so that very large step sizes are now possible.

The equilibration process does not have a clearly defined termination, due to the asymptotic nature of the approach to the chemical equilibrium state. Since equilibrium values of temperatures and species concentration are easily determined a priori by a Gibbs function minimization scheme (Gordon and McBride 1971), the end of the equilibration period can be defined as the time at which all of the mole numbers and the temperature are within (say) one percent of their chemical equilibrium values.

A number of single-step algorithms were considered, including:

(i) implicit Euler rule,
(ii) implicit midpoint rule,
(iii) trapezoidal rule,
(iv) exponential-fitted trapezoidal rule: "Liniger-Willoughby No. 1" (Liniger and Willoughby 1970),
(v) trapezoidal rule with end correction: cubic spline, or "Obreschkoff (2,2)" (Wanner 1977),
(vi) exponential-fitted trapezoidal rule with end correction: "Liniger-Willoughby No. 3" (Liniger and Willoughby 1970),
(vii) locally-exact or exponential solutions.

Variations and combinations of the seven basic algorithms were explored; for example, Young and Boris' SAIM (Selected Asymptotic Integration Method), which utilizes a combination of locally-exact ("asymptotic") solutions and an explicit Euler-trapezoidal rule predictor-corrector method.

Of the seven algorithms and their variants tested, it was found that the most useful algorithms were variations of "Liniger-Willoughby No. 1," the exponential-fitted trapezoidal rule.

It was found necessary to develop two distinct versions of the algorithm to accommodate the widely differing characteristics of the physical/computational regimes of induction, heat release, and equilibration.

Consider an approximate solution of (5.24) based on the "theta method" variation of the trapezoidal rule:

$$\sigma_i^{n+1} = \sigma_i^n + h[U_i f_i^{n+1} + (1 - U_i) f_i^n], \qquad i = 1, \ldots, N, \tag{5.48}$$

where U_i is a degree-of-implicitness or "tuning" factor such that $U_i = 0$ recovers the explicit Euler approximation, $U_i = 1$ recovers the implicit Euler approximation, and $U_i = 1/2$ recovers the trapezoidal rule.

Liniger and Willoughby introduced the concept of "exponential-fitting" the parameter U_i to the locally exact solutions (5.36) or (5.46):

$$\sigma_i^{n+1} = \sigma_i^n + hf_i^n \left(\frac{\exp(\theta_i h) - 1}{\theta_i h}\right), \qquad i = 1, \ldots, N, \tag{5.49}$$

where θ_i denotes either $-L_i$ or Z_i, as appropriate.

Equation (5.49) resulted from integration of a locally linearized form of the rate equation (5.24), as given for example by (5.35) or (5.44), suitably generalized to give

$$f_i = (f_i^n - \theta_i \sigma_i^n) + \theta_i \sigma_i \,, \tag{5.50}$$

where $\theta_i = -L_i$ or Z_i, as appropriate.

Applying (5.50) to the entire time step h,

$$f_i^{n+1} = (f_i^n - \theta_i \sigma_i^n) + \theta_i \sigma_i^{n+1}. \tag{5.51}$$

Substituting (5.50) for the term f_i^{n+1} in (5.48), eliminating σ_i^{n+1} between (5.48) and (5.49), and solving explicitly for U_i, there results

$$U_i = \frac{1}{\theta_i h} + \left(\frac{1}{1 - \exp(\theta_i h)}\right). \tag{5.52}$$

However, in order to maintain the A-stability of (5.48), it is necessary to restrict U_i to the interval (0.5, 1.0). Thus, whenever U_i is less than 0.5, which occurs only for Z_i greater than zero, (5.52) must be modified so that $U_i = 0.5$; thus (5.48) defaults to the second-order-accurate trapezoidal rule. However, for the entire L-formulation, and for negative Z_i in the Z-formulation, (5.48) and (5.52) together are equivalent to the locally-exact or exponential solution (5.49). Brandon (1974) has shown that the equivalent polynomial accuracy of (5.49) is of order six to eight. Thus (5.48) and (5.52), with the restriction ($0.5 \leqslant U_i \leqslant 1$), constitute an *exponential-fitted trapezoidal rule*, a method which is A-stable, and has polynomial-order accuracy of at least two, and as great as six to eight. It is interesting to note that, using Liniger and Willoughby's terminology, Treanor's method would be called an "exponential-fitted, fourth-order Runge–Kutta" method.

We now explain a predictor–corrector method for combustion kinetic problems that uses these ideas.

In the induction regime and the first half of the heat-release regime, the eigenvalues of the Jacobian matrix for (5.24) are predominantly positive real numbers. For the Z-formulated equations, the basic algorithm defaults to the trapezoidal rule except for those few equations which may have negative real eigenvalues. Therefore, it is necessary to take small time steps for reasons of accuracy and stability. It is therefore desirable to avoid the computational work associated with evaluation of both the Jacobian matrix elements and eigenvalues, and with matrix inversion. It follows that a predictor–corrector method with Jacobi iteration or simple successive substitution is an appropriate method of convergence.

A "filtered, explicit-Euler" version of (5.49) is used as a predictor:

$$\sigma_i^{n+1} = \sigma_i^n + hf_i^n \left(\frac{\exp(\theta_i h) - 1}{\theta_i h}\right), \qquad i = 1, \ldots, N, \tag{5.53}$$

where $\theta_i = -L_i^n$ or Z_i^n for the L- or Z-formulations, respectively. In the Z-formulation, the term in brackets in (5.53) is set equal to unity whenever Z_i^n is positive.

A Jacobi-Newton form of (5.48) is chosen for the corrector,

$$\Delta\sigma_i^{n+1} = \frac{\sigma_i^n - \sigma_i^{n+1} + h[U_i f_i^{n+1} + (1 - U_i) f_i^n]}{1 + hU_i L_i^{n+1}}, \qquad i = 1, \ldots, N. \tag{5.54}$$

The loss coefficients L_i $(= -\,\partial f_i/\partial\sigma_i)$ are used in the denominator of (5.54) for both L- and Z-formulations. Because the L_i are always positive, the denominator in (5.54) acts as a self-scaled damping factor, allowing greater step sizes than directly using (5.48) as a simple or Jacobi iteration corrector.

In (5.54), $\Delta\sigma_i^{n+1}$ is the current Newton iterate correction, and U_i is restricted to the interval (0.5, 1.0). U_i is related to either L_i or Z_i according to (5.52). When the Z-formulation is used, $\theta_i = \bar{Z}_i$ in (5.52) is determined by a central-difference approximation to (5.41),

$$\bar{Z}_i \approx \frac{f_i^{n+1} - f_i^n}{\sigma_i^{n+1} - \sigma_i^n}. \tag{5.55}$$

After each iteration of (5.54), (5.52), and (5.55), the mole numbers are corrected without acceleration,

$$\sigma_i^{n+1} = \sigma_i^{n+1} + \Delta\sigma_i^{n+1}, \tag{5.56}$$

and iteration of (5.54–5.56) is continued until all of the current iterate corrections satisfy a convergence criterion:

$$|\Delta\sigma_i/\sigma_i| < \epsilon \approx 10^{-3}, \qquad i = 1, \ldots, N. \tag{5.57}$$

After all of the species corrector equations are converged, the temperature is updated by one or two Newton iterations of (5.32), as follows

$$\Delta T = \frac{H_0 - \sum_{i=1}^{N} \sigma_i h_i}{\sum_{i=1}^{N} \sigma_i C_{pi}} \quad \text{and } T \leftarrow T + \Delta T. \tag{5.58}$$

In the last half of the heat release regime and throughout the equilibration regime, the eigenvalues of the Jacobian matrix of (5.24) have predominantly negative real values, so that it is possible to use very large time steps, limited only by the accuracy of the integration method. However, when using very large time steps, it is necessary to use a Newton-Raphson scheme with full Jacobian calculation and matrix inversion to converge the equations.

Equation (5.53) is again used as a predictor for species mole numbers. The predicted values of σ_i $(i = 1, \ldots, N)$ are then used in (5.58) to obtain a predicted value for the temperature T.

A Newton-Raphson corrector equation is used, with log-variable corrections,

$$\sum_{k=1}^{N} \frac{\partial F_i}{\partial \ln \sigma_k} \Delta \ln \sigma_k + \frac{\partial F_i}{\partial \ln T} \Delta \ln T = -F_i, \qquad i = 1, \ldots, N, \tag{5.59}$$

where the Newton-Raphson functional F_i is defined from (5.48) by

$$F_i \equiv \frac{\sigma_i^{n+1} - \sigma_n^n}{U_i h} - \left(\frac{1 - U_i}{U_i}\right) f_i^n - f_i^{n+1}. \tag{5.60}$$

The partial derivatives in (5.59) are given by

$$\frac{\partial F_i}{\partial \ln \sigma_k} = \frac{\delta_{ik}\sigma_i}{U_i h} + \rho^{-1} \sum_{j=1}^{J} (a'_{ij} - \alpha''_{ij})(R_j \alpha'_{kj} - R_{-j}\alpha''_{kj}), \quad i = 1, \ldots, N, \tag{5.61}$$

and

$$\frac{\partial F_i}{\partial \ln T} = \rho^{-1} \sum_{j=i}^{J} (\alpha'_{ij} - \alpha''_{ij}) A_j, \tag{5.62}$$

where

$$A_j = [R_j(B_j + E_j/RT - n'_j + 1) - R_{-j}(B_{-j} + E_{-j}/RT - n''_j + 1)],$$

and where n'_j and n''_j are the molecularities of the forward and reverse reactions, respectively. The partial derivatives of the reciprocal molecular weight, σ_m, are assumed to be negligible compared to other terms. The "tuning factors" U_i are determined by (5.52) with $\theta_i = Z_i^n$. Note the computational efficiency that results from using the Jacobian elements both for *accuracy* (5.52), and for *convergence* (5.59–5.62), an advantage of this method pointed out by Brandon (1974).

The Newton-Raphson functional for temperature is determined from (5.27),

$$F_T = H_0 - \sum_{i=1}^{N} \sigma_i h_i, \tag{5.63}$$

and the corresponding log-variable correction equation is

$$\sum_{k=1}^{N} h_k \sigma_k \, \Delta \ln \sigma_k + T \sum_{i=1}^{N} C_{pi}\sigma_i \, \Delta \ln T = -F_T. \tag{5.64}$$

Equations (5.59) and (5.64) are solved simultaneously by LU decomposition and back-substitution, and the resulting log-variable corrections are used to update the current iterate values of σ_i and T by the approximate equations

$$\sigma_i^{n+1} \approx \sigma_i^{n+1} [1 + (\Delta \ln \sigma_i)^{n+1}], \tag{5.65}$$

and

$$T^{n+1} = T^{n+1} [1 + (\Delta \ln T)^{n+1}]. \tag{5.66}$$

The objective of the calculation is to solve the following initial value problem: Given a set of initial conditions σ_i^0, $(i = 1, \ldots, N)$, T^0, and P solve and return the values of $(\sigma_i, i = 1, \ldots, N)$ and T at the end of a prescribed time interval Δt. The overall steps taken to accomplish this are:

(i) filter the initial conditions by integrating over a small time step $h = \min(L_i^{-1})$, using an L-formulated predictor (5.36) and an implicit Euler corrector

(5.54) ($U_i^n = 1$ for all $i = 1, \ldots, N$) with Jacobi-Newton iteration to convergence;

(ii) adjust the step size for the rate of convergence at step (i); use an explicit-Euler predictor and a Jacobi-Newton trapezoidal-rule corrector (5.54) with $U_i^n = 0.5$ for all $i = 1, \ldots, N$), iterate to convergence;

(iii) estimate local truncation error from step (ii) and therefrom the maximum step size permissible to stay within desired error tolerance, h_{accy}; from the observed convergence rate of J-N iteration, estimate the maximum step size permissible to stay within a desired range of convergence rates, h_{iter};

(iv) if ($h_{\text{accy}} \sim h_{\text{iter}}$), continue Jacobi-Newton iteration with new step size $h = \min(h_{\text{accy}}, h_{\text{iter}})$. If ($h_{\text{accy}} \gg h_{\text{iter}}$), go to Newton-Raphson iteration (5.59) with new step size $h = h_{\text{accy}}$;

(v) continue steps (iii) and (iv) until the end of the prescribed interval Δt is reached; return the solution.

Convergence rate is measured by monitoring the largest of the iterate corrections (5.57); the ratio of largest corrections for two or more iterations is defined as the convergence rate, R_{conv}. A range of desirable convergence rates is the interval (0.2, 0.6); computational efficiency suffers if convergence is either too fast or too slow. Therefore,

$$h_{\text{iter}} = h_{\text{old}}(0.6/R_{\text{conv}})^{1/2}, \tag{5.67}$$

and step size is adjusted up or down by the square-root relation of (5.67) if R_{conv} falls outside (0.2, 0.6). (The square root is due to the super-linear convergence of Jacobi-Newton iteration.)

The local truncation error (lte) estimate is obtained as follows:

(i) from the converged solution, determine the average value of $\bar{Z}_i$ for each $i = 1, \ldots, N$ from (5.55);

(ii) calculate $\bar{U}_i = U_i(\bar{Z}_i h)$ from (5.52);

(iii) the local truncation error is then given by (Liniger and Willoughby 1970)

$$\text{lte} = h \max_i \left| \frac{f_i^n(\bar{U}_i - U_i^n)\bar{Z}_i h}{1 - \bar{U}_i \bar{Z}_i h} \right|, \tag{5.68}$$

where $U_i^n \equiv U_i(Z_i^n h)$ is the integrating factor used in solving either (5.54) or (5.60). Of course, if $Z_i^n > 0$, then $U_i^n = 0.5$ was used.

Because both the U_i functions in (5.68) depend nonlinearly on the step size h according to (5.52), analysis shows that lte $\sim h^3$ when $Z_i h \to 0$ (trapezoidal rule), lte $\sim h^2$ up to about $|Z_i h| \sim 5$, and lte $\sim h$ when $|Z_i h| \to \infty$. Since the maximum of lte over all $i = 1, \ldots, N$ variables is always of interest, (5.68) is inverted to estimate h_{accy} as follows:

$$h_{\text{accy}} \sim h\epsilon_0/(\text{lte}), \tag{5.69}$$

where ϵ_0 is the prescribed error tolerance.

The computational work of evaluating logarithms and exponentials may be almost entirely eliminated by judicious use of approximating functions. For example, the term $(e^x - 1)/x$ in (5.31) is evaluated in the code by means of a (2,2) diagonal Pade (rational function) approximation for e^x,

$$e^x_{(2,2)} = \frac{1 + x/2 + x^2/12}{1 - x/2 + x^2/12}, \qquad |x| \leqslant 1, \tag{5.70}$$

which results in

$$\left(\frac{e^x - 1}{x}\right)_{(2,2)} = \frac{1}{1 - x(1/2 - x/12)}, \qquad |x| \leqslant 1. \tag{5.71}$$

Similarly, the "tuning factor" U_i of (5.51) can be evaluated with sufficient accuracy by means of a (3, 3) diagonal Pade approximation for e^x, to give

$$\frac{1}{x} + \frac{1}{1 - e^x_{(3,3)}} = \frac{1}{2} - \frac{x}{12 + x^2/5}, \qquad |x| \leqslant 1. \tag{5.72}$$

The right-hand sides of (5.71) and (5.72) require only five operations to evaluate and do not exhibit the singularity at $x = 0$ of the left-hand sides.

Another significant reduction in computational work was achieved by locally linearizing the fifth-order polynomial approximations used in evaluating the species-i enthalpy h_i and specific heat capacities C_{pi}, which appear in (5.58) (5.63), and (5.64).

Finally, it should be noted that while log-variable corrections are used in the Newton–Raphson version of the algorithm, (5.59), evaluation of logarithms of the variables themselves is avoided by use of (5.65). The only evaluations of exponentials in the code are those necessarily required for the Arrhenius rate expressions, (5.2a, b), and there are no logarithm evaluations in the code at all.

A working version of the present code, denoted CREK-1D (for Combustion Reaction Kinetics—1-Dimensional), has been tested against LSODE by Radhakrishnan (1982) and described in Pratt (1979a). Preliminary results indicate that the present version of CREK-1D is about five times faster than LSODE, using what appears to be the fastest combination of LSODE options. The current version of CREK-1D does not contain sophisticated numerical or coding techniques and therefore is probably capable of some modest improvements in execution speed.

The most significant differences in performance between the two codes are:

(i) In the induction regime, both codes automatically select approximately the same step sizes. Since CREK-1D does not compute Jacobian elements or perform matrix inversion in this regime, it is much faster than LSODE.

(ii) In the post-induction regimes, both CREK-1D and LSODE calculate Jacobian elements and do matrix inversion; however, CREK-1D automatically selects step sizes which are three to four times larger than those selected by LSODE, while maintaining comparable accuracy and the same number of iterations per step for convergence (one to three). This difference in performance is apparently due to the unusual suitability of exponentials as approximating functions for transient equilibration of chemical species.

CREK-1D also was tested on a very stiff, isothermal kinetics test problem considered by Edsberg (1974). The test problem was readily solved by CREK-1D; however, no direct comparison with Edsberg's reported solution was possible. Detailed performance comparisons of CREK-1D against LSODE will be performed and published sometime in 1983.

A major conclusion of this work is that the L-formulated equations, widely used at present for modelling both atmospheric chemistry and combustion processes, should be employed *only* in the initial induction period, when some mole numbers may be very small, and then only in combination with step sizes scaled on the larger of the L- or Z-formulation estimates. In all other regimes, use of the Z-formulated equations completely obviates the need to use asymptotic or quasi-steady-state assumptions to resolve the near-equilibrium stiffness problem uniquely associated with the L-formulated equations. For atmospheric chemistry problems (for which regime II is nonexistent), the present algorithm without the energy constraint should constitute a very fast and robust method for calculation of isothermal, batch-homogeneous gas-phase kinetics.

The success of the present, two-part algorithm stems from recognition of the fact that the approximating equations describing the induction processes are *stability-limited*, whereas the corresponding post-induction equations are only *accuracy-limited*. In the former case, low-accuracy methods with the least possible computational work per iteration are indicated due to the necessarily restricted step size. However, in the post-induction processes, where stability is not a problem, step sizes may be as large as the accuracy and convergence radius of the approximating equation permits. The use of exponentials as approximating functions satisfies the accuracy requirement, and Newton iteration gives an infinite convergence radius. The added computational work of Jacobian evaluation and matrix inversion required for Newton iteration is offset by very large step sizes and quadratic convergence.

5.2.4 *Integration of the ill-conditioned system of boundary valued ODEs for the laminar, opposed jet diffusion flame* (*Hahn and Wendt*)†

The laminar opposed jet diffusion flame, Figure 5.4 is a combustion configuration that possesses a number of useful attributes. First, it allows a flat flame to be established in space, without direct influences of burners or possible catalytic surfaces; second, it allows investigation of kinetic mechanisms occurring in hot fuel rich regions and over a wide range of conditions, which due to flammability limitations, cannot be established in the premixed mode; third, it may be thought to represent a prototype model of laminar flamelets, each of which is strained in its own plane, and which represent the reaction zones important in turbulent diffusion flames. Its usefulness as a tool to test and corroborate detailed kinetic mechanisms rests on its ability to be modelled and on the power of numerical procedures to enable accurate solution of the ensuing ill-conditioned system of ordinary boundary valued differential equations to be achieved.

Unlike previous work (Fendell 1965; Krishnamurthy and Williams 1974) which has focused on analytical solutions of models utilizing simplified combustion chemistry with either infinitely rapid chemical reactions or a simple reaction with a very high activation energy (Peters 1978), we examine here numerical procedures necessary to solve the problem for an arbitrary large set of reactions describing the free-radical combustion chemistry pertinent to this flame. The

† The authors would like to thank W. Steven Lanier for his suggestion to attempt the modified discretization described in this work. Additional insight contributed by J. Heinrich and C. J. Kau is also much appreciated. This work was supported by the U.S. Environmental Agency under Grant R806685–01 to the University of Arizona.

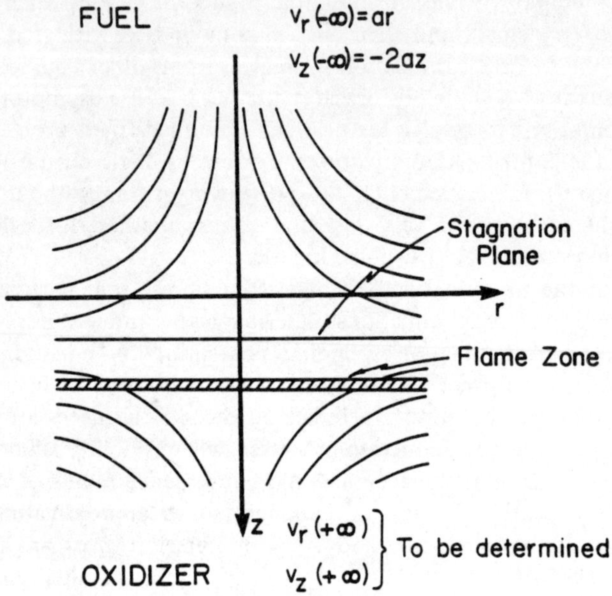

Fig. 5.4 Schematic of the laminar opposed jet diffusion flame.

equations describing the laminar opposed jet diffusion flame comprise a severe test of any numerical integration scheme. The problem is a boundary valued problem with concentrations and temperature set at $\pm\infty$. Moreover, this is one of the cases in nature where there is direct coupling between the energy and the momentum balances through the changes in gas density ρ. An accurate prediction of the velocity profile is essential since this has a first-order influence on all other profiles. Inclusion of the effects of viscosity and other dynamic effects on the overall concentration profiles, together with chemical reaction, distinguishes this problem from that of the one-dimensional premixed flat flame, where the velocity profile can be calculated from kinematic considerations alone.

The reaction zone thickness depends on a parameter, ϵ, the rate of stretching, and very steep spatial gradients in concentrations, temperature, and reaction rates are usual. A detailed combustion chemistry mechanism typically consists of many free-radical reactions with characteristic times differing by over thirty orders of magnitude at a point in space. This naturally introduces computational difficulties, and special techniques must be implemented to overcome them. The purpose of this subsection is to survey the problem qualitatively and to present the approaches used to solve it. These approaches address both the problem of ill-conditioning and that associated with the steep gradients occurring in the spatial domain near the reaction zone.

A complete description of the system shown in Figure 5.4 involves the simultaneous solution of the following equations in cylindrical coordinates:

(i) continuity:

$$\frac{\partial \rho}{\partial t} + \frac{1}{r}(\rho r v_r) + \frac{\partial}{\partial z}(\rho v_z) = 0, \tag{5.73}$$

(ii) r momentum:

$$\rho\frac{\partial v_r}{\partial t} + \rho v_r\frac{\partial v_r}{\partial r} + \rho v_z\frac{\partial v_r}{\partial z} = -\frac{\partial P}{\partial r} - \frac{1}{r}\frac{\partial}{\partial r}(r\tau_{rr}) - \frac{\tau_{\theta\theta}}{r} + \frac{\partial \tau_{rz}}{\partial z}, \tag{5.74}$$

(iii) z momentum:

$$\rho\frac{\partial v_z}{\partial t} + \rho v_r\frac{\partial v_z}{\partial r} + \rho v_z\frac{\partial v_z}{\partial z} = -\frac{\partial P}{\partial z} - \frac{1}{r}\frac{\partial}{\partial r}(r\tau_{rz}) + \frac{\partial \tau_{zz}}{\partial z}, \tag{5.75}$$

(iv) energy:

$$cC_{\mathrm{p}}\left(\frac{\partial T}{\partial t} + v_r\frac{\partial T}{\partial r} + v_z\frac{\partial T}{\partial z}\right) = +\frac{1}{r}\frac{\partial}{\partial r}\left(rk\frac{\partial T}{\partial r}\right) + \frac{\partial}{\partial z}\left(k\frac{\partial T}{\partial z}\right) + \sum_{j=1}^{n_{\mathrm{r}}} r_j\Delta h_j, \tag{5.76}$$

(v) species:

$$\frac{\partial(cx_A)}{\partial t} + \frac{1}{r}\frac{\partial}{\partial r}(crx_Av_r) + \frac{\partial}{\partial z}(cx_Av_z) = \frac{1}{r}\frac{\partial}{\partial r}\left(rcD_A\frac{\partial x_A}{\partial z}\right) + \frac{\partial}{\partial z}\left(cD_A\frac{\partial x_A}{\partial z}\right) + R_A,$$

$$(A = 1, \ldots, N_{\mathrm{S}}), \tag{5.77}$$

(vi) Ideal Gas Law:

$$P/c = RT.$$

Table 5.2 contains nomenclature for these and subsequent equations.

In the above equations, pseudo binary diffusion has been assumed and radiative heat losses and thermal diffusion have been neglected following Field, Gill, Morgan, and Hawksley (1967). The time dependence has been retained since we wish to solve these equations from an initial guess at $t = 0$ to the steady state at $t \to \infty$.

The boundary conditions to (5.73-5.77) are:

$$\left.\begin{aligned} v_z &= -2\epsilon z, \\ v_r &= \epsilon r, \end{aligned}\right\} \quad z = -\infty \text{ for all } r, \tag{5.78}$$

$$P = P_\infty, \qquad z = \pm\infty \text{ for all } r, \tag{5.79}$$

$$T = T_\infty, \qquad z = +\infty \text{ for all } r, \tag{5.80}$$

$$T = T_\infty, \qquad z = -\infty \text{ for all } r, \tag{5.81}$$

$$x_A = x_{A1}, \qquad z = +\infty \text{ for all } r, \tag{5.82}$$

$$x_A = x_{A2}, \qquad z = -\infty \text{ for all } r. \tag{5.83}$$

It can be shown that (5.73-5.77) can be reduced to a system of ODEs by assuming self-similar solutions of the following forms:

$$v_r = \epsilon r\psi(z)\,; v_z = v(z), \tag{5.84}$$

TABLE 5.1
Nomenclature

A	Block tridiagonal matrix
A_i	Variable coefficient
B_i	Variable coefficient
c	Molar density (g-mol cm^{-3})
C_i	General dependent variable
C_p	Heat capacity (cal g-mol^{-1} K^{-1})
D_A	Pseudobinary diffusion coefficient of species A in gaseous mixture
D_i	Variable coefficient
$F_{i,p}^k$	Forcing function evaluated at grid p and time k
F_i	Block vector of forcing functions
h	Mesh size for modified discretization
k	Thermal conductivity of gaseous mixture (cal cm^{-1} s^{-1} K^{-1})
m	Number of grid points
W_M	Molecular weight of mixture
n	Number of gases in mixture
n_r	Number of chemical reactions
N_S	Number of species in reaction set
N	$N_\mathrm{S} + 2$
P	Pressure (atm)
r	Radial distance (cm)
r_j	Rate of reaction j (g-mol cm^{-3} s^{-1})
R_A	Rate of formation of species A (g-mol cm^{-3} s^{-1})
R_i	Forcing function
t	Time (s)
T	Temperature (K)
v	Axial velocity (cm s^{-1})
$\boldsymbol{v}$	Velocity vector (cm s^{-1})
v_r	Radial component of $\boldsymbol{v}$ (cm s^{-1})
v_z	Axial component of $\boldsymbol{v}$ (cm s^{-1})
$W_{i,p+j}$	Weighting factor for modified discretization ($i = 1, 2; j = -1, 1$)
x_A	Mole fraction of species A
z	Axial distance (cm)
Greek Symbols	
$\alpha_{i,p}$	Auxiliary variable
$\beta_{i,p}$	Auxiliary variable
$\gamma_{i,p}$	Auxiliary variable
ΔC	Block vector of corrections
$\Delta C_{j,p}$	Correction to variable j at point p
Δh_j	Heat of reaction j (cal g-mol^{-1})
Δz	Increment of grid
ϵ	Stretching rate (s^{-1})
θ	Dimensionless temperature $= (T - 298)/298$
μ	Viscosity of mixture (g cm^{-1} s^{-1})

ρ	Density (g cm^{-3})
ψ	Similarity transformation

Subscripts

∞	Refers to distance from flame zone
i	Refers to dependent variable i
p	Refers to grid point

Superscripts

k	Refers to time step

$$\frac{\partial P}{\partial r} = f(r); \quad \frac{\partial P}{\partial z} = g(z), \tag{5.84}$$

$$\rho = \rho(z); \quad T = T(z); \quad x_A = x_A(z). \tag{5.86}$$

Furthermore, since

$$\frac{\partial P}{\partial r} = f(r) = \epsilon^2 \rho_\infty r,$$

it can be shown (Hahn 1979) that the proposed self-similar solution is consistent with the equations, which now become totally independent of r. It is convenient to define a dimensionless temperature as $\theta = (T - T_0)/T_0$, to yield the following differential equations to be solved:

(i) Mass:

$$\frac{\partial \rho}{\partial t} = -\frac{\partial}{\partial z}(\rho v) + 2\epsilon\phi\psi, \tag{5.87}$$

(ii) r momentum:

$$\frac{\partial \psi}{\partial t} = \frac{\mu}{\rho}\frac{\partial^2 \psi}{\partial z^2} + \left(\frac{1}{\rho}\frac{\partial \mu}{\partial z} - v\right)\frac{\partial \psi}{\partial z} - \epsilon\psi^2 + \frac{\rho_\infty}{\rho}\epsilon, \tag{5.88}$$

(iii) Energy:

$$\frac{\partial \theta}{\partial t} = \frac{k}{cC_p}\frac{\partial^2 \theta}{\partial z^2} + \left(\frac{1}{cC_p}\frac{\partial k}{\partial z} - v\right)\frac{\partial \theta}{\partial z} + \frac{1}{T_0 c C_p}\sum_{j=1}^{n_r} r_j \Delta h_j, \tag{5.89}$$

(iv) Species:

$$\frac{\partial x_A}{\partial t} = D_A \frac{\partial^2 x_A}{\partial z^2} + \left(\frac{1}{c}\frac{\partial}{\partial z}(cD_A) - v\right)\frac{\partial x_A}{\partial z} + \left(-\frac{1}{c}\frac{\partial c}{\partial t} - 2a\psi - \frac{1}{c}\frac{\partial (cv)}{\partial z}\right) \times$$

$$\times x_A + \frac{1}{c}R_A, \qquad (A = 1, \ldots, N_S). \tag{5.90}$$

The z momentum equation plays no role other than to calculate $\partial P/\partial z$.

Of interest is the solution as $t \to \infty$ when the left-hand sides of these equations are equal to zero. This is the desired one-dimensional steady-state solution. It should be noted that all physical and transport properties are allowed to vary with position. Simplifying assumptions with respect to diffusion can be made, depending on which level of complication for $\tilde{D}_A$, the pseudobinary diffusion coefficient, is appropriate.

Equations (5.88-5.90) were soived numerically in a time-dependent, fully implicit, and fully coupled mode. These equations have the form

$$\frac{\partial c_i}{\partial t} = A_i(z)\frac{\partial^2 c_i}{\partial z^2} + B_i(z)\frac{\partial c_i}{\partial z} + D_i(z)\, C_i + R_i(z), \qquad i = 1, \ldots, N_S + 2, \tag{5.91}$$

where the coefficients are given in Table 5.2.

TABLE 5.2
Coefficients for (5.91)

	A_i	B_i	D_i	R_i
x_A	$\tilde{D}_A$	$\frac{1}{c}\frac{\partial}{\partial z}(c\tilde{D}_A) - v$	$-\frac{1}{c}\frac{\partial c}{\partial t} - 2\epsilon\psi - \frac{1}{c}\frac{\partial (cv)}{\partial z}$	$\frac{1}{c}R_A$
θ	$\frac{k}{cC_p}$	$\frac{1}{cC_p}\frac{\partial k}{\partial z} - v$	0	$\frac{1}{T_0 cC_p}\Sigma\, r_j \Delta h_j$
ψ	$\frac{\mu}{\rho}$	$\frac{1}{\rho}\frac{\partial \mu}{\partial z} - v$	0	$-\epsilon\psi^2 + \frac{\rho_\infty}{\rho}\epsilon$

If the domain in which a solution is sought is given by

$$-l/2 \leqslant z \leqslant +l/2,$$

the boundary conditions for all the x_A and θ values are specified, while for ψ:

$$\psi = 1, \qquad z = -l/2,$$
$$\psi = \sqrt{(\rho_{-\infty}/\rho_\infty)} \qquad z = +l/2.$$

Continuity (5.87) is solved separately in the spatial domain

$$\frac{\partial}{\partial z}(\rho v) = -2\epsilon\phi\psi - \frac{\partial \rho}{\partial t}, \tag{5.92}$$

where

$$\frac{\partial \rho}{\partial t} = \frac{(W_M)P}{R}\frac{\partial}{\partial t}\left(\frac{1}{T}\right). \tag{5.93}$$

Ill-conditioning in the system arises through the term $R_i(z)$ in (5.91) which is the forcing function due to detailed combustion chemical reactions. The coefficient $B_i(z)$ changes sign in the neighborhood of the stagnation plane, causing a regular first-order turning point.

An explicit approach to solve (5.91) is to evaluate A_i, B_i, and R_i using values of a previous iteration and solve for C_i, thus decoupling the system of equations and leading to inversion of a series of tridiagonal matrices. This was attempted but failed. Tyson (1964) developed a technique for solution of systems of stiff initial valued problems in which the term $R_i(z)$ is expanded about the backward time-step. This idea was later used by Wendt, Martinez, Lilley, and Corley (1979) to solve ill-conditioned systems of boundary-valued ODEs and forms the centerpiece for the solution method utilized in this work. Thus

$$R_i^{k+1}(z) = R_i^k + \sum_{j=1}^{N} \left.\frac{\partial R_i}{\partial C_j}\right|^k (C_j^{k+1} - C_j^k) \tag{5.94}$$

with $N = N_S + 2$ and k refers to the time step. This expansion is performed at each grid point p.

Both a collapsing and an equally spaced grid were used with standard finite difference discretization. The appropriate formulae are

$$\frac{dC_i}{dz} = \frac{\Delta z_{p-1}}{\Delta z_{p+1}(\Delta z_{p+1} + \Delta z_{p-1})} C_{i,p+1}^{k+1} - \frac{\Delta z_{p-1} - \Delta z_{p+1}}{\Delta z_{p-1}\Delta z_{p+1}} C_{i,p}^{k+1} - \frac{\Delta z_{p+1}}{\Delta z_{p-1}(\Delta z_{p+1} + \Delta z_{p-1})} C_{i,p-1}^{k+1}, \tag{5.95}$$

and

$$\frac{d^2C_i}{dz^2} = \left(\frac{2}{\Delta z_{p+1} + \Delta z_{p-1}}\right) \frac{C_{i,p+1}^{k+1} - C_{i,p}^{k+1}}{\Delta z_{p+1}} - \frac{C_{i,p}^{k+1} - C_{i,p-1}^{k+1}}{\Delta z_{p-1}}. \tag{5.96}$$

The finite difference form of (5.91) now becomes

$$\left[\alpha_{i,p}, \beta_{ip} - \frac{1}{\Delta t}, \gamma_{i,p}\right] \begin{bmatrix} \Delta C_{i,p-1} \\ \Delta C_{i,p} \\ \Delta C_{i,p+1} \end{bmatrix} + \left[\left.\frac{\partial R_i}{\partial C_j}\right|_p^k\right] \begin{bmatrix} \Delta C_{1,p} \\ \Delta C_{2,p} \\ \vdots \\ \Delta C_{N,p} \end{bmatrix} = -F_{i,p}^k, \tag{5.97}$$

with

$$\alpha_{i,p} = \frac{1}{\Delta z_{p-1}(\Delta z_{p+1} + \Delta z_{p-1})}(2 - \Delta z_{p+1}A_{i,p}), \tag{5.98}$$

$$\beta_{i,p} = -\left[\frac{2}{\Delta z_{p+1} + \Delta z_{p-1}}\left(\frac{1}{\Delta z_{p+1}} + \frac{1}{\Delta z_{p-1}}\right) + \frac{(\Delta z_{p-1} - \Delta z_{p+1})A_{i,p}}{\Delta z_{p+1}\Delta z_{p-1}} - B_{i,p}\right], \tag{5.99}$$

$$\gamma_{i,p} = \frac{1}{\Delta z_{p+1}(\Delta z_{p+1} + \Delta z_{p-1})}(2 + \Delta z_{p-1}A_{i,p}), \tag{5.100}$$

and

$$\Delta C_{j,p} = C_{j,p}^{k+1} - C_{j,p}^{k}. \tag{5.101}$$

$F_{i,p}^{k}$ gathers all remaining terms known at the kth time step and $\alpha_{i,p}$, $\beta_{i,p}$, and $\gamma_{i,p}$ as well as $\partial R_i/\partial C_j$ are also evaluated at the known kth step. If m is the number of grid points (5.97) represents a set of (N.m) simultaneous algebraic equations for the values of the increments $\Delta C_{i,p}$ for a time step Δt. The system can be rearranged as follows:

$$\boldsymbol{A} \cdot \Delta \boldsymbol{C} = -\boldsymbol{F} \tag{5.102}$$

where $\boldsymbol{A}$ is a $(m \times N) \times (m \times N)$ block tridiagonal matrix, i.e. an $m \times m$ matrix in which each element is a $N \times N$ matrix. The diagonal $N \times N$ blocks are full, nondiagonally dominant (depending on Δt) matrices while the off-diagonal blocks are diagonal matrices.

Classical block tridiagonal matrix inversion logic (Wendt, Martinez, Lilley, and Corley 1979) was not successful in solving the flame problems attacked in this work. However, use was made of the fact that $\boldsymbol{A}$ is also a band matrix and therefore the solution to (5.102) can be obtained employing LU decomposition. which was found to be very efficient for this type of problem (Peaceman 1977). This method also saves core storage, since only the band of dimension $2N \times (m \times N)$ need be treated.

With the equally spaced grid, $\Delta z_{p-1} = \Delta z_{p+1}$, a large number of grid points must be used when the reaction zone is thin and gradients are steep. For example, with a CO flame, 8 species and 7 reactions and with $\epsilon = 10\ \text{sec}^{-1}$, 101 grid points were employed to obtain a solution, while 35 grid points were insufficient for proper resolution of the peaks.

Since systems involving 35 species or more are common in this type of problem it was clear that a discretization allowing fewer grid points was essential. This forced utilization of a collapsing grid.

The collapsing grid is shown in Figure 5.5 and was automatically adjusted as follows. The location of the peak in the temperature profile was determined and a grid was created to the left with a specified expansion factor. The expansion factor for the right side was then determined such that both intervals next to the peak temperature location were equal and such that an equal number of grid points were located on each side. This method allowed the CO opposed jet diffusion flame to be solved with 35 grid points, provided that the stretching rate ϵ, was less than 7 sec^{-1}. In these cases the equally spaced grid with 35 grid points was too coarse in the neighborhood of the peak temperature to allow convergence. However, at $\epsilon \geqslant 7\ \text{sec}^{-1}$ the collapsing grid was also prone to problems due to nonsymmetric numerical diffusion of errors, and this led to oscillations and an inability to satisfy the boundary conditions. This is described in detail below.

The burner spacing, l, or domain in which the problem was being solved, was automatically adjusted so that there was no net flux of heat or any species into either of the burners. Thus the distance l could be increased by 50 percent if, at any iteration, any net flux had the incorrect sign at $z = +l$ or $z = -l$. Likewise, l could be automatically decreased by two-thirds if the temperature profile indicated

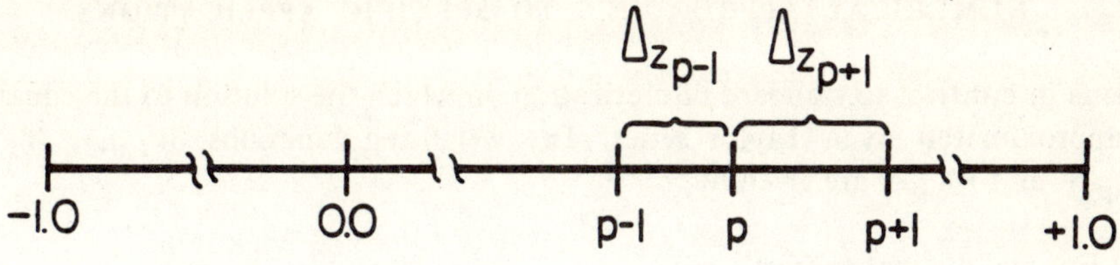

Fig. 5.5 Collapsing grid utilized for finite difference solution.

that greater resolution was appropriate. Whenever the grid had to be adjusted, the new values of all variables were obtained by linear interpolation between values at the previous grid points.

The original initial estimate of profiles at $t = 0$ was from the classical Burke-Shumann flame with an infinitesimally thin reaction zone, together with arbitrary free radical profiles to facilitate ignition. Much computer time can be saved if intermediate nonconverged or converged results are saved and used as the initial guesses for the next run. The adjustable time step was controlled entirely by the temperature profile, allowing changes of one percent at any point in any time interval. At convergence to steady state, determined by relative changes of *all* variables, the time step had become very large, greater than one second. This typically occurred, from a 'reasonable' initial guess within 200 to 400 iterations, from an initial $\Delta t = 10^{-6}$ sec.

When solving the model corresponding to the opposed jet diffusion flame with a high rate of stretching, numerical oscillations were observed in the temperature profile, the ψ profile and the reactant profiles. These undulations occur close to the boundaries of the domain and do not satisfy the boundary conditions and thus prevent convergence. These instabilities originated from the collapsing gird and diffusion of numerical errors towards the boundaries. Yet, a collapsing grid is essential in order to handle the sharp temperature profiles in the center with less than 35 grid points overall. One solution was to use 100 equally spaced grid points and this appeared to allow convergence. However, this large number of grid points constrained the program to solution of relatively small kinetic problems (due to core storage limitations) and did not allow investigation of hydrocarbon plus fuel nitrogen chemistry together. Since the latter was the key aspect of modelling in this work, these results put us on the horns of a dilemma.

This dilemma was ultimately resolved by developing a modified central difference discretization in which the first and second derivations are given by (for equally spaced grids):

$$\frac{dC_i}{dz} = \frac{W_{1,p+1}C_{i,p+1} - W_{1,p-1}C_{i,p-1}}{2h}, \tag{5.103}$$

$$\frac{d^2C_i}{dz^2} = \frac{W_{2,p+1}C_{i,p+1} - 2C_{i,p} + W_{2,p-1}C_{p-1}}{h^2}, \tag{5.104}$$

with the weighting functions $W_{1,p+1}$, $W_{1,p-1}$, $W_{2,p+1}$, and $W_{2,p-1}$ calculated from the roots, α_1 and α_2 of the auxiliary equation for the second-order ODE with constant coefficients used to approximate the true equations within a mesh interval.

This is in contrast to standard discretization, in which the solution to the equations is approximated by a Taylor series. The weighting functions, $W_{1,p+1}$, $W_{1,p-1}$, $W_{2,p+1}$, and $W_{2,p-1}$ are given by:

(i) For real distinct roots α_1, α_2:

$$W_{1,p+1} = 1 - \frac{e^{-\alpha_2 h}[\sinh(\alpha_1 h) - \alpha_1 h]}{\sinh[(\alpha_1 - \alpha_2)h]} - \frac{e^{-\alpha_1 h}[\sinh(\alpha_2 h) - \alpha_2 h]}{\sinh[(\alpha_2 - \alpha_1)h]}, \quad (5.105a)$$

$$W_{1,p-1} = 1 - \frac{e^{\alpha_2 h}[\sinh(\alpha_1 h) - \alpha_1 h]}{\sinh[(\alpha_1 - \alpha_2)h]} - \frac{e^{\alpha_1 h}[\sinh(\alpha_2 h) - \alpha_2 h]}{\sinh[(\alpha_2 - \alpha_1)h]}, \quad (5.105b)$$

$$W_{2,p+1} = 1 - \frac{e^{-\alpha_2 h}[\cosh(\alpha_1 h) - 1 - \frac{1}{2}\alpha_1^2 h^2]}{\sinh[(\alpha_1 - \alpha_2)h]} - \frac{e^{-\alpha_1 h}[\cosh(\alpha_2 h) - 1 - \frac{1}{2}\alpha_2^2 h^2]}{\sinh[(\alpha_2 - \alpha_1)h]}, \quad (5.105c)$$

$$W_{2,p-1} = 1 + \frac{e^{\alpha_2 h}[\cosh(\alpha_1 h) - 1 - \frac{1}{2}\alpha_1^2 h^2]}{\sinh[(\alpha_1 - \alpha_2)h]} + \frac{e^{\alpha_1 h}[\cosh(\alpha_2 h) - 1 - \frac{1}{2}\alpha_2^2 h^2]}{\sinh[(\alpha_2 - \alpha_1)h]}. \quad (5.105d)$$

(ii) For complex roots: $\alpha_1 = \alpha_R - i\alpha_I$, $\alpha_2 = \alpha_R + i\alpha_I$:

$$W_{1,p+1} = 1 - \frac{2e^{-\alpha_R h}(M_1 + M_2)}{\sin(2\alpha_I h)}, \quad (5.106a)$$

$$W_{1,p-1} = 1 + \frac{2e^{\alpha_R h}(M_1 - M_2)}{\sin(2\alpha_I h)}, \quad (5.106b)$$

with

$$M_1 = \cos(\alpha_I h)\,[\alpha_I h - \cosh(\alpha_R h)\sin(\alpha_I h)],$$

$$M_2 = \sin(\alpha_I h)\,[\alpha_R h - \sinh(\alpha_R h)\cos(\alpha_I h)],$$

and

$$W_{2,p+1} = 1 + \frac{2e^{-\alpha_R h}(N_1 - N_2)}{\sin(2\alpha_I h)}, \quad (5.106c)$$

$$W_{2,p-1} = 1 - \frac{2e^{\alpha_R h}(N_1 + N_2)}{\sin(2\alpha_I h)}, \quad (5.106d)$$

with

$$N_1 = \cos(\alpha_I h)\,[\alpha_R \alpha_I h^2 - \sinh(\alpha_R h)\sin(\alpha_I h)],$$

$$N_2 = \sin(\alpha_I h)\,[\cosh(\alpha_R h)\cos(\alpha_I h) - 2 - \alpha_R^2 h^2 + \alpha_I^2 h^2].$$

(iii) For equal roots, α,

$$W_{1,p+1} = 1 - e^{-\alpha h}[\sinh(\alpha h) - \cosh(\alpha h) - \alpha h - 1], \qquad (5.107a)$$

$$W_{1,p-1} = 1 + e^{\alpha h}[\sinh(\alpha h) - \cosh(\alpha h) - \alpha h + 1], \qquad (5.107b)$$

$$W_{2,p+1} = 1 - e^{-\alpha h}[\cosh(\alpha h) + \sinh(\alpha h) - \tfrac{1}{2}\alpha^2 h^2 - \alpha h - 1], \qquad (5.107c)$$

$$W_{1,p-1} = 1 - e^{\alpha h}[\cosh(\alpha h) - \sinh(\alpha h) - \tfrac{1}{2}\alpha^2 h^2 + \alpha h - 1]. \qquad (5.107d)$$

Thus it can be seen that this modified finite difference formulation is exact, with no truncation error, if the general solution for the real problem behaves like that for linear ODEs locally. The roots α_1 and α_2 are calculated from the actual homogeneous equation at position p. The particular integral is also taken to be a constant locally. Since the equations for the opposed jet diffusion flame fit into this category it would appear that the new formulation might allow a drastic reduction in the number of grid points.

Implementation of the modified discretization resulted in solution of the opposed jet diffusion flame model, with CO kinetics and with a stretching rate $\epsilon = 10\ s^{-1}$, with 35 grid points. This was not possible with standard discretization with either collapsing or uniform grids, in which cases no convergence could be achieved. With uniform standard discretization, convergence was obtained with 101 grid points. Therefore the modified discretization allowed the minimum number of mesh points to be reduced from 101 to 35 and this allowed development of a general purpose laminar opposed jet diffusion flame code.

The program also allowed the laminar opposed jet diffusion flame model for a $CH_4/O_2/N_2$ flame to be solved. This involved 16 species and 39 (reversible) reactions. Core limitations required the use of not more than 35 grid points for this case. Neither the evenly spaced grid nor collapsing grid with standard discretization allowed a convergent solution to be found. The modified discretization did converge, indicating again the necessity of this approach.

The general purpose code developed allows an arbitrary reaction set containing up to 150 reactions and 70 species to be interpreted and solved for the laminar opposed jet flame configuration. Diffusion coefficients, thermal conductivities, viscosities, and thermochemical properties are supplied in the program. These are calculated from the proper Lennard–Jones and Stockmayer intermolecular potential parameters and from the NASA equilibrium tape. Appropriate mixing rules including the bifurcation approximation for multicomponent diffusion are incorporated. Results presented herein utilized a detailed reaction set from a recent review of $CO/H_2/O_2$ kinetics by Corley and Bowman (1983). The data can be listed in any order and a free format input is utilized. The program requires the stretching rate as an input parameter and adjusts the burner spacing automatically such that there is no net flux of any species back into either of the burners. Appropriate velocities are calculated and this is used in the design of the experimental configuration.

An objective of the model was to determine at the outset if and under which conditions it is possible to obtain a laminar opposed jet diffusion flame with a reaction zone of sufficient thickness to allow for detailed probing for species profiles. The conditions examined correspond to fairly low stretching rates because of limitations in gas flow rates for the experiment proposed. The question addressed is whether it is possible to obtain a sufficient number of sampled points to properly define gradients, and even second derivatives in the reaction zone. This question always arises when flat flame data is utilized to extract net rates of formation or destruction of a particular species.

Figures 5.6-5.8 profile temperature, velocity, and the similarity function ψ, of major and of intermediate species for a laminar opposed jet diffusion flame with $CO/H_2O/N_2$ as the fuel and O_2/N_2 as the oxidant. The stretching rate was 3.62 s^{-1}. Note how density changes influence the axial velocity profile and the location of the stagnation place (Figure 5.6). The flame thickness is predicted to be approximately 1 cm, which is barely sufficient to allow for flame probing. Figure 5.8 shows that the free radical profiles are present in only a similarly very narrow region.

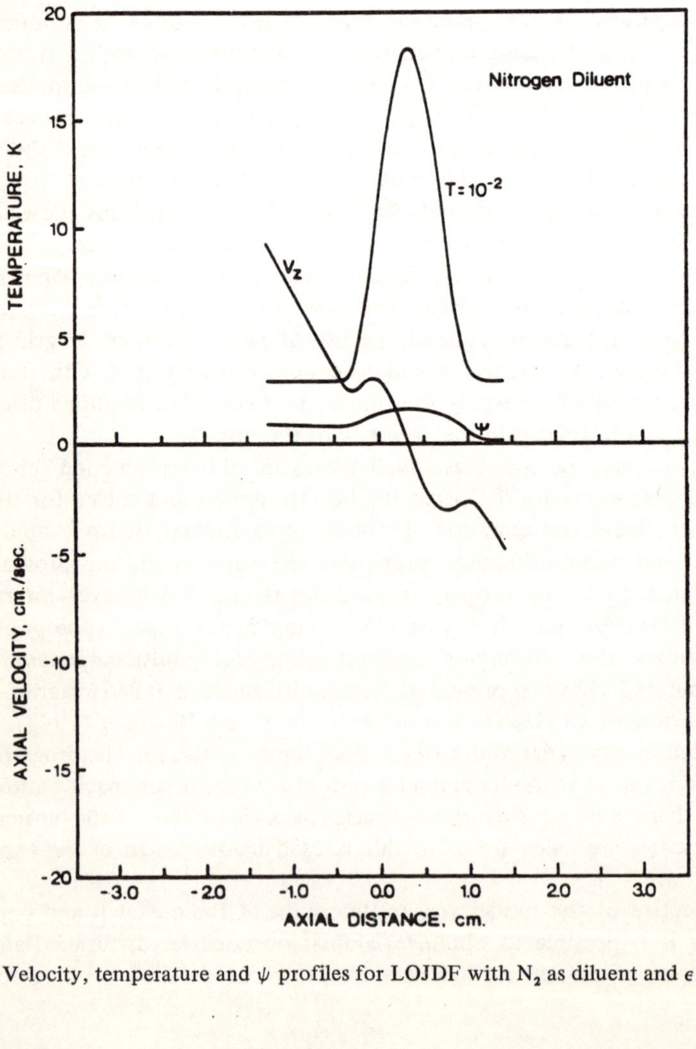

Fig. 5.6 Velocity, temperature and ψ profiles for LOJDF with N_2 as diluent and $\epsilon = 3.62\ s^{-1}$.

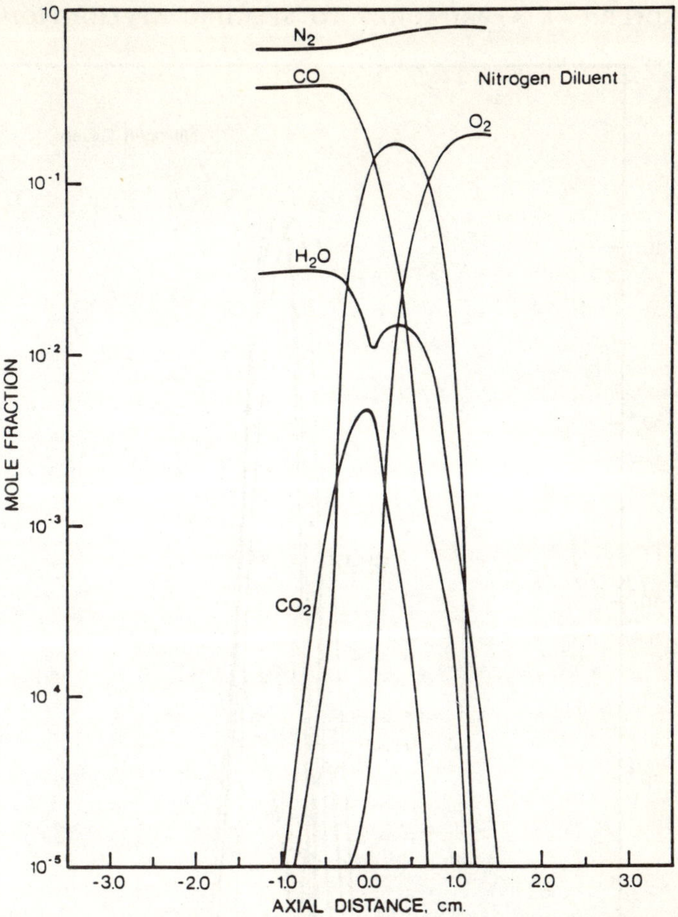

Fig. 5.7 Major species profiles ($\epsilon = 3.62\ s^{-1}$).

The laminar opposed jet diffusion flame can be modelled in detail and used to corroborate and investigate combustion kinetic mechanisms under a wide range of conditions. The ensuing equations to be solved involve ill-conditioning through the reaction rate terms, very steep spatial gradients because of diffusion, and a velocity which changes sign in the domain of interest. These equations can best be solved numerically in a fully coupled, fully implicit manner, utilizing a modified discretization in the spatial domain that approximates the equation within a mesh interval rather than its solution. This approach may be relevant to many other problems involving multiple reactions, diffusion and convection, and future work might be directed towards improving the discretization to allow better approximations for the forcing function within a mesh interval.

5.3 Process analysis (Seider, White III, Prokopakis)

Increasingly, engineers and scientists are reporting the importance of physical insights in significantly reducing computation times; in some cases, permitting

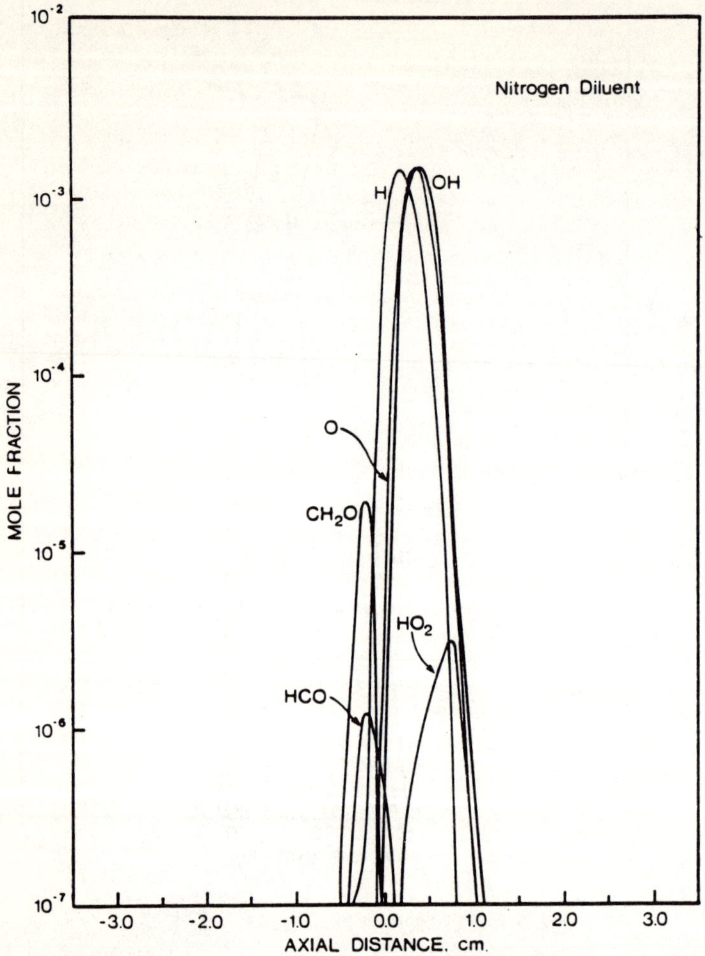

Fig. 5.8 Free radical species profiles ($\epsilon = 3.62\ s^{-1}$).

solutions that otherwise could not be obtained with available computing resources. It is noteworthy, however, that some approaches, presented in the name of eliminating stiffness, reduce the computation time for nonstiff systems.

In the next few sections, we consider a number of approaches for reducing computation times.

5.3.1 *Linearization*

For systems that are mildly nonlinear, linearization can offer an efficient solution method. The Jacobian, $\boldsymbol{J}$, is computed as well as its eigenvalues and their corresponding eigenvectors. Then, the analytical solution is used for $t_i < t \leqslant t_{i+1}$, where the step size, $h = t_{i+1} - t_i$, is determined by the nonlinearities. Mah, Michaelson, and Sargent (1961) applied this repeated linearization with some success for integration of the mass balances in distillation. However, the approach has not been widely adopted, primarily due to difficulties in adjusting the step size and the complexity of calculations to determine $\boldsymbol{J}$ and its eigenvalues and eigenvectors.

5.3.2 *Classification by rate of change*

For systems that exhibit widely spread rates of change (usually not stiff, unless $|\mathrm{Re}(\lambda)|_{max}$ is sufficiently large to require that a nonstiff integrator reduce its step size below that necessary to accurately track the fastest variables), significant savings in integration time can often be achieved by classifying the variables according to rate of change. Let the m ODEs be listed in order of descending magnitude of rate of change:

$$\dot{y}_1 = f_1(t, \boldsymbol{y}),$$

$$\dot{y}_2 = f_2(t, \boldsymbol{y}),$$

$$\vdots$$

$$\dot{y}_m = f_m(t, \boldsymbol{y}),$$

where $|\dot{y}_1| \gg |\dot{y}_m|$. Some authors refer to $y_1, \ldots, y_n$ (where n is arbitrary) as the stiff variables (Aiken and Lapidus 1974; Prokopakis and Seider 1981) with $y_{n+1}, \ldots, y_m$ as nonstiff variables. Since the system may not be stiff, we prefer to call these "fast" and "slow."

In some cases, it may not be necessary to accurately track the fast variables. If not, their derivatives can be approximated, say, with a first-order finite difference or as zero (the steady-state approximation (ssa), see §7.2.5) resulting in algebraic equations. The resulting ODEs (with the slower rates of change) permit larger step sizes to accurately track the slower variables. However, with rates of change in closer proximity, they are more likely to be stiff, requiring an implicit or semi-implicit algorithm with evaluations of the Jacobian $\boldsymbol{J}$. Normally, the increase in h reduces computation time far more than that added to evaluate $\boldsymbol{J}$.

Shampine and Gear (1979) state: "Current codes for stiff differential equations are sufficiently efficient that there is no need to consider such model changes for most problems for reasons of cost, and there are excellent reasons of convenience and theoretical support for not changing the model. To be sure, there are exceptions because we are discussing general purpose codes." We believe that the equations of distillation constitute such an exception.

In other cases, the slowest variables may be changing too slowly to require integration or when the rate of change is significant, it may be possible to integrate their ODEs with a nonstiff integrator (if $|\mathrm{Re}(\lambda)|_{max}$ is sufficiently small). Such decompositions are recommended by Prokopakis and Seider (1982) for integration of the distillation equations and by Edsberg (1976) for integration of the mass balances in reaction systems.

For pyrolysis, reforming, and combustion models, where the need to include the fast reactions involving free-radicals is generally acknowledged, there is question concerning how to apply an ssa for radical concentrations. Blakemore and Corcoran (1969) postulate a free-radical mechanism for the pyrolysis of n-butance and integrate the mass balances with and without this approximation. After 10 msec in a batch reactor at 519 °C they show negligible difference in the radical concentrations, with no changes in the radical concentrations thereafter. With the ssa, the step size for their nonstiff integrator is increased significantly, presumably because the resulting ODEs are nonstiff.

However, others question the suitability of the ssa. Sundaram and Froment (1978) studied the pyrolysis of propane with a free-radical mechanism at 800 °C and found significant errors. Furthermore, Edelson, and Allara (1973) report that, using the model of Herriott, Eckert, and Albright (1972) for propane pyrolysis, a conventional ssa does not seem to work.

It seems clear that the conventional ssa is appropriate for some reaction systems and not for others. Unfortunately, the rates of change of the fast forward and reverse reactions, or any other measure (to our knowledge), does not indicate its appropriateness. Its suitability can be determined by numerical experimentation. Since this involves integrating the unabridged ODEs, it is pointless to make the assumption unless repeated integrations are planned with small changes in the parameters.

5.3.3 *Chemical equilibrium*

In chemical reaction systems, it may be possible to reduce computation times if some of the reactions are in local equilibrium (i.e., have equivalent forward and back reaction rates); for example, if the reaction:

$$H + O_2 \underset{k_{rev}}{\overset{k_{for}}{\rightleftharpoons}} O + OH$$

is in local equilibrium, the forward and reverse rates of reaction are equal:

$$r_{for} = k_{for}c_H c_{O_2} = r_{rev} = k_{rev}c_O c_{OH}\ .$$

Sorensen and Stewart (1980) present a method to determine an independent set of mass balances when a subset of the reactions can be taken at equilibrium. White and Seider (1981) apply this method to a system of 21 reactions for combustion of H_2 and CO in air, with seven reactions taken at equilibrium. The eight independent mass balances are reduced to four, since there are four independent reactions at equilibrium, and have the form:

$$\frac{d\zeta}{dt} = f(r_n), \tag{5.112}$$

where ζ is a linear combination of concentrations (e.g. $\zeta_1 = c_{NO} + c_H + c_{H_2} - c_{O_2}$, $\zeta_2 = c_{OH} + 2c_{O_2} + c_H + 2c_O, \ldots$) and $\boldsymbol{r}_n$ is a vector of net rates of reaction for reactions not taken at equilibrium. White and Seider use the GEAR integrator to give ζ across a time step. Then, the concentrations, $\boldsymbol{c}$, are computed using the Rand method (Newton's method) to minimize the Gibbs free energy given ζ. For this system, computation times are not reduced primarily because the reactions not taken at equilibrium are fast as well as slow. They conclude that computation times should be reduced when all reactions not at equilibrium are relatively slow, reducing the rates of change, $d\zeta/dt$, and probably the stiffness of (5.112). This should permit large time steps and possibly the use of a nonstiff integrator. This has been demonstrated for a very small system.

5.3.4 *Near-analytical integration*

Often the responses of process variables, such as temperatures and compositions, have a familiar form, approaching exponential variations, for example. In these cases, it may be possible to obtain algebraic equations that contain this functionality through analytical integration with the appropriate numerical approximations and, in this manner, avoid numerical integration with small time steps in tracking these fast variables.

As an example, consider the following model for reacting polydispersed particles in an entrained bed reactor to devolatilize coal (Kayihan 1980). Figure 5.9 is a schematic of the reactor into which a slurry of pulverized coal in carrier gas (at temperature T_{G0}, gas flow rate G_0, and mass flow rates of particles in K discrete size ranges, $m_1, \ldots, m_K$) is mixed with hot gas (at temperature, T_{F0}, and flow rate, F_0). As the coal particles move through the reactor, they are heated by convection and devolatilization occurs.

The model assumes a uniform density and heat capacity in all solid particles and uniform heat capacity in the gas phase. The energy balance for particles in each size range, j, is

$$m_j c \frac{\mathrm{d}T_j}{\mathrm{d}t} = a_{sj} h (T_{\text{F}} - T_j) + \frac{\mathrm{d}m_j}{\mathrm{d}t} \left[(C - c) T_j + \Delta H_{\text{R}} \right]. \qquad (5.113)$$

where

T_j = temperature (does not vary with particle radius), °C

c = heat capacity of solid particles, cal g K^{-1}

C = heat capacity of gas, cal g^{-1} K^{-1}

ΔH_{R} = heat of reactions, cal g^{-1}

a_{sj} = area for convective heat transfer (surface area of particle in size range j), cm^2

h = convective heat transfer coef., cal s^{-1} cm^{-2} K^{-1}

m_j = mass flow rate of particles in size range j, g s^{-1}.

It is assumed that N first-order reactions remove the volatile species from each size range:

$$\frac{\mathrm{d}v_{ij}}{\mathrm{d}t} = k_{0i}(v_i^* - v_{ij}) \exp(-E_i/RT_j), \qquad i = 1, \ldots, N, \quad j = 1, \ldots, K, \qquad (5.114)$$

where v_i^* is the maximum volatiles produced by reaction i in grams per gram of solid initially and v_{ij} is related to m_j by the mass balances

$$m_j = m_{0j} \left(1 - \sum_i v_{ij} \right), \qquad j = 1, \ldots, K. \qquad (5.115)$$

For devolatilization of Montana lignite, Kayihan uses kinetic parameters of Suuberg (1977) for 15 parallel reactions to produce 8 volatile species. With $K = 10$, (5.113-5.115) give 160 ODEs to be solved with the algebraic mass and energy balances in the gas phase. The particle temperatures range from 500 °K (for the large particles) to 1000 °K and with a wide range of reaction rates the differential equations are expected to be stiff. Kayihan estimates an SR equal to 10^{24}.

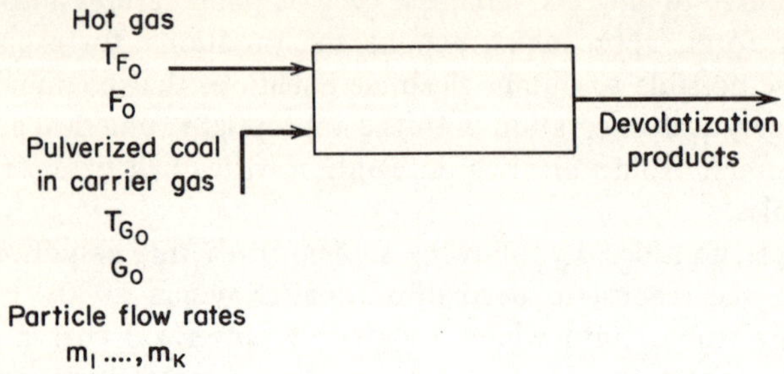

Fig. 5.9 Schematic of entrained bed devolatilization reactor.

This led him to use the DVOGER stiff ODE integration routine in the IMSL Library which uses the GEAR algorithm. The 160 stiff ODEs were integrated, using analytical expressions for the Jacobian elements, over 1/100 of the residence time for the largest particles and projected to require over 10 hours on a CDC CYBER 73 computer. This was primarily due to the excessive operations with zero elements in LU factorization of a sparse 160 × 160 Jacobian matrix and the subsequent elimination and substitution steps. A significant reduction in operations would be expected using GEARS with its routines for efficient inversion of sparse matrices.

Instead, he integrated (5.113) and (5.114) analytically with numerical approximations and obtained algebraic equations that preserve the exponential variations in temperature and conversion with time.

The energy balances for particles in size range j, (5.113) are rewritten:

$$\alpha_j \frac{dT_j}{dt} + T_j = T_F + \beta_j \frac{dm_j}{dt},$$

where

$$\alpha_j = \frac{m_j c}{a_{sj} h}, \quad \beta_j = \frac{(C-c)T_j + \Delta H_R}{a_{sj} h}.$$

An integrating factor, $\exp(t/\alpha_j)$, is introduced:

$$d[\exp(t/\alpha_j)T_j] = \frac{\exp(t/\alpha_j)}{\alpha_j}\left(T_F + \beta_j \frac{dm_j}{dt}\right) dt. \tag{5.116}$$

This is integrated analytically from t_{n-1} to t_n assuming $T_F = (T_{Fn-1} + T_{Fn})/2$, $\alpha_j = (\alpha_{j,n-1} + \alpha_{j,n})/2$, $\beta_j = (\beta_{j,n-1} + \beta_{j,n})/2$, and a linear variation of $m_j \exp(t/\alpha_j)$ with t to give:

$$T_j(t_n) \doteq T_j(t_{n-1})\exp(-\Delta t/\alpha_j) + T_F(t_n)\,[1 - \exp(\Delta t/\alpha_j)]$$
$$+ \frac{\beta_j}{\alpha_j}\left[m_j(t_n)\left(1 - \frac{\Delta t}{2\alpha_j}\right) - m_j(t_{n-1})\left(1 - \frac{\Delta t}{2\alpha_j}\right)\exp(-\Delta t/\alpha_j)\right]. \tag{5.117}$$

Similarly, (5.114) are integrated analytically:

$$\nu_{ij} = \nu_i^* \left[1 - \exp\left(-k_{oi}\int_0^t \exp(-E_i/RT_j)dt\right)\right]$$

or

$$v_{ij} = v_i^* \left[1 - \exp\left(-k_{0i} \sum_{m=1}^{n} \phi_{ij}(t_m)\right)\right] ,$$

where

$$\phi_{ij}(t_m) = \int_{t_{m-1}}^{t_m} \exp(-E_i/RT_j)\mathrm{d}t,$$

and the integration is performed assuming linear variation of T_j between t_{m-1} and t_m. The resulting algebraic equations are solved simultaneously with (5.117) and the mass and energy balances for the gas phase to compute profiles of all unknowns in time. The equations are decoupled with guesses for the profiles of particle temperatures to avoid operations with large sparse matrices.

Using a uniform Δt (= 0.01 $t_{\max}$), four to five iterations gave three significant figures in less than thirty seconds of CDC CYBER 73 computer time. Whereas the stiff integrator is very convenient to use, this is illustrative of the many problems for which physical insights lead to specialized algorithms that are more efficient, in some cases permitting a solution that could otherwise not be obtained with available computing resources.

5.4 Nuclear power plants (Girijashankar, Hetrick, Keepin, Paluskinski, Bailey)

The simulation of large engineering systems such as power plants of both nuclear and fossil type, aircraft systems, etc., has been an important activity for more than a decade. System modelling, using the conservation laws of mass, momentum and energy in the case of mechanical and hydraulic systems and Kirchoff's voltage and current laws in the case of electrical systems, is a first step in the simulation scheme. These laws when used to model systems such as pressurized water reactors, for example, result in a set of PDEs. Using some form of lumping, these are reduced to a set of coupled nonlinear ODEs. Conventional integration schemes such as Runge–Kutta, EPISODE, GEAR, etc., while generally satisfactory, are highly unsuitable for some classes of problems. We will direct our attention to one such class of problems and refer to it as two-time-scale and oscillatory.

The above class of problems having highly oscillatory solutions for a small subset of system variables requires a small step size for the solution of all system equations when using the above integration techniques. Since the number of system variables that contribute to the oscillatory part of the solution is generally small compared to the total number of system variables, significant advantage could be gained by partitioning the system into (a) those variables which produce highly oscillatory solutions necessitating a small step size for integration and (b) the remaining variables which can be integrated by using a relatively large step size. We define the first subsystem as the fast subsystem, the second as the slow subsystem, and the scheme of integration of such partitioned systems as multirate integration. It must be pointed out that a system which is not highly oscillatory at the start of integration could become so during integration, and that a system which has been partitioned in one way may need repartitioning during the course of integration.

We present two examples illustrating the above features of partitioning and multirate integration for the cases of (a) a thirteenth-order autopilot model and (b) a thirtieth-order pressurized-water-reactor system. Also, we will present a proposed scheme for monitoring the system to be used as an indicator in the partitioning of the system.

The scheme for identifying the slow and fast components of a linear system is due to O'Malley and Anderson (1979) and has been extended to nonlinear systems below.

Consider a dynamic system described by

$$\dot{Z} = F(Z, U, t), \tag{5.118}$$

where Z and U are n and m-dimensional vectors respectively, and F denotes the vector dependence between Z and U. In the control literature, these are referred to as the state and control vectors. Around any operating trajectory (Z_0, U_0), the system equations could be written as

$$\dot{z} = Az + Bu, \tag{5.119}$$

where z and u are perturbations around (Z_0, U_0) and A and B are Jacobians with respect to Z and U, respectively. We assume that the system (5.118) can be adequately represented by the linearized version (5.119) for the purposes of determining the fast and slow parts of the system. The homogeneous part of the solution (5.119) is governed by A while the forcing part is dictated by B. The total solution, therefore, is a summation of the homogeneous and its forcing part. If u, the forcing function, is slowly varying as compared to the fast components of the homogeneous solution, we can assume that the fast and slow components of the solution can be identified by an analysis of A.

Let us assume also that the eigenvalues of the system matrix A can be separated into two sets W and S of order N_1 and N_2, respectively, such that eigenvalues $f_i \in W$ and eigenvalues $s_j \in S$ and $|s_j| \ll |f_i|$. We refer to such systems as two-time-scale; the set W is the fast set of eigenvalues and set S is the slow set. Later, we will further assume that the fast set contains at least one complex conjugate pair with relatively large imaginary parts. This class of problems is referred to as two-time-scale and highly oscillatory. The scheme presented below however, works whether oscillations are present or not. We can rewrite (5.119) as

$$\begin{bmatrix} \dot{z}_1 \\ \dot{z}_2 \end{bmatrix} = \begin{bmatrix} A_{11} & A_{12} \\ A_{21} & A_{22} \end{bmatrix} \begin{bmatrix} z_1 \\ z_2 \end{bmatrix} + \begin{bmatrix} B_1 \\ B_2 \end{bmatrix} u, \tag{5.120}$$

where z_1 and z_2 are n_1- and n_2-dimensional subvectors of z and A_{11}, B_1, etc. are submatrices of A and B. We can cast the A matrix into its Jordan form using the eigenvector matrix M and its inverse Q, thus,

$$J = QAM. \tag{5.121}$$

It can easily be seen that if the transformation $P = Qz$ is used, (5.119) will be completely decoupled in the new state variable vector P, and the homogeneous part of the solution is governed by the eigenvalues in the Jordan form of A. Thus, while accomplishing total decoupling the above transformation requires the evaluation of the complete set of eigenvalues and its eigenvectors. But as Golub and Wilkinson (1976) point out, the smallest eigenvalues of a large ill-conditioned matrix cannot be calculated accurately and hence their eigenvectors are approximate at best. We use a slightly different approach which achieves partial decoupling and is all that is needed for partitioning. Only a very brief outline of the method is given below and the reader may refer to O'Malley and Anderson (1979) for details. We will rewrite (5.121) as follows:

$$\begin{bmatrix} J_1 & 0 \\ 0 & J_2 \end{bmatrix} = \begin{bmatrix} Q_{11} & Q_{12} \\ Q_{21} & Q_{22} \end{bmatrix} \begin{bmatrix} A_{11} & A_{12} \\ A_{21} & A_{22} \end{bmatrix} \begin{bmatrix} M_{11} & M_{12} \\ M_{21} & M_{22} \end{bmatrix}, \tag{5.122}$$

where J_1 and J_2 are submatrices of J containing the fast and slow eigenvalues, respectively, and $(M_{11}M_{21})^{\mathsf{T}}$ corresponds to the eigenvectors of the smaller set of n_1 eigenvalues that has been identified for partial decoupling. Various other submatrices in (5.122) can be easily identified.

Let us define the matrices L and T by the relations

$$L = -M_{21}M_{11}^{-1}, \tag{5.123}$$

and

$$T = \begin{bmatrix} I_1 & 0 \\ L & I_2 \end{bmatrix}. \tag{5.124}$$

We will further define a transformation of the $(z_1, z_2)^{\mathsf{T}}$ state space into $(x, y)^{\mathsf{T}}$ space by

$$\begin{bmatrix} x \\ y \end{bmatrix} = T \begin{bmatrix} z_1 \\ z_2 \end{bmatrix}. \tag{5.125}$$

Using the transformation defined by (5.125), it can be shown that (5.120) will be transformed into

$$\begin{bmatrix} \dot{x} \\ \dot{y} \end{bmatrix} = \begin{bmatrix} \tilde{A}_{11} & A_{12} \\ 0 & \tilde{A}_{22} \end{bmatrix} \begin{bmatrix} x \\ y \end{bmatrix} + \begin{bmatrix} B_1 \\ LB_1 + B_2 \end{bmatrix} u, \tag{5.126}$$

where

$$\begin{aligned} \tilde{A}_{11} &= A_{11} - A_{12}L, \\ \tilde{A}_{22} &= A_{22} + LA_{12}. \end{aligned} \tag{5.127}$$

Note that $\tilde{A}_{11}$ and $\tilde{A}_{22}$ contain the fast and slow eigenvalues, respectively. Also, it has been shown that L satisfies the algebraic Riccati equation (ARE)

$$LA_{11} + A_{21} - LA_{12}L - A_{22}L = 0. \tag{5.128}$$

It may be observed by looking at (5.126) in the (x, y) state space there is partial decoupling between the fast and slow parts of the system. The evaluation of L needs only the eigenvectors corresponding to the fast n_1 eigenvalues. Even if this

L is not accurate, the algebraic matrix Riccati equation could be used to formulate an iterative procedure to improve its accuracy. If order reduction were the main consideration then (5.126) could be used towards this end as has been done by O'Malley and Anderson. However, since our aim is to integrate the original nonlinear system, we will follow a slightly different approach.

We have assumed in the evaluation of the transformation matrix that the linearized system represents adequately the original nonlinear system. This is true at least in a small neighborhood of the operating point (A_0, u_0) if the nonlinearities are severe, otherwise it will be valid over a wider range. Thus, defining a transformation of the nonlinear variables

$$V = TZ, \tag{5.129}$$

we can write (5.118) as

$$\dot{V} = TF(T^{-1}V, U, t), \tag{5.130}$$

where $V(t_0) = TZ(t_0)$.

Denoting the first n_1 components of V by X and the remaining n_2 components by Y, we have

$$V^{\mathrm{T}} = (X, Y)^{\mathrm{T}},$$

and (5.118), in terms of X and Y will become

$$\begin{bmatrix} \dot{X} \\ \dot{Y} \end{bmatrix} = \begin{bmatrix} G(X, t, Y, U) \\ H(Y, t, X, U) \end{bmatrix}, \tag{5.131}$$

where $X(0) = X_0$, $Y(0) = Y_0$ and G and H correspond to the first n_1 and n_2 right-hand-side components of the nonlinear equation. Using the analogy of the linear system presented earlier, we can denote the X and Y as fast and slow variables respectively. Once the fast and slow subsystems have been identified, these partitioned subsystems can be integrated using a multirate integration scheme.

As mentioned earlier, the nonlinearities produce a continually changing Jacobian matrix A as the solution unfolds. Two schemes for monitoring this change have been developed. One is a perturbation technique for estimating a residual in terms of A at a point in time and in terms of the nonlinear functions as they change for subsequent times. When this residual has grown beyond some pre-set value, the new A may be computed and a new partitioning determined as needed. Another scheme monitors the "zero block" in the matrix A. Nonlinearities cause the entries in this block to drift away from zero, and repartitioning can be instituted when the norm of this sub-matrix exceeds some appropriate value. These results can be used in devising an automated repartitioning scheme, and this is a subject for continuing research.

We have for the slow variable, using an s-Stage Runge-Kutta method,

$$Y_{n+1} = Y_n + h \sum_{r=1}^{s} c_r k_r, \qquad \sum_{r=1}^{s} c_r = 1, \tag{5.132}$$

$$k_1 = H(Y_n, t_n, X_n), \tag{5.133}$$

$$k_r = H\left(Y_n + h\sum_{i=1}^{r-1} b_{ri}k_i, t_{n+a_r}, X_{n+a_r}\right), \tag{5.134}$$

$$a_r = \sum_{i=1}^{r-1} b_{ri}, \quad r = 2, 3, \ldots, s. \tag{5.135}$$

Computing k_r successively requires X_{n+a_r}; hence we need to integrate the fast variable. The Runge-Kutta-Merson scheme has been used in the integration of the fast variable and with interpolation of $Y(t)$ based on values $Y_n, Y_{n+a_2}, \ldots, Y_{n+a_{r-1}}$ which are computed successively using imbedded Runge-Kutta schemes of highest-possible order ϵ (Vernar 1979). Note that different integration schemes could be used to integrate the two subsystems.

For example, in order to compute

$$k_2 = H(Y_n + ha_2k_1, t_{n+a_2}, X_{n+a_2}), \tag{5.136}$$

we have to determine X_{n+a_2}. Since k_1 is known, we can estimate using the Euler formula

$$Y_{n+a_2} = Y_n + a_2hk_1,$$

and this suggests the use of a linear polynomial

$$Y(t) = Y_n + (t - t_n)k_1, \quad t_n \leqslant t \leqslant t_n + a_2h,$$

to integrate the fast subsystem. The resulting X_{n+a_2} can now be used to evaluate k_2 by (5.136). Since k_2 is known, we can improve the estimate of Y_{n+a_2} using the improved Euler formula. Estimate Y_{n+a_3} using the linear polynomial between t_{n+a_2} and t_{n+a_3} as above. Construct a second order polynomial based on $Y_n, Y_{n+a_2}, Y_{n+a_3}$ to be valid for $t_{n+a_2} \leqslant t \leqslant t_{n+a_3}$. Since X_{n+a_2} is known, the fast variable can be integrated from t_{n+a_2} to t_{n+a_3}. We can use X_{n+a_3} to compute k_3. The above scheme can be repeated to continue the integration. Thus, the previously evaluated $k_1, k_2, \ldots, k_{r-1}$ have been used in updating $Y_{n+a_2}, Y_{n+a_3}, \ldots, Y_{n+a_{r-1}}$ and hence the scheme is known as imbedded Runge-Kutta multirate integration. The above scheme can be used continuously to march ahead in time.

We will demonstrate the usefulness of the scheme for two-time-scale and highly oscillatory problems with the help of two examples. The first is an autopilot model having 13 differential equations while the second is a pressurized water reactor having 30 differential equations. In the first case, partitioning was done at time $t = 0$. In the second case the system was partitioned at 93 sec. In both cases, there was considerable saving in computer time.

The autopilot model is composed of four major subsystems, the vertical sensing unit, the amplifier, the servo-motor along with its pressure stabilizer, and the flight dynamics of the aircraft. The system equations and other details can be found in Girijashankar, Hetrick, Keepin, and Palusinski (1980). It consists of 13 differential equations and a table look-up. The eigenvalues of the system are as shown in Table 5.3. Since an analytical solution was not available for this system, a benchmark solution was obtained numerically by solving the system with very small values of relative error control parameter. Three different methods, (a)

TABLE 5.3
Eigenvalues of the autopilot system

No.	Real part	Imaginary part
1	−0.1235 E+05	
2	−0.4999 E+04	
3	−0.8401 E +03	−0.8998 E+04
4	−0.8401 E +03	+0.8998 E+04
5	−0.1000 E+03	
6	−0.2437 E+02	−0.3200 E+02
7	−0.2437 E+02	+0.3200 E+02
8	−0.8828 E+01	
9	−0.7687 E+00	−0.4807 E+01
10	−0.7687 E+00	+0.4807 E+01
11	−0.1809 E+00	−0.3299 E+00
12	−0.1809 E+00	+0.3299 E+00
13	−0.7640 E−14	

Runge-Kutta-Merson, (b) Gear's stiffly stable, and (c) Hindmarsh's EPISODE were used to yield agreement in solutions to a four-digit accuracy. For ease of reference we will denote the multirate integration scheme by P_0 and the methods of Runge-Kutta-Merson, Gear, and Hindmarsh by P_1, P_3, and P_{11} respectively. This notation corresponds to M_1, M_3, and M_{11} respectively, used in DAREP (Korn and Wait 1977) simulation software for these three integration methods.

The error criterion for all the schemes was selected so that it was not more than one percent during the entire run. The computer times involved for different schemes are shown in Table 5.4. Two representative fast variables and two representative slow variables from the multirate integration scheme are shown in Figure 5.10. From the results of Table 5.4, it may be concluded that considerable savings could be obtained by partitioning and using multirate schemes. One more interesting conclusion that can be drawn is that the Runge-Kutta-Merson method yields much shorter solution time as compared to either GEAR or the EPISODE. This may be explained from the fact that stiff routines are good only when there is stiffness caused by a spread of real eigenvalues without fast oscillatory components.

The simulation of nuclear power reactors has gained more attention and importance since the Three Mile Island accident at Harrisburg, Pennsylvania. The PWR models are based on the conservation laws of mass, momentum, and energy. These equations, together with the equations of state for water and steam, give a complete set of equations. The integration of such systems may not readily submit to efficient numerical treatment using readily available routines, as we will see for this example.

The PWR model is based on the data for St Lucie Unit 2, a Combustion Engineering plant in Florida. In building the reactor model, numerous assumptions have been made. It may be pointed out that while some of these assumptions limit the

TABLE 5.4

Comparison of computer CPU *times for the autopilot model using different integration schemes* (TMAX = 0.5 secs)

Method	Error Control Parameters	Cyber 175 Run Time (sec)	Maximum percentage errors					
			x	y	x_1	x_2	x_3	TH
P_1	EMAX = 10^{-6}	15.6	8.49×10^{-5}	2.94×10^{-6}	0.126	0.877	4.37×10^{-3}	3.28×10^{-4}
P_3	EMAX = 10^{-8}	33.4	8.79×10^{-5}	2.94×10^{-6}	0.129	0.84	4.53×10^{-2}	3.26×10^{-4}
P_{11}	EMAX = 10^{-8}	44.6	8.79×10^{-5}	3.29×10^{-6}	0.124	0.884	4.54×10^{-2}	3.25×10^{-4}
P_0	EMAX = 10^{-5}	5.07	0.90	9.4×10^{-3}	0.11	0.839	5.03×10^{-2}	0.476

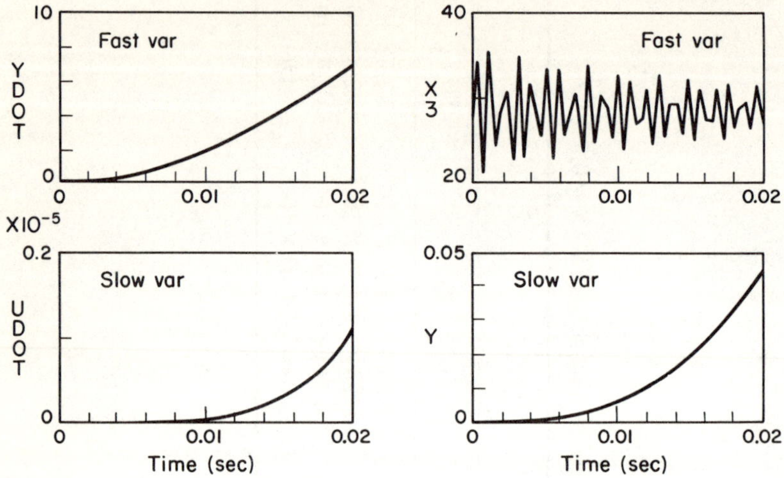

Fig. 5.10 Two representative fast and slow variables of the autopilot system.

use of the model to small transients only, the model should still retain the eigenvalue–eigenvector structure in the neighborhood of any operating point as compared to more detailed models. The plant as modelled is a one-loop representation of a two-loop plant. It consists of a core, a pressurizer, a steam generator and a recirculating pump. The core has been modelled by a representative average channel, and the primary coolant is assumed to be under pressure such that no boiling is allowed in the model. The steam generator is simulated by a representative U-tube and the primary flow is divided into three sectors. The first sector extends from the inlet to the level of the boiling interface, the second sector from this interface to the boiling interface on the exit side, and the third sector from this interface on the exit side to the exit itself. Each sector is represented by an average temperature node and an outlet temperature node. No boiling is permitted for the primary liquid. The nonboiling region of the secondary side of the steam generator is similarly represented by an outlet temperature, while the downcomer exit temperature is determined by treating the downcomer as a tank problem. The steam production rate is calculated from the overall heat transfer coefficient and the difference between the average temperature in the second region and the saturation temperature. The pressurizer has been modelled to represent four states for the water in the bottom portion and steam in the top portion of the pressurizer. These states are (a) top superheated, bottom subcooled, (b) top superheated, bottom boiling, (c) top condensing, bottom subcooled, and (d) top condensing, bottom boiling. The recirculation pump (centrifugal type) has been modelled along the traditional lines of nondimensional homologous curve. This treatment allows the pump to traverse any of the four possible quadrants of pump operation. The final model consists of thirty differential equations and a large number of algebraic equations and table look-ups for the steam properties. A digital simulator using these equations has been constructed and maintained as an input file to the DAREP translator. The transient response appears to be consistent with results for similar plants.

Attempts were made to carry out the integration for the above transient initiator up to 100 seconds. For all of the three methods of integration mentioned earlier, the first 93 seconds of real time needed 1 second of computed CPU time for each second of real time. After 93 seconds, nearly 40 seconds were needed for each second of real time. The first four largest eigenvalues of the Jacobian at two times are shown in Table 5.5. It can be seen that the imaginary part of the complex pair has increased by 25 fold in the interval. The equations have been partitioned into fast and slow parts with 3 and 27 differential equations in the fast and slow systems, respectively. The three variables that have been selected for treatment as fast variables using the approach given above are neutron power, delayed neutron precursors, and the recirculation flow rate. The simulation was restarted at 93 seconds using both the original and partitioned systems. The original system took 80 seconds of CPU time for 2 seconds of system time, while the partitioned system reached 15 seconds of system time in the same CPU time, indicating a six-fold saving of CPU time. The relative error for the most sensitive variable (neutron power) was less than 0.002 percent at 95 seconds.

TABLE 5.5
Four largest eigenvalues of the Jacobian at 11.0 seconds and 93.0 seconds

Time (sec)	Eigenvalue No.	Eigenvalue Real Part	Eigenvalue Imaginary Part
11.0	1	−29.12	0.0
	2	−100.68	4.07
	3	−100.68	−4.07
	4	−18.46	0.0
93.0	1	−146.99	0.0
	2	−61.41	108.49
	3	−61.41	−108.49
	4	−60.76	0.0

5.5 An advanced simulation package for large chemical reaction systems (Bader, Nowak, Deuflhard)†

5.5.1 *Introduction*

The development of a new stiff integrator on an extrapolation basis by Bader and Deuflhard (1981) has led to the development of a software package designed to handle large chemical kinetics problems. Details about this package LARKIN have been presented by Deuflhard, Bader, and Nowak (1981) and in Bader, Nowak, and

† The authors wish to thank H. J. Ederer and G. Isbarn for extensive discussions concerning the chemist's point of view. They are indebted to F. W. Schneider and M. Heinrichs for ceding the RNA polymerization example to them. Further, they are grateful to P. Kunkel for his programming assistance. The work of the authors has been supported by the Deutsche Forschungsgemeinschaft.

Deuflhard (1982). This early version of LARKIN, however, was portable only on IBM computers. In addition, part of that code was written in PL/1 (whereas most of it was written in FORTRAN). These disadvantages of the previous code were certainly prohibitive for a wider distribution.

It is our present purpose to report on progress made in the development of the package LARKIN since then. The present code which is exclusively written in FORTRAN IV, should be portable on a wide range of computers. Moreover, it permits a user to attack much larger problems. This feature has been essentially achieved by the implementation of a new "chemical compiler system" and a new sparse solver (Duff and Nowak 1983).

A general outline of the structure and some special properties of LARKIN are given in §5.5.2. Input/output facilities are described in §5.5.3. §5.5.4 contains details of the "chemical compiler system" that automatically generates the ODE right-hand side together with the associated Jacobian matrix and its sparse pattern from the input in terms of chemical reaction equations and rate constants. A short presentation of the selected stiff integrator of extrapolation type follows in §5.5.5, where also a brief comparison with the competing BDF-type integrator is made.

The philosophy behind the development of this code has been to allow any chemical user to concentrate on the modelling aspects of his chemical reaction system under consideration. In view of this aim, the package has been developed in direct and steady contact with a local group of physical chemists who have been using the system successfully for quite a while (Isbarn, Ederer, and Ebert 1981). Therefore, the software package LARKIN, as it now stands, should be a useful and reliable tool in other chemical research environments.

5.5.2 *General description of the package*

In developing LARKIN, the main emphasis has been put into the implementation of an interactive dialogue system. In order to be able to use this system, a user is only required to know details about his chemical model (and certainly not details about the computer system or about numerical analysis). In particular, the question-answer dialogue contains only either self-explanatory or chemical terms. There is no command language to be memorized by the user.

The package LARKIN as a whole is written in a "reasonably portable" FORTRAN. The FORTRAN compiler needs to be called essentially once, when the load module is produced from the source program (typical service of a computing center). Further calls of the FORTRAN compiler are necessary only if changes in the size of the allocated storage are made. All model changes (of chemical reactions, initial values, rate constants, etc.) can be easily and quickly performed within the frame of the dialogue system.

A structure diagram of the package LARKIN is represented in Figure 5.11. Details concerning the individual main blocks are as follows:

EDIT — input dialogue block: accepts chemical input and transfers it to the database CHEMIN; the contents of CHEMIN are available in EDIT (for possible modifications). For more details see §5.5.3.

COMPILER — "Chemical compiler" block: reads contents of data base CHEMIN;

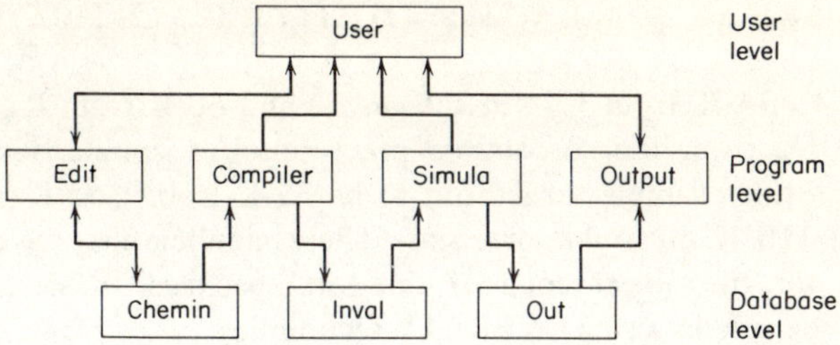

Fig. 5.11 Structural diagram of LARKIN.

checks syntax of chemical input; produces all pertinent information, which is necessary for the numerical simulation; stores this information in binary database INVAL. For more details see §5.5.4.

SIMULA — numerical integrator block: performs numerical simulation of chemical reaction system using information from database INVAL. In the course of the integration, data of chemical interest (usually not only chemical species concentrations) are updated in database OUT. For details concerning the stiff integrator see §5.5.5.

OUTPUT — output dialogue block: gives (possibly selected) output information in either graphical or numerical form as requested by the user; requires interface to plot software of individual computing centers. For more details see §5.5.3.

The whole package LARKIN applies to the numerical solution of mass action kinetics problems, which involves the numerical integration of large stiff ODE systems with polynomial right-hand sides. The standard formulation is in terms of molar concentrations (changes to mass or mole fractions are possible; compare §5.5.3). The additional treatment of diffusion or the handling of differential-algebraic systems is presently not possible (in preparation). Generally speaking, the modularized form of the package facilitates changes of individual blocks.

In view of the possible solution of very large systems, storage economy has had a high priority in the development of the package. With the notation

NSP: number of species,
NCEQ: number of chemical equations (reversible reactions count as 2),
NZE: number of nonzero elements in Jacobian matrix,
NLU: maximum number of nonzero elements in LU-decomposition,

the total integer (I) and real (R) workspace required can be estimated as follows:

$$I \approx NLU + NZE + 6\ NCEQ + 15\ NSP,$$

$$R \approx NLU + NZE + 3\ NCEQ + 15\ NSP.$$

Upon referring to the rough considerations of Deuflhard, Bader, and Nowak (1981), the following rules of thumb are obtained;

$$I_{est} \approx 6\ NCEQ + 36\ NSP,$$

$$R_{est} \approx 3\ NCEQ + 36\ NSP.$$

For the largest example (RNA-polymerization) presented in this paper, these numbers lead to 34 **K** from I_{est} (in integer*2) and 60 **K** from R_{est} in single precision, or 120 **K** from R_{est} in double precision. For comparison, the required array storage in this example turned out to be 32 K from I, 55 K from R in single precision, and 110 K in double precision. These numbers may be compared with about 100 K for the object code of the total package. A considerable saving, however, can be made by using OVERLAY techniques.

As a consequence, the present version of LARKIN should be helpful in simulating comparatively large chemical reaction systems on moderate size computers.

5.5.3 *Input/output facilities*

This section deals with details of the two dialogue blocks EDIT and OUTPUT, as far as they are relevant from the user's point of view. A complete description of the syntax of the chemical input would be beyond the scope of this paper. Rather, part of the documentation of problem HEXAN1 of §5.5.5 is inserted for illustration purposes. First, the block EDIT will be described in a sequential order. Note, however, that the user input may be given in arbitrary order.

Head (optional): sub-block may contain a short documentation of the problem (up to 5 lines).

```
    ISBARN, EDERER, EBERT:
            THE THERMAL DECOMPOSITION OF N-HEXANE,
            KINETICS, MECHANISM, AND SIMULATION.
                      (DATE OF LATEST CHANGE: JAN. 21, '82)
```

Elements (optional): list of chemical element names, each of which may be followed by its atomic weight. The names are used in the subsequent stoichiometric test. The atomic weights are necessary for a possible change from molar concentrations to mass or mole fractions in the output.

```
    H     1.008
C     12.011
```

Species (optional): list of chemical species names, each of which may be followed by the element composition (for a stoichiometric test).

```
H                             H  1
CH3                           C  1      H  3
    3-M-C4H8                            C   5  H    10
      E-BENZOL                    C  8     H 10
    STYROL                    H     8             C    8
```

The order of appearance of the names defines the order in the output. Further species names appearing only in the block reaction are automatically added. However, it is strongly advised to use the optional stoichiometric test for the possible detection of input flaws.

Reaction: complete list of irreversible or reversible chemical reactions, each of which is followed by the associated kinetic parameters (or (ln A, Δ E) for Arrhenius law).

```
CH4                 ⇒   H +    CH3      (15.3,104000)
C2H6                    ⇒ H +  C2H5                (16.0, 98000)
C2H6                ⇔   CH3 + CH3       (16.6, 88D3)  (10.7,0)
C2H5 + 1-M-AL ⇒ 3-M-C5H10  (9.0,       0.0)
C2H5 + 1-E-AL ⇒                           ( 9,  0)

02– + CS+       ⇒ CS  +   02                  (0.5D-7)
CS+ + E–        ⇒ CS                          (0.1D-11)
CS    ⇒     CS– + E–                          (0.324D-2)
02–   ⇒     02    + E–                        (0.4D0)
02    +   CS   +    M  ⇒     CS02  +   M            (0.1D-30)
02    +   02   +    E– ⇒     02–   +   02           (0.124D-29)
02    +   E–   +    N2 ⇒      02     +   N2           (0.1D-30)
```

As can be seen from the above special example, the equations may be written in a standard chemical nomenclature. The above insertions have been deliberately presented in a rather irregular shape to demonstrate the robustness of the package.

(v) Initial concentrations: list of nonzero initial concentrations in the form

C6H14   0.0021

(vi) Temperature (optional, fixed value): only required if either Arrhenius or modified Arrhenius law is used for the rate constants; may be skipped if rate constants are prescribed as numbers.

(vii) Numerical input: list of additional information which is necessary in the simulation, such as required relative accuracy, user prescribed output points (including initial and final point; output points, which are automatically selected by the integrator are independently stored in database OUT for print or plot purposes), and print option parameter.

The aforementioned sub-blocks comprise the block EDIT. Every (even linewise) change of the input is automatically updated in database CHEMIN.

All output information is available from database OUT, for further use either inside or outside the package LARKIN. The dialogue block OUTPUT offers possibly selected results of the simulation either in the form of numerical data or in graphic form. The numerical data in database OUT contain at least the concentrations at all intermediate points (both integrator selected and possibly user prescribed), if not suppressed by the above print option. From the experience of the authors, additional numerical information like reaction rates, integrated reaction rates, pressure, etc., are helpful in the process of modelling and are therefore available from OUT. The graphic output contains several options including either single or double logarithmic scale and phase diagrams. Only minor emphasis has been put into the development of the plot software—the adaptation of the graphic part of the package requires a special interface to the local plotting facilities.

In view of any adaptations, the database OUT has been especially well structured to facilitate the access.

5.5.4 *Chemical compiler system*

The above input in terms of chemical reactions, rate constants, and initial concentrations uniquely defines an associated ODE initial-value problem,

$$y' = f(y), \quad y(0) = y_0,$$

with time t as independent variable. The actual generation of the ODE system is usually done by a so-called "chemical compiler" or "interpreter" (see Curtis 1979; Edelson 1976b; Kee, Miller, and Jefferson 1980; Stabler and Chesick 1978). There is the following principal alternative: (a) generation of explicit FORTRAN subroutines f and f_y (required by standard stiff integrators) with the consequence of a necessary FORTRAN compilation step after each model change, (b) symbolic generation of f and f_y, which are evaluated by means of formal routines contained in the chemical kinetics package.

Advantages of (a) are:

(i) lower evaluation time for f and f_y,
(ii) slightly more transparent program structure.

Advantages of (b) are

(i) model changes do not require additional calls of the FORTRAN compiler,
(ii) smaller storage for object module.

This implies that (b) is preferable for large problems and is certainly more convenient from the user's point of view even for smaller problems. In the standard version of LARKIN, version (b) is realized; version (a) can be produced by a few minor changes. The limitations of version (a) may be clearly seen from its performance in the following example: RNA polymerization (NCEQ = 770, NSP = 352) (Schneider, Heinrichs, and Poll 1980).

Version (a): The listing of the source code of f and f_y is about 70 printed pages. After about ten minutes compilation time, the FORTRAN compiler gives the message: "source program too large."

Version (b): A listing of the symbolic representation of f and f_y would be about one printed page. For comparison, note that about two seconds are needed to generate f and f_y and less than one minute is needed to simulate the whole chemical system (see also §5.5.5).

The symbolic generation of f uses ideas that have been suggested by Stabler and Chesick (1978) in their realization of the package HAVCHM. Unlike HAVCHM, however, but similar to the package CHEMKIN by Kee, Miller, and Jefferson (1980), LARKIN accepts a familiar chemical language formulation with only few format restrictions. This user-oriented structure of the input leads to the necessary implementation of a scanning procedure. All species names appearing in the optional species list or in the chemical reactions are identified and internally listed in an integer representation. The integer position in this list (called NAME) represents the internal coding. This coding is then used to update the structure of the reaction scheme line by line in two integer vectors. In the course of this scanning process, the rate constants (following each reaction) are simultaneously read and analyzed. The evaluation of f is done in a standard way as in comparable packages, e.g., HAVCHM. The build-up of the list NAME needs a large number (about 4 NCEQ) of table look-ups. In order to avoid this look-up time in very large systems, a special hash technique is used (for details see Knuth 1973).

The symbolic generation of the Jacobian f_y would then be comparatively trivial if a full-mode updating scheme were applied (where each entry of f_y is characterized by its row and column index). However, in view of large systems a sparse storage scheme seems to be crucial, which means that in addition to the nonzero elements of f_y, the associated sparse pattern of $I - hf_y$ has to be provided. The pattern representation and the storage scheme in LARKIN are on one hand rather concise and on the other hand general enough to permit the use of any of the presently available sparse solvers. For more details see Deuflhard, Bader, and Nowak (1981). This side condition implies that the evaluation of f_y is technically more complicated than in packages that just use full-mode representation.

By means of scanning routines similar to the one described above, further information from data base CHEMIN is picked up, analyzed, and transferred to the binary file INVAL. In the course of the compilation, a series of error checks are made: syntax monitor for each line, check on input inconsistencies, and optional stoichiometric test.

If any of these checks is not passed, self-explanatory warning or error messages are issued.

The priorities in the construction and implementation of the chemical compiler system have been storage economy, portability, and ease of handling from the user's point of view. These priorities lead to a lot of rather technical but important details which are not suited for presentation here. These details are best documented by the code itself.

5.5.5 *Numerical simulation*

The actual numerical simulation, i.e., integration of the ODE system, is performed in the block SIMULA, which uses information only from database INVAL. Part of the information which is not used is just passed over to database OUT. Another part of the information is preprocessed before the call of the stiff integrator (e.g., computation of the rate constants by virtue of the Arrhenius law). In the course of the numerical integration, the intermediate results are transferred to database OUT by an extra routine. Once again, note that this interface structure facilitates a possible exchange of sub-blocks, such as the integrator.

The stiff integrator used in LARKIN is METASC, a special version of METAN 1 (Bader and Deuflhard 1981) adapted to large chemical kinetics problems. The code is based on the semi-implicit midpoint rule with polynomial extrapolation. The special adaptation of METASC refers to:

(i) a scaling technique, which uses further information from mass action kinetics (Deuflhard, Bader, and Nowak 1981);
(ii) the additional computation of quantities of chemical interest; and
(iii) the sparse solver MA 30 in the version developed by Duff and Nowak (1983).

Nearly all chemical kinetics packages use a stiff integrator of BDF-type for the numerical simulation. The question to be discussed now is whether the new stiff integrator of extrapolation type is competitive. A number of numerical comparisons already have been documented in Bader and Deuflhard (1981) and Deuflhard, Bader, and Nowak (1981), including standard test problems and chemical kinetics examples. In order to give some impression about the performance of the present version of LARKIN, a few further chemical kinetics examples are presented here using either the standard extrapolation integrator or the BDF-integrator by Gear

TABLE 5.6
Comparison of simulation times (seconds)

	METASC	GEARS maxord = 5	GEARS maxord = 3
HEXAN 1			
1.E–3	3.98	Fail*	Fail*
1.D–3	3.46	4.09	3.66
1.D–6	7.50	7.14	8.83
HEXAN 2			
1.E–3	10.63	12.05	12.02
1.D–3	12.30	11.14	11.10
1.D–6	22.62	21.15	24.22
RNA			
1.E–3	48.28	38.70	41.86
1.D–3	55.68	44.46	48.20
1.D–6	114.36	106.59	147.23

* Fail run also with maxord = 2

TABLE 5.7
Comparison of A(α)*-stability properties (for discretization of order* k) *Hairer, Bader, and Lubich (1981)*

BDF discretization:						
k	1	2	3	4	5	6
α	90°	90°	86°	76°	50°	16°
semi-implicit midpoint discretization:						
k	1	3	5	7	9	11
α	90°	90°	90°	87°	86°	86°

(1971a) and Hindmarsh (1974b). Of course, both integrators have been endowed with common scaling and sparse techniques. All comparison runs have been made on the IBM 370/168 at the University of Heidelberg using the FORTRAN optimizing compiler with OPT = 3. The following three problems are selected as typical:

(i) HEXAN1, thermal decomposition of *n*-hexane (Isbarn, Ederer, and Ebert 1981) NCEQ = 240, NSP = 59, NZE = 523;

(ii) HEXAN2, extended model of *n*-hexane (Isbarn 1981) NCEQ = 705, NSP = 166, NZE = 1270;

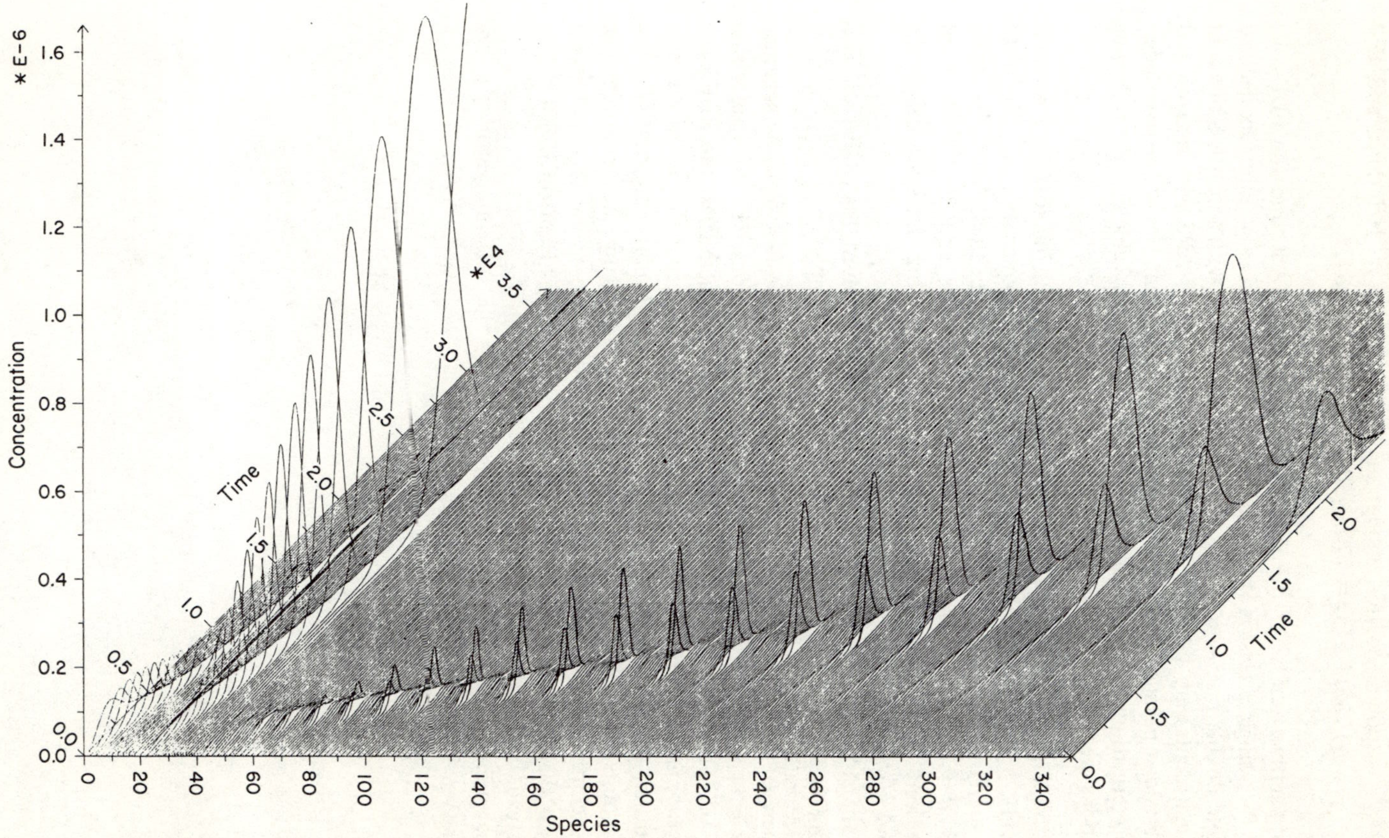

Fig. 5.12 RNA problem: graphical output for 349 (out of 352) species.

(iii) RNA polymerization model of RNA (Schneider, Heinrichs, and Poll 1980) NCEQ = 770, NSP = 352, NZE = 2642.

The comparative results for the extrapolation code METASC and the BDF-code GEARS for a single precision (tolerance 1.E–3 with a mantissa of about six decimal digits) and double precision (tolerance 1.D–3 and 1.D–6 with a mantissa of about 16 decimal digits) are listed in Table 5.5. The column GEARS with maxord = 3 has been added because in this case the A(α)-stability properties of the two underlying discretizations are comparable (see Table 5.7). From Table 5.6, the following observations are worth discussing:

(i) In the largest example (RNA), the code GEARS with maxord = 5 is about twenty percent faster than METASC. A possible explanation of this might be that the dynamic behavior of that chemical system turns out to be rather regular (compare Figure 5.12). This property favors multistep methods of this kind. In the other successful runs, the computing times are nearly the same.

(ii) In the single precision version, METASC is more robust than GEARS. This property seems to be typical over a wider class of examples.

In view of storage economy (compare §5.5.3) and the fact that low accuracy requirements are often sufficient for chemical applications, the single-precision version of LARKIN seems to be preferable. For low accuracies, the BDF code usually supplies enough intermediate output points to permit the use of a simple plot routine, whereas the extrapolation code tends to supply comparatively few output points. However, the few output points obtained by the new order and step-size control (Deuflhard 1980a) reflect the dynamic behavior of the chemical system so that a more sophisticated plot routine (such as Rentrop 1980) can be used instead.

These points seem to justify the decision to use the stiff integrator of extrapolation type in the simulation package LARKIN.

6

THEORY

6.1 Listing review (Aiken)

For a listing of some available work with stiff theory (1974–1982), refer to Appendix I. This is only meant to be a guide to the literature and is not exhaustive or definitive.

6.2 Some comments on stability and error analysis for stiff nonlinear differential systems (Dahlquist)†

6.2.1 *Basic notions*

We shall discuss error propagation in the numerical integration of initial-value problems for differential systems,

$$\mathrm{d}y/\mathrm{d}t = f(t, y), \qquad f: \mathbb{C} \times \mathbb{C}^s \to \mathbb{C}^s. \tag{6.1}$$

The classical theory of numerical methods for nonlinear problems is excellently summarized in Henrici (1962). It is based on the assumption that the step size h is chosen so that $\|hf_y'(t, y)\|$ is small. This is inappropriate for stiff problems. Moreover, the error bounds found in most texts contain factors like $\exp[L(t-t_0)]$, where the *Lipschitz constant* L is an upper bound of the norm of the Jacobian in some convex domain $\Omega \subseteq \mathbb{C}^s$ (or $\mathbb{R}^s$):

$$L \geqslant \sup_{y \in \Omega} \|f_y'(t, y)\|.$$

Such factors are inappropriate also for nonstiff problems with converging integral curves.

A more adequate concept is a one-sided Lipschitz condition, based on the *logarithmic matrix norm* $\mu(A)$,

$$\mu(A) = \lim_{\epsilon \downarrow 0} (\|I + \epsilon A\| - 1)/\epsilon, \tag{6.2}$$

where $\| \ldots \|$ is the matrix norm induced by the vector norm chosen. The most important properties of the logarithmic norm, in the general case are:

(i) $\mu(\alpha A + cI) = \alpha\,\mu(A) + \operatorname{Re} c, \quad \text{if } \alpha \geqslant 0, \quad c \in \mathbb{C},$

† Part of this work was carried out during the author's visit at Stanford University, spring 1982, sponsored by the NSF Grant No. MCS-7811985. The author is also grateful to North Holland Publishing Company for permission to use some of the material of Dahlquist and Söderlind (1982).

(ii) $\mu(\alpha A + \beta B) \leqslant \alpha\,\mu(A) + \beta\,\mu(B)$, if $\alpha \geqslant 0$, $\beta \geqslant 0$, (sub-additivity),

(iii) $\mu(A) \leqslant \|A\|$,

(iv) $\mu(A) \geqslant \operatorname{Re}\lambda$, for all eigenvalues λ of A,

(v) $\|(aI - A)^{-1}\| \leqslant [a - \mu(A)]^{-1}$, if $\mu(A) < a \in \mathbb{R}$,

(vi) $\|\exp(tA)\| \leqslant \exp[t\mu(A)]$, if $t \geqslant 0$.

The subadditivity can be extended to an arbitrary number of terms and to integrals, etc. For an inner-product norm it can be shown that $\mu(A)$ maximizes a Rayleigh quotient; in fact

$$\mu(A) = \sup\{\operatorname{Re}\langle \boldsymbol{x}, A\boldsymbol{x}\rangle / \langle \boldsymbol{x}, \boldsymbol{x}\rangle : \boldsymbol{x} \in \mathbb{C}^s\}. \tag{6.3}$$

For the usual l_2-norm $\mu(A)$ equals the largest eigenvalue (the spectral abscissa) of $\frac{1}{2}(A + A^{\mathrm{H}})$.

Desoer and Haneda (1972) apply this concept in an analysis of the backward Euler method for stiff nonlinear systems.

The one-sided Lipschitz constant m is an upper bound for the logarithmic norm of the Jacobian,

$$m \geqslant \sup_{y \in \Omega} \mu[f_y'(t, y)]. \tag{6.4}$$

The condition $\mu(A) \leqslant m$ is also referred to as a half-plane condition, since it can be shown (by a similarity transformation of A to an appropriate canonical form) that for any matrix, the spectrum of which is contained in the half-plane $\{z \in \mathbb{C} : \operatorname{Re} z < m\}$ one can find an inner-product norm (unfortunately dependent on A) such that $\mu(A) < m$. It is advantageous also to utilize a more general condition, called disk condition, which will be introduced in §6.2.3.

The study of the linear model problem,

$$y' = \lambda y, \tag{6.5}$$

where λ is a complex constant, tells us that when $|h\lambda|$ is not small, a numerical solution may in the long run behave quite differently from the exact solution of the ODE.

The region of (absolute) stability, S, of a numerical method is defined as the set of $h\lambda \in \overline{\mathbb{C}}$, such that the numerical solution of the model problem $y' = \lambda y$ with fixed step size h and arbitrary initial conditions, will remain bounded when $n \to \infty$. A useful collection of stability regions has been given by Sand and Østerby (1979).

For the ODE itself the stability region is the half-plane

$$\mathbb{C}^{-} = \{\lambda : \operatorname{Re}\lambda \leqslant 0\}.$$

$S = \mathbb{C}^{-}$ only in rare cases, such as the methods based on diagonal Padé approximants to the exponential function, e.g. the trapezoidal method. If the method is not A-stable, i.e. if S does not contain $\mathbb{C}^{-}$, we have numerical instability for some $\lambda \in \mathbb{C}^-$ unless h is small enough. This is the case for Euler's method, the Adams multistep methods and many other classical methods (designed for nonstiff problems). On the other hand, if S contains a large portion of $\mathbb{C}^+$, the method may be called super-stable. This is the case for the BDF-methods; it is actually an unfortunate consequence of the desirable property that the negative real axis should belong to S. In the writer's opinion super-stable methods should not be avoided, but the user should be aware that, in a nonlinear problem, it may happen that the numerical solution approaches an unstable stationary solution of the ODE, without any warning, if the step-size control is inappropriate (e.g. due to inadequate scaling) or if the tolerance is high; see examples in Dahlquist, Edsberg, Sköllermo, and Söderlind (1982).

There is work in progress about the handling of such situations; see the end of §6.2.3. For linear systems $\boldsymbol{y}' = A\boldsymbol{y}$, where A is a constant matrix, a numerical solution will be bounded, if the spectrum of A is interior to S. It is generally believed that for nonlinear systems the numerical solution will be stable, if the spectrum of the (time-dependent) Jacobian is always interior to S, at least if the nonlinearity is not too wild. The intuitive basis for this belief is that if $\boldsymbol{y}(t) + \boldsymbol{x}$ is the solution of a perturbed system,

$$\mathrm{d}[\boldsymbol{y}(t) + \boldsymbol{x}]/\mathrm{d}t = \boldsymbol{f}[\boldsymbol{y}(t) + \boldsymbol{x}, t] + \boldsymbol{p}(t), \tag{6.6}$$

then $\boldsymbol{x}$ satisfies the "pseudo-linear" variational equation,

$$\mathrm{d}\boldsymbol{x}/\mathrm{d}t = \boldsymbol{J}(t, \boldsymbol{x})\boldsymbol{x} + \boldsymbol{p}(t), \tag{6.7a}$$

where $\boldsymbol{J}(t, \boldsymbol{x})$ is a neighborhood average of the Jacobian,

$$J(t, \boldsymbol{x}) = \int_0^1 f_y' [\boldsymbol{y}(t) + \boldsymbol{x}\theta, t]\, \mathrm{d}\theta. \tag{6.7b}$$

Logarithmic norms yield handy bounds for the propagation of perturbations, which can be quite realistic, if the norm is appropriately chosen. By (6.7),

$$\begin{aligned} \|\boldsymbol{x}(t+h)\| &\leqslant \|\boldsymbol{I} + h\boldsymbol{J}(t, \boldsymbol{x})\| \cdot \|\boldsymbol{x}(t)\| + h\|\boldsymbol{p}(t)\| + O(h), \\ \|\boldsymbol{x}(t+h)\| &\leqslant [1 + h\mu \boldsymbol{J}(t, \boldsymbol{x})]\, \|\boldsymbol{x}(t)\| + h\|\boldsymbol{p}(t)\| + O(h). \end{aligned} \tag{6.8a}$$

Letting $h \to 0$ we obtain the basic differential inequality,

$$\mathrm{d}\|\boldsymbol{x}(t)\|/\mathrm{d}t \leqslant \mu[\boldsymbol{J}(t, \boldsymbol{x})]\, \|\boldsymbol{x}(t)\| + \|\boldsymbol{p}(t)\|. \tag{6.8b}$$

One can derive various global bounds from this, e.g. according to the following lemma.

LEMMA 6.1. Suppose that f satisfies the one-sided Lipschitz condition (6.4) and that $\boldsymbol{y}(t) \in \Omega$, $\|\boldsymbol{p}(t)\| \leqslant \hat{p}$. Then $\|\boldsymbol{x}(t)\| \leqslant \xi(t)$ where $\xi(0) \geqslant \|\boldsymbol{x}(0)\|$ and

$$\mathrm{d}\xi/\mathrm{d}t = m\xi + \hat{p}, \quad t \geqslant 0,$$

as long as the following global bound guarantees that $\boldsymbol{y}(t) + \boldsymbol{x}(t)$ stays in Ω.

$$\xi(t) = \hat{p}(\mathrm{e}^{mt} - 1)/m + \mathrm{e}^{mt}\xi(0) = -\hat{p}/m + \mathrm{e}^{mt}(\xi(0) + \hat{p}/m).$$

It is important that these results hold also when $m < 0$. Note that, in this case, $\xi(t) \to -\hat{p}/m$ when $t \to \infty$. When $m = 0$, l'Hospital's rule yields the result $\xi(t) = \hat{p}t + \xi(0)$.

The phrase "as long as . . ." is essential in nonlinear initial-value problems. When the condition is no longer true, anything can happen. There are several unpleasant "nonlinear phenomena" in ODEs and difference equations, such as

(i) $\boldsymbol{x}(t)$ may become infinite in finite time;
(ii) chaotic behavior;
(iii) there may be doubts about the existence and uniqueness of the solution of the nonlinear system to be solved at each step when an implicit method is used.

A similar neighborhood average can also be defined for the study of perturbations in a difference equation, see e.g. the arguments leading to Theorem 6.4.

It is reasonable to require that the local perturbations are small enough and the predictor is good enough so that, for a scheme that is stable for linear problems with time-dependent coefficients, the perturbation $\boldsymbol{x}$ will not grow so large that we have to worry about such serious nonlinear difficulties. In fact, an approximate solution ceases to be relevant several steps before such nonlinear phenomena occur. Note, however, that an adequate step-size control is necessary for these requirements to be satisfied, see again Dahlquist et al. (1982).

However, also linear problems with *variable* coefficients can in important respects be different from those with *constant* coefficients. Consider e.g. the difference equation

$$\boldsymbol{y}_{n+1} = \boldsymbol{A}_n \boldsymbol{y}_n, \quad n = 0, 1, 2, \ldots,$$

where

$$\boldsymbol{A}_{2m} = \begin{bmatrix} 0.5 & 0 \\ 1.5 & 0.1 \end{bmatrix}, \qquad \boldsymbol{A}_{2m+1} = \boldsymbol{A}_{2m}^{\mathrm{T}}, \qquad m = 0, 1, 2, \ldots,$$

$$\boldsymbol{A}_{2m+1}\boldsymbol{A}_{2m} = \begin{bmatrix} 2.5 & 0.15 \\ 0.15 & 0.01 \end{bmatrix}.$$

The spectral radius of $\boldsymbol{A}_{2m+1}\boldsymbol{A}_{2m}$ is 2.5090. Hence there are exponentially increasing solutions of the difference equation, in spite that the spectral radius of $\boldsymbol{A}_n$ is equal to 0.5 for all n.

There are analogous examples for ODEs. In fact, consider

$$\mathrm{d}\boldsymbol{y}/\mathrm{d}t = \boldsymbol{B}(t)\boldsymbol{y}, \qquad \boldsymbol{y}(0) = \boldsymbol{y}_0,$$
$$\boldsymbol{B}(t) = \ln \boldsymbol{A}_n, \qquad n \leqslant t < n+1, \qquad n = 0, 1, 2, \ldots,$$

where $\boldsymbol{A}_n$ is the same as before. Then $\boldsymbol{y}(1) = \boldsymbol{A}_0\boldsymbol{y}(0) = \boldsymbol{y}_1$. More generally, $\boldsymbol{y}(n) = \boldsymbol{y}_n$ which grows exponentially despite the eigenvalues of $\boldsymbol{B}(t)$ being strictly negative for all t. In fact

$$\boldsymbol{B}(2m) = \begin{bmatrix} -0.6931 & 0 \\ 6.0354 & -2.3026 \end{bmatrix}, \qquad \boldsymbol{B}(2m+1) = \boldsymbol{B}^{\mathrm{T}}(2m).$$

Evidently one can modify $\boldsymbol{B}(t)$ a little to make it smooth, and still there will be exponentially increasing solutions. There are also in the literature examples of equations with smooth coefficients of the form

$$y'' + ay' + (1 + b\cos 2\pi t)y = 0,$$

with unbounded solutions (expressible in terms of Mathieu functions) although $a > 0$, $|b| < 1$, so that the criterion for asymptotic stability is satisfied for the equation, if the coefficients have been frozen at an arbitrary time. (There is no difficulty in handling such problems numerically if $\|h\boldsymbol{B}(t)\|$ is small enough. There will be essentially the same sort of instability in the numerical as in the exact solution.)

These examples indicate that one has to be careful in the generalization of stability results from constant to variable coefficients (in particular if $\|h\boldsymbol{B}(t)\|$ is not small), and to variable step-size, see Brayton and Conley (1972), Sand (1982b), Nevanlinna and Liniger (1979).

Nevertheless there are several results, which can be described as generalizations of stability results from the linear model problem to large classes of stiff nonlinear systems, also when hL is large. In addition to Desoer and Haneda (1972), two other early writers should be mentioned. Dahlquist (1963) analyzed the trapezoidal method for nonlinear systems, which satisfy the half-plane condition (6.4) with $m = 0$ for an inner-product norm. This is generalized in several directions for multistep methods, see §6.2.3 and §6.2.4. Sandberg (1967) introduced a technique for deriving global l_2-estimates for the error of multistep methods, which proved to be very useful several years later, see §6.2.7. About 1974 a lively activity was started, and the early results are now particular cases of theories, which are more general, though still fairly incomplete. Most of this section is concerned with multistep methods, but in §6.2.12 there are some results for Runge–Kutta methods.

We shall now introduce the formalism for multistep methods. Two k-step methods are generated by a pair of polynomials with real coefficients,

$$\rho(\zeta) = \sum_{j=0}^{k} \alpha_j \zeta^j, \qquad \sigma(\zeta) = \sum_{j=0}^{k} \beta_j \zeta^j. \tag{6.9}$$

Let E be the translation operator, $Ey_n = y_{n+1}$. The *linear k-step method* is for a constant time-step h defined by the formula:

$$\rho(E)\hat{y}_n = h\sigma(E)f(t_n, \hat{y}_n). \tag{6.10a}$$

Similarly, the *one-leg method* is defined by the formula:

$$\rho(E)y_n = hf[\sigma(E)t_n, \sigma(E)y_n]. \tag{6.10b}$$

We shall mainly discuss the one-leg method, because the formulation of stability results for stiff nonlinear systems are simpler. It is, however, well known, see e.g. Dahlquist (1975c) that there is a close relation between the methods. E.g. if $[y_n]$ satisfies the one-leg equation (6.10b), then (for constant step-size) the sequence $[\hat{y}_n]$, where

$$\hat{y}_n = \sigma(E)y_n,$$

satisfies the linear multistep equation (6.10a). This is true for arbitrary nonlinear systems. For linear autonomous systems the two equations are identical (with $\hat{y}_n = y_n$). A disadvantage of the one-leg methods compared to the linear multistep methods is that there are more constraints on the coefficients for obtaining a certain order of consistency for variable-step-size computations, see Dahlquist (1979a). We shall not go into details. We only note that nevertheless, the class of one-leg methods contains methods with good stability and accuracy properties e.g. the BDF methods and that the memory requirements are moderate. No f-values are stored. For the linear model problem,

$$\mathrm{d}y/\mathrm{d}t = \lambda y, \qquad \lambda \in \mathbb{C},$$

we obtain the difference equation,

$$\rho(E)y_n = q\sigma(E)y_n, \qquad (q = \lambda h).$$

Introduce the vector:

$$Y_n = (y_n, y_{n+1}, \ldots, y_{n+k-1})^{\mathrm{T}}.$$

The difference equation (6.10b) can then be written in vector-matrix notation,

$$Y_{n+1} = C(q)Y_n,$$

where $C(q)$ is the companion matrix, associated with the polynomial $\rho - q\sigma$, i.e.

$$C(q) = \begin{bmatrix} 0 & 1 & 0 & \ldots & 0 \\ 0 & 0 & 1 & \ldots & 0 \\ \ldots & & & \ldots & \\ 0 & 0 & 0 & \ldots & 1 \\ c_1 & c_2 & c_3 & \ldots & c_k \end{bmatrix}, \tag{6.11a}$$

where

$$c_{\nu+1} = c_{\nu+1}(q) = -(\alpha_\nu - \beta_\nu q)/(\alpha_k - \beta_k q). \tag{6.11b}$$

Note that the point α_k/β_k is a pole of $C(q)$.

For the model problem (6.5), our analysis is based on an inner product in $\mathbb{C}^k$,

$$\langle u, v\rangle_G = u^{\mathrm{H}}Gv, \quad G \doteq [g_{ij}] = \text{positive-definite Hermitian}, \quad \|u\|_G^2 = \langle u, u\rangle_G.$$

The related norm, $\|u\|_G = \langle u, u\rangle_G^{\frac{1}{2}}$ will be called a G-norm. The square of the induced matrix norm equals

$$\|A\|_G^2 = \sup_x \langle Ax, Ax\rangle_G / \langle x, x\rangle_G,$$

i.e. the largest root κ of the generalized eigenvalue equation,

$$\det(A^{\mathrm{H}}GA - \kappa G) = 0. \tag{6.12}$$

For systems of s differential equations this is extended to an inner product of the form $\langle \ldots, \ldots \rangle_{G\otimes H}$, where the Hermitian matrix $\boldsymbol{H}$ defines an inner product in $\mathbb{C}^s$, appropriate for the differential system (6.1) and $\otimes$ denotes the Kronecker product, i.e., if we set $\boldsymbol{G} = [g_{ij}]$, then

$$\boldsymbol{G} \otimes \boldsymbol{H} = [g_{ij}\boldsymbol{H}]. \tag{6.13}$$

The basic rules for Kronecker products are (with obvious generalizations to an arbitrary number of factors)

$$(\boldsymbol{AB}) \otimes (\boldsymbol{CD}) = (\boldsymbol{A} \otimes \boldsymbol{C})(\boldsymbol{B} \otimes \boldsymbol{D}), \tag{6.14}$$

$$(\boldsymbol{A} \otimes \boldsymbol{B})^{-1} = \boldsymbol{A}^{-1} \otimes \boldsymbol{B}^{-1}. \tag{6.15}$$

If vector norms (in the three spaces involved) are defined such that $\|\boldsymbol{x} \otimes \boldsymbol{y}\| = \|\boldsymbol{x}\| \cdot \|\boldsymbol{y}\|$, for all $\boldsymbol{x}, \boldsymbol{y}$ the same relation holds for the induced matrix norms, i.e.

$$\|\boldsymbol{A} \otimes \boldsymbol{B}\| = \|\boldsymbol{A}\| \cdot \|\boldsymbol{B}\|. \tag{6.16}$$

The advantage of the use of norms and contractivity regions depends on the fact that a matrix norm is submultiplicative, while the spectral radius is not. Local inequalities for the norms of matrices of the form $\boldsymbol{C}(q)$ or $\boldsymbol{C}(A)$, where A is an $s \times s$ matrix, can therefore be applied to yield estimates for solutions or errors in computations with variable coefficients or variable step size. In the case of variable step size we must, however, remember that the coefficients α_j, β_j depend on n.

In §6.2.2 the basic theory is developed for the model problem, while the logical structure of the extension to systems is given in §6.2.3.

In §6.2.4 there will be a brief discussion of alternative approaches and analogous questions for other methods. Roughly speaking, the changes of the results in the extension from the model problem (6.5) to nonlinear systems (6.1) are smaller for one-leg multistep methods than for, say, Runge-Kutta methods. The extension from constant to variable step size is, of course, easier for Runge-Kutta methods.

In the following there will be frequent references to DS 82, which means Dahlquist and Söderlind (1982). We quote many results which are proved there.

6.2.2 *Stability and contractivity for the linear model problem*

Let $\overline{\mathbb{C}}$ be the complex plane, closed by the point ∞. We define two families of regions in $\overline{\mathbb{C}}$,

$$S(\theta) = \{q \in \overline{\mathbb{C}} : (C(q)/\theta)^n \text{ is uniformly bounded}, \quad n = 1, 2, 3, \ldots\}, \tag{6.17}$$

$$K_G(\theta) = \{q \in \overline{\mathbb{C}} : \|C(q)\|_G \leqslant \theta\}. \tag{6.18}$$

$S(1)$ is the familiar region of (absolute) stability of the method. $K_G(1)$ is called the contractivity region, Nevanlinna and Liniger (1978). A-stability and A-contractivity mean that $\overline{\mathbb{C}}$ is contained in, respectively, $S(1)$ and $K_G(1)$. Clearly,

$$K_G(\theta) \subseteq S(\theta). \tag{6.19}$$

A set in $\overline{\mathbb{C}}$, which is either a closed disc, the exterior of an open disc, a closed half-plane or a point is called a *circular domain*. It is called symmetric when it is symmetric with respect to the real axis.

We shall first find transformations, so that questions concerning the relation of a circular domain D to the regions $S(\theta)$ or $K_G(\theta)$ can be reduced to the particular case when $D = \mathbb{C}^-$, $\theta = 1$, i.e. the case of A-stability or A-contractivity.

Note that for any symmetric circular domain D we can find a real quadruple $(\alpha, \beta, \gamma, \delta)$ such that

$$D = \left\{ q \in \bar{\mathbb{C}} : \operatorname{Re} \frac{\alpha q + \beta}{\gamma q + \delta} \leqslant 0 \right\}. \tag{6.20}$$

The following lemmas show how a general situation can be transformed into the A-stability situation, where $D = \mathbb{C}^-$, $\theta = 1$. They are proved by straightforward calculation, see DS 82.

LEMMA 6.2. Let (ρ, σ) be a given method. Define another method (ρ^*, σ^*) by the polynomials,

$$\rho^* = \alpha\rho + \beta\sigma, \tag{6.21a}$$

$$\sigma^* = \gamma\rho + \delta\sigma. \tag{6.21b}$$

Set

$$q^* = (\alpha q + \beta)/(\gamma q + \delta), \tag{6.22}$$

and let $C^*(q^*)$ be the companion matrix associated with the polynomial $\rho^* - q^*\sigma^*$. Then

$$C(q) = C^*(q^*). \tag{6.23}$$

This is true also if q is a matrix (with appropriate interpretation of singular cases).

LEMMA 6.3. Let (ρ, σ) be a given method, and set

$$\rho_\theta(\zeta) = \rho(\theta\zeta), \qquad \sigma_\theta(\zeta) = \sigma(\theta\zeta), \qquad (\theta > 0). \tag{6.24}$$

Let $C_\theta(q)$ be the companion matrix associated with the polynomial $\rho_\theta - q\sigma_\theta$. Then

$$\theta \cdot C_\theta(q) = \Theta^{-1} C(q) \Theta, \tag{6.25}$$

where $\Theta = \operatorname{diag}(1, \theta, \ldots, \theta^{k-1})$.

This is true also if q is a matrix (with $\Theta \otimes I$ instead of Θ).

LEMMA 6.4. Let G be any positive-definite symmetric matrix, and set

$$G(\theta) = \Theta^{-H} G \Theta^{-1}.$$

Then,

$$\|C(q)\|_{G(\theta)} = \theta \|C_\theta(q)\|_G = \theta \|C_\theta^*(q^*)\|_G. \tag{6.26}$$

This is also valid if q is a matrix (with $\Theta \otimes I$ instead of Θ).

Note that if D is defined by (6.20), then

$$q \in D \Leftrightarrow q^* \in \mathbb{C}^-, \tag{6.27a}$$

$$D \subseteq S_{\rho,\sigma}(\theta) \Leftrightarrow (\rho_\theta^*, \sigma_\theta^*) \text{ is A-stable}, \tag{6.27b}$$

where ρ_θ^*, σ_θ^* means the polynomials obtained by combining the transformations given in the lemmas.

We are now in a position to formulate the main theorems.

THEOREM 6.1. To any symmetric circular subdomain D of $S(\theta)$ one can find a G-norm such that $D \subset K_G(\theta) \subset S(\theta)$.

The proof of this result in Dahlquist (1978b) is constructive. By the previous lemmas it is sufficient to consider the A-stability case ($D = \mathbb{C}^-, \theta = 1$); "$G$-stability is equivalent to A-stability." A-stability is characterized by the condition,

$$\text{Re}\ \rho(\zeta)/\sigma(\zeta) > 0 \text{ for } |\zeta| > 1.$$

An algorithm resembling a continued fraction expansion is developed which decomposes $\text{Re}\rho(\zeta)/\sigma(\zeta)$ into a sum of squares. It is shown how to calculate a G-matrix from the coefficients produced in this decomposition. A Fortran program exists, see LeVeque (1979).

The next result was first given by Dahlquist (1979b); see DS 82 for a complete proof.

THEOREM 6.2. $K_G(\theta)$ is, if it is not empty, a symmetric circular domain, which does not contain the point α_k/β_k.

COROLLARY. The real function,

$$\gamma(q) = \|C(q)\|_G, \qquad q \in \bar{\mathbb{R}},$$

defines $\|C(q)\|_G$ in the whole of $\bar{\mathbb{C}}$, see Figure 6.1.

$\gamma(q)$ has several pleasant properties, e.g. it is piece-wise analytic on $\bar{\mathbb{R}}$. Only those values of q, for which the largest root of (6.12) (with $A = C(q)$) is a multiple root, can be nonanalyticities. It is also a convex function of q', where

$$q' = q/(1 - q\beta_k/\alpha_k), \tag{6.28}$$

since we can write,

$$C(q) = C(0) + q' e_k \cdot \nu^{\mathsf{T}}, \tag{6.29}$$

$$\nu^{\mathsf{T}} = (\nu_1, \nu_2, \ldots, \nu_k), \qquad \nu_{\nu+1} = (\beta_\nu \alpha_k - \alpha_\nu \beta_k)/\alpha_k^2. \tag{6.30}$$

The following result follows easily from Theorem 6.2, and the facts that $K_G(1) \subset S(1)$ and, for a consistent method, $S(1)$ cannot contain $\mathbb{R}^+$.

LEMMA 6.5. Let q_1, q_2 be two points in $\mathbb{R} \cap K_G(\theta)$. Then $K_G(\theta)$ contains either the disc, for which the interval (q_1, q_2) is a diameter, or the exterior of that disc. In particular, if the method is consistent and $\gamma(\infty) \leqslant \gamma(0) = 1$, then $K_G(1) \supseteq \mathbb{C}^-$, which is possible for A-stable methods only.

Let D be that symmetric circular domain which intersects $\mathbb{R}$ at the points q_1 and q_2 and does not contain the pole α_k/β_k. Then

$$\max_{q \in D} \|C(q)\|_G = \max[\gamma(q_1), \gamma(q_2)].$$

LEMMA 6.6. If $\gamma(0) = 1$, $\gamma(\infty) < 1$ then the order of consistency cannot exceed 1.

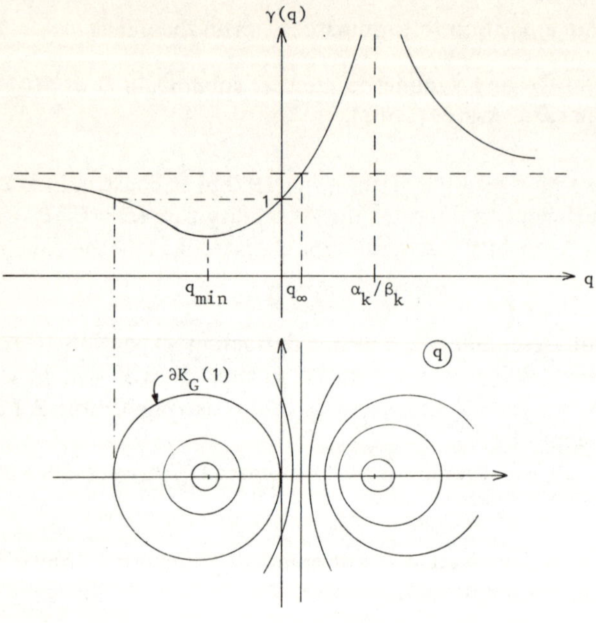

Fig. 6.1 Typical graph of $\lambda(q)$ if $q_{\min} < q_\infty < \alpha_k/\beta_k$ and the corresponding level curves of $\|C(q)\|_G$ in the q-plane.

Proof. Let the order of consistency be p, $p \geqslant 1$. If $\gamma(0) = 1$, it follows from the consistency that $\gamma(q) > 1$ for $q \in (0, \epsilon)$. Since $\gamma(\infty) < 1$, $\gamma(2R) = 1$ for some positive R. It follows that $\{q : |q - R| > R\} = K_G(1) \subseteq S(1)$. This is impossible for $p \geqslant 2$, because it is well known, e.g. Sanz-Serna (1980), that the curvature of the boundary locus $q = \rho(\zeta)/\sigma(\zeta)$ is zero at $\zeta = 1$, when $p \geqslant 2$.

There are other ways to construct $\boldsymbol{G}$-norms. E.g. if $\boldsymbol{C}(q_0) = \boldsymbol{TAT}^{-1}$, where $\|\boldsymbol{A}\|_2$ equals the spectral radius of $\boldsymbol{C}(q_0)$, then $\boldsymbol{G} = (\boldsymbol{TT}^{\mathsf{H}})^{-1}$ is useful for q in the vicinity of q_0 on the Riemann sphere. In fact, by (6.30) and the triangle inequality,

$$\boldsymbol{C}(q) = \boldsymbol{C}(q_0) + \boldsymbol{e}_k \boldsymbol{v}^{\mathsf{T}}(1 - q_0\beta_k/\alpha_k)^{-1}(q - q_0)(1 - q\beta_k/\alpha_k)^{-1}, \tag{6.31}$$

$$\|\boldsymbol{C}(q)\| \leqslant \|\boldsymbol{C}(q_0)\| + \|\boldsymbol{e}_k \boldsymbol{v}^{\mathsf{T}}\| \cdot |(1 - q_0\beta_k/\alpha_k)^{-1}(q - q_0)| \cdot |(1 - q\beta_k/\alpha_k)^{-1}|. \tag{6.32}$$

In particular for $q_0 = \infty$ (if $\beta_k \neq 0$),

$$\|\boldsymbol{C}(q)\| \leqslant \gamma(\infty) + \|\boldsymbol{e}_k \boldsymbol{v}^{\mathsf{T}}\| \cdot |\alpha_k/\beta_k| \cdot |(1 - q\beta_k/\alpha_k)^{-1}|. \tag{6.33}$$

These formulae have been written so that they are easily generalized to the matrix case.

6.2.3 *Stability and contractivity for systems*

We shall first give a different algebraic formulation (Dahlquist 1975a) for the contractivity condition,

$$\|\boldsymbol{C}(q)\|_G \leqslant 1 \text{ for all } q \leqslant 0, \tag{6.34}$$

where q does not occur! This will then be generalized to the system case. Let $X_n \in \mathbb{R}^k, X \in \mathbb{R}^{k+1}$, be defined thus,

$$X_n = (x_n, x_{n+1}, \ldots, x_{n+k-1})^{\mathrm{T}}, \qquad X = (x_0, x_1, \ldots, x_k)^{\mathrm{T}},$$

and introduce the quadratic forms

$$Q(X) = \rho x_0 \cdot \sigma x_0, \tag{6.35}$$

$$P_0(X) = \| X_0 \|_G^2 - \| X_1 \|_G^2, \tag{6.36}$$

and denote the corresponding symmetric matrices by $\boldsymbol{Q}, \boldsymbol{P}_0$. We need the following lemma, proved in DS 82.

LEMMA 6.7. If (6.34) holds, there exists a $\lambda > 0$, such that $\lambda \boldsymbol{Q} + \boldsymbol{P}_0$ is positive—semidefinite (6.37).

This condition was earlier called the G-stability condition, but that concept is no longer needed.

LEMMA 6.8. Let $C(q)$ be defined by (6.11), and assume that

$$\alpha_k/\beta_k > 0, \quad \| C(0) \|_G = 1, \quad \| C(\infty) \|_G \leqslant 1, \tag{6.38}$$

and let $\boldsymbol{H}$ be a positive definite $s \times s$ matrix. Then

$$\| C(A) \|_{G \otimes H} \leqslant 1, \tag{6.39}$$

for all $s \times s$ complex matrices $\boldsymbol{A}$, such that $\mu_H(A) \leqslant 0$, i.e. if

$$\operatorname{Re}\langle \boldsymbol{x}, A\boldsymbol{x} \rangle_H \leqslant 0, \quad \text{for all } \boldsymbol{x} \in \mathbb{C}^s. \tag{6.40}$$

($\mu_H(\ldots)$ is the logarithmic matrix norm induced by the $\boldsymbol{H}$-norm.)

Proof. Condition (6.34) holds by Lemma 6.5. Hence, by Lemma 6.7, there exists a $\lambda > 0$ such that $\lambda \boldsymbol{Q} + \boldsymbol{P}_0$ is positive-semidefinite. It is well known, see e.g. Bellman 1960, p. 227, that then the Kronecker product $(\lambda \boldsymbol{Q} + \boldsymbol{P}_0) \otimes \boldsymbol{H}$ is also positive-semidefinite for any positive-definite $\boldsymbol{H}$; i.e., with the notation introduced in §6.2.1, there exists a $\lambda > 0$ such that for $\forall \boldsymbol{X} \in \mathbb{C}^{s \cdot (k+1)}$

$$\lambda \operatorname{Re}\langle \rho \boldsymbol{x}_0, \sigma \boldsymbol{x}_0 \rangle_H + \| \boldsymbol{X}_0 \|_{G \otimes H}^2 - \| \boldsymbol{X}_1 \|_{G \otimes H}^2 \geqslant 0. \tag{6.41}$$

In particular if $\rho \boldsymbol{x}_0 = A \sigma \boldsymbol{x}_0$, then $\boldsymbol{X}_1 = C(A) \boldsymbol{X}_0$, and we obtain, by (6.40) and (6.41),

$$\| \boldsymbol{X}_0 \|_{G \otimes H}^2 - \| C(A) \boldsymbol{X}_0 \|_{G \otimes H}^2 \geqslant 0, \quad \forall \boldsymbol{X}_0 \in \mathbb{C}^{sk},$$

which is equivalent to (6.39).

Let, as before, D be the set

$$D = \{ q \in \overline{\mathbb{C}} : \operatorname{Re}[(\alpha q + \beta)/(\gamma q + \delta)] \leqslant 0 \}. \tag{6.41a}$$

We now define $D(H)$ to be the corresponding set of matrices:

$$D(H) = \{ A \in \mathbb{C}^{s \times s} : \operatorname{Re}\langle (\alpha A + \beta I)\boldsymbol{x}, \quad (\gamma A + \delta I)\boldsymbol{x} \rangle_H \leqslant 0, \quad \forall \boldsymbol{x} \in \mathbb{C}^s \}. \tag{6.41b}$$

We say that the matrix $\boldsymbol{A}$ satisfies a disk condition if $\boldsymbol{A} \in D(H)$, and that the differential system satisfies a *disk condition*, with the step size h, in $\Omega \in \mathbb{C}^s$, if $hf_y'(t, \boldsymbol{y}) \in D(H)$ for all $\boldsymbol{y} \in \Omega$. Note that

$$A \in D(H) \Leftrightarrow (\alpha A + \beta I)(\gamma A + \delta I)^{-1} \in \overline{\mathbb{C}}(H),$$

and that the condition $A \in \mathbb{C}^{-}(H)$ is equivalent to the condition (6.40), which in §6.2.1 was called a half-plane condition, with $m = 0$. If the spectrum of A is interior to D, one can choose H, such that $A \in D(H)$. For time-dependent $A, A = hJ(t)$, this detail is discussed in DS 82.

A circle-bounded set can be defined by inequalities different from (6.41a). One can e.g. show the following equivalences:

$$D = \{q : |q - b| \leqslant r\} \Leftrightarrow D(H) = \{A : \|A - bI\|_H \leqslant r\}, \tag{6.42a}$$

$$D = \{q : |q - b| \geqslant R\} \Leftrightarrow D(H) = \{A : \|(A - bI)^{-1}\|_H \leqslant R^{-1}\}, \tag{6.42b}$$

$$D = \{q : \operatorname{Re} q \leqslant m\} \Leftrightarrow D(H) = \{A : \mu_H(A) \leqslant m\}. \tag{6.42c}$$

By the combination of Lemma 6.4 and Lemma 6.8, we obtain the following result; see also Söderlind and Dahlquist (1981).

THEOREM 6.3. If $D \subseteq K_G(\theta)$ and $A \in D(H)$ then

$$\| C(A)\|_{G\otimes H} \leqslant \theta.$$

Then, by Lemma 6.5, we obtain:

COROLLARY: If $A \in D(H)$, $\alpha_k/\beta_k \notin D$, then

$$\|C(A)\|_{G\otimes H} \leqslant \max\{\gamma(q_1), \gamma(q_2)\},$$

where $\{q_1, q_2\} = \partial D \cap \mathbb{R}$, $\gamma(q) = \| C(q)\|_G$.

The application of the one-leg method (6.4b) with variable step size to the *nonautonomous linear system*,

$$\mathrm{d}y/\mathrm{d}t = J(t)y + p(t), \quad t \in T \subseteq \mathbb{R},$$

yields the difference equation,

$$Y_{n+1} = C_n(h_nJ_n)Y_n + e_k \otimes h_nM_np_n, \tag{6.43a}$$

where

$$J_n = J(\sigma(E)t_n), \quad M_n = (\alpha_kI - \beta_kh_nJ_n)^{-1}, \quad p_n = p(\sigma(E)t_n). \tag{6.43b}$$

We write $C_n(\ldots)$ here, since the coefficients α_j, β_j depend on n, when h does so. If $G \otimes H$ is a constant matrix, we obtain our basic inequality, analogous to (6.8b),

$$\| Y_{n+1}\|_{G\otimes H} \leqslant \|C_n(h_nJ_n)\|_{G\otimes H} \, \|Y_n\|_{G\otimes H} + h_n\|e_k\|_G \, \|M_np_n\|_H. \tag{6.44}$$

We shall now derive a similar inequality for the study of the error of the variable step-size application of a one-leg method to the *nonlinear system* (6.1). Consider the difference equation

$$\rho(E)y_n = h_nf(\sigma(E)t_n, \sigma(E)y_n) + h_n\epsilon_n, \tag{6.45}$$

where the coefficients of ρ and σ depend on n. Let (y_n^{I}), (y_n^{II}) be solutions of this equation, when the perturbation term ϵ_n is equal to, respectively $p_n^{\mathrm{I}}, p_n^{\mathrm{II}}$. Set

$$P_n = p_n^{\mathrm{I}} - p_n^{\mathrm{II}}, \quad z_n = y_n^{\mathrm{I}} - y_n^{\mathrm{II}}, \quad Z_n = (z_n^{\mathrm{T}}, z_{n+1}^{\mathrm{T}}, \ldots, z_{n+k-1}^{\mathrm{T}})^{\mathrm{T}}. \tag{6.46}$$

It was shown by Dahlquist (1978b) that Z_n satisfies the equation

$$Z_{n+1} = C_n^* Z_n + e_k \otimes h_n M_n^* p_n, \tag{6.47}$$

where the matrices C_n^*, M_n^* are averages of C_n, M_n between y_n^{II} and y_n^{I}, i.e.

$$C_n^* = \int_0^1 C_n[h_n J_n(\kappa)]\, d\kappa, \quad M_n^* = \int_0^1 [\alpha_k I - \beta_k h_n J_n(\kappa)]^{-1} d\kappa, \tag{6.48a}$$

$$J_n(\kappa) = f_y' [\sigma(E)t_n, \phi(\kappa)], \tag{6.48b}$$

where $\phi(\kappa)$ is a function that interpolates between $\sigma(E)y_n^{\mathrm{II}}$ and $\sigma(E)y_n^{\mathrm{I}}$ in the following way. Let

$$b = \beta_k/\alpha_k, \quad \tau(\zeta) = \sigma(\zeta) - b\rho(\zeta), \tag{6.49}$$

and note that $\tau(E)z_n$ does not depend on z_{n+k}.

$$\phi(0) = \sigma(E)y_n^{\mathrm{II}},$$

$$\phi(\kappa) - hbf[\sigma(E)t_n, \phi(\kappa)] = \phi(0) - hbf[\sigma(E)t_n, \phi(0)] + \kappa[\tau(E)z_n + hbp_n]. \tag{6.50}$$

It can be verified by straightforward computation that

$$\phi(1) = \sigma(E)y_n^{\mathrm{I}},$$

$$\sigma(E)z_n = \int_0^1 \phi'(\kappa) d\kappa = \alpha_k M_n^* [\tau(E)z_n + hbp_n],$$

and that the last equation is equivalent to (6.47). More details are found in Dahlquist (1978b).

By Theorem 6.3 and the application of the triangle inequality etc. to (6.47) and (6.48), we then obtain the following result:

THEOREM 6.4. Let $[y_n^{\mathrm{II}}]$ be a "known" sequence, e.g. such that $\sigma(E)y_n^{\mathrm{II}} = y(\sigma(E)t_n)$, where $y(t)$ is a solution of the differential system (6.1). Let $[D_n]$ be a sequence of circular domains such that $D_n(H)$ contains a neighborhood of the matrix $h_n f_y'(\sigma(E)t_n, \sigma(E)y_n^{\mathrm{II}})$. Set

$$\theta_n = \max_{q \in D_n} \|C_n(q)\|_G.$$

Then

$$\|Z_{n+1}\|_{G \otimes H} \leqslant \theta_n \|Z_n\|_{G \otimes H} + h_n \|e_k\|_G \cdot \|M_n^* p_n\|_H, \tag{6.51}$$

as long as the bounds obtained by the recursive application of this inequality guarantee that the matrices $h_n J_n(\kappa)$, $0 \leqslant \kappa \leqslant 1$, (defined by (6.48b)) belong to $D_n(H)$, for $n = 0, 1, \ldots$.

Comments

(i) The last sentence "as long as . . ." is typical for nonlinear problems where assumptions are made in a subset of the phase space.

(ii) Note that p_n is multiplied by $h_n M_n^*$, hence the local truncation error per step is multiplied by M_n^*, i.e. there is a considerable compression in the subspace spanned by the eigenvectors belonging to numerically large eigenvalues of $h_n J_n$. By (6.45) $\sigma(E)y_n$ is determined by an algebraic equation, which according to (6.49), can be written thus

$$\alpha_k \sigma(E) y_n - \beta_k h_n f[\sigma(E) t_n, \sigma(E) y_n] = \beta_k h_n \epsilon_n + \alpha_k \tau(E) y_n, \tag{6.52}$$

where $\tau(E)y_n$ is independent of y_{n+k}. Therefore

$$\sigma(E) z_n \approx \alpha_k M_n \tau(E) z_n + \beta_n h_n M_n \epsilon_n.$$

This shows that the component of the global error in the subspace mentioned above (which we call the retired component), is influenced more by the coupling to the error of the other (active) components than by its own local errors. This fact has implications for the measurement of the local error in computer programs, see also Dahlquist (1981).

(iii) For A-stable methods we choose, by Theorem 6.1, G so that $K_G(1) \subsetneq \mathbb{C}^-$. Then we can take $\theta_n = 1$, $D_n = \mathbb{C}^-$ for any constant step-size and any problem satisfying the monotonicity condition

$$\langle f(t,x) - f(t,y), x - y \rangle_H \leqslant 0, \quad \forall x, y \in \Omega,$$

where $\Omega \subseteq \mathbb{R}^s$ is some neighborhood of the orbit.

Unfortunately, due to Lemma 6.5, this theorem is not very useful for stiff problems, if the method is not A-stable. We shall indicate how to get around this difficulty. It is beyond the scope of this article to go into details. The author hopes to deal with them in a paper under preparation.

We need a few assumptions and notations. Note that A-stability is not assumed. The assumptions concerning the one-leg method ρ, σ, are:

(i) The method is consistent.
(ii) All zeros of $\rho(\zeta)$ except $\zeta_0 = 1$ are strictly inside the unit circle.
(iii) $0 < \alpha_k/\beta_k \leqslant \infty$.
(iv) All zeros of $\sigma(\zeta)$ are strictly inside the unit circle, and the method is $A(0)$ stable.

Let D_R denote the following circular domain, whose boundary is tangent to the imaginary axis. $D_R(H)$ is then given by (6.42).

$$D_R = \begin{cases} \{q \in \mathbb{C} : |q + R| \leqslant R\} & \text{if } R > 0, \\ \mathbb{C}^- & \text{if } R = \infty, \\ \{q \in \overline{\mathbb{C}} : |q + R| \geqslant -R\} & \text{if } R < 0. \end{cases} \tag{6.53}$$

The following lemma is essentially a combination of Lemma 2.3 and Theorem 3 of Söderlind and Dahlquist (1981). It tells that in a nonstiff problem, i.e. if $\|hJ\|^2 \ll 1$, the errors in the numerical solution propagate similarly to perturbations in the ODE itself, provided that the method satisfies the first three asumptions. (Recall the introductory criticism to the classical error bounds.) We shall then see that, if the fourth assumption is also satisfied, it can be applied also to a class of stiff problems, because the error propagation is in that case determined by the "active component," for which an assumption of the form $\|hD_{mm}\| \ll 1$ is reasonable, where D_{mm} is a certain submatrix in a factorized Jacobian.

LEMMA 6.9. Assume that the method (ρ, σ) satisfies assumptions (i), (ii), and (iii). Then there exists a G-norm such that, for some $R \in [-\infty, \infty]$, $K_G(1) = D_R$, and $\gamma(q)$ is analytic at the origin. Then, for $|\mu + i\nu| \ll 1$,

$$\|C(\mu + i\nu)\|_G = \exp\{\mu + \tfrac{1}{2}\nu^2/R + \tfrac{1}{2}[\gamma''(0) - 1]\mu^2 + O(|\mu + i\nu|^3)\}.$$

We associate a complex number $\mu + i\nu$ to the matrix hJ, as follows,

$$\mu = \mu_H(hJ), \quad \nu^2 = \|hJ\|_H^2 - \mu^2,$$

and assume that $\mu < \alpha_k/\beta_k$. Then

$$\|C(hJ)\|_{G \otimes H} \leqslant \begin{cases} \|C(\mu + i\nu)\|_G, & \text{if the level circle of Figure 6.1, which passes through } \mu + i\nu \text{ has its center in } \mathbb{R}^- \\ \gamma(\mu), & \text{if } \gamma(\mu) \geqslant \gamma(\infty). \end{cases}$$

This lemma is no surprise, if one thinks of Figure 6.1 with the circular level curves for $\|C(q)\|_G$, the labelling of which is determined by the piecewise analytic function $\gamma(q)$, $q \in \mathbb{R}$. For any $D_R \subseteq S$, the G-norm determined by the program of LeVeque, Dahlquist, and Andree (1979) can be used, unless 1 is a multiple root of equation (6.12) for $A = hJ$, $q = 0$. We have never seen the multiple root case occur. In the proof of Lemma 2.3 of Söderlind and Dahlquist 1981, it is shown, by a different construction, that in fact this exceptional case can be avoided.

First, consider the special case where the Jacobian is a block-triangular matrix, and a corresponding partitioning of Z and y

$$J(t) = \begin{bmatrix} J_{11} & J_{12} & \cdots & J_{1m} \\ 0 & J_{22} & \cdots & J_{2m} \\ \vdots & & & \vdots \\ 0 & 0 & \cdots & J_{mm} \end{bmatrix}, \quad Z = \begin{bmatrix} Z_1 \\ Z_2 \\ \vdots \\ Z_m \end{bmatrix}, \quad y = \begin{bmatrix} y_1 \\ y_2 \\ \vdots \\ y_m \end{bmatrix}. \tag{6.54a}$$

$C(hJ)$ is, by a similarity permutation, equivalent to a blocktriangular matrix with the matrices $C(hJ_{ii})$ in the diagonal. The partitioning is assumed to be done so that all eigenvalues of hJ_{11} have large moduli, while $\|hJ_{mm}\|^2 \ll 1$. The spectra of J_{ii}, $2 \leqslant i \leqslant m - 1$, are somewhere between. We also assume, roughly speaking, that for some positive constant c of moderate size and for suitable norms,

$$\mu(J_{ii}) + c \sum_{j, j \neq i} \|J_{ij}\| < 0, \quad 1 \leqslant i \leqslant m - 1. \tag{6.54b}$$

This is a block generalization of the "essentially negative dominant" matrices of Kreiss (1978), which we cannot work out further in this paper. See related ideas in Dahlquist (1983a). y_m contains the active components in the interval considered. The other components are retired; their differential equations are more or less close to being algebraic equations, in the time-interval (epoch) E under consideration.

We assume that, for each i, we can find a circular domain $D_i \subseteq S$, such that the spectrum Σ_i of hJ_{ii} is strictly inside D_i. For $i < m$, we assume in addition that $\Sigma_i \subseteq \mathbb{C}^-$. Then one has a fairly good chance to find a matrix H_i so that $hJ_{ii}(t) \in D_i(H_i)$, $i = 1, 2, \ldots, m$, $t \in E$, and

$$\| C(hJ_{ii})\|_{G_i \otimes H_i} \leqslant \rho < 1, \quad i = 1, 2, \ldots, m-1, \tag{6.55a}$$

for some matrix G_i. We can then define weights w_i in a norm,

$$\|Z\| = \max_i \|Z_i\|_{G_i \otimes H_i}/w_i, \quad i = 1, 2, \ldots, m, \tag{6.55b}$$

such that the induced norm of $C(hJ)$ satisfies an inequality like

$$\|C(hJ)\| \leqslant \max[\tfrac{1}{2}(1+\rho), \|C(hJ_{mm})\|_{G_m \otimes H_m}]. \tag{6.55c}$$

G_m should be a norm of the type mentioned in Lemma 6.9, so that

$$\|C(hJ_{mm})\|_{G_m \otimes H_m} \leqslant \exp[h\mu_{H_m}(J_{mm}) + O(\|hJ_{mm}\|^2_{H_m})]. \tag{6.55d}$$

The interpretation of this result is that, for a one-leg method satisfying the four assumptions given above, the error propagation in systems of this block-triangular form, is essentially the same as the perturbation propagation in the ODE itself, if the latter can be adequately discussed in terms of the logarithmic norm, induced by the vector norm

$$\|y\| = \max_i \|y_i\|_{H_i}/w_i. \tag{6.56}$$

A much more general class of Jacobians can be brought to a block triangular system of this type by a time-dependent, though fairly slowly varying, similarity transformation $T(t)$. The following general theorem shows what correction factors one has to apply in order to compensate for the variation of T. See also §6.2.6 concerning the handling of additive modifications of J.

The condition number of a matrix is denoted by cond(. . .), i.e.

$$\text{cond}(\mathbf{A}) = \|A\| \cdot \|A^{-1}\|.$$

THEOREM 6.5a. Consider the difference equation,

$$y_{n+1} = C_n y_n + r_n.$$

Then

$$\|y_\nu\| \leqslant \eta_\nu,$$

where $\eta_0 \geqslant \text{cond}(T_0)\|y_0\|$, and η_ν is defined by the difference equation,

$$\eta_{n+1} = \text{cond}(T_{n+1}^{-1}T_n)\|T_n^{-1}C_nT_n\|\eta_n + \text{cond}(T_{n+1})\|r_n\|. \tag{6.57a}$$

Proof. Set $y_\nu = T_\nu z_\nu$. Then

$$z_{n+1} = (T_{n+1}^{-1}T_n)(T_n^{-1}C_nT_n)z_n + T_{n+1}^{-1}r_n.$$

Then $\|z_\nu\| \leqslant \zeta_\nu$, $\zeta_0 = \|T_0^{-1}\| \cdot \|y_0\|$, where

$$\zeta_{n+1} = \|T_{n+1}^{-1}T_n\| \cdot \|T_n^{-1}C_nT_n\|\zeta_n + \|T_{n+1}^{-1}\| \cdot \|r_n\|.$$

Set $\zeta_\nu = \|T_\nu\|^{-1}\tilde{\eta}_\nu$, and note that $\|y_\nu\| \leqslant \|T_\nu\| \cdot \zeta_\nu = \tilde{\eta}_\nu$. Then

$$\tilde{\eta}_{n+1} = \|T_{n+1}\| \cdot \|T_n\|^{-1}\|T_{n+1}^{-1}T_n\| \cdot \|T_n^{-1}C_nT_n\|\tilde{\eta}_n + \mathrm{cond}(T_{n+1})\|r_n\|,$$

$$\tilde{\eta}_0 = \|T_0\|\zeta = \mathrm{cond}(T_0)\|y_0\|.$$

Since $\|T_{n+1}\| \cdot \|T_n\|^{-1} \leqslant \|T_n^{-1}T_{n+1}\| = \|(T_{n+1}^{-1}T_n)^{-1}\|$, it follows that $\tilde{\eta}_\nu \leqslant \eta_\nu$ where η_ν is defined by (6.57a).

This result has many applications. $T_n^{-1}C_nT_n$ is not necessarily block-triangular. The following continuous analog is proved in the same way.

THEOREM 6.5b. Consider the differential system,

$$\mathrm{d}y/\mathrm{d}t = J(t)y + p(t),$$

and let $T = T(t)$ be a differentiable nonsingular transformation. Then $\|y(t)\| \leqslant \eta(t)$, where $\eta(0) \geqslant \mathrm{cond}[T(0)]\|y(0)\|$, and $\eta(t)$ satisfies the differential equation,

$$\mathrm{d}\eta/\mathrm{d}t = [\mu(T^{-1}JT) + \mu(T^{-1}\dot{T}) + \mu(-T^{-1}\dot{T})]\,\eta + \mathrm{cond}(T)\|p(t)\|, \quad (6.57\mathrm{b})$$

where $\mu(\ldots)$ denotes the logarithmic matrix norm, induced by the vector norm $\|\ldots\|$.

Comment: Evidently these theorems are a fortiori true if the coefficients in (6.57) are replaced by simpler upper bounds. It follows, for example that if

$$\mathrm{cond}(T_{n+1}^{-1}T_n)\|T_n^{-1}C_nT_n\| \leqslant \theta, \quad \mathrm{cond}(T_{n+1})\|r_n\| < \hat{r},$$

then

$$\|y_n\| \leqslant \eta_n \leqslant \frac{1-\theta^n}{1-\theta}\,\hat{r} + \theta^n\eta_0 = \frac{\hat{r}}{1-\theta} + \theta^n\left(\eta_0 - \frac{\hat{r}}{1-\theta}\right). \quad (6.57\mathrm{c})$$

There are a lot of alternative nonrecursive bounds, dependent on the choice of assumptions for the coefficients. It is left to the reader to work out this and also to derive analogous bounds related to (6.51) and to compare analogous bounds for the continuous case with bounds obtained for $\theta = \exp(\hat{\mu}h)$ (say). Bounds where θ or $\hat{\mu}$ depend on time are also of interest. It is a mean of θ_n or $\hat{\mu}(t)$ that determines the bound; $\lim\sup\|y_n\|$ can be finite, even if θ_n sometimes exceeds unity. Also note that we do not obtain very important improvements in the bounds below $\theta = 1/2$ (say).

Note that if $T_n = T(t_n)$ then $T_{n+1}^{-1}T_n \approx T_{n+1}^{-1}(T_{n+1} - h_n\dot{T}_{n+1}) = I - hT^{-1}\dot{T}$, hence $\|T_{n+1}^{-1}T_n\| \approx 1 + h\mu(-T^{-1}\dot{T})$ and similarly $\|T_n^{-1}T_{n+1}\| \approx 1 + h\mu(T^{-1}\dot{T})$, hence $\mathrm{cond}(T_{n+1}^{-1}T_n) \approx 1 + h\mu(T^{-1}\dot{T}) + h\mu(-T^{-1}\dot{T})$, which shows the similarity of (6.57a) and (6.57b). A particular case is worth mentioning. If T is orthogonal, and if l_2-norm is used then $\mathrm{cond}(T_{n+1}^{-1}T_n) = 1$. One can show that in this case $T^{-1}\dot{T}$ is skew-symmetric, hence $\mu(T^{-1}\dot{T}) = \mu(-T^{-1}\dot{T}) = 0$.

For a general one-leg method T_n has a more complicated relationship to the transformation $T(t)$ of the continuous case. We conjecture that the correction factors needed to compensate for the variability of the transformation will, nevertheless, be approximately the same.

In many applications (e.g. chemical kinetics) the Jacobian is, after appropriate scaling of the variables a "graded matrix," which is a generalization of the singular perturbation form, see Theorem 6.6. The next theorem shows that for such matrices $T^{-1}T$ is very small. This is in our opinion very satisfactory, because we feel that the introduction of a correction factor is a little disturbing, even if it is bounded, if the bound is not expressed in easily estimated terms.

THEOREM 6.6. Let $E = \text{diag}(\epsilon_i)$, where $\epsilon_i/\epsilon_{i+1} \leqslant \eta \ll 1, i = 1, 2, \ldots, m-1$, and consider the graded matrix,

$$J = E^{-1}A,$$

where $A(t)$ can be factorized in the form,

$$A = (I + L)D(I + U),$$

where $L(U)$ are strictly lower (upper) block triangular matrices, and D is block-diagonal. We assume that L, D, U, and their time-derivatives are bounded uniformly with respect to t and η.

Then $A(t)$ can be brought to block upper triangular form $T^{-1}A'T = R$ where T is block lower triangular of the form,

$$T = I + L' + O(\eta^2),$$

$$(L')_{ij} = L_{ij} \cdot \epsilon_j/\epsilon_i, \quad j < i.$$

Note that

$$T^{-1}\dot{T} \approx \mathrm{d}L'/\mathrm{d}t = O(\eta),$$

so that the correction factor for the variation of T is negligible if $\eta \ll 1$; actually "the stiffer, the better". The proof will be given in Dahlquist (1983c).

Our discussion of the block triangular form (6.54) is to be applied to subintervals, *epochs,* of the whole run, in this case Epoch m. The active variables of one epoch are normally the retired of the next. In the early epochs there are also variables, *infants*, which we have ignored, because they are practically constant, but they will become active in a later epoch. Within each epoch the step size should vary fairly slowly; it typically grows by a factor of 20 in 50 steps (say). At the end of Epoch i, a quasisteady state is reached, if $\epsilon_{i+1} > 20\epsilon_i$ (say), which can be determined with a much smaller error than the maximum global error during the preceding computation. Therefore *each epoch can be treated almost independently in the error analysis*. Between two epochs there is a "generation shift," when the step size may grow from ϵ_i to $\epsilon_{i+1}/20$ (say), a growth by powers of 10, e.g. in problems of chemical kinetics, while very little happens to the solution. This rapid increase of step size sometimes may cause trouble with some present multistep codes (teenager problems?).

In addition to this "normal" development, there are cases when assumption (6.54b) ceases to be valid, i.e. *the retired ask to return to active state*. This happens e.g. in the simulation of stiff nonlinear oscillators. A fast reduction of step size is then needed, and the active return to infancy. It may happen, see Dahlquist Edsberg, Sköllermo, and Söderlind (1982), that a code does not detect such a situation and produces a completely erroneous solution with no warning. For systems with graded Jacobians it seems possible to design a strategy for the safe and easy handling of this, but we cannot yet report any experience.

6.2.4 *One-leg methods based on several pairs of polynomials* (ρ_i, σ_i)

Söderlind and Dahlquist (1981) study error propagation for the case where two different one-leg methods (ρ_1, σ_1), (ρ_2, σ_2), with the same collocation point i.e. $\sigma_1 t_n = \sigma_2 t_n$, are applied to, respectively, $\boldsymbol{x}_1$ and $\boldsymbol{x}_2$ in a singular perturbation example,

$$\epsilon \mathrm{d}\boldsymbol{x}_1/\mathrm{d}t = J_{11}(t)\boldsymbol{x}_1 + J_{12}(t)\boldsymbol{x}_2, \qquad J(t) = \begin{bmatrix} J_{11} & J_{12} \\ J_{21} & J_{22} \end{bmatrix},$$
$$\mathrm{d}\boldsymbol{x}_2/\mathrm{d}t = J_{21}(t)\boldsymbol{x}_1 + J_{22}(t)\boldsymbol{x}_2.$$

For brevity, we write $\rho \boldsymbol{x}_n$, etc., instead of $\rho(E)\boldsymbol{x}_n$, here and later.

The difference equation reads,

$$\epsilon h^{-1}\rho_1 \boldsymbol{x}_{1n} = J_{11}(\sigma t_n)\sigma_1 \boldsymbol{x}_{1n} + J_{12}(\sigma t_n)\sigma_2 \boldsymbol{x}_{2n},$$
$$h^{-1}\rho_2 \boldsymbol{x}_{2n} = J_{21}(\sigma t_n)\sigma_1 \boldsymbol{x}_{1n} + J_{22}(\sigma t_n)\sigma_2 \boldsymbol{x}_{2n}.$$

A generalization to graded matrices with more than two blocks suggests itself. A block LU-decomposition of $J(t)$ is performed instead of the similarity transformation $\boldsymbol{T}_n$ employed by us in §6.2.3. The Schur complement, $J_s = J_{22} - J_{21}J_{11}^{-1}J_{12}$, becomes the relevant concept. Consider an $\boldsymbol{H}$-norm for the variable $\boldsymbol{x}_2$. A bound of the following type is obtained for the growth or decay of the norm of the error in one step of a stiff interval.

$$\exp\left[h\mu_H(J_s) + O(\|hJ_s\|_H^2 + \epsilon)\right],$$

if $\|hJ_s\|_H \ll 1$ and $\epsilon/h \ll 1$. Although the latter condition is harder than we needed when the case $(\rho_1, \sigma_1) = (\rho_2, \sigma_2)$ was discussed with the similarity transformation approach, we strongly believe in the use of different methods for different variables for problems with a graded Jacobian. The reason is, of course, that we overcome the difficulty of combining high accuracy with a large stability or contractivity region. After the first epoch high accuracy is not needed for (ρ_1, σ_1) and a large contractivity region is not needed for (ρ_2, σ_2). If ϵ is very small, (ρ_1, σ_1) does not even need to be consistent; one can put $\epsilon = 0$. D is of the form (6.42b) for $\boldsymbol{x}_1$ and (6.42a) for $\boldsymbol{x}_2$.

6.2.5 *Double-step contractivity*

Another approach to overcoming the difficulty is the double-step contractivity investigated by Söderlind (1981); see also DS 82. Here the original difference equation is augmented by a copy of the difference equation shifted k steps, i.e.

$$Y_{n+1} = C_n(h_n J_n)Y_n + \boldsymbol{e}_k \otimes h_n M_n \boldsymbol{p}_n,$$
$$Y_{n+1-k} = C_{n-k}(h_{n-k} J_n)Y_{n-k} + \boldsymbol{e}_k \otimes M_{n-k}\boldsymbol{p}_{n-k}$$

is considered. In order to use the presented theory, this system has to be transformed to a difference equation with a single companion matrix. Since the new companion matrix is no longer linear in q' as in (6.29), the proof of Theorem 6.2, and its corollary (Lemma 6.5), do not hold, and the contractivity region is no longer circular (see Figure 6.2).

For BDF-methods "almost contractivity," i.e. a bound for the error growth in one step is derived,

$$\theta_n = 1 + Ch^2 \sup \|\dot{J}(t)\|.$$

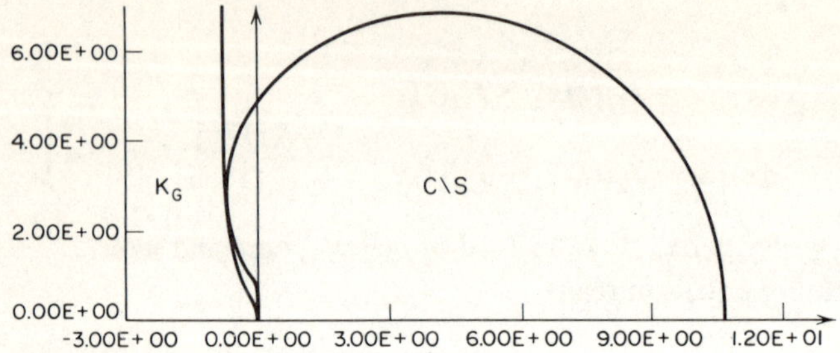

Fig. 6.2 Double-step contractivity and stability regions of the BDF4.

Unlike our previous results, we here have a condition on the rate of change of the Jacobian.

6.2.6 *Error bounds of Gronwall's type*

Consider the application of a one-leg method to the system

$$dy/dt = A(t)y + g(t, y). \tag{6.58}$$

Let (y_n^{I}) and (y_n^{II}) be two solutions of the difference equation with different initial conditions. Set

$$A_n = A(\sigma t_n), \quad J_n = J(\sigma t_n, \sigma x_n), \quad x_n = y_n^{\text{I}} - y_n^{\text{II}},$$

where $J(. . . . , . . .)$ is the neighborhood average of the Jacobian, introduced in (6.7a, b). Then (x_n) satisfies the difference equation,

$$\rho x_n = h_n A_n \sigma_n + h J_n \sigma x_n.$$

Set

$$X_n = (x_n^{\text{T}}, x_{n+1}^{\text{T}}, \ldots, x_{n+k-1}^{\text{T}})^{\text{T}}, \quad U_n = C_n(h_n A_n),$$
$$V_n(X_n) = [C_n(h_n A_n + hJ_n) - C_n(h_n A_n)] \cdot h_n^{-1}.$$

The difference equation reads, in matrix form,

$$X_{n+1} = [U_n + h_n V_n(X_n)]\, X_n. \tag{6.59}$$

By a matrix version of (6.31),

$$V_n(X_n) = e_k v^{\text{T}} \otimes (I - hA_n\beta_k/\alpha_k)^{-1} J_n [I - h_n(A_n + J_n)\beta_k/\alpha_k]^{-1}.$$

Let $k_1(t_n)$ be an upper bound of $\|V_n(X_n)\|$. If we use matrix norms such that $\|B \otimes C\| = \|B\| \cdot \|C\|$, we have if u.b. means upper bound,

$$k_1(t_n) = \|e_k v^{\text{T}}\| \cdot \text{u.b. } \|(I - hA_n\beta_k/\alpha_k)^{-1} J_n\|$$
$$\cdot \text{ u.b.}(1 - h\mu(A + g_y'))^{-1}. \tag{6.60}$$

We shall use the following lemma of Gronwall's type, see e.g., Gear and Tu (1974).

LEMMA 6.10. If $W_n = U_n + h_n V_n$, $t_{n+1} - t_n = h_n$, and there exist constants k_0 and k_1 such that for all $m \in [n, N]$, $n \in [0, N]$,

$$\|V_n\| \leqslant k_1, \quad \max(1, \|U_{m-1}U_{m-2} \ldots U_n\|) \leqslant k_0,$$

then

$$\|W_{m-1}W_{m-2}\ldots W_n\| \leqslant k_0 \exp[k_0 k_1(t_m - t_n)].$$

This can be used to estimate solutions of the pseudolinear difference equation (6.59), because $X_m = W_{m-1}W_{m-2} \ldots W_n X_n$, and it does not matter that W_j depends on X_j.

LEMMA 6.11. Let X_n be determined by (6.59) for $n = 0, 1, 2, \ldots, N$. If $\|V_n(X_n)\| \leqslant k_1$ for $\|X_n\| \leqslant k_2(t_n)$, $n = 0, 1, 2, \ldots, N$, then

$$\|X_m\| \leqslant k_0 \exp[k_0 k_1(t_m - t_n)] \cdot \|X_n\|, \quad m \geqslant n.$$

as long as this bound remains less than $k_2(t_m)$. There is a continuous analog to this.

LEMMA 6.12. Assume that all solutions of the system,

$$\mathrm{d}x/\mathrm{d}t = A(t)x \tag{6.61}$$

satisfy the inequality (i.e. uniform stability):

$$\|x(t)\| \leqslant k_0\|x(\tau)\|, \quad \forall t \geqslant \tau \geqslant a. \tag{6.62}$$

Assume that $\|J(t, x)\| \leqslant k_1, \forall t \geqslant a$ if $\|x\| \leqslant k_2(t)$. Then all solutions of the pseudolinear system,

$$\mathrm{d}x/\mathrm{d}t = [A(t) + J(t, x)]\,x, \tag{6.63}$$

satisfy

$$\|x(t)\| \leqslant k_0\|x(\tau)\| \cdot \exp[k_0 k_1(t - \tau)], \quad t \geqslant \tau \geqslant a,$$

as long as this inequality guarantees that $\|x(t)\| \leqslant k_2(t)$.

Lemma 6.12 can be derived either by a passage to the limit from the discrete case or by writing (6.63) in the equivalent integral equation form,

$$x(t) = W(t)x(a) + \int_a^t W(t)W(\tau)^{-1}J[\tau, x(\tau)]\,x(\tau)\,\mathrm{d}\tau,$$

where $W(\ldots)$ is the fundamental matrix solution of (6.61) with the initial condition $W(a) = I$. Then (6.62) is equivalent to the condition, $\|W(t)W(\tau)^{-1}\| \leqslant k_0$, hence $\|x(t)\|$ satisfies the inequality,

$$\|x(t)\| \leqslant k_0\|x(a)\| + k_0 \int_a^t k_1(\tau)\|x(\tau)\|\,\mathrm{d}\tau,$$

if we allow k_1 to depend on time. The result then follows from Gronwall's lemma, which we quote from Coppel (1965).

GRONWALL'S LEMMA. Let $\lambda(t)$ be a real continuous function and $k(t)$ be a non-negative continuous function on the interval $[a, b]$. If a continuous function $y(t)$ has the property that

$$y(t) \leqslant \lambda(t) + \int_a^t k(s)y(s)\,\mathrm{d}s$$

for $a \leqslant t \leqslant b$, then on the same interval

$$y(t) \leqslant \lambda(t) \int_a^t \lambda(s) k_1(s) \exp\left(\int_a^t k(\tau) \mathrm{d}\tau \right) \mathrm{d}s.$$

In particular if $\lambda(t) \equiv \lambda$ is a constant,

$$y(t) \leqslant \lambda \exp\left(\int_a^t k(s) \mathrm{d}s \right).$$

We used the particular case.

Comments:

(i) It is correct to let k_1 be time-dependent also in the preceding lemmas; one has just to replace $k_1(t_m - t_n)$ by

$$\int_{t_n}^{t_m} k_1(s) \mathrm{d}s.$$

This is of interest if $k_1(s)$ is a decreasing function, so that one can obtain a reasonable bound also over a large interval.

(ii) Another extension is when k_0 is replaced by respectively $k_0 \psi(t_m)/\psi(t_n)$ and $k_0 \psi(t)/\psi(\tau)$, where ψ is a positive function, for example an increasing or decreasing exponential.

(iii) The assumptions contain the contractivity assumption as a particular case, namely $k_0 = 1$. In this case the lemmas are easily obtained by the triangle inequality, e.g.

$$\| W_j \| \leqslant \| U_j \| + h_j \| V_j \| \leqslant 1 + h_j k_1 \leqslant \exp(h_j k_1), \quad \text{if } k_0 = 1.$$

The triangle inequality can also be used in the recursive error bounds of §6.2.3. In the continuous case,

$$k_0 = 1 \Leftrightarrow \mu[A(t)] \leqslant 0, \quad \forall t.$$

In this case a sharper result can be obtained by the one-sided Lipschitz condition $\mu[J(\ldots, \ldots)] \leqslant k_1$ (also with $k_1 < 0$) instead of $\|J(\ldots, \ldots)\| \leqslant k_1$. This exemplifies that a weak point of the Gronwall approach is that it does not use directional information in the perturbations. Note that $k_1(t)$ is assumed to be non-negative, while no corresponding assumption is needed about the logarithmic norm $\mu(J)$ in the basic differential inequality (6.8b). On the other hand, the Gronwall approach is more efficient than approaches based on logarithmic norms and (generalized) contractivity, when it is difficult to assume the existence of a norm such that the bound for $\Pi \| U_j \|$ is not much larger than the bound for $\| \Pi U_j \|$. (How often is this the case?)

(iv) If A_n and J_n are graded matrices, it is important that we use $\|(I - hA)^{-1} J\|$ instead of $\|(I - hA)^{-1}\| \cdot \|J\|$ in (6.60). Usually, there are eigenvalues of A and J of small as well as of large modulus. Then $\|(I - hA)^{-1}\|$ is determined by the small eigenvalues (it is $1 - O(h)$) while $\|J\|$ depends on the large eigenvalues.

(v) It has been suggested to use (6.58) with a constant matrix A as a sufficient model for stiff non-linear problems. In our experience it is relevant in particular cases only. The basic difficulty of the whole theory is in fact related to the non-commutativity of the Jacobians at different times, see §6.2.1. In the final phase of each epoch, when a quasi-steady state is approached, it can be used. It is the standard technique in the classical stability theory of ODEs.

It also has been suggested to use a model with a piecewise constant $\boldsymbol{A}$. Note, however, that the factors k_0 belonging to all the preceding pieces have to be multiplied, and they are never less than unity! We prefer instead the use of Theorem 6.5a or 6.5b with a smoothly changing $\boldsymbol{T}$. It is also possible to use piecewise constant $\boldsymbol{T}$. If in Theorem 6.5b, $\boldsymbol{T}$ changes discontinuously from $\boldsymbol{T}_-$ to $\boldsymbol{T}_+$, $\dot{\boldsymbol{T}}$ becomes a delta function, but it can be shown that η at such a point is to be multiplied by $\text{cond}(\boldsymbol{T}_+^{-1}\boldsymbol{T}_-)$, which is not surprising, if one compares with (6.57a) for the discrete case. This factor is likely to be less than k_0, since it is a condition number for the relative change of $\boldsymbol{T}$, while k_0 is more like the condition number for $\boldsymbol{T}$ itself.

6.2.7 *Multiplier techniques and l_2-estimates*

There is another approach to the study of linear multistep methods for stiff non-linear problems, where one does not estimate the local error growth, but aims directly to a bound for the global error. Let ρ and σ be the usual generating poly-nomials (see §6.2.1), and let γ be the coefficient sequence in the expansion,

$$\rho(\zeta)/\sigma(\zeta) = \sum_0^\infty \gamma_j \zeta^{-j}. \tag{6.64}$$

The equation governing the propagation of error x_n can be written in the form

$$\gamma * x_n - h[f(y_n + x_n) - f(y_n)] = hq_n, \tag{6.65}$$

where the star means convolution of sequences, and $\boldsymbol{q}$ is related to the local errors. Liniger and Odeh (1977) and Nevanlinna (1977) scalar-multiplied this equation with $\boldsymbol{e}_n$, and obtained error bounds for A-stable schemes and problems satisfying the half-plane condition with $m = 0$. This idea can be thought of as an adaptation of Popov's work in control theory. Its application to the study of numerical methods for ODEs can be traced back to Sandberg (1967), whose important paper was unnoticed for many years.

Nevanlinna and Odeh (1981) have found a powerful modification of this approach through the introduction of a multiplier sequence $\mu = (\mu_0, \mu_1, \mu_2, \ldots, \mu_M)$ which gives access to stability results for A(α)-stable methods (also with $\alpha < \pi/2$) for certain stiff nonlinear systems. Let $\boldsymbol{J}(t, \boldsymbol{x})$ be the neighborhood average defined by (6.7a, b). By (6.65),

$$\sum_{n=p}^N \langle \mu * x_n, \gamma * x_n \rangle - \sum_{n=0}^N \langle \mu * x_n, J(t_n, x_n) x_n \rangle = \sum_{n=0}^N \langle \mu * x_n, q_n \rangle h. \tag{6.66}$$

Let

$$\hat{\mu} = \sum_{n=0}^M \mu_n e^{-in\tau},$$

and define $\hat{x}$ analogously. By Parseval's relation the real part of the first term equals

$$\frac{1}{2\pi}\int_{-\pi}^{\pi} \hat{\mu}(\tau)\rho(e^{i\tau})/\sigma(e^{i\tau}) \cdot |\hat{x}(\tau)|^2 d\tau.$$

They choose $\hat{\mu}(\tau)$ so that the integrand becomes non-negative, where

$$|\arg \hat{\mu}(\tau)| \leqslant \pi/2 - \alpha.$$

This is possible, only if the method (ρ, σ) is A(α)-stable, and they show that, for any $\epsilon > 0$, it is indeed possible, for any A($\alpha - \epsilon$)-stable method. For BDF-methods with $k \leqslant 4$ they construct multipliers with $M = 1$ and ϵ fairly small; for $k = 5$, $\alpha = 51° \, 50'$, $\alpha - \epsilon = 35° \, 19'$.

Concerning the second sum on the left-hand side of (6.66) they point out that if J is a constant matrix such that $|\pi - \arg \lambda| < \alpha$ for all eigenvalues, then one can find an inner product such that it becomes negative. They show that the sum is negative also in many cases with variable coefficients, e.g. they define two measures K_1, K_2 of, respectively, asymmetry and variability (due to nonlinearity or time-dependency), such that

$$|\langle u, -J(t, x)v\rangle| \leqslant \tfrac{1}{2}K_1(\langle u, J(t, x)u\rangle + \langle v, J(t, x)v\rangle), \quad \forall u, v,$$

$$\langle u, -J(t, x)u\rangle \leqslant (1 + 2K_2 h)\langle u, -J(s, y)u\rangle, \quad \forall u.$$

Here $|t - s| = O(h)$, $\|x - y\| = O(h)$. Then they show that the second sum is nonpositive if

$$\mu_0^{-1} \sum_{j \geqslant 1} |\mu_j| < K_1^{-1} \cdot (1 + K_2 h)^{-1}.$$

Hence if $f(t, y) + \alpha y$ satisfies such conditions for some $\alpha > 0$, one can obtain by Schwarz inequality a bound for

$$\left(\sum_{n=0}^{N} \|x_n\|^2\right)^{1/2}.$$

By an interesting combination of this approach with G-norm theory (Theorem 6.1) they obtain error bounds of the form

$$\|x_n\| \leqslant C\left(\max_{0 \leqslant j < k} (\|x_j\| + h\|J(\ldots)x_j\|) + h \sum_{j=k}^{n} \|\sigma(E)q_j\|\right).$$

Note that the initial data are multiplied by hJ here, which was not the case for one-leg methods. This is related to the lack of D-stability of linear multistep methods. For the time being this theory is limited to fixed step size, but Nevanlinna (personal communication) indicates that an extension to variable step seems possible.

6.2.8 *Contractivity in other norms*

Nevanlinna and Liniger (1979) discuss contractivity in maximum norm. The contractivity regions are no longer circular domains, and there exist A_0-contractive methods (i.e. methods such that the contractivity set contains $\mathbb{R}^-$), which are not A-stable. They show that, for any p and any $\alpha < \pi/2$, there exists a pth order A(α)-contractive method with $\rho(\zeta) = \zeta^k - \zeta^{k-1}$. The widely used backward differentiation methods are, however, not A_0-contractive (in fact not even contractive at the origin) in maximum norm, for $k \geqslant 2$. For a class of A_0-contractive methods they prove a boundedness result for the "pseudolinear" system $\mathrm{d}\boldsymbol{y}/\mathrm{d}t = \boldsymbol{J}(t,\boldsymbol{y})\boldsymbol{y} + \boldsymbol{p}(t)$, when $\boldsymbol{J}(t, \boldsymbol{y})$ satisfies a certain condition of negative diagonal-dominance type. The first use of the max-norm in this context is due to Liniger (1957). See also §6.3.

6.2.9 *The approach of Kreiss*

Conditions of negative diagonal-dominance type are used also by Kreiss (1978) in a profound analysis of the smoothness properties of the solutions of stiff non-autonomous linear systems, and the stability properties of linear multistep methods in their application to such systems with piecewise constant step size. His terminology differs from ours, and he does not use the contractivity concept. We shall briefly outline one of his main results. He considers the class of systems of the form, $\mathrm{d}\boldsymbol{y}/\mathrm{d}t = \boldsymbol{A}(t)\boldsymbol{y} + \boldsymbol{F}(t)$, where the basic assumptions are that

$$\min[|a_{ii}|^{-1}, 1]\ (|\mathrm{d}^\nu a_{ij}/\mathrm{d}t^\nu| + |\mathrm{d}^\nu \boldsymbol{F}/\mathrm{d}t^\nu|) \leqslant K, \quad \nu = 0, 1,$$

that the matrix is essentially negative dominant and that the spectrum of $h\boldsymbol{A}(t)$ belongs to a certain star-shaped subdomain of the stability region S(1). The existence of this subdomain is proved. Then he proves the existence of constants K', γ such that for any $m, n, m \leqslant n$, the numerical solution for every problem in this class will satisfy the inequality

$$\|Y_n\|/\|Y_m\| \leqslant K' \cdot \exp[\gamma(t_n - t_m)],$$

where

$$\|Y_n\| = \|y_n\|_\infty + \|y_{n+1}\|_\infty + \ldots + \|y_{n+k-1}\|_\infty .$$

Note that Kreiss allows a restrained exponential growth of the numerical solution, also when there is no such growth for the solutions of the differential system.

As we have remarked earlier, we try to avoid that as much as possible. The norm introduced in (6.55) is a kind of synthesis of the approach of Kreiss and our $\boldsymbol{G} \otimes \boldsymbol{H}$ norm approach.

6.2.10 *The approach of van Veldhuizen*

Van Veldhuizen (1981) considers discretizations of the form

$$Y_{i+1} = G(t_i, h) Y_i,$$

for linear systems of the form

$$\mathrm{d}\boldsymbol{y}/\mathrm{d}t = \boldsymbol{T}(t)\boldsymbol{A}(t)\boldsymbol{T}^{-1}(t), \quad t_0 \leqslant t \leqslant t_{\mathrm{end}},$$

where

$$\boldsymbol{A}(t) = \begin{bmatrix} \lambda(t)/\epsilon & 0 \\ 0 & \mu(t) \end{bmatrix}, \quad 0 < \epsilon \leqslant \epsilon_0, \quad \mathrm{Re}\,\lambda(t) \leqslant \lambda < 0,$$

with smoothness assumptions for all functions involved, uniformly with respect to t and ϵ. He calls a method D-stable if, for some M which depends only on h_0, $\|G(t, h)\| \leqslant M$ for all $h < h_0$. In particular, M should be independent of ϵ. He emphasizes that this concept is concerned with just one step and has no relation to the propagation of errors. (We would therefore have preferred the term D-robustness (say), since "stability" names should be reserved to descriptions of desirable behavior in the long run of a solution and to sufficient algebraic conditions for obtaining such a behavior in the long run. Moreover the same word has been used by other authors with an entirely different content.) He shows that one-leg methods are "D-stable" if $\beta_k/\alpha_k > 0$, while linear multistep methods are not (except for BDF methods). He discusses why this does not contradict the fact that these methods are equivalent for constant step size (Dahlquist 1975c). We want to comment that his argument does not apply to linear multistep methods with appropriate smoothing of the initial data, similar to the one proposed by Lindberg (1971a) for the trapezoidal method. For variable step size or for rough functions, however, some sort of repeated smoothing would be needed, and one-leg methods are more robust, and are therefore to be preferred in stiff problems, as suggested by Nevanlinna and Liniger (1978). One must, however, take into account their different orders of consistency, e.g. Dahlquist (1983b). The "D-stability" condition, which is less demanding than contractivity and independent of the choice of norm, describes, in our opinion, a desirable robustness, but it is far from being a sufficient condition on a method.

6.2.11 *Variable step-size and order*

Since the use of variable step and order or, more generally, variable formula, is an essential feature of stiff computation, the analysis of fixed-step computation is to be considered as a preliminary study. That the extension is nontrivial is seen from the following results, quoted from Sand (1981). Set

$$h_n = t_{n+k} - t_{n+k-1}; \quad \delta_n = h_n/h_{n-1} - 1; \quad \Delta_n = (\delta_n, \delta_{n-1}, \ldots, \delta_{n-k+2})^{\mathrm{T}}. \tag{6.67}$$

The k-step BDF method ($2 \leqslant k \leqslant 4$) is zero-stable for a fixed $\delta_n = \delta$ iff δ is less than the following bounds

$$\begin{array}{cccc} k & 2 & 3 & 4 \\ \delta & \sqrt{2} & 0.6180\ldots & 0.2807\ldots \end{array} \tag{6.68a}$$

For $k = 5, 6$, the formula is zero-stable for δ at the left endpoint and zero-unstable at the right endpoint of the following intervals,

$$\begin{aligned} k = 5 &: [0.127, 0.128] \\ k = 6 &: [0.044, 0.045]. \end{aligned} \tag{6.68b}$$

(In view of this we have doubts about the practical use of BDF6, unless extremely high accuracy is demanded.)

Most of the early articles on variable step-size and variable order, Piotrowski (1969), Gear and Tu (1974), Gear and Watanabe (1974), and also the more recent work of Zlatev (1981), are concerned with zero-stability. An exception is the study of BDF methods by Brayton and Conley (1972), who considered A(α)-stability and used contractivity in polyhedral norms and scaled max-norms. Contractivity in such norms and also in the usual max-norm has been established for several methods by Liniger and Nevanlinna (1978, 1979), Brayton and Tong (1978), Nevanlinna (1979), Dahlquist, Liniger and Nevanlinna (1980), including A-contractivity results for a class of two-step one-leg methods with arbitrary step-size sequences. We shall not penetrate this subject more, since it is well covered by Liniger in §6.3. Sand (1981, 1982b) has done a thorough study of the subject with many new results.

We shall instead indicate how the discussion in §6.2.3 of systems with graded (or "gradable") Jacobians might be extended to variable step formulae. With G-norms we have the complication that the matrix G obtained by the general algorithm of LeVeque et al. will usually vary from step to step. In principle this can be accounted for by Theorem 6.5, but it seems easier to use piecewise constant matrices $G_i \otimes H_i$. Changes of G_i are mainly needed for $i = m$. We expect that a small number of pieces are usually sufficient for each epoch.

The matrix version of (6.31) reads, for $q \cdot = \infty$,

$$C(hJ) = C(\infty) \otimes I + e_k v^{\mathrm{T}} \alpha_k / \beta_k \otimes (I - hJ\beta_k/\alpha_k)^{-1}. \tag{6.69}$$

$C(\infty)$ is the companion matrix of the polynomial σ. Let α_{kn}, β_{jn}, $C(q; \Delta_n)$ be, respectively, the coefficients and the companion matrix of the variable step method at $t = t_n$; and let α_j, β_j, $C(q)$ be the corresponding quantities for $\Delta_n = 0$ ("frozen step size"). Δ_n is defined in (6.67).

$$h_n \beta_{kn}/\alpha_{kn} = \tilde{h}_n \beta_k/\alpha_k. \tag{6.70}$$

The assumptions for variable step method are:

(a) The coefficients of the variable-step one-leg formula are smooth functions of Δ_n, if $\| \Delta_n \| \leqslant \kappa_0$.

(b) The system satisfies the half-plane conditions $\mu_H[J_{ii}(t)] \leqslant 0$, $1 \leqslant i \leqslant m-1$. Then

$$\|(I - hJ_{ii}(t)\beta_k/\alpha_k)^{-1}\|_H \leqslant 1. \tag{6.71}$$

(c) The four assumptions (i)-(iv) formulated after Theorem 6.4 hold for the method for every Δ_n, $\|\Delta_n\| \leqslant \kappa_0$. The consistency relations read,

$$\sum_{j=0}^{k} \alpha_{jn} = 0, \quad \sum_{j=0}^{k} \beta_{jn} = 1, \quad \sum_{j=0}^{k} \alpha_{jn} t_{n+j} = h_n.$$

Then the assumptions and (6.70) imply an inequality for J_n $(= J_{ii,n})$,

$$\|C(h_n J_n \, ; \Delta_n) - C(\tilde{h}_n J_n)\| \leqslant K_0 \cdot \|\Delta_n\|. \tag{6.72}$$

K_0 depends only on the coefficients of the method, and on κ_0 and the choice of norm.

Then by (6.55), and the triangle inequality,

$$\|C(h_n J_{ii,n}; \Delta_n)\|_{G_i \otimes H_i} \leqslant \rho + K_0 \|\Delta_n\|, \quad i = 1, 2, \ldots, m-1.$$

Hence, if $\|\Delta_n\|$ is small enough,

$$\|C(h_n J_{ii,n};\Delta_n)\|_{G_i \otimes H_i} \leqslant \rho' < 1. \tag{6.73}$$

For J_{mm} we proceed differently. By consistency, the spectral radius of $C(0;\Delta)$ equals 1, hence

$$\min_{\Delta} \|C(0;\Delta)\|_G = \|C(0;\boldsymbol{0})\|_G = 1, \tag{6.74}$$

if G defines a norm of the type mentioned in Lemma 6.9. Then $\|C(q;\Delta)\|_G$ is smooth in the neighborhood of $(0;\boldsymbol{0})$, and by Taylor's formula, there are constants $a \in R$, $\boldsymbol{b} \in R^{k-1}$, such that

$$\|C(q;\Delta)\|_G = 1 + a\text{Re}q + \boldsymbol{b}^{\mathsf{T}}\Delta + O(|q|^2 + \|\Delta\|^2), \quad q \to 0, \quad \Delta \to 0,$$

by Lemma 6.9. By (6.74) $\boldsymbol{b}^{\mathsf{T}} = \boldsymbol{0}$. The following result can then be derived from Lemma 6.9, see also Dahlquist (1981b).

LEMMA 6.13. Let the G-norm be determined for $C(q;\boldsymbol{0})$ according to Lemma 6.9. Then there exists a constant K_1, such that, if $\|\Delta\|$ and $|q|$ are small enough,

$$\|C(q;\Delta)\|_G \leqslant \exp[\text{Re}\, q + K_1(|q|^2 + \|\Delta\|^2)], \tag{6.75a}$$

$$\|C(hJ;\Delta)\|_{G\otimes H} \leqslant \exp[\mu_H(hJ) + K_1(\|hJ\|^2 + \|\Delta\|^2)]. \tag{6.75b}$$

By combination of Lemma 6.13 for $J = J_{mm,n}$ and (6.73), we find, under similar assumptions as in §6.2.3 an inequality for $\|C(hJ_n, \Delta_n)\|$ similar to (6.55c). Towards the end of the epoch $\|hJ_{mm}\|$ may be so large that it is advantageous to treat J_{mm} just like J_{ii} for $i < m$.

We assumed above that $\|\Delta_n\|$ is small enough. By the asymptotic theory, developed in Dahlquist (1981b), $\|\Delta_n\| = O(N^{-1})$ and $\|hJ_{mm}\| = O(N^{-1})$, as $N \to \infty$ where N is the number of steps required for the computation of the epoch. Therefore, if the error tolerance is small enough this assumption will be valid, in the epoch, almost up to the "retirement" of the active variables.

In the (normal) generation shifts, between two epochs, the situation is different. It is more like a restart. There a rapid increase of step size is demanded, which, by (6.68), cannot be handled by high-order BDF methods. As long as the step size has not grown to the adequate size for the next epoch, the one-leg version of a first-order θ-method with $\theta \approx 0.5$ can be used, after which the order is allowed to grow again, when criteria based on the variable step size stability analysis permit a more rapid increase of step size than the local error tolerance allows for the first-order method. (This exemplifies how we consider the application of this theory in practice. The G-matrices themselves should probably never appear in the integration routines.) The details are to be worked out for each proposed variable step size-variable formula method. Another interesting possibility, at least for systems with graded Jacobians, is the use of a Runge–Kutta restarter that allows the presence of algebraic side conditions. The equations for the retired variables may be treated as such, with a simple backwards difference approximation for the derivative if necessary.

We mentioned above that G may change discontinuously at a point t', say from G_- to G_+. This also includes a point where the order (formula) is changed. By Theorem 6.5, the error bound is to be multiplied by the square root of the condition number of $G_+^{-1} \cdot G_-$. A change of norm can, namely, also be interpreted as a transformation of variable; note that $\|Y\|_G = \|T^{-1}Y\|_2$ where $G^{-1} = TT^{\mathrm{T}}$.

6.2.12 *Runge-Kutta methods*

Burrage and Butcher (1980) have introduced a general concept called (k, l)-algebraic stability, which is applicable to the contractivity analysis for a general class of methods including both Runge-Kutta and multistep methods. We are, unfortunately, not able to summarize their results, without a lengthy introduction of their notions.

We shall, however, say a few words about contractivity questions for Runge-Kutta methods, when the same H-norm is used for the solutions of the differential system as for the numerical solution, i.e. when $G = 1$. As in the case of multistep methods the theory started with the following question. Consider all systems, $y' = f(t, y)$, for which the distance (in H-norm) between any two solutions $y^{\mathrm{I}}(t)$, $y^{\mathrm{II}}(t)$ never grows. For what methods does the numerical solution have the analogous property, i.e. the contractivity in H-norm,

$$\|y_{n+1}^{\mathrm{I}} - y_{n+1}^{\mathrm{II}}\|_H \leqslant \|y_n^{\mathrm{I}} - y_n^{\mathrm{II}}\|_H? \tag{6.76}$$

The class of ODEs is characterized by the half-plane condition

$$\mu_H(f') \leqslant 0,$$

or in other words,

$$\mathrm{Re}\langle f(y^{\mathrm{I}}) - f(y^{\mathrm{II}}), y^{\mathrm{I}} - y^{\mathrm{II}}\rangle_H \leqslant 0, \quad \forall y^{\mathrm{I}}, \quad y^{\mathrm{II}} \in \mathbb{C}^s. \tag{6.77}$$

A-stability is a necessary condition for the validity of (6.76) because the equations of the form $y' = \lambda y$, $\lambda \in \mathbb{C}^-$, obviously belong to this class.

For one-leg multistep methods we know, by Theorem 6.1 and Theorem 6.4, (in the case $D_n = \mathbb{C}^-$, $\theta_n = 1$, $p_n = 0$) that A-stability is also sufficient for the existence of a G-norm such that

$$\|Y_{n+1}^{\mathrm{I}} - Y_{n+1}^{\mathrm{II}}\|_{G\otimes H} \leqslant \|Y_n^{\mathrm{I}} - Y_n^{\mathrm{II}}\|_{G\otimes H},$$

whenever (6.77) is satisfied.

For Runge-Kutta methods, Butcher (1975) pointed out that A-stability is not sufficient for what he called B-stability, i.e. for the validity of (6.76) in the whole class of systems defined by (6.77). This is due to the presence of several values of f in each step. In fact, the main obstacle occurs already in the attempt to generalize results from the constant coefficient equation, $y' = \lambda y$, to the scalar variable coefficient equation, $y' = \lambda(t)y$, Burrage and Butcher (1979). In the same paper they gave necessary and sufficient conditions (on the coefficients of the Runge-Kutta method) for B-stability; see also Crouzeix (1979).

The conditions of Burrage and Butcher were later formulated for more general disk conditions by Dahlquist and Jeltsch (1979). Consider the Runge-Kutta method,

$$Y_i = y_n + h \sum_{j=1}^{k} a_{ij} f(t_n + c_j h, Y_j), \quad i = 1, 2, \ldots, k,$$

$$y_{n+1} = y_n + h \sum_{j=1}^{k} b_j f(t_n + c_j h, Y_j).$$

Let $\boldsymbol{E}$ be the matrix of ones, i.e. $e_{ij} = 1, 1 \leqslant i, j \leqslant k$, and set

$$\boldsymbol{A} = [a_{ij}], \quad \boldsymbol{B} = \text{diag}(b_1, b_2, \ldots, b_k).$$

THEOREM 6.7. Let the circular domain D_R be defined by (6.53). Then there exists an $R \in [-\infty, \infty]$, such that the condition $f'_y \in D_R(\boldsymbol{H})$ implies contractivity in $\boldsymbol{H}$-norm, if and only if

$$b_j \geqslant 0, \quad j = 1, 2, \ldots, k.$$

R is to be chosen so that $\boldsymbol{BA} + \boldsymbol{A}^{\mathrm{T}}\boldsymbol{B} - \boldsymbol{BEB} + \boldsymbol{B}/R$ becomes positive-definite.

Burrage and Butcher treated the case $R = \infty$, called algebraic stability. The reader is referred to the proof of an equivalent result in Dahlquist and Jeltsch (1979). It is also shown there, for example, that no D_R can exist for an explicit Runge–Kutta method that is more than fourth-order accurate. We are not yet sure whether the right conclusion of this is that one should be careful with the use of such methods for strongly nonlinear problems, or that it is too much to ask for this kind of contractivity.

Finally we mention that Hundsdorfer (1981) has recently done a nonlinear analysis for a Rosenbrock method.

6.3 Contractivity of multistep and one-leg methods with variable steps (Liniger)

6.3.1 *Introduction*

A linear multistep formula,

$$\sum_{j=0}^{k} a_j x_{n-j} - h \sum_{j=0}^{k} b_j \dot{x}_{n-j} = 0, \tag{6.78}$$

normalized by

$$\sum_{j=0}^{k} b_j = 1, \tag{6.79}$$

is said to be *stable* at $q = h\lambda$, if all solutions $\{x_n\}$ of the difference equation

$$\sum_{j=0}^{k} (a_j - qb_j) x_{n-j} = 0, \tag{6.80}$$

generated by applying (6.78) to the test equation

$$\dot{x} = \lambda x, \quad \lambda \text{ const., complex}, \tag{6.81}$$

on the uniform grid $\{t_n : t_n = nh,\ n = 0, 1, \ldots\}$ are bounded for a given $h > 0$ as

$n \to \infty$. The set of all q at which (6.78) is stable is called the stability region S of the formula. On the uniform grid and for constant formula coefficients, we have stability at q iff the "root condition" is satisfied.

In practical applications, multistep formulae are used to solve nonlinear systems of differential equations, or linear equations with variable coefficients. Furthermore, they are normally implemented with variable steps, and under any or all of these circumstances it is in general hard to find necessary and sufficient conditions for asymptotic stability of the difference equation. The theory of contractivity, part of which is summarized here, gives sufficient conditions for stability in some of these cases, notably (a) for dissipative (monotone negative) nonlinear systems of differential equations

$$\dot{\boldsymbol{x}} = \boldsymbol{f}(t, \boldsymbol{x}), \tag{6.82}$$

i.e. systems satisfying

$$\langle \boldsymbol{x}-\boldsymbol{y}, \boldsymbol{f}(t, \boldsymbol{x})-\boldsymbol{f}(t, \boldsymbol{y})\rangle \leqslant \mu \|\boldsymbol{x}-\boldsymbol{y}\|^2, \tag{6.83}$$

for some scalar product $\langle . , . \rangle$ and some $\mu \leqslant 0$, where $\|x\|^2 \equiv \langle x, x \rangle$; and (b) for the variable coefficient test equation

$$\dot{x} = \lambda(t)x, \quad \lambda(t) \text{ complex}, \tag{6.84}$$

with arbitrary $\lambda(t)$ in the negative left-half plane or in some subset of it. Furthermore, contractivity results can be obtained for variable (in some cases, arbitrary) step sequences $\{h_n\}$ as well as for constant steps. For variable steps, the formula usually has variable coefficients,

$$\sum_{j=0}^{k} a_{j,n} x_{n-j} - h_n \sum_{j=0}^{k} b_{j,n} \dot{x}_{n-j} = 0, \tag{6.85}$$

normalized again by

$$\sum_{j=0}^{k} b_{j,n} = 1. \tag{6.86}$$

In general, however, stability results for variable steps are obtained only if the formula (6.85) is implemeneted as a one-leg (OL) method

$$\rho_n x_n - h_n f(\sigma_n t_n, \sigma_n x_n) = 0, \tag{6.87}$$

where

$$\rho_n x_n \equiv \sum_{j=0}^{k} a_{j,n} x_{n-j}, \qquad \sigma_n x_n \equiv \sum_{j=0}^{k} b_{j,n} x_{n-j}, \tag{6.88}$$

rather than in the usual multistep (MS) form

$$\rho_n x_n - h_n \sigma_n f(t_n, x_n) = 0. \tag{6.89}$$

In studying contractivity of the discrete solutions $\{x_n\}$ of the test equation (6.81) or (6.84) (respectively, the nonlinear system (6.82)), we ask whether $\{x_n\}$ (respectively, the difference $\{x_n\}$ of any two solutions $\{y_n\}$ and $\{y_n + x_n\}$) is non-increasing in an appropriate sense (rather than bounded as in discussing stability)? More precisely, we ask whether $\|X_{n-1}\| \geqslant \|X_n\|$ in some given norm $\|\cdot\|$ (independent of n), where $X_n := (x_{n-k+1}, x_{n-k+2}, \ldots, x_n)$? If this is the case for all $n \geqslant k$, then $\|X_{k-1}\| \geqslant \|X_n\|$ and we have stability. When applied to the constant coefficient test equation (6.81) with fixed steps, both the MS- and the OL-implementations of (6.78) lead to the difference equation (6.80). In this case, the formula (6.78) is said to be *contractive* (Nevanlinna and Liniger 1978) at $q = h\lambda$ if $\{x_n\}$ is non-increasing in the above sense for any solution of (6.80) computed with constant step size h. The set of all q at which the formula is contractive (in the given norm) is called the contractivity region, K. Clearly, contractivity at q implies stability at q and thus $K \subseteq S$ for any norm. By analogy to the corresponding stability concepts, we say that a formula is A-contractive, A_0-contractive, etc. if, respectively, the left-half q-plane, the negative real q-axis, etc., are contained in K. Every one of these contractivity properties implies the corresponding stability property.

When applied to (6.81), both implementations of the variable-step formula (6.85) give

$$\sum_{j=0}^{k} (a_{j,n} - q_n b_{j,n}) x_{n-j} = 0, \tag{6.90}$$

where $q_n = h_n\lambda$. Note that $q_n \in K_n$ for all n is sufficient for stability, i.e. q need not be constant. When implemented as an OL-method and applied to the variable-coefficient test equation (6.84), the formula (6.85) still generates a difference equation of the form (6.90) but with $q_n = h_n\lambda(\sigma_n t_n)$, and again we have stability if $\{q_n\} \subseteq K_n$. By contrast, the MS-method applied to (6.84) produces a difference equation of a more complicated type,

$$\sum_{j=0}^{k} (a_{j,n} - q_{j,n} b_{j,n}) x_{n-j} = 0,$$

one whose contractivity depends on $(k + 1)$ complex parameters $q_{j,n} = h_n\lambda(t_{n-j})$, rather than on only one of them, and this is the case even for constant steps.

Most of the existing work on contractivity falls into one of two categories. In the first, contractivity in inner product norms of OL-solutions of dissipative nonlinear systems (G-stability) was introduced in Dahlquist (1975) and studied, for example, in Nevanlinna (1976) and Nevanlinna and Liniger (1978, 1979). Contractivity of Runge–Kutta methods for the same problem class (B-stability) was studied in Butcher (1975), Wanner (1976), and Burrage and Butcher (1979).

The second approach, the one which is mainly surveyed here, is to investigate contractivity in the max norm or related norms for the test equation (6.81), respectively (6.84). For constant steps, contractivity in this sense was first studied in Liniger (1957) for classical Adams multistep and Runge–Kutta methods applied to (6.84) with $\lambda(t) \leqslant 0$. In Nevanlinna and Liniger (1978, 1979) many aspects of the contractivity theory were developed systematically; §6.3.2 summarizes that work. For non-uniform grids, contractivity of BDFs in certain "polygonal" norms were analyzed in Brayton and Conley (1972) and Brayton and Tong (1979).

Some results on max-norm contractivity of variable-step Adams-type OL-methods were given in Nevanlinna and Liniger (1979). A-Contractivity was studied in Dahlquist, Liniger, and Nevanlinna (1983) for the class of all second-order ($p = 2$) two-step ($k = 2$) formulae, a two-parameter family. It was shown that, for any given ratio of the two steps, all formulae of this class which are A-contractive, in either the max-norm or in constant inner product norms, are members of one and the same one-parameter sub-family of all $p = k = 2$-formulae. The OL-implementations of these formulae are A-stable with arbitrary step sequences for any dissipative nonlinear system and for (6.84) with any $\lambda(t)$, $\operatorname{Re} \lambda(t) \leqslant 0$. Finally it was shown in Liniger (1983) that for any $k \geqslant 1$ and for arbitrary step sequences the set of all $p = 2$, k-step formulae which are A-contractive in the max norm represent a $(k - 1)$-parameter sub-family of the $(2k - 2)$-parameter family of all $p = 2$, k-step formulae. The OL-implementation of these formulae are A-stable with respect to (6.84) for any $\lambda(t)$, $\operatorname{Re} \lambda(t) \leqslant 0$, for arbitrary variable steps. The A-contractivity work is summarized in §6.3.3.

6.3.2 *General discussion of contractivity*

The contractivity region K has a simple geometry. With respect to any norm, K is closed and connected by arcs of circles (Nevanlinna and Liniger 1978, Th. 2.1). For explicit formulae, K is convex. In contrast to this, the stability region S in general consists of several disconnected components, as illustrated in Nevanlinna and Liniger (1977) for a formula given in Grigorieff and Schroll (1977). With respect to the max norm, we have (Nevanlinna and Liniger 1978, Th. 3.1)

$$K = \left\{ q \in \bar{\mathbb{C}} : F(q) \equiv \sum_{j=1}^{k} |a_j - qb_j| - |a_0 - qb_0| \leqslant 0 \right\}. \tag{6.91}$$

The boundary of K is $\partial K = \{q\colon F(q) = 0\}$. It is smooth except possibly at its intersection points with the real axis. Examples of contractivity curves are plotted in Nevanlinna and Liniger (1979) in comparison with the corresponding root-locus curves.

For any BDF (defined by $b_0 = 1$, $b_1 = b_2 = \ldots = b_k = 0$) the max-norm contractivity region is a circle of center a_0 and radius

$$\sum_{j=1}^{k} |a_j|.$$

In inner product norms, all formulae have contractivity regions which are either half-planes or the inside or outside of circles (Dahlquist, 1979). It was shown (Nevanlinna and Liniger 1978) that the formula (6.78) is contractive at $q = 0$ iff

$$a_0 > 0, \quad a_j \leqslant 0, \quad j = 1, \ldots, k. \tag{6.92}$$

It is contractive† at $q = \infty$ iff

$$\gamma = b_0 - \sum_{j=1}^{k} |b_j| \geqslant 0, \tag{6.93}$$

† From here on, unless otherwise stated, contractivity is to be understood in the max-norm.

and A_∞-contractive (strictly contractive at $q = \infty$) iff $\gamma > 0$. The formula is A_0-contractive iff both (6.92) and (6.93) hold. (The corresponding statement for A_0-stability is, of course, not true in general.) Finally, the formula (6.78) is A-contractive iff (Nevanlinna and Lingier 1978) it is A_0-contractive and

$$\sum_{j=1}^{k} [(a_j^2 + b_j^2\eta)/(a_0^2 + b_0^2\eta)]^{1/2} \leqslant 1, \quad 0 \leqslant \eta \leqslant +\infty. \tag{6.94}$$

For a given η, the inequality (6.94) is necessary and sufficient for contractivity at $q = \mathrm{i}y$, y real, with $\eta = y^2$.

For any $\delta > 0$ and any $\psi < \pi/2$ there exist A(α)-contractive formulae of Adams type (i.e. with $a_0 = 1, a_1 = -1, a_2 = \ldots = a_k = 0$) and of any order of accuracy p such that γ defined by (6.93) satisfies $\gamma \geqslant 1 - \delta$ and $\alpha \geqslant \psi$. It should be noted, however, that these formulae exist only for $k \geqslant k_{\min}(p)$ where $k_{\min}(p)$ is a rather rapidly increasing function (empirically, $k_{\min} \gtrsim 3p$). Furthermore, formulae with $p > 2$ which are A(α)-contractive (and thus A(α)-stable) with α near $\pi/2$ cannot be accurate, i.e. they must have large error constants (Jeltsch and Nevanlinna 1980). A sequence of A(α)-contractive formulae of Adams type and of orders $p \leqslant 6$ was given in Nevanlinna and Liniger (1979). Among them is the $p = k = 2$-formula

$$x_n - x_{n-1} - \frac{h}{4}(3\dot{x}_n + \dot{x}_{n-2}) = 0, \tag{6.95}$$

which is characterized in §6.3.3 hereafter and which is A-stable and A(α)-contractive with $\alpha = 70.5°$, and a formula with $p = 3$ and $k = 4$,

$$x_n - x_{n-1} - \frac{h}{12}(8\dot{x}_n + 5\dot{x}_{n-2} - \dot{x}_{n-4}) = 0, \tag{6.96}$$

which is A(α)-stable with $\alpha = 78°$ and A(α)-contractive with $\alpha = 41°$. Variable-step extensions of (6.95) and (6.96) were defined uniquely by requiring that they, too, have $a_{2,n} = b_{1,n} = 0$, respectively $a_{2,n} = a_{3,n} = b_{1,n} = b_{3,n} = 0$, and it was shown that the variable-step extension of (6.95),

$$x_n - x_{n-1} - h_n\left(\frac{2 + r_n}{2 + 2r_n}\dot{x}_n + \frac{r_n}{2 + 2r_n}\,\dot{x}_{n-2}\right) = 0 \tag{6.97}$$

remains A_0-contractive for arbitrary step ratios $r_n \equiv h_n/h_{n-1}$. The familiar BDF with $p = k \geqslant 2$ are not contractive in the max-norm at $q = 0$ (nor are they, therefore, A_0- A(α)-, or A-contractive in this norm). The two-step second-order BDF,

$$\tfrac{3}{2}x_n - \tfrac{4}{2}x_{n-1} + \tfrac{1}{2}x_{n-2} - h\dot{x}_n = 0, \tag{6.98}$$

is A-stable and thus (Dahlquist 1978a) A-contractive in an inner-product norm (G-stable). However, as shown in Dahlquist, Liniger, and Nevanlinna (1983) this contractivity does not naturally extend to variable grids.

In Nevanlinna and Liniger (1979), a boundedness result was given for the discrete solution of a nonlinear $m \times m$-system of the form

$$\dot{\boldsymbol{x}} = \boldsymbol{A}(t, \boldsymbol{x})\boldsymbol{x} + \boldsymbol{d}(t, \boldsymbol{x}), \tag{6.99}$$

where $A = [a_{ij}]$ with $a_{ii}(t, x) < 0$, $i, j = 1, \ldots, m$, and $d(t, x)$ is bounded. A somewhat sharper result is the following (Odeh and Liniger 1980): Assume $\{x_n\}$ is a discrete solution of (6.99) generated by a one-leg method which is both A_0- and A_∞-contractive in the max-norm, and let $B = [b_{ij}]$ with $b_{ii} = -a_{ii}\gamma > 0$ and

$$b_{ij} = -\left(\sum_{j=0}^{k} |b_j|\right) \sup_{x,\, t} |a_{ij}(\sigma t_n, \sigma x_n)| \leqslant 0.$$

If we let $\xi = (\xi_i)$, $\delta = (\delta_i)$, $\xi_i \equiv \|x_i\|_\infty$ and $\delta_i \equiv \|d_i\|_\infty$ where $\|u\|_\infty$ denotes the max-norm of a scalar sequence $\{u_n\}$, then we have

$$B\xi \leqslant \delta, \quad \xi \geqslant 0, \quad \delta \geqslant 0, \tag{6.100}$$

and if B is an M-matrix (which is the case if A has a sufficient amount of diagonal dominance) then we have input-output stability,

$$\|\xi\|_\infty \leqslant \kappa \|\delta\|_\infty, \tag{6.101}$$

for some κ because B has a bounded inverse.

6.3.3 *A-contractivity of two-step formulas with variable steps*

In Dahlquist, Liniger, and Nevanlinna (1983) contractivity in inner product norms (G-stability) for dissipative nonlinear systems and A-contractivity in the max-norm were analyzed for the three-parameter family of all methods with $p = 1$ and $k = 2$ and/or the two-parameter family of all $p = k = 2$-methods. We summarize the results of this theory.

For variable-step G-stability analysis it turns out to be useful to write the two-step formula

$$a_0 x_n + a_1 x_{n-1} + a_2 x_{n-2} - \hat{h}_n(b_0 \dot{x}_n + b_1 \dot{x}_{n-1} + b_2 \dot{x}_{n-2}) = 0, \tag{6.102}$$

in terms of the step

$$\hat{h}_n \equiv a_0 h_n - a_2 h_{n-1}, \tag{6.103}$$

where $h_n = t_n - t_{n-1}$ and $h_{n-1} = t_{n-1} - t_{n-2}$. With this choice of the step, the coefficients of the consistent ($p = 1$) two-step formulae can be written in terms of three parameters a, b, c as

$$\begin{aligned} a_0 &= \tfrac{1}{2}(c+1), & b_0 &= \tfrac{1}{4}[1 + b + (a+c)], \\ a_1 &= -c, & b_1 &= \tfrac{1}{2}(1-b), \\ a_1 &= \tfrac{1}{2}(c-1), & b_2 &= \tfrac{1}{4}[1 + b - (a+c)]. \end{aligned} \tag{6.104}$$

We characterize the non-uniformity of the steps by the grid-parameter

$$\epsilon = \epsilon_n \equiv (h_n - h_{n-1})/(h_n + h_{n-1}), \quad |\epsilon| < 1, \tag{6.105}$$

where $\epsilon \gtreqless 0$ iff $h_n \gtreqless h_{n-1}$. Then $h_n = \bar{h}_n(1 + \epsilon)$, $h_{n-1} = \bar{h}_n(1 - \epsilon)$, and $\hat{h}_n = \bar{h}_n(1 + \epsilon c)$, where $\bar{h}_n := \tfrac{1}{2}(h_n + h_{n-1})$ is the average step. For equal steps ($\epsilon = 0$), $\hat{h}_n = \bar{h}_n = h_n = h_{n-1}$. In general, ϵ is related to the step ratio $r = r_n := h_n/h_{n-1}$ via

$$r = (1+\epsilon)/(1-\epsilon) \leftrightarrow \epsilon = (r-1)/(r+1). \tag{6.106}$$

It is easy to verify that the formula whose coefficients are specified by (6.104) is second-order ($p = 2$) accurate iff

$$a = \epsilon\left(\frac{1-c^2}{1+\epsilon c} - b\right). \tag{6.107}$$

Thus for $\epsilon = 0$ we have $p = 2$ iff $a = 0$.

Global contractivity,

$$\|X_n\| \leqslant \|X_{k-1}\|, \quad n \geqslant k, \tag{6.108}$$

which implies A-stability, is assured for all discrete solutions of dissipative nonlinear systems generated by variable-step one-leg methods provided these methods are (locally) G-stable, i.e. $\|X_i\| \leqslant \|X_{i-1}\|$ for all i, $i = k, k+1, \ldots, n$, in a G-norm $\|\ldots\|$ which is constant with respect to n, i.e. independent of the step changes. It was shown that the only second-order two-step formulae for which a constant G-norm exists are defined by

$$\begin{aligned} a(\epsilon) &= \epsilon^2 c(1-c^2)/(1+\epsilon c)^2, \\ b(\epsilon) &= (1-c^2)/(1+\epsilon c)^2, \\ c(\epsilon) &= c = const., \ 0 \leqslant c \leqslant 1. \end{aligned} \tag{6.109}$$

They represent a one-parameter subfamily of the two-parameter family of all $p = k = 2$-formulae defined by (6.104), subject to the constraint (6.107).

In analyzing A-contractivity in the max-norm for $k = 2$, the inequality (6.94) was squared twice to obtain an equivalent rational condition in the coefficients. Subject to the constraints for second-order accuracy and A_0-contractivity, the latter turns out to define the same one-parameter subfamily (6.109) of the two-parameter family of all $p = k = 2$-formulae as does the contractivity analysis in inner product norms (except that, from the viewpoint of A-contractivity in the max-norm $c(\epsilon)$ need not be constant).

In Nevanlinna and Liniger (1978) the particular formula (6.95) was derived by minimizing, over all A_0-contractive $p = k = 2$-formulae, a bound for the global error generated by applying the OL-implementation of any such formula to $\dot{x} = \lambda(t)x$, $\lambda(t) \leqslant -a$, for some arbitrarily small $a > 0$. The objective function whose minimization led to (6.95) was $|c_3|/\gamma$, where c_3 is the error constant and γ is defined by (6.93). As an example of an A-contractive formula, we give the one minimizing the same objective function over all A-contractive $p = k = 2$-formulae:

$$\tfrac{5}{6}x_n - \tfrac{4}{6}x_{n-1} - \tfrac{1}{6}x_{n-2} - h(\tfrac{5}{9}\dot{x}_n + \tfrac{2}{9}\dot{x}_{n-1} + \tfrac{2}{9}\dot{x}_{n-2}) = 0. \tag{6.110}$$

The formula (6.110) is associated with $a = 0$, $b = \frac{2}{9}$ and $c = \frac{2}{3}$. Its variable-step extension (6.102) is defined by $c \equiv \frac{2}{3}$,

$$\hat{h}_n = \tfrac{5}{6}h_n + \tfrac{1}{6}h_{n-1}, \tag{6.111}$$

and by (6.104) with

$$\begin{aligned} a &= a(\epsilon_n) = 10\epsilon_n^2/[3(3+2\epsilon_n)^2], \\ b &= b(\epsilon_n) = 5/(3+2\epsilon_n)^2, \end{aligned} \tag{6.112}$$

and with ϵ_n defined by (6.105). The error constant of this variable-step formula (with respect to $\hat{h}_n$) can be computed from the general expression

$$c_3 = c_{3,n} = \tfrac{1}{6}(-u^3\alpha_0 + v^3\alpha_2) - \tfrac{1}{2}(u^2\beta_0 + v^2\beta_2), \tag{6.113}$$

which is valid for any $p = k = 2$-formula; here

$$\begin{aligned} u = u_n &:= h_{n-1}/\hat{h}_n = (1-\epsilon)/(1+\epsilon c), \\ v = v_n &:= h_n/\hat{h}_n = (1+\epsilon)/(1+\epsilon c), \end{aligned} \tag{6.114}$$

and again $c = 2/3$ for (6.110).

For uniform steps, the $p = k = 2$ BDF (6.98) is A-stable and thus G-contractive as stated above. However, it is not a member of the family (6.109) for $\epsilon = 0$ and its variable-step extension (normally defined by setting $b_{1,n} = b_{2,n} = 0$) is not contractive in any fixed G-norm. In fact, the BDF was shown (Dahlquist, Liniger, and Nevanlinna, 1983, and Odeh and Liniger, 1980) to become unstable for problems of type (6.84) with marginally stable, oscillatory solutions of increasing period and for geometrically increasing step sequences with a fixed number of steps per period. For similar problems with decreasing period and steps, the BDF was stable but overdamped. By contrast, the OL-method associated with the variable-step version of (6.110) was not only stable for both problems but gave a much more accurate amplitude response than the BDF.

6.3.4 *A-contractivity of second-order multistep formulae with variable steps*

Consider the k-step formula (6.85) written in terms of the leading step, h_n, and let

$$\theta_j = \theta_{j,n} \equiv (t_n - t_{n-j})/h_n, \quad j = 0, 1, \ldots, k,$$

so that $\theta_0 = 0$ and $\theta_1 = 1$ by definition. For $k = 2$, $\theta_2 = 2/(1+\epsilon)$ where ϵ is defined by (6.105). For uniform grids, $\theta_j = j$. In terms of the weighted moments

$$\begin{aligned} A_m = A_{m,n} &\equiv \sum_{j=0}^{k} \theta_j^m(-a_j), \\ B_m = B_{m,n} &\equiv \sum_{j=0}^{k} \theta_j^m b_j, \end{aligned} \tag{6.115}$$

where m is any integer $\geqslant 0$, the constraints for pth-order accuracy of (6.85) are (Dahlquist 1979):

$$A_0 = 0, \quad A_m = mB_{m-1}, \quad m = 1, \ldots, p, \tag{6.116}$$

and the error constant (with respect to h_n) is given by

$$c_{p+1} = \frac{(-1)^p}{(p+1)!}\,[A_{p+1} - (p+1)B_p]. \tag{6.117}$$

The squaring technique used in Dahlquist, Liniger, and Nevanlinna (1980) to analyze max-norm A-contractivity for $k = 2$ is impractical for $k > 2$. But another approach given in Liniger (1980) is applicable to arbitrary k: subject to (6.92) and for $a_j \neq 0$ one can write (6.94) in the form

$$F(\eta) \equiv \sum_{j=0}^{k} a[1 + (b_j^2\eta/a_j^2)]^{1/2} \geq 0. \tag{6.118}$$

By consistency, $F(\eta) = \frac{1}{2}F\eta + O(\eta^2)$, and thus

$$F \equiv \sum_{j=0}^{k} (b_j^2/a_j) \geq 0 \tag{6.119}$$

is necessary for (6.118) to hold. But it is easy to prove that second-order accuracy implies $F \leq 0$. Thus at best, $F = 0$. For any given fixed a_j (satisfying the consistency relations), F is a quadratic function of the b_j which takes a unique maximum,

$$F_{\max} = 0, \tag{6.120}$$

identically in the a_j, for

$$b_j = (\theta_j - \tfrac{1}{2}A_2)(-a_j), \quad j = 0, 1, \ldots, k. \tag{6.121}$$

Thus the formulae defined by (6.121) do satisfy the necessary condition (6.119) and for $a_j \neq 0$ no other formulae will.

One finally proves that, subject to A_0-contractivity and second-order accuracy, (6.119) is also sufficient for (6.118) to hold, and thus for A-contractivity. This is done by squaring (6.118) once and by majoring the remaining irrational terms, which are geometric means of any two of the radicands, by the corresponding arithmetic means. The result is (6.119).

From the consistency relations and normalization, $A_0 = 0$ and $A_1 = B_0 = 1$, it follows that

$$a_0 = 1 - \sum_{j=2}^{k} (\theta_j - 1)(-a_j),$$

$$a_1 = -\left(1 - \sum_{j=2}^{k} \theta_j(-a_j)\right), \tag{6.122}$$

and thus the contractivity condition (6.92) at $q = 0$ defines a simplex in the $(k-1)$-dimensional space of the parameters a_j, $j = 2, \ldots, k$, whose vertices are the origin and the intersection points $\{-a_j = 1/\theta_j, a_i = 0, i \neq j, j = 2, \ldots, k\}$ of the a_j-axes with the plane $a_1 = 0$. One immediately verifies that these extreme points represent respectively the trapezoidal rules with step lengths $t_n - t_{n-j} = \theta_j h_n, j = 1, \ldots, k$.

From the above it follows that all $p = 2$, k-step formulae which are A-contractive in the max-norm are members of the $(k-1)$-parameter sub-family (6.120) of the $(2k-2)$-parameter family of all $p = k = 2$-formulae with the parameters $\{a_2, \ldots, a_k\}$ representing any point in the above mentioned simplex. A particularly interesting case is $p = 2$, $k = 3$, in view of the fact that, for evaluating the local error of a second-order method, at least three backward data must be available at every integration step. In this case, the above construction provides a two-parameter family of A-contractive formulae for arbitrary step sequences.

7

ANALYSIS OF DETAILED DYNAMIC SYSTEMS

7.1 Systems approach (Aiken)

Physical and chemical phenomena become understood through experiment and analysis. The best procedure for obtaining this understanding incorporates both experiment and analysis in an iterative program; an analysis of an experiment leads to the design of another experiment for use in the analysis, and so on until an acceptable level of knowledge is reached. Clear definition of the objective for this knowledge search is quite important as it affects each step of the process.

This sequential evolution of knowledge is illustrated in Figure 7.1. Figure 7.1(a) shows the feedback loop entered initially through experiment, passed through analysis (often mathematics combined with intuition), leading to a conjecture. The conjecture might be very doubtful indeed with few iterations; but it could suggest the design of additional experiments. In Figure 7.1(b) the iterations continue until the quest is completed (note that the first four letters of *C*onjective, *D*esign, *E*xperimental, *A*nalysis represent the first four notes of the song, "Getting to Know You"; this courtesy of Hunter 1979). We are concerned here with the analysis stage and how it relates to objectives. Figure 7.2(a) shows analysis as a further sequence. Aside from the intuitive, mathematical analysis begins with a model; by *model* we mean here a mathematical formulation that attempts actually to describe fundamentals, rather than be merely correlative or data-fitting equations. Next, the parameters specific to the experiments performed need to be given values. These parameters have some qualitative physical or chemical meaning, at least within the definition of the model. The model may then be solved under operating conditions other than those used to obtain the parameters, the simulation step. In this sense, it is predictive; the need to obtain good predictors constitutes a typical industrial objective for process design and control.

In the great majority of situations, progression through analysis involves a different individual or group of individuals for each stage. Very often their efforts are neither coordinated nor focused on the ultimate objective. Figure 7.2(b) indicates some connections. The model should be only as detailed as is necessary to reach the objective. The appropriate detail may not be evident on first pass, but *a priori* simplicity is often possible and improves with additional experimentation. Parameter estimation should be performed on the basis of attainment of the objective, rather than on a performance index specific to the particular experiments performed (§7.4.7). It could occur that the objective is simply to obtain and report to the literature values of certain physical and chemical parameters, but it is more practical to consider the context of how these parameters will be

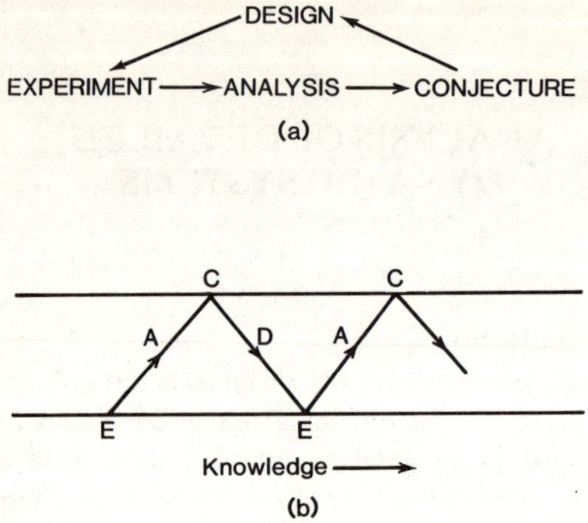

Fig. 7.1 Sequential nature of knowledge acquisition (from Box, Hunter, and Hunter 1978, with permission)

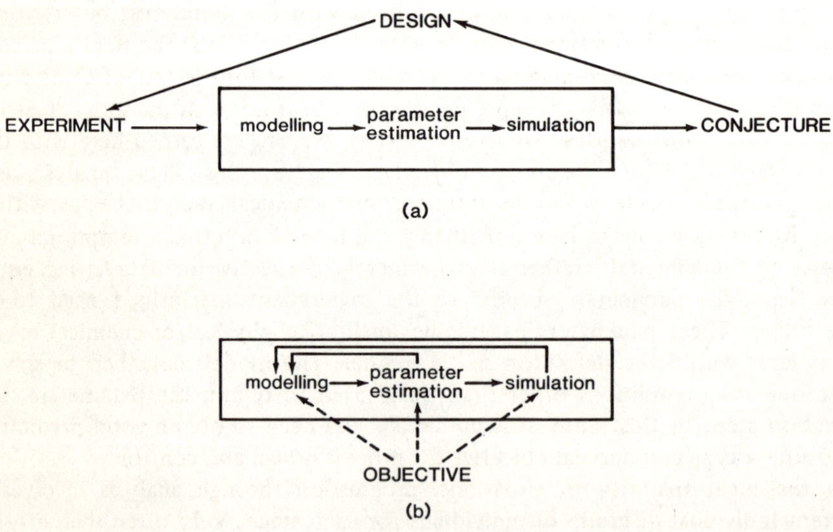

Fig. 7.2 Schematic of systems approach to analysis; (a) breakdown of analysis step in Fig. 7.1; (b) interrelations of analysis step

used. The accuracy and extent of the simulation, as well as the amount and form of the output are also strongly related to the objective. Then, too, parameter estimation should be kept in mind when formulating the model. If the experiments were not sensitive to certain parameters or their associated variables, then the model should not include them. Finally, note that the simulation accuracy need not exceed the accuracy of the data on which parameter estimates are based.

7.2 Stiff model approximation† (Aiken)

From a strictly experimental standpoint, not every possible variable will be measurable with available equipment. The modeller must be sensitive to this fact and not produce a model for which it is critical to have data on unmeasurables. State-of-the-art measuring devices may be incapable of measuring a variable because it is too small, is too similar to other measured variables, or the rate of change is too fast to be followed.

If a model has been created that is too detailed, mathematical approximation can guide the simplification process. For dynamic models, these mathematical approximations include:

(i) elimination of a variable by setting its derivative to zero and solving the resulting algebraic equation;
(ii) setting a parameter to zero.

Approximation (i) is for systems with too strong a coupling, approximation (ii) for too weak a coupling; (i) may be handled by singular perturbation techniques, (ii) by regular perturbations. Approximation (i) is called here the steady-state approximation; it is given a mathematical basis for the general stiff system in §7.2.1, 7.2.2. Application to general stiff systems is discussed in §7.2.5–7.2.8. Approximations (i) and (ii) can be used in a single approximation (§7.2.5).

7.2.1 *Equivalence of stiffness and singular perturbations*

Consider a two-variable set of first-order ODEs with a small parameter multiplying the derivative of one of the variables, the type of system for which singular perturbation methods have been developed:

$$\begin{aligned} \frac{\mathrm{d}x}{\mathrm{d}t} &= f(x, y, \epsilon), \quad x(0) = \zeta, \\ \epsilon\frac{\mathrm{d}y}{\mathrm{d}t} &= g(x, y, \epsilon), \quad y(0) = \eta, \end{aligned} \tag{7.1}$$

where $f(0, 0, \epsilon) = g(0, 0, \epsilon) = 0$. If (7.1) is linearized along its trajectory, it may be expressed as

$$\begin{bmatrix} \dfrac{\mathrm{d}x}{\mathrm{d}t} \\ \\ \dfrac{\mathrm{d}y}{\mathrm{d}t} \end{bmatrix} = \begin{bmatrix} f_x & f_y \\ \\ g_x/\epsilon & g_y/\epsilon \end{bmatrix} \begin{bmatrix} x \\ \\ y \end{bmatrix}, \quad \begin{bmatrix} x(0) \\ \\ y(0) \end{bmatrix} = \begin{bmatrix} \zeta \\ \\ \eta \end{bmatrix}. \tag{7.2}$$

Examination of the Jacobian eigenvalues indicates that their spread increases the smaller ϵ becomes: one approaches zero while the other grows larger in absolute value. The occurrence of an eigenvalue relatively large in absolute value indicates a stiff problem is likely; thus (7.2) can be regarded as the linearized representation of a typical stiff system with widely separated eigenvalues. It then follows that such a stiff system and the singular perturbation form of 7.1 are at least locally equivalent.

† §7.2.1–7.2.4 is based on Aiken and Lapidus (1974); §7.2.5–7.2.8 on Aiken and Lapidus (1975a); §7.2.9, 7.2.10 on Aiken and Lapidus (1975b).

7.2.2 *Formal perturbation method*

Consider initial value problems of the form (7.1) with the perturbing parameter ϵ small. First assume the solution may be approximated by the simple first-order expansion in the (outer) variables, obtained by truncation from the full infinite power series ϵ expansion:

$$\begin{aligned} x^{**} &= x_0 + \epsilon x_1 \\ y^{**} &= y_0 + \epsilon y_1 \end{aligned} \tag{7.3}$$

Substitute (x^{**}, y^{**}) into (7.1) for (x, y) and expand about $(x_0(t), y_0(t))$. Matching terms with like powers of ϵ results in

$$\frac{dx_0}{dt} = f(x_0, y_0, \epsilon), \quad x_0(0) = \zeta, \tag{7.4a}$$

$$0 = g(x_0, y_0, \epsilon), \tag{7.4b}$$

and

$$\frac{dx_1}{dt} = f_x(x_0, y_0)x_1 + f_y(x_0, y_0)y_1, \quad x_1(0) = (0), \tag{7.5a}$$

$$\frac{dy_0}{dt} = g_x(x_0, y_0)x_1 + g_y(x_0, y_0)y_1. \tag{7.5b}$$

An inconsistency can arise when (7.4b) is not satisfied by (ζ, η). To alleviate this, additional (inner) variables are introduced which are particularly important in the very early stages of the transient. Expanding these variables to first order in ϵ and adding them to (7.3) gives the new composite solution approximation (x^*, y^*):

$$\begin{aligned} x^*(t) &= x_0(t) + \epsilon x_1(t) + X_0(t/\epsilon) + \epsilon X_1(t/\epsilon) \\ y^*(t) &= y_0(t) + \epsilon y_1(t) + Y_0(t/\epsilon) + \epsilon Y_1(t/\epsilon) \end{aligned} \tag{7.6}$$

A boundary-layer type characteristic is required of the inner variables:

$$\lim_{t/\epsilon \to \infty} X_0 = X_1 = Y_0 = Y_1 = 0 \tag{7.7}$$

Let $\tau = t/\epsilon$ and make this variable change in (7.1):

$$\begin{aligned} \frac{dx}{d\tau} &= \epsilon f(x, y, \epsilon), \quad x(0) = \zeta, \\ \frac{dy}{d\tau} &= g(x, y, \epsilon), \quad y(0) = \eta, \end{aligned} \tag{7.8}$$

a set of regularly perturbed equations. Now substitute (x^*, y^*) into (7.8) and expand about $[x_0(\epsilon\tau) + X_0(\tau), y_0(\epsilon\tau) + Y_0(\tau)]$. Matching terms with like powers in ϵ results in the following as $\epsilon \to 0$:

$$\frac{dX_0}{d\tau} = 0,$$

$$\frac{dY_0}{d\tau} = g[x_0(0) + X_0(\tau), y_0(0) + Y_0(\tau)], \tag{7.9}$$

and

$$\frac{dX_1}{d\tau} = f[x_0(0) + X_0(\tau), y_0(0) + Y_0(\tau)] - f[x_0(0), y_0(0)],$$

$$\frac{dY_1}{d\tau} = g_x X_1(\tau) + g_y y_1(0) + g_y Y_1(\tau) - g_x(x_0, y_0)x_1(0) - g_y(x_0, y_0)y_1(0) + g_x x_1(0). \tag{7.10}$$

Equations (7.4) and (7.9) share the initial conditions

$$x_0(0) + X_0(0) = \zeta,$$
$$y_0(0) + Y_0(0) = \eta, \tag{7.11}$$

while (7.5) and (7.10) share

$$x_1(0) + X_1(0) = 0,$$
$$y_1(0) + Y_1(0) = 0. \tag{7.12}$$

Note that as a result of (7.7) and (7.9), $X_0(\tau) = 0$. The condition under which (7.6) may be expected to be a valid solution representation over the domain of interest may be found in Hoppensteadt (1971).

7.2.3 *Singular perturbation numerical method for stiff equations*

It will now be shown that the perturbation method of the last section is entirely applicable to the general stiff system for which the perturbing parameter ϵ may not be directly identifiable. This leads to a novel approach to the numerical integration of such systems. Miranker (1972) was the first to recognize and detail the possibilities of such an approach. It will be demonstrated in the examples following this section that knowledge of these inner terms, thus the length of the boundary layer, is of paramount concern.

Consider the system:

$$\frac{dx}{dt} = f(x, y), \quad x(0) = \zeta, \tag{7.13a}$$

$$\frac{dy}{dt} = w(x, y), \quad y(0) = \eta, \tag{7.13b}$$

in which $w(x, y)$ is very large and thus stiff. For convenience, introduce the small parameter ϵ into (7.13b):

$$\epsilon \frac{dy}{dt} = g(x, y), \quad y(0) = \eta, \tag{7.14}$$

where $g \equiv \epsilon_w$ is of the same order as f in (7.13a). Equations (7.13a) and (7.14) are now in the same form as (7.1).

Equation (7.4) may be written as

$$\frac{dx_0}{dt} = f(x_0, y_0), \quad x_0(0) = \zeta, \tag{7.15a}$$

$$0 = w(x_0, y_0). \tag{7.15b}$$

Equation (7.15) can be written since no assumption has been made on the magnitude of ϵ. Equation (7.15), the zeroth-order outer expansion, corresponds to the steady-state approximation often used in chemical kinetics, but now written for the general stiff system. The common procedure of setting ϵ to zero, if this quantity is identified, may mechanically produce the steady-state approximation, but no such restriction was placed on the magnitude of ϵ in deriving (7.15). This may have interesting consequences (Ray 1969). Equation (7.15) may be simultaneously solved to yield $y_0(0)$, $x_0(h)$, and $y_0(h)$, where h is the chosen time-step size. This will require an extraneous nonstiff numerical integration routine for (7.15a).

From (7.9):

$$\frac{dY_0}{d\tau} = g(x_0, y_0 + Y_0), \quad Y_0(0) = \eta - y_0(0)$$

$$\approx g(x_0, y_0) + g_y(x_0, y_0) Y_0(\tau),$$

a linear equation with the solution

$$Y_0(\tau) = Y_0(0)\exp[g_y(x_0, y_0)\tau]$$

or (7.16)

$$Y_0(h/\epsilon) = Y_0(0)\exp[w_y(x_0, y_0)h].$$

This linearization should be a good approximation since the boundary layer will decay with a large exponential over most of $Y_0(\tau)$. Equation (7.16) may be visualized as the linearization of (7.13b) for the scalar w linear in y, where $x_0(0) = \eta$ and $y_0(0)$ does not appear in w_y; the interpretation for the vector version where $y_0(0)$ appears in the Jacobian w_y is not as obvious.

Rewrite (7.5) as

$$\epsilon \frac{dx_1}{dt} = f_x(x_0, y_0)\epsilon x_1 + f_y(x_0, y_0)\epsilon y_1, \quad \epsilon x_1(0) + \epsilon X_1(0) = 0, \tag{7.17a}$$

$$\epsilon \frac{dy_0}{dt} = g_x(x_0, y_0)\epsilon x_1 + g_y(x_0, y_0)\epsilon y_1. \tag{7.17b}$$

Take the total derivative of (7.15b) to be used in solving (7.17b) for ϵy_1. After substitution into (7.17a):

$$\epsilon \frac{dx_1}{dt} = \left(f_x(x_0, y_0) - \frac{f_y(x_0, y_0) w_z(x_0, y_0)}{w_y(x_0, y_0)} \right) \epsilon x_1 - \frac{f_y(x_0, y_0) w_z(x_0, y_0)}{w_y(x_0, y_0)^2}$$

$$\equiv \alpha \epsilon x_1 - \beta,$$

a linear equation with the solution

$$\epsilon x_1(h) = \{[\alpha \epsilon x_1(0) - \beta]\, \mathrm{e}^{\alpha h} + \beta\}/\alpha, \tag{7.18a}$$

and thus

$$\epsilon y_1(h) = -\left(\frac{w_z(x_0, y_0)}{w_y(x_0, y_0)}\, \epsilon x_1(h) + \frac{w_z(x_0, y_0)}{w_y(x_0, y_0)^2} f \right). \tag{7.18b}$$

Now since

$$\epsilon x_1(0) + \epsilon X_1(0) = 0,$$

and

$$\lim_{t \to \infty} X_1(t) = 0,$$

then

$$\begin{aligned}
\epsilon x_1(0) &= \epsilon \int_0^\infty \frac{\mathrm{d}X_1}{\mathrm{d}\tau}\, \mathrm{d}\tau, \\
&= \epsilon \int_0^\infty \{ f[x(0), y_0(0) + Y_0(\tau)] - f[x(0), y_0(0)] \}\, \mathrm{d}\tau, \\
&= \epsilon f_y [x(0), y_0(0)] \int_0^\infty Y_0(\tau)\, \mathrm{d}\tau.
\end{aligned}$$

If $Y_0(\tau)$ is assumed linear over τ, with slope defined by $\dot{Y}_0(0) = \epsilon w(x(0), y(0))$,

$$\epsilon x_1(0) = f_y [x(0), y_0(0)]\, Y_0^2(0)/2w(x(0), y(0)). \tag{7.19}$$

This allows rather simple computation of $\epsilon x_1(h)$, $\epsilon y_1(h)$, and $\epsilon y_1(0)$.

We note from (7.10) that $\dot{X}_1(\tau)$ is functionally dependent only on known arguments, so that immediately

$$\epsilon X_1(h/\epsilon) = \epsilon X_1(0) + h \dot{X}_1(h/\epsilon), \quad \epsilon X_1(0) = -\, \epsilon x_1(0). \tag{7.20}$$

From (7.10):

$$\epsilon \dot{Y}_1(\tau) = g_y \epsilon Y_1(\tau) + \gamma,$$

where

$$\gamma = g_x \epsilon X_1(\tau) + [g_x - g_x(x_0, y_0)]\, \epsilon x_1(0) - [g_y - g_y(x_0, y_0)]\, \epsilon y_1(0),$$

a linear equation with the solution

$$\epsilon Y_1(\tau) = \{[g_y \epsilon Y_1(0) + \gamma] \exp(g_y \tau) - \gamma\}/g_y,$$

which yields

$$\begin{aligned}
\epsilon Y_1(h/\epsilon) = \{&[w_x \epsilon X_1(h/\epsilon) + (w_x - w_x(x_0, y_0))\epsilon x_1(0) \\
&- (w_y - w_y(x_0, y_0))\epsilon y_1(0) + w_y \epsilon Y_1(0)] \exp(w_y h) \\
&- w_x \epsilon X_1(h/\epsilon) + (w_x - w_x(x_0, y_0))\epsilon x_1(0) \\
&- (w_y - w_y(x_0, y_0))\epsilon y_1(0)\}/w_y.
\end{aligned} \tag{7.21}$$

Finally, we may join all of the above results for the numerical approximation $(x^*(h), y^*(h))$ to $(x(h), y(h))$,

$$x(h) \sim x_0(h) + \epsilon x_1(h) + X_0(h/\epsilon) + \epsilon X_1(h/\epsilon),$$

$$y(h) \sim y_0(h) + \epsilon y_1(h) + Y_0(h/\epsilon) + \epsilon Y_1(h/\epsilon).$$

A step-by-step summary of the algorithm with comments appears in Table 2.1.

Thus, the parameter ϵ is introduced into the general stiff system (7.13) merely as a bookkeeping device to group systematically terms of comparable degrees of accuracy. Use of the artificial parameter ϵ has demonstrated its convenience in many common solution approximation schemes of classical implicit function theory (Bellman 1964). Thus (7.1) should be viewed as identically equivalent to the general stiff system form of (7.13).

TABLE 7.1
Summary of singular perturbation algorithm sequence

Variable	Equation	Comments
$x_0(h), y_0(h)$	(7.15a) (7.15b)	Simultaneous solution by any accurate integration and root-finding procedures
$X_0(h/\epsilon)$	(7.7) (7.9)	Always equal to zero
$Y_0(h/\epsilon)$	(7.16)	Boundary layer importance to accuracy indicated by $Y_0(0) = \eta - y_0(0)$; linear integration
$\epsilon x_1(h)$	(7.18a)	$\epsilon x_1(0)$ from (7.19); w_y may not be singular; linear integration
$\epsilon y_1(h)$	(7.18b)	Follows directly from $\epsilon x_1(h)$
$\epsilon X_1(h/\epsilon)$	(7.20)	$\epsilon X_1(0)$ from (7.20); simple function evaluations for the integration
$\epsilon Y_1(h/\epsilon)$	(7.21)	$\epsilon Y_1(0) = - \epsilon y_1(0)$ obtained from (7.18b) with $h = 0$; w_y may not be singular; linear integration

7.2.4 *Example of perturbation numerical method*

This problem was used by Bowen, Acrivos, and Oppenheim (1963) as an important system in singularly perturbed form with which to derive an analytical solution approximation via matched asymptotic expansions. The accepted kinetic steps involved for a dilute ozone-oxygen mixture are

$$O_3 + O_2 \underset{k_2}{\overset{k_1}{\rightleftharpoons}} O + 2O_2$$

$$O_3 + O \overset{k_3}{\rightarrow} 2O_2$$

If the following dimensionless variables are defined:

$$x = [O_3]/[O_3]_0, \quad y = [0]/\epsilon[O_3]_0,$$
$$\kappa = 2k_3[O_2]_0/k_1, \quad \epsilon = k_1[O_2]_0/2k_3[O_3]_0,$$

and the time scale divided by $2/k_1[O_2]_0$, then the transient behavior may be described by

$$\frac{dx}{dt} = -x - xy + \epsilon\kappa y, \quad x(0) = 1$$

$$\epsilon\frac{dy}{dt} = x - xy - \epsilon\kappa y, \quad y(0) = 0,$$

with $\epsilon = 1.0/98.0$, $\kappa = 3.0$.

The authors expended considerable effort in obtaining the inner, outer, and far-out solutions which, when combined, appear as in Figure 7.3. Also drawn in the figure is the accurate Runge-Kutta-Gill fourth-order algorithm (RKG4) numerical solution and the present perturbation algorithm. The analytical approximation is seen to be considerably in disagreement once past the initial boundary layer. Since the analytical solution is accurate only to $O(\epsilon)$ in the far-out portion of the curve, while the inner and outer expansions are accurate to $O(\epsilon^3)$ and since the RKG4 and the perturbation algorithm agree, the conclusion is that the analytical approximation is in error for $t > 0.12$.

Table 7.2 contains a detailed account of all inner and outer terms for both variables, along with the exact solution. The inner terms are seen to make a substantial contribution up to $t \approx 0.07$, corresponding to the length of the boundary layer, after which their effect is much smaller. The first-order outer terms are seen to contribute as much as 10 percent for the stiff variable but are always very small for the non-stiff variable; in fact with the small variation within the boundary-layer, the zeroth-order outer approximation x_0 is adequate for 0.1 percent accuracy through the entire transient.

The maximum step size that could be run with RKG4 for 0.1 per cent accuracy was 0.01; for the perturbation routine, step sizes as large as 1.0 could be taken for an agreement within 1 percent of RKG4. This gave a decrease in computation time by a factor of 25. A step of 1.0 actually leaps over the boundary layer so that, in this case, the outer terms would suffice for the same accuracy. If a step is taken within the boundary layer with only the outer terms, however, errors incurred there can propagate and considerably lower accuracy. For example, a step size of 0.01 using outer terms only can cause as much as a 70 percent deviation from the inner plus outer solution within the boundary layer and 35 percent deviation thereafter.

7.2.5 *The steady-state approximation for general stiff equations*

In §7.2.3 we presented a singular-perturbation-based numerical solution to stiff equations partitioned into stiff (y) and non-stiff (x) parts. The solution was given in terms of a zeroth-order inner (X_0, Y_0) and outer (x_0, y_0) and first-order inner and outer (X_1, Y_1, x_1, y_1) terms

$$\begin{aligned} x &\approx X_1 + \epsilon X_1 + x_0 + \epsilon x_1, \\ y &\approx Y_0 + \epsilon Y_1 + y_0 + \epsilon y_1, \end{aligned} \tag{7.22}$$

TABLE 7.2

Ozone: exact, and first-order solutions

(a) non-stiff variable

t	x	X_0	ϵX_1	x_0	ϵx_1	x^*
0.01	0.986615	0	−0.001231	0.980787	0.004559	0.984115
0.02	0.969778	0	−0.000450	0.965216	0.001665	0.966430
0.04	0.933960	0	−0.000062	0.930135	0.000225	0.930298
0.07	0.881443	0	−0.000003	0.877917	0.000005	0.877919
0.10	0.831826	0	7×10^{-7}	0.828503	−0.000008	0.828495
0.50	0.389468	0	0.000005	0.387841	−0.000022	0.387824
0.90	0.189700	0	0.000020	0.188700	−0.000045	0.188675
5.00	0.006656	0	−0.000028	0.006470	0.000213	0.006655
10.00	0.002582	0	−0.000007	0.002657	0.000069	0.002719
20.00	0.001148	0	−0.000002	0.001257	0.000019	0.001275
50.00	0.000428	0	-3×10^{-7}	0.000499	0.000003	0.000502
100.00	0.000208	0	-7×10^{-8}	0.000251	9×10^{-7}	0.000252

(b) stiff variable

t	y	Y_0	ϵY_1	y_0	ϵy_1	y^*
0.01	0.692814	0	–0.002720	0.969733	0.000685	0.614298
0.02	0.837530	0	–0.000578	0.969260	0.000620	0.837778
0.04	0.949530	0	–0.000245	0.968137	0.000619	0.949664
0.07	0.965869	0	–0.000275	0.966306	0.000681	0.965779
0.10	0.965161	0.000216	–0.000322	0.964368	0.000759	0.965021
0.50	0.929962	0.001792	–0.001955	0.926847	0.002965	0.929647
0.90	0.869663	0.006708	–0.007525	0.860417	0.009351	0.868952
5.00	0.192618	0.003348	–0.003362	0.174480	0.021907	0.196373
10.00	0.080713	0.000892	–0.001058	0.079862	0.006025	0.085720
20.00	0.036792	0.000224	–0.000288	0.039471	0.001607	0.041014
50.00	0.013872	0.000037	–0.000050	0.016050	0.000278	0.016316
100.00	0.006787	0.000009	–0.000013	0.008143	0.000072	0.008212

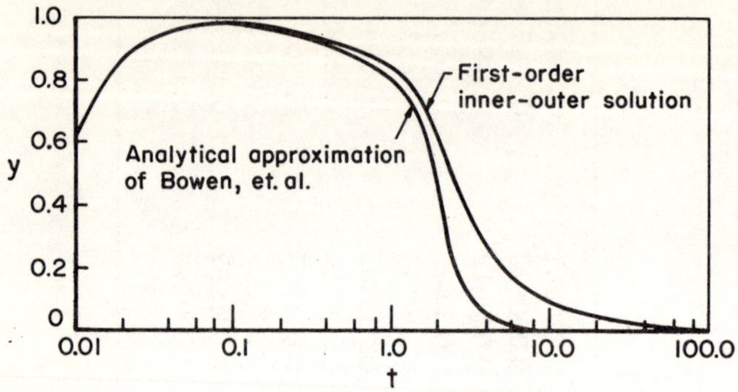

Fig. 7.3 Transient behavior of dimensionless oxygen atom concentration.

where ϵ is an artificial bookkeeping indication of the degree of stiffness, defined by

$$\epsilon \frac{dy}{dt} = g(x, y), \tag{7.23}$$

where $g \equiv \epsilon w$. This parameter need not actually exist or be identified for the solution. The outer terms are of more interest than the inner terms, which are important only within a relatively small boundary layer region of the transient. For systems stiff enough to require special integration techniques, the zeroth-order outer approximation often is sufficiently accurate:

$$x_0 = f(x_0, y_0), \quad x_0(0) = x(0), \tag{7.24a}$$

$$0 = g(x_0, y_0) = w(x_0, y_0). \tag{7.24b}$$

The last equality in (7.24b) is made since ϵ is not zero. This is what properly has been referred to as the *steady-state approximation* (*ssa*).

Note that this could be different from the zeroth-order outer solution if the parameter ϵ is identified. Thus in §7.2.2 the ϵ term on the right-hand side of (7.4b) must be dropped. This changes the transient solution slightly but perhaps more importantly it changes the steady state. This can be disturbing to, for example, a chemist who demands an accurate mass balance. Under these circumstances we should refer to a *pseudo steady-state approximation* (*pssa*).

The conditions for the validity of (7.24), or for regular degeneracy to the low-order solution, require that the initial conditions $x(0) = \zeta$, $y(0) = \eta$ be within the region of asymptotic stability of

$$\frac{dy}{d\tau} = g(\alpha, y), \tag{7.25}$$

where $\tau \equiv t/\epsilon$, and x is replaced by some constant vector α at each instant (Tikhonov 1952).

The ssa has been used for some time to permit an analytical solution or to aid data gathering and parameter fitting of ODEs, as Bodenstin (1924) first demonstrated for chemical kinetic schemes. The ssa has been used extensively in more recent years to aid numerical computation of stiff equations although the theoretical basis for this was not clarified. For example, Luss and Amundson (1968) examined a simplified model for the dynamics of a catalytic fluidized bed in which the concentration and temperature of the interstitial fluid is assumed in equilibrium with the particle. Gelinas (1972) applied the ssa to a large set modelling photochemical smog; Herriott (1972) to a large set of the pyrolysis of propane. In nuclear reactor kinetics, the ssa is called the *zero power approximation* (Birkhoff 1966), obtained by setting to zero the small perturbing prompt neutron lifetime. Aiken (1982b) applied it to combustion kinetics where the application is quite non-intuitive. A few recent interesting applications of the *ad hoc* version of the approximation are: Chen, Douglas, and Georgakis (1979); Devooght and Mund (1980); Kao (1980); Farrow and Graedel (1977); Aronowitz, Nategeli, and Glassman (1977); Warner (1977); Cho and Joseph (1979); and sophisticated versions in O'Malley and Anderson (1979); and O'Malley and Flaherty (1980).

In spite of all the applications of the ssa, there is some confusion as to its interpretation and validity. The zeroth-order outer expansion is derived without the assumption of the perturbing parameter's value. As such, it is not true that the ssa is defined as the original model with ϵ equal to zero. For systems in the singular perturbed form, with ϵ identified, the ssa may be mechanically obtained by setting ϵ to zero. For systems not obviously in singular perturbed form, the ssa may also be obtained mechanically by setting the time derivations of the stiff variables to zero. Note, however, that this does not imply that the stiff variables are necessarily invariant over any domain, as is often believed to be required for it to be valid. Thus Edelson and Allara (1973), examining the ssa applied to a stiff model of propane pyrolysis, report agreement with an accepted solution as surprisingly good even though the stiff variables do not settle to a steady state during most of the transient.

7.2.6 *Estimate of the boundary layer thickness*

There are two main difficulties in the use of a valid ssa to the partitioned stiff system: *a priori* knowledge of the applicable region of the domain and the accuracy within this region. Hirschfelder (1957) suggested that the length of the boundary layer alone be used as a qualitative gauge of the stiffness and so of the accuracy of the ssa, although he gives no practical method of arriving at this. Johnson (1966) uses chemical rate constants related to the stiffness of certain kinetic schemes to guide application of the ssa. Snow (1966) has written a computer package which compares results from integration of the full equations to those from an ssa, then uses only the ssa when the agreement is close; Keneshea (1967) continued periodically to examine this agreement. Ponzo and Wax (1972) derived a criterion for the accuracy and region of applicability of the ssa for the particular form of the Lienard equation. Their criterion, however, is not generally practical to use. Heinecken, Tsuchya, and Aris (1967) explain that the Monod kinetics can be expected to be more valid for very low initial enzyme concentrations. This can be shown to diminish the magnitude of the stiff variables within the boundary layer and thus require less accuracy for simulation of the large negative eigenvalue.

Consideration of the ssa as the zeroth-order outer approximation reveals that the region of applicability corresponds to the region where the outer variables are much more dominant than the inner ones. The inner variables are then important only within a narrow initial boundary layer and thus can be used to define this region. The zeroth-order inner term for the stiff variable is by far the most dominant:

$$Y_0(t) = Y_0(0) \exp[\partial w/\partial y(x_0, y_0)t] \tag{7.26}$$

where $Y_0(0) = y(0) - y_0(0)$. Experience has indicated that (2.26) is capable of providing an *a priori* estimate of the boundary layer for linear and nonlinear applications. In this way the boundary layer is defined as a fractional decay of the zeroth-order stiff inner variable. The effective boundary layer length for the vector version of (7.26) is given by t_I:

$$t_I = \frac{|\ln[Y_0(t)/Y_0(0)]|}{\left\| \frac{\partial w}{\partial y} [x_0(0), y_0(0)] \right\|}, \tag{7.27}$$

where $\|\cdot\|$ is a suitable matrix norm. Since for any matrix A,

$$\|A\| > \rho(A),$$

where $\rho(A)$ is the spectral radius of A,

$$\rho(A) \equiv \max |\lambda_i|$$

(λ_i the eigenvalues of A), a conservative estimate of the boundary layer is thus provided by the use of the spectral radius for the matrix norm: that is, if the first step can be taken greater than this boundary layer estimate, the ssa is applicable.

7.2.7 *Accuracy of the steady-state approximation*

Since the first-order terms may easily be evaluated periodically at any time during the solution, a convenient upper bound on the error of using the ssa for many common systems is $\epsilon x_1(t)/x_0(t)$ or $\epsilon y_1(t)/y_0(t)$. If these ratios are less than say 0.001, the accuracy of the ssa is indicated to be better than 0.1 percent.

7.2.8 *Numerical experiment on ssa accuracy and region of applicability*

As a simple illustration of the effectiveness of the boundary layer and accuracy estimates, consider the linear system ($\lambda_2 \ll \lambda_1 < 0$)

$$\dot{x} = \lambda_1 x, \; x(0),$$

$$\dot{y} = \alpha x + \lambda_2 y, \; y(0),$$

with the solution

$$\begin{aligned} x(t) &= x(0) \exp(\lambda_1 t), \\ y(t) &= y^i(t) + y^0(t), \\ y^i(t) &= [y(0) - \alpha x(0)/(\lambda_1 - \lambda_2)] \exp(\lambda_2 t), \\ y^0(t) &= [1/(\lambda_1 - \lambda_2)] \alpha x(0) \exp(\lambda_1 t), \end{aligned}$$

where y^i and y^0 represent contributions important in the inner and outer regions, respectively. Using (7.26):

$$Y_0(t) = [y(0) + \alpha x(0)/\lambda_2] \exp(\lambda_2 t).$$

Comparison of Y_0 with $y(t)$ shows exact agreement of decay rates and quite good agreement of magnitudes, the agreement being better the stiffer the system. The ssa gives

$$y_0(t) = [\alpha x(0)/-\lambda_2] \exp(\lambda_1 t).$$

From (7.18b)

$$\epsilon y_1(t) = [-\alpha\lambda_1 x(0)/\lambda_2^2] \exp(\lambda_1 t).$$

Thus the second-order approximation is

$$y_0 + \epsilon y_1 = -[(1 + \lambda_1/\lambda_2)/\lambda_2] \, \alpha x(0) \exp(\lambda_1 t).$$

It can be shown that the accuracy of inclusion of the first-order term is given by

$$\frac{y - (y_0 + \epsilon y_1)}{y} = \left(\frac{\lambda_1}{\lambda_2}\right)^2,$$

while the accuracy of the ssa is

$$\frac{y - y_0}{y} = \left(\frac{\lambda_1}{\lambda_2}\right),$$

and the contribution of the first-order term in comparison to the zeroth-order term, the indication of accuracy recommended earlier, is

$$\epsilon y_1/y_0 = \lambda_1/\lambda_2,$$

which agrees exactly with the real accuracy in this simple linear example.

Hence the solution $y_0 + \epsilon y_1$ is always more accurate than just y_0, the difference being greater the stiffer the system (that is, the smaller λ_1/λ_2). This first-order contribution can thus become large enough to require inclusion in the solution for sufficient accuracy; the system is then not stiff enough for the ssa. Note that in this analysis a perturbing ϵ was not identified.

7.2.9 *Choice of initial condition to eliminate stiffness*

If a problem does contain stiff variables, the stiff contribution dies out early in the solution, after which the transient will be dictated entirely by the remaining eigenvalues. Thus, the initial conditions on these variables have virtually no effect on the solution and may be chosen freely. The best choice for the initial condition is that which eliminates any contribution from the stiff eigenvalues, so that high accuracy is not required. This choice results from applying the ssa initially. Applying it to the stiff variables initially gives:

$$0 = g[x(0); y_0(0)], \tag{7.28}$$

where $x(0)$ are given initial conditions, and $y_0(0)$ are the zeroth-order initial conditions that result from the ssa. We take

$$y(0) = y_0(0). \tag{7.29}$$

Equation (7.29) eliminates the contribution from the inner variable Y_0 in (7.11) to yield a non stiff problem. This procedure may be preferred over use of the ssa throughout the transient when (7.28) is difficult to solve. It also allows standard integration packages to be used without modification.

To illustrate the principles involved, consider first the constant coefficient linear system

$$\frac{dx}{dt} = \lambda_1 x, \qquad \text{with } x(0) \text{ given,}$$
$$\frac{dy}{dt} = ax + \lambda_2 y, \quad \text{with } y(0) \text{ given,} \tag{7.30}$$

and with the analytic solution ($\lambda_2 \ll \lambda_1 < 0$)

$$x(t) = x(0)\exp(\lambda_1 t),$$
$$y(t) = C_1 \exp(\lambda_1 t) + C_2 \exp(\lambda_2 t),$$

where

$$C_1 = ax(0)/(\lambda_1 - \lambda_2),$$
$$C_2 = y(0) - C_1.$$

The ssa applied initially yields

$$0 = ax(0) + \lambda_2 y_0(0);$$

so rather than use the given $y(0)$ we use

$$y_0(0) = -ax(0)/\lambda_2.$$

With this choice of initial condition, the coefficient of the large eigenvalue, C_2, becomes $C_1(\lambda_1/\lambda_2)$. Therefore, the stiffer the system, the smaller will be the contribution from the stiff eigenvalue and the less the accuracy requirement for numerical integration.

The A-stable trapezoidal rule

$$z(n+1) = z(n) + (h/2)[F(n+1) + F(n)]$$

was applied to (7.30) with $a = 1$, $\lambda_1 = -1$, $\lambda_2 = 1{,}000$, $x(0) = 1$, $y(0) = 1$, only a moderately stiff problem. Equation (7.31) gives the alternative ssa initial condition $y_0(0) = 10^{-3}$. The results of using the original initial conditions are shown in Figure 7.4. The original initial condition causes severe accuracy problems for $h = 0.05$ (oscillating sawtooth curve), although it eventually converges to the steady state, as the trapezoidal rule must. The use of (7.31), however, gives better than 0.01 percent agreement with the exact solution for times greater than the boundary-layer length (approximately $t = 0.01$, as computed by (7.26) when $Y_0(t)/Y_0(0) = 10^{-3}$) and step sizes of 0.1. In order to achieve similar accuracy with the original initial conditions, step sizes of not larger than 0.001 are required, two orders of magnitude slower than using (7.31).

The second example is a nonlinear problem in reaction kinetics used first by Liniger and Willoughby (1970):

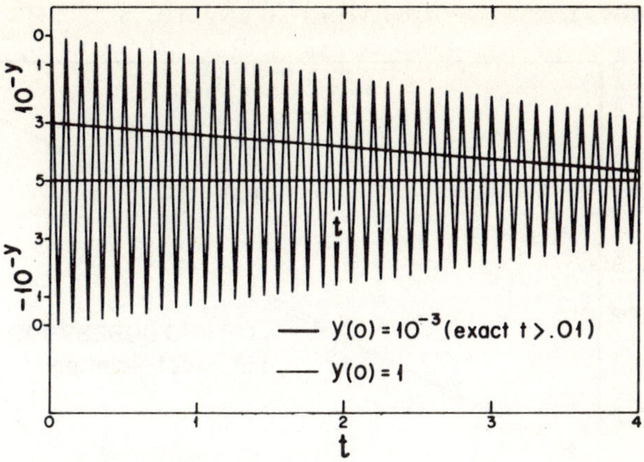

Fig. 7.4 Problem approximation to nonlinear kinetics example.

$$\frac{dx}{dt} = 0.01 - (1 + x^2)(0.01 + x + y), \quad x(0) = 0, \tag{7.32a}$$

$$\frac{dy}{dt} = 0.01 - [1 + (y + 10^3)(y + 1)](0.01 + x + y), \quad y(0) = 0. \tag{7.32b}$$

The initial ratio of eigenvalues is $1{:}10^7$; later, at $t = 80$, the ratio is about $1{:}10^5$, and at $t = 100$ the eigenvalues vary rapidly. Application of the ssa initially to the derivative of the stiff variable (9b) gives

$$0 = y_0(0)^3 + [1001.01 + x(0)]\, y_0(0)^2 \\ + [1011.01 + 1001\, x(0)]\, y_0(0) + 1001\, x(0) + 10\ . \tag{7.33}$$

Because of the nonlinearities, a choice of the roots for $y_0(0)$ in (7.33) must be made. However, the only stable root (for which $\partial y/\partial t < 0$) is

$$y_0(0) = -0.009989909, \tag{7.34}$$

Although this ssa-derived initial condition is close to that of the original initial condition, a substantial increase in accuracy results over most of the solution domain with $y(0) = y_0(0)$, as shown in Figure 7.5. There, solutions for each initial condition are drawn as solved by the trapezoidal rule with the Newton–Raphson method to solve the implicit equation and steps of $h = 1.0$ Use of (7.34) with this step size gives better than 0.01 percent accuracy with two Newton–Raphson iterations per step. A first step of 1.0 jumps over the boundary-layer region (here $t = 10^{-2}$, as computed by (7.26)) in which the ssa is not valid. Use of the original initial condition yields an increasingly inaccurate solution which actually enters the domain of influence of another of the three steady states!

7.2.10 *Problem approximation for two-point boundary-value problems*

The initial-value approximation for stiff equations is also of use for two-point boundary-value problems (TPBVP) with sharp transients. It is possible to provide as many new equations for the missing conditions as there are fast-acting variables

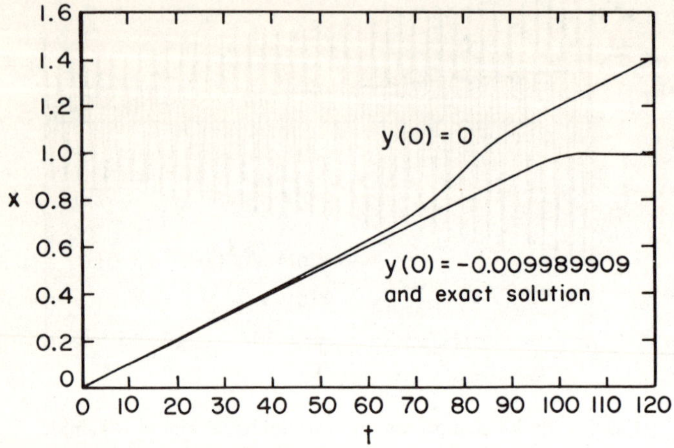

Fig. 7.5 Problem approximation to linear example.

in the problem. Consider for illustration the classical dimensionless equations for the isothermal steady-state tubular reactor

$$\frac{1}{\mathrm{Pe}}\left(\frac{d^2x}{ds^2}\right) = 2\frac{dx}{ds} + R(x), \tag{7.35a}$$

with

$$\frac{dx(0)}{ds} = 2\mathrm{Pe}\,[x(0) - 1], \quad \text{and}\ \frac{dx(1)}{ds} = 0, \tag{7.35b}$$

where differentiation is with respect to axial distance from entrance to exit ($s =$ 0 to 1), x is related to concentration, Pe is the Peclet number, and R is the reaction rate. The solution of (7.35) for realistic Pe (>20) requires a large amount of computation with conventional methods. The iterative nature of most methods for TPBVP magnifies this difficulty. Although some special singular perturbation approaches have been devised (O'Malley 1970; Ray and Ajinkya 1973), there exists no general approach. As will be shown, this problem can be particularly simple to solve by the initial-value approximation, with (7.35) written in the partitioned form:

$$\frac{dx}{ds} = y, \tag{7.36a}$$

$$\frac{1}{\mathrm{Pe}}\left(\frac{dy}{ds}\right) = 2y + R(x), \tag{7.36b}$$

$$y(0) = 2\mathrm{Pe}\,[x(0) - 1], \quad \text{and}\ y(1) = 0. \tag{7.37}$$

Note that there is only one initial-value relation for this two-variable set. As the local Jacobian of (7.36) has positive eigenvalues, the equations are usually integrated from the end of the reactor back to the front to take advantage of the higher stability regions of most accurate methods when applied to equations with negative eigenvalues. By defining $t = 1 - s$, (7.36) and (7.37) become

$$\frac{dx}{dt} = -y, \tag{7.38a}$$

$$\frac{1}{\text{Pe}}\left(\frac{dy}{dt}\right) = -2y - R(x), \tag{7.38b}$$

$$y(1) = 2\text{Pe}\,[x(1) - 1] \text{ and } y(0) = 0. \tag{7.39}$$

Equations (7.38) have the same characteristics as discussed previously for the initial value problem (7.1); that is, there is a small region near $t = 0$, over which y varies rapidly, causing the severe accuracy and stability limitations. As has been indicated, the initial condition on y can be set to any value within the domain of influence of the desired solution. Because the only initial condition provided (7.39) is on the stiff variable, however, no additional condition is provided by the ssa. From consideration of (7.38), the ssa is expected to apply to (7.38b) beyond the boundary layer up to and including the bed entrance. Thus, we may use (7.38) with

$$0 = 2y(0) + R\,[x(0)]\,, \tag{7.40}$$

and (7.37), to determine the initial conditions. These yield a good approximation to the actual initial conditions, since the ssa is valid at this point. The equation may then be solved by any method stable for large eigenvalues and with moderate accuracy. The solution from the reactor entrance need be continued only up to the boundary layer at the end of the reactor (as can be estimated by (7.26)), since beyond that point the smoother concentration variable is approximately constant.

As an example, (7.36) was solved with Pe $= 1000$ and $R = 10x^2$. Equations (7.37) and (7.40) give $x(0) = 0.9975124$ and $y(0) = -4.97520$. The solution obtained with the trapezoidal rule and steps of $h = 0.1$ was at least 0.1 percent accurate over $0 \leqslant s \leqslant 1$. The computation time for the problem as formulated is very small compared with other TPBVP methods and is significantly easier than applying special singular perturbation methods.

7.3 Case study of stiff model approximations: enzyme kinetics (Aiken and Venkateshwaran)†

Enzyme kinetic modelling has been based for over half a century on an approximation to a more detailed model, usually referred to as the Michaelis-Menten approximation. Its success lies not only in an initial concentration difference that is set experimentally but also in a certain difference among kinetic constants, and this is not widely recognized. Other differences among these constants lead to additional model approximations, often much simpler than that of Michaelis-Menten. Under the conditions where these alternate approximations apply, the Michaelis-Menten approximation gives misleading parameter estimates. This variety of approximate models will be examined in the next few subsections with the objective of fitting data with well-determined parameters and obtaining an efficient numerical solution.

† This section supported in part by NIH Biomedical Research Support Grant No. RR07092.

7.3.1 *Enzyme model*

The simplest enzyme kinetic model is (Henri 1902):

$$e^* + s^* \underset{k_{-1}}{\overset{k_1}{\rightleftharpoons}} c^* \overset{k_2}{\rightarrow} p^* + e^*.$$

This symbolic kinetic scheme represents elementary successful collisions of correctly oriented enzyme (e^*) and substrate (s^*) to form an enzyme-substrate complex (c^*) that can either react to form product (p^*) or disassemble back into unreacted constituents. By use of mass action kinetic theory the equations for the rate of change of each concentration are:

$$\frac{ds^*}{dt^*} = -k_1 s^* e^* + k_{-1} c^*, \tag{7.41}$$

$$\frac{de^*}{dt^*} = -k_1 s^* e^* + (k_{-1} + k_2) c^*, \tag{7.42}$$

$$\frac{dc^*}{dt^*} = k_1 s^* e^* - (k_{-1} + k_2) c^*, \tag{7.43}$$

$$\frac{dp^*}{dt^*} = k_2 c^*. \tag{7.44}$$

The experimental conditions at the start of the reaction ($t^* = 0$) are:

$$s^* = s^*(0), \ e^* = e^*(0), \ c^* = p^* = 0.$$

Since the sum of the right-hand sides of (7.42) and (7.43) is zero,

$$e^* + c^* = e^*(0). \tag{7.45}$$

We need only two equations. We eliminate e^* using (7.45) and consider

$$\frac{ds^*}{dt^*} = -k_1 e^*(0) s^* + (k_1 s^* + k_{-1}) c^*, \tag{7.46a}$$

$$\frac{dc^*}{dt^*} = k_1 e^*(0) s^* - (k_1 s^* + k_{-1} + k_2) c^*. \tag{7.46b}$$

This system is nonlinear. Commonly, two independent conditions make (7.46) stiff. One of these conditions is the well-known requirement (Heineken, Tsuchiya, and Aris 1967) that $e^*(0) \ll s^*(0)$. The other condition, not so well known, requires that $k_1 s^*(0), k_{-1} \gg k_2$. The point becomes clear if we examine the so-called stiffness ratio of the system:

$$\mathrm{SR} = \frac{|\mathrm{Re}(\lambda)|_{\min}}{|\mathrm{Re}(\lambda)|_{\max}} \tag{7.47}$$

Our stiff system has a "small" value of SR. For a nonlinear system of ODEs the eigenvalues, corresponding to those of the Jacobian, will vary in magnitude over the range of integration. The Jacobian of (7.46) is

$$J = \begin{bmatrix} -k_1[e^*(0)-c^*] & k_1s^* + k_{-1} \\ k_1[e^*(0)-c^*] & -(k_1s^* + k_{-1} + k_2) \end{bmatrix}.$$

The eigenvalues are given by:

$$\lambda = \frac{-b \pm (b^2 - 4ax)^{1/2}}{2},$$

where

$$a = k_1k_2,$$

$$x = e^*(0) - c^*,$$

and

$$b = k_1(s^* + x) + k_{-1} + k_2.$$

Table 7.3 gives the stiffness ratio of (7.46) evaluated at $t^* = 0$, for four different sets of rate constant values and initial conditions. Comparing SR for cases 1 and 2, we see that $e^*(0) \ll s^*(0)$ definitely makes (7.46) stiff. Comparison of the SR values for cases 2 and 3 indicates that $k_1s^*(0)$, $k_{-1} \gg k_2$ can also make (7.46) stiff. These two conditions are independent of each other. Case 4 examines a combination of these two conditions which makes the SR very small.

7.3.2 *Conventional approximate solutions*

We follow the excellent treatment of Lin and Segel (1974) which is based on that of Heineken, Tsuchiya, and Aris (1967). First (7.46) is made dimensionless by the introduction of scales, resulting in new dimensionless variables:

$$t = t^*/t_0, \quad x = s^*/s_0 \quad \text{and} \quad y = c^*/c_0, \tag{7.48}$$

where t_0, s_0, and c_0 are as yet unknown scales for time, substrate, and complex respectively, that result in a solution with roughly the same magnitude in x and y over a normalized time range. Substitution of (7.48) into (7.46) gives the dimensionless equations:

$$\frac{1}{t_0k_1e^*(0)}\dot{x} = -x + \frac{c_0}{e^*(0)}xy + \frac{k_{-1}c_0}{k_1e^*(0)s_0}y, \tag{7.49}$$

$$\frac{c_0}{t_0k_1e^*(0)s_0}\dot{y} = x - \frac{k_1s_0c_0}{k_1e^*(0)s_0}xy - \frac{(k_{-1}+k_2)c_0}{k_1e^*(0)s_0}y, \tag{7.50}$$

where $\dot{}$ denotes d/dt. The initial conditions are $t = 0$, $x = s^*(0)/s_0$, and $y = 0$. Let us take $t_0 = 1/k_1e^*(0)$. At least initially the substrate concentration, assumed much larger than the enzyme concentration, will be on the order of its initial concentration; so take $s_0 = s^*(0)$. The complex concentration starting at zero value will become nonzero soon and would reach a maximum value of about $e^*(0)$ while the substrate is being processed; so take $c_0 = e^*(0)$. With the choice of these scales, (7.49) and (7.50) become

$$\dot{x} = -x + (x + K - \lambda)y, \tag{7.51}$$

TABLE 7.3
Stiffness ratio of (7.47) at $t^* = 0$

Case	k_1	k_{-1}	k_2	$z = e^*(0)/s^*(0)$	SR
1	1.0 E+ 05	1.0 E+ 00	1.0 E+ 00	1.0 E− 03	1.12 E− 04
2	1.0 E+ 05	1.0 E+ 00	1.0 E+ 00	1.0 E+ 00	7.19 E−01
3	1.0 E+ 05	1.0 E+ 00	1.0 E− 03	1.0 E+ 00	1.12 E− 04
4	1.0 E+ 05	1.0 E+ 00	1.0 E−02	1.0 E− 03	1.24 E− 06

$$\epsilon\dot{y} = x - (x + K)y, \qquad x(0) = 1 \text{ and } y(0) = 0, \tag{7.52}$$

where $\epsilon = e^*(0)/s^*(0)$, $K = (k_{-1} + k_2)/k_1 s^*(0)$ and $\lambda = k_2/k_2 s^*(0)$. The zeroth-order outer singular perturbation solution of (7.52) is obtained by setting the right-hand side equal to zero as

$$y = \frac{x}{x + K},$$

which in dimensional terms results in

$$\frac{dp^*}{dt^*} = \frac{k_2 e^*(0)s^*}{K_M + s^*}, \tag{7.53}$$

where $K_M = (k_{-1} + k_2)/k_1$, the so-called Michaelis constant.

The degree of stiffness of the system represented by (7.51) and (7.52) is dependent upon the magnitude of ϵ; the smaller the magnitude of ϵ, the stiffer the system. Equation (7.53) is only approximate for such stiff systems, but it seems to work rather well even for systems which are only moderately or slightly stiff. This is so because for many practical systems, the product formation is slow compared to the other reactions. In fact, this latter point was the sole qualitative basis for the original reasoning of Michaelis and Menten (1913).

Michaelis and Menten obtained (7.53) by assuming that the binding of e^* and s^* goes to rapid equilibrium. This leads to the following condition on the rate constants:

$$k_1 s^*(0), \quad k_{-1} \gg k_2.$$

If the time scale $t_0 = 1/k_2$ (the slower time) is used and $s_0 = s^*(0)$, then the characteristic concentration of the enzyme-substrate complex is

$$c_0 = k_1 e^*(0) s^*(0)/k_{-1}.$$

Equations (7.49) and (7.50) reduce to

$$\dot{x} = -\left(\frac{k_1 e^*(0)}{k_2}\right)x + \left(\frac{k_1 e^*(0)}{k_2}\right)\left(\frac{k_1 s^*(0)}{k_{-1}}\right)xy + \left(\frac{k_1 e^*(0)}{k_2}\right)y, \tag{7.54}$$

and

$$\epsilon\dot{y} = x - \frac{k_1 s^*(0)}{k_{-1}} xy - y - \epsilon y, \tag{7.55}$$

where $\epsilon = k_2/k_{-1} \ll 1$. Again the order of magnitude of ϵ is indicative of the stiffness of the system. For a stiff system, the zeroth-order outer approximation on (7.55) is

$$y = \frac{x}{1 + \dfrac{k_1 s^*(0)}{k_{-1}} x}. \tag{7.56}$$

In dimensional form,

$$\frac{\mathrm{d}p^*}{\mathrm{d}t^*} = \frac{k_2 e^*(0) s^*}{K_\mathrm{M} + s^*}. \tag{7.57}$$

This is of the same form as (7.53), but there is an important difference. Substitution of (7.56) into (7.54) gives

$$\dot{x} = 0,$$

which in dimensional form is

$$s^* = s^*(0). \tag{7.58}$$

Equations (7.57) and (7.58) together with the initial conditions give a linear relationship between product concentration and time:

$$P^* = \frac{k_2 e^*(0) s^*(0)}{K_\mathrm{M} + s^*(0)} t^*. \tag{7.59}$$

The slope of product with time is very small, which is to be expected since not much reactant is consumed. This is the correct mathematical approximation based on Michaelis and Menten's physical reasoning.

If we add here the condition $e^*(0) \ll s^*(0)$, it happens that $y \gg \dot{y}$ or $\epsilon y \gg \epsilon\dot{y}$. Retention of ϵy in (7.55) leads to the form of (7.53). We offer no scaling scheme that recognizes both types of stiffness (concentration and kinetic origins together). As mentioned in §7.2.5, often in practice a steady-state approximation is achieved by simply setting a derivative equal to zero without neglecting appropriately small terms on the right-hand side. While this has no mathematical basis, it does eliminate processing into singular perturbation form and retains the exact equilibrium condition.

Most practical systems in enzyme kinetics are modelled fairly well by (7.53) because of a combination of these *two* factors. Typically, the substrate is more plentiful than the enzyme and the product formation reaction is relatively slower than the other reactions. Thus a system only moderately stiff in $\epsilon = k_2/k_{-1}$ or $\epsilon = e^*(0)/s^*(0)$ still may be well approximated by (7.53).

7.3.3 *Higher-order perturbation solution*†

It has been pointed out that the Michaelis–Menten approximation is the zeroth-order outer solution. The inner zeroth-order solution has to be appropriate to the

† The main results in this subsection appear in Lin and Segel (1974).

first few moments of the reaction and must satisfy the original initial conditions. This involves rescaling the variables to make the terms appropriate in the boundary layer. A uniform approximation is obtained by adding the inner and outer approximations while subtracting their common parts. The zeroth-order uniform approximation is

$$S_0^{(u)}(t) = S_0(t), \tag{7.60a}$$

$$C_0^{(u)}(t) = \frac{S_0(t)}{S_0(t)+K} - \frac{\exp[-(K+1)t/\epsilon]}{K+1}. \tag{7.60b}$$

It is possible to estimate all three parameters using the zeroth-order solution (7.60). The quality of the estimates will depend on how well (7.60) approximates the exact model, by the extent of experimental error in the complex concentration data, and by the location of this data.

It is possible to improve on these approximations by assuming a series solution to the inner and outer equations. The uniform first-order approximation for S is

$$S_1^{(u)}(t, \epsilon) = S_0(t) + \epsilon S_1(t) + \frac{\epsilon(1+K-\lambda)\exp[-(K+1)t/\epsilon]}{(K+1)^2}. \tag{7.61}$$

where

$$S_1(t) = \frac{S_0}{S_0+K}\left[\frac{K-\lambda}{K}\ln\left(\frac{K+S_0}{(1+K)S_0}\right) - \frac{K-\lambda+s_0}{K+S_0}\right].$$

All the parameters appear in this solution, too. It is more attractive from the parameter estimation standpoint to have an equation for the substrate rather than the complex, because it is much easier to measure. The above expression for $S^{(u)}(t, \epsilon)$ is an asymptotic approximation to the exact solution.

Figures 7.6a and 7.6b compare the approximate solution to the exact solution for $\epsilon = 0.1$. As the system becomes stiffer, the approximate solution becomes more exact. The zeroth-order solution for c (the stiff variable) is more accurate than that for s (the nonstiff variable).

7.3.4 *Other approximations*

Case I. $k_2 \gg k_{-1}, k_1 s^*(0)$.

This is the reverse kinetic condition of the last section. Take as the scales $t_0 = 1/k_1 e^*(0)$, $s_0 = s^*(0)$ and $c_0 = k_1 e^*(0) s^*(0)/(k_{-1} + k_2)$. This will give a characteristic complex concentration much less than $e^*(0)$. Equation (7.49) and (7.50) now become

$$\dot{x} = -x + \epsilon_1 xy + \epsilon_2 y, \tag{7.62}$$

$$\epsilon_3 \dot{y} = x - \epsilon_1 xy - y, \tag{7.63}$$

where

$$\epsilon_1 = \frac{k_1 s^*(0)}{(k_{-1}+k_2)}, \quad \epsilon_2 = \frac{k_{-1}}{(k_{-1}+k_2)} \quad \text{and } \epsilon_3 = \frac{k_1 e^*(0)}{(k_{-1}+k_2)}.$$

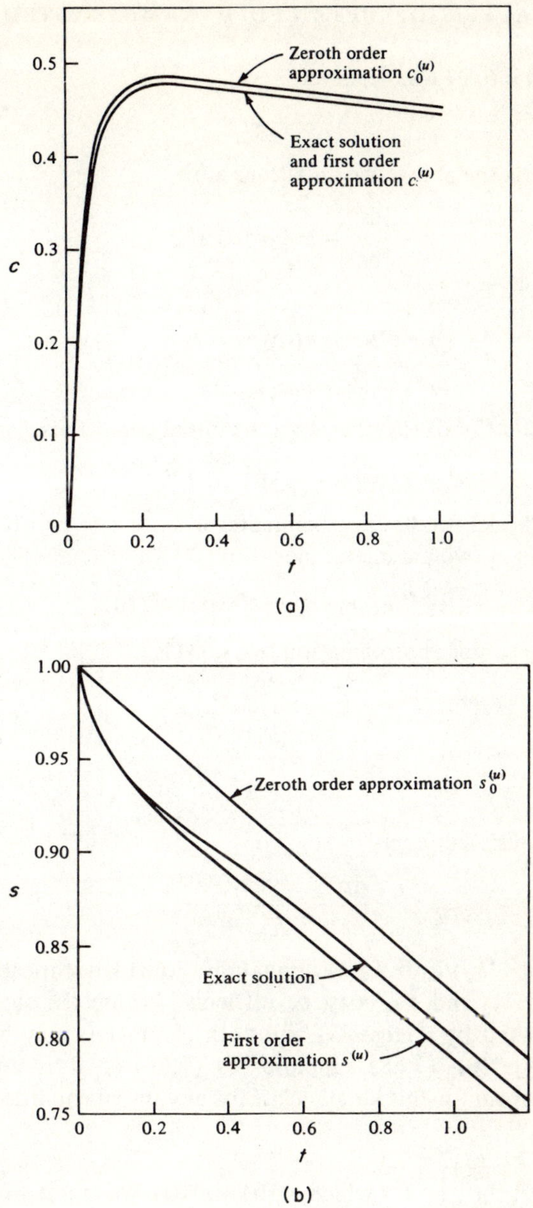

Fig. 7.6 Approximations to enzyme model for $\epsilon = 0.1$; (a) substrate (b) complex (from Heineken, Tsuchiya, and Aris 1967, with permission).

If $e^*(0) \ll s^*(0)$ then $\epsilon_3 \ll \epsilon_1, \epsilon_2$; otherwise the three parameters are of the same order of magnitude. Let us consider the two possibilities separately:

A. $$\epsilon_1 \sim \epsilon_2 \sim \epsilon_3 \text{ (i.e. } e^*(0) \sim s^*(0)).$$

The zeroth-order outer approximation of (7.63) is

$$x = y. \tag{7.64}$$

Substituting (7.64) into (7.62) we have

$$\dot{x} = x.$$

In dimensional terms the above two equations are

$$\frac{ds^*}{dt^*} = -k_1 e^*(0) s^*, \tag{7.65}$$

and

$$\frac{dp^*}{dt^*} = \frac{k_1 k_2 e^*(0) s^*}{(k_{-1} + k_2)} \approx k_1 e^*(0) s^*. \tag{7.66}$$

Equations (7.65) and (7.66) together with the initial conditions give

$$p^* = s^*(0)\{1 - \exp[-k_1 e^*(0) t^*]\}. \tag{7.67}$$

A one-parameter model results; this parameter is k_1, a rate constant not estimable from the conventional two-parameter model of (7.53).

B. $\epsilon_3 \ll \epsilon_1, \epsilon_2$ (i.e. $e^*(0) \ll s^*(0)$).

The zeroth-order outer approximation of (7.63) is

$$y = \frac{x}{1 + \dfrac{k_1 s^*(0) x}{k_{-1} + k_2}},$$

leading to the dimensional form

$$\frac{dp^*}{dt^*} = \frac{k_2 e^*(0) s^*}{K_M + s^*}, \tag{7.68}$$

where $K_M = (k_{-1} + k_2)/k_1 \approx k_2/k_1$. Equation (7.68) is identical to (7.53). However, estimation of k_2 and K_M may be difficult because the sigmodial character of s^* may be masked by a large K_M and both parameters may be ill-determined. But since $k_2 \gg k_1 s^*(0)$, (7.68) simplifies to (7.66) so it is unwise to use the Michaelis–Menten form for this kinetic class for any initial conditions.

Case II. $k_1 s^*(0) \gg k_2, k_{-1}$.

A. Consider first the case for which $e^*(0) \sim s^*(0)$. We take $t_0 = 1/k_2$, $c_0 = s^*(0)$, $s_0 = k_2 s^*(0)/k_1 e^*(0)$. Equations (7.49) and (7.50) reduce to:

$$\epsilon \dot{x} = -x + \frac{s^*(0)}{e^*(0)} xy + \frac{k_{-1}}{k_2} y, \tag{7.69}$$

$$\dot{y} = x - \frac{s^*(0)}{e^*(0)} xy - \left(\frac{k_{-1} + k_2}{k_2}\right) y, \tag{7.70}$$

where $\epsilon = k_2/k_1 e^*(0) \ll 1$. So x is the stiff variable rather than y. With the use of $s^* + c^* + p^* = s^*(0)$, it can be shown that in dimensional form

$$\frac{dp^*}{dt^*} \approx \frac{k_2 s^*(0) - k_2 p^* - \dfrac{k_2 s^*(0)^2}{e^*(0)}\left(1 - \dfrac{p^*}{s^*(0)}\right)^2}{1 - \dfrac{s^*(0)}{e^*(0)} + \dfrac{p^*}{e^*(0)}}. \tag{7.21}$$

Only one parameter, the rate constant k_2, can be evaluated from (7.71).

B. Now consider the case $e^*(0) \ll s^*(0)$. With $t_0 = 1/k_2$, $c_0 = e^*(0)$ and $s_0 = s^*(0)$, Equations (7.49) and (7.50) become

$$\dot{x} = \frac{-k_1 e^*(0)}{k_2} x + \frac{k_1 e^*(0)}{k_2} xy + \frac{k_1 e^*(0)}{k_2 s^*(0)} y, \tag{7.72}$$

$$\epsilon_1 \dot{y} = x - xy - \epsilon_2 y, \tag{7.73}$$

where

$$\epsilon_1 = \frac{k_2}{k_1 s^*(0)} \text{ and } \epsilon_2 = \frac{(k_2 + k_{-1})}{k_2 s^*(0)}.$$

The zeroth-order outer approximation of (7.73) gives

$$y = 1.$$

In dimensional terms, we have

$$\frac{dp^*}{dt^*} = k_2 e^*(0),$$

or

$$p^* = k_2 e^*(0) t^*. \tag{7.74}$$

Equation (7.74), a linear approximation, will give one well determined rate constant, k_2. Use of (7.53) may not give a good estimate of K_M because that parameter is masked by s^*.

Case III. $k_{-1}, k_2 \gg k_1 s^*(0)$.

This is the opposite of the previous case. Take $t_0 = 1/k_1 e^*(0)$, $s_0 = s^*(0)$ and $c_0 = k_1 e^*(0) s^*(0)/k_{-1}$. We have:

$$\dot{x} = -x + \epsilon_1 xy + y, \tag{7.75}$$

$$\epsilon_2 \dot{y} = x - \epsilon_1 xy - \left(\frac{k_{-1} + k_2}{k_{-1}}\right) y, \tag{7.76}$$

where

$$\epsilon_1 = \frac{k_1 s^*(0)}{k_{-1}} \text{ and } \epsilon_2 = \frac{k_1 e^*(0)}{k_{-1}}.$$

A. If $e^*(0) \sim s^*(0)$, then $\epsilon_1 \sim \epsilon_2$ and the zeroth-order outer approximation on (7.76) gives in dimensional terms

$$\frac{dp^*}{dt^*} = \frac{k_1 k_2 e^*(0)}{k_{-1} + k_2} s^* = \frac{k_2 e^*(0)}{K_M} s^*, \tag{7.77}$$

and

$$\frac{ds^*}{dt^*} = \frac{-k_1 k_2 e^*(0)}{k_{-1} + k_2} s^* = \frac{-k_2 e^*(0)}{K_M} s^*. \tag{7.78}$$

Equations (7.77) and (7.78), along with the initial conditions, give

$$p^* = s^*(0) \left[1 - \exp\left(\frac{-k_2 e^*(0)}{K_M} t^* \right) \right]. \tag{7.79}$$

B. If $e^*(0) \ll s^*(0)$, then $\epsilon_2 \ll \epsilon_1$ and the zeroth-order outer approximation of (7.76) gives in dimensional terms

$$\frac{dp^*}{dt^*} = \frac{k_2 e^*(0) s^*}{K_M + s^*}, \tag{7.80}$$

$$\frac{ds^*}{dt^*} = \frac{k_2 e^*(0) s^*}{K_M + s^*}. \tag{7.81}$$

Equation (7.80) is in the conventional form of (7.53). Here again K_M could mask the sigmoidal character of s^* (since $K_M = (k_{-1} + k_2/k_1 \sim k_{-1}/k_1$ and $k_{-1} \gg k_1 s^*(0))$ leading to ill-determined parameters in the model. If the relative magnitudes of the kinetic constants are recognized, (7.81) reduces to (7.77).

7.3.5 *Parameter estimation of approximate models*

Simulated data was used to carry out the parameter estimation. The product concentration versus time data over the entire time range was simulated by solving (7.44), (7.46a), and (7.46b) with LSODE. Up to five percent simulated experimental error was introduced into the product concentration in a random manner. The various simulation runs are classified into two sets used to study the effect of initial concentration differences and kinetic constant differences.

Table 7.4 lists the simulation runs used to estimate the parameters in (7.53). In runs 1 through 4, the value of $e^*(0)/s^*(0)$ varies progressively from moderately stiff ($e^*(0)/s^*(0) = 1.0\text{ E }-3$) to non stiff ($e^*(0)/s^*(0) = 1.0$) conditions. For each run, the other independent condition introducing stiffness, k_2/k_{-1}, is either 1.0 E−2 or 1.0 E−1. In run 4, k_2/k_{-1} is either 1.0 E −3 or 1.0 E −2. This data set was used to study the effect of degree of stiffness on the estimation of parameters.

Table 7.5 lists the cases used to estimate parameters in the appropriate approximations developed in the previous section for various kinetic constant differences. The numbering of cases is consistent with the notation used in the previous section.

The parameter estimation was carried out using the nonlinear regression program available on the BMDP (1981) statistical software package. For the estimation of parameters in dynamic models, this package uses a Runge-Kutta method for the numerical integration of the dynamic equations (Ralston, Jennrich, Sampson, and Uho 1980). Hence it is inefficient for moderately stiff systems. Such cases were handled using another package developed by Chandler, Hill, and Spivey (1972).

TABLE 7.4

Parameter values used for simulation of data set to estimate parameters in equation (7.53)

Runs		k_1	k_{-1}	k_2	$e^*(0)/s^*(0)$
1	a	1.0E8	1.0E3	1.0E1	1.0E–3
	b	1.0E8	1.0E3	1.0E2	
2	a	1.0E8	1.0E3	1.0E1	1.0E–2
	b	1.0E8	1.0E3	1.0E2	
3	a	1.0E8	1.0E3	1.0E1	1.0E–1
	b	1.0E8	1.0E3	1.0E2	
4	a	1.0E8	1.0E3	1.0	1.0
	b	1.0E8	1.0E3	1.0E1	

TABLE 7.5

Parameter values used for simulation of data set to estimate parameters in other approximations

Case		k_1	k_{-1}	k_2	$e^*(0)/s^*(0)$
I	A	1.0E5	1.0	1.0E3	1.0
	B	1.0E5	1.0	1.0E3	1.0E–2
II	A	1.0E8	1.0	1.0	1.0
	B	1.0E8	1.0	1.0	1.0E–2
III	A	1.0E5	1.0E3	1.0E3	1.0
	B	1.0E5	1.0E3	1.0E3	1.0E–2

For the simulation runs lised in Table 7.4, the parameters of (7.53) were estimated. The results are summarized in Table 7.6.

In runs 4a and 4b stiffness is introduced only by the condition $k_2/k_{-1} \ll 1$. The other condition which introduces stiffness is inactive because $e^*(0)/s^*(0) = 1.0$. The estimates of the parameters are reasonable. In runs 3a and 3b, both the conditions capable of introducing stiffness are active; but one of the conditions, $e^*(0)/s^*(0) = 0.1$, introduces only a very slight degree of stiffness. Yet the parameter estimates are very good. Runs 1 and 2 clearly illustrate that (7.53) is a good approximation even for moderately stiff systems.

Thus two conditions, one requiring that the substrate be initially more plentiful than the enzyme and the other requiring that the product formation be relatively slower than the other reactions, can indeed independently make (7.41)–(7.44) stiff. In many practical systems, both conditions are active to some extent. So the zeroth-order approximation of (7.53) works very well even for only slightly stiff systems.

TABLE 7.6

*Results for runs in Table 7.3**

Runs		1a	1b	2a	2b	3a	3b	4a	4b
Exact parameter	$k_2e^*(0)$	1.0 E –7	1.0 E –6	1.0 E –6	1.0 E –5	1.0 E –4	1.0 E –4	1.0 E –5	1.0 E –4
value	K_M	1.01 E –5	1.1 E –5	1.01 E –5	1.1 E –5	1.01 E –5	1.1 E –5	1.001 E –5	1.01 E –5
Initial guess	$k_2e^*(0)$	1.0 E –6	1.0 E –5	1.0 E –5	1.0 E –4	1.0 E –4	1.0 E –3	1.0 E –4	1.0 E –3
	K_M	1.0 E –4	1.0 E –4	1.0 E –4	1.0 E –4	1.0 E –4	1.0 E –4	1.0 E –4	1.0 E –4
Estimated value	$k_2e^*(0)$	0.872 E –7	0.894 E –6	1.133 E –6	.804 E –5	1.207 E –5	1.131 E –4	2.16 E –5	1.7 E –4
	K_M	0.798 E –5	0.823 E –5	1.144 E –5	.722 E –5	1.327 E –5	1.304 E –5	4.15 E –5	3.22 E –5
Std. deviation	$k_2e^*(0)$	2.32 E –8	1.69 E –7	3.492 E –7	1.72 E –6	4.15 E –6	3.77 E –5	*	2.44 E –4
	K_M	3.6 E –6	2.64 E –6	5.245 E –6	2.74 E –6	6.5 E –6	6.31 E –6	*	5.4 E –5

* Run 4a was handled using the package developed by Chandler, Hill, and Spivey (1972). The standard deviation of fit was 0.3096 E –6.

Case I. $k_2 \gg k_{-1}, k_1 s^*(0)$.

A. $e^*(0) \sim s^*(0)$. The approximation given by (7.67) was fitted to the simulated data of case IA. The results are summarized in Table 7.7a. This model is appropriate and allows a good data fit and the determination of the single parameter in the model, k_1.

B. $e^*(0) \ll s^*(0)$. An attempt to evaluate $k_2 e^*(0)$ and K_M from (7.68) using simulated data of case IB yielded results which are summarized in Table 7.7b. The estimates are inaccurate because K_M masks the sigmoidal character of s^*. Note that the attempt to estimate the unestimable constant here destroys the ability to find even one constant because the two constants are dependent. The Michaelis–Menten form should not be used.

Case II. $k_1 s^*(0) \gg k_{-1}, k_2$.

A. $e^*(0) \sim s^*(0)$. For the particular case when $e^*(0) = s^*(0)$, (7.71) reduces to

$$\frac{dp^*}{dt^*} = k_2 [s^*(0) - p^*] .$$

The data of case IIA was fitted to this equation. The results are summarized in Table 7.8a. The fit and the signifiance of the estimates are excellent.

B. $e^*(0) \ll s^*(0)$. The linear approximation given by (7.74) was fitted to the simulated data of case IIB. The results are summarized in Table 7.8b. Again the one-parameter fit is excellent.

An attempt to fit the same data to (7.53) yielded results which are summarized in Table 7.8c. One parameter, k_2, is well estimated; the other, K_M, is not.

When the relative order of magnitude of the rate constants is such as to satisfy $k_1 s^*(0) \gg k_{-1}$, k_2, a linear approximation indicated by (7.74) is the easiest way to evaluate k_2. K_M cannot be well estimated and use of the two-parameter model is not justified. Thus, once again the Michaelis–Menten equation would lead to a fortuitous fit and an insignificant parameter estimate—misleading information on the kinetics.

Case III. $k_{-1}, k_2 \gg k_1 s^*(0)$.

A. $e^*(0) \sim s^*(0)$. The approximation given by (7.79) was fitted to simulated data of case IIIA. The results are summarized in Table 7.9. The one-parameter model gives significant results.

B. $e^*(0) \ll s^*(0)$. Table 7.9b and Figure 7.7 summarize the results of an attempt to estimate the parameters of (7.80) using the data of case IIIB. Once again indiscriminate use of the Michaelis–Menten type of equation leads to trouble although the fit appears adequate. For these kinetics, the appropriate approximation is the one-parameter model (7.79) regardless of the initial concentration.

7.3.6 *Concluding comments*

The enzyme model and its Michaelis–Menten approximation have been used and analyzed for many years and yet fundamental aspects of rate constant disparity effects have not been recognized. Depending upon these differences, a number of simpler yet more significant approximations are indicated—regardless of the initial conditions. These can lead to well determined values of individual rate constants in the enzyme model.

TABLE 7.7

(a) Results for Case IA

Parameter	k_1
Exact value	1.0 E 5
Initial guess	1.0 E 4
Estimated value	1.09 E 5
Std. deviation	3.57 E 3

(b) Results for Case IB

Parameter	$k_2e^*(0)$	K_M
Exact value	1.0 E –4	1.0 E –2
Initial guess	1.0 E –3	1.0 E –1
Estimated value	1.01 E –5	9.55 E –4
Std. deviation	2.13 E –4	2.02 E –2

TABLE 7.8

(a) Results for Case IIA and (7.71)

Parameter	k_2
Exact value	1.0
Initial guess	10.0
Estimated value	1.074
Std. deviation	3.7 E –2

(b) Results for Case IIB and (7.74)

Parameter	$k_2e^*(0)$
Exact value	1.0 E –7
Estimated value	0.95 E –7
Std. reg. coeff.	0.996

(c) Results for Case IIB and (7.53)

Parameter	$k_2e^*(0)$	K_M
Exact value	1.0 E –7	2.0 E –8
Initial guess	1.0 E –6	1.0 E –7
Estimated value	1.0 E –7	4.9 E –7
Std. deviation	5.76 E –9	3.01 E –7

TABLE 7.9

(a) Results for Case IIIA

Parameter	$\dfrac{k_2 e^*(0)}{K_M}$
Exact value	0.5
Initial guess	1.0 E −3
Estimated value	0.515
Std. deviation	2.139 E −2

(b) Results for Case IIIB

Parameter	$k_2 e^*(0)$	K_M
Exact value	1.0 E −4	2.0 E −2
Initial guess	1.0 E −5	2.0 E −3
Estimated value	2.2 E −7	3.56 E −5
Std. deviation	1.48 E −7	2.8 E −5

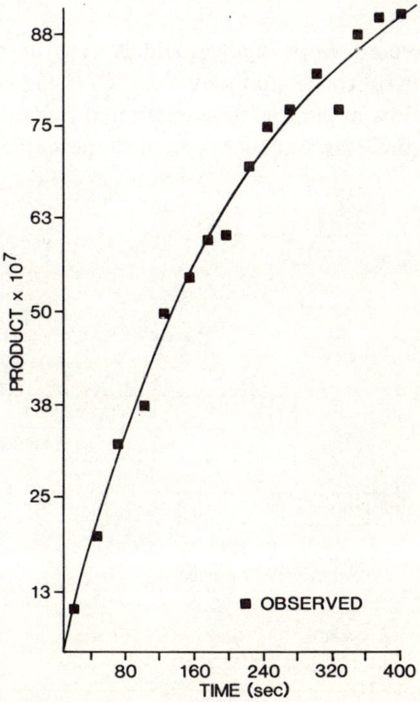

Fig. 7.7. Fit of Michaelis–Menten equation containing poorly determined parameters.

7.4 Stiff estimation (Aiken)

We define informally the idea of *stiff estimation* as estimation of the parameters in stiff differential equations. We refer to the parameters directly involved in producing sharp transients, or large absolute-valued eigenvalues, as ***stiff parameters***. It is assumed that the detail in the model leading to stiff computation is necessary.

7.4.1 *Measurement limitations*

The variable with sharp initial transients, the stiff variable, can present two major measurement problems: one is associated with the speed of the transient and the other with the potentially small magnitude of the stiff variable.

Consider chemical-kinetics reaction-rate measurement. A number of techniques exist for measuring fast individual reactions, illustrated in Figure 7.8. These are roughly classified as flow methods and relaxation methods. The continuous flow method was the first fast-reaction technique, developed by Hartridge and Roughton (1923) who measured the rate of binding carbon monoxide with hemoglobin. Flow methods rapidly mix two reagent streams: adjustment of flow velocities for a constant detector upstream location will vary the reaction rate. Relaxation methods perturb an equilibrium system with any variable capable of being changed rapidly enough. Relaxation methods can be used for reaction rates as fast as 10^{-10} sec, as occur in electron or proton transfer reactions and do not require the quantities of reagent that flow methods do. However, only perturbations about an equilibrium can be studied.

Chance (1943) pioneered work that provided experimental evidence for the existence of an enzyme-substrate complex (see §7.3.1). He follow its spectral change with a stopped-flow apparatus he constructed for fast reactions in solution. For example, consider the reaction of horseradish peroxidase (E) with hydrogen peroxide (S),

$$E + S \underset{k_{-1}}{\overset{k_1}{\rightleftharpoons}} ES, \tag{7.82}$$

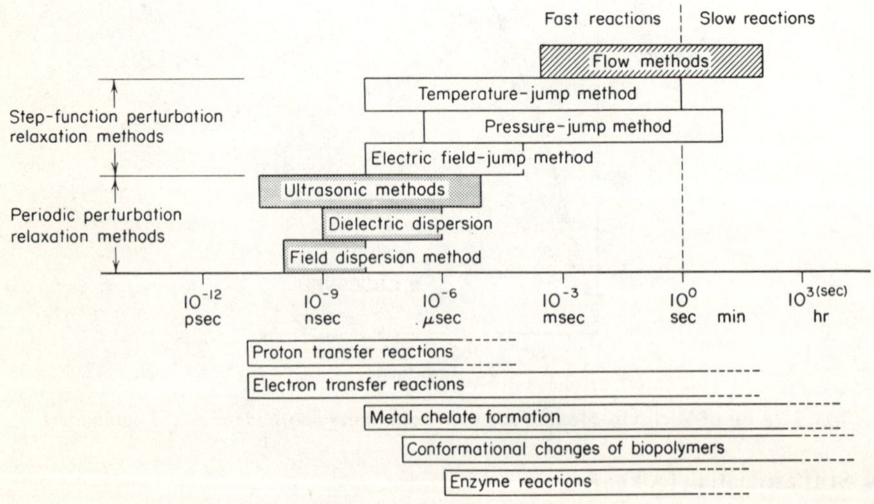

Fig. 7.8 Time ranges of fast reactions in solution and methods for studying them. (Adapted from Hiromi, Kinetics of Fast Enzyme Reactions, Halsted Press 1979, by permission.)

$$ES + AH_2 \overset{k_2'}{\rightarrow} E + A + 2H_2, \tag{7.83}$$

where the mechanism of §7.3.1 is modified by the presence of a second substrate, leuco-malachite green (AH_2), and A is the product malachite green. The rate constant k_2 of §7.3.1 is retained if we take

$$k_2 = k_2'[AH_2].$$

This will be approximately constant if this substrate's concentration is much larger than the enzyme's initial concentration. The transient of the complex (ES) is shown in Figure 7.9, from which data was extracted by Chance as shown in Figure 7.10a; similarly, data for product (A) is given in Figure 7.10b. The complex was monitored by absorbance changes at 400 nm, the product at 610 nm.

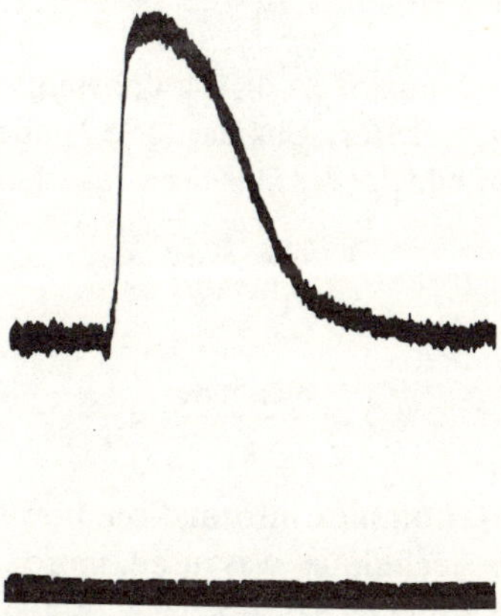

Fig. 7.9 Stiff complex variable as measured by Chance (1947) (by permission).

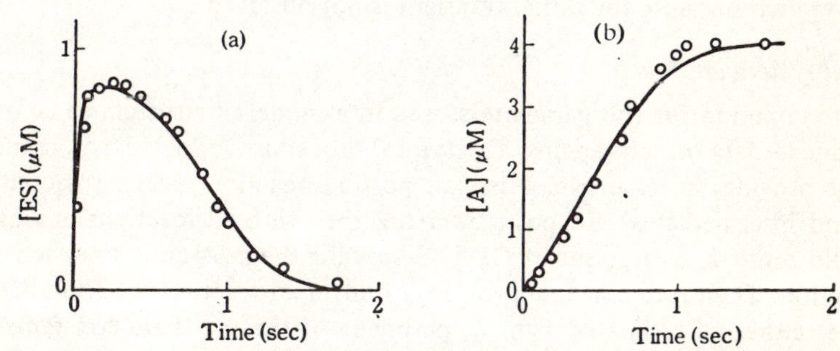

Fig. 7.10 Data chosen from Figure 7.9 by Chance (1947, by permission): (a) complex; (b) product.

Final concentrations were: enzyme 1 μM, hydrogen peroxide 4 μM, leuco-malachite green 15 μM. This rather high enzyme concentration (initially 1 μM compared with initial substrate hydrogen peroxide 8 μM) is necessary to see the transient for this only mildly stiff problem. Chance did not perform a transient analysis for all three constants. Instead, he used the integrated area under the complex curve, Figure 7.10a, to obtain k_2:

$$\frac{dA}{dt} = k_2[\text{ES}], \tag{7.84}$$

$$k_2 = [\text{A}]_t \bigg/ \int_0^t [\text{ES}]\, dt. \tag{7.85}$$

Since $[AH_2]$ is known, k_2' was obtained as $3.0 \times 10^5\ M^{-1}\ s^{-1}$ ($k_2 = 4.5\ s^{-1}$). The value of k_1 was directly given by observing the formation of ES from E and S in the absence of AH_2, $k_1 = 1.2 \times 10^7\ M^{-1}s^{-1}$. Finally k_{-1} was estimated from the dissociation constant of ES (static spectrophotometric titration):

$$K_s = \frac{k_{-1}}{k_1} = 2 \times 10^{-8}\ \text{M},$$

as $k_{-1} = K_s/k_1 = 0.2s^{-1}$. Comparison of the constants gives $k_1[S(0)] = 100\ s^{-1}$, $k_{-1} = 0.2\ s^{-1}$, $k_2 = 4.5\ s^{-1}$. Later, Chance (1957) adapted a modified approach to rate costant evaluation in which (7.85) is used together with the assumption

$$k_2 \gg k_{-1},$$

to give k_1 from

$$K_M = \frac{k_{-1} + k_2}{k_1} \approx \frac{k_2}{k_1},$$

and the known value of k_2. Complex information, however, is not required for this two-constant model. This technique was used more recently, for example, by Ishimura, Nozaki, and Hayaishi (1970). A typical oscillograph trace of the complex (*l*-tryptophan is substrate, tryptophan pyrrolase is enzyme) is shown in Figure 7.11. The ordinant is 1 percent transmission (at 415 mμ) per division, the abscissa is 1 msec per division; note the initial transient is not observed.

7.4.2 *Subsystem approach*

It is quite common for the parameters used in a model of a system to be obtained from reduced data on subsystems. The typical procedure followed by a combustion kineticist provides an example. First, all possible reaction species (reactants, products, and intermediates) are postulated together with all reactions among them; this would result in a large set of ODEs describing the mass action of each elementary reaction. Then rate constants for each individual reaction are found from the literature, either as estimated from experiment or theory; these constants can be incorrect by orders of magnitude, particularly if derived from theory or correlation. The constants are then inserted into the full nonlinear equation set which may then be solved numerically. If there is an interest in reducing the number of equations, some equations are omitted and new results obtained. If these new results agree with those when the full set was used, the simplification is considered justified.

Thus, in combustion kinetic modelling, experimental data is virtually never taken on all species of interest in a single set of experiments. This is because of the difficulties in measuring several species simultaneously with their greatly differing time scales and concentrations. Resulting mathematical models are often quite

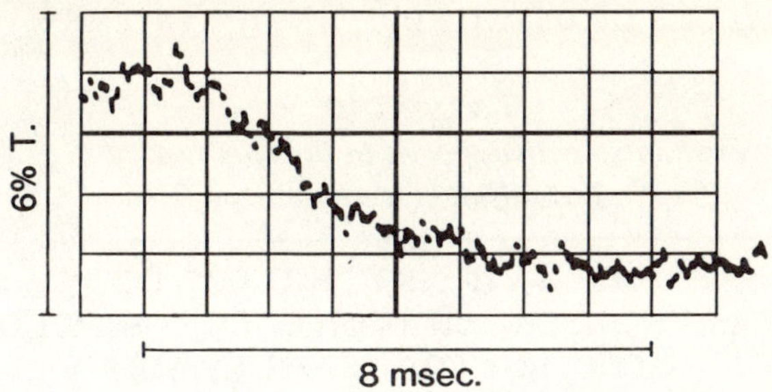

Fig. 7.11 Typical oscillograph trace of the changes in optical transmissions at 415 mμ of the complex l-tryptophan. (Ishimura, Nozaki, and Hayaishi, 1970, by permission).

poor in their predictive capabilities. Even if the postulated mechanism is slightly incorrect, there is no room for data-fitting the model to the particular experiment at hand, i.e. no adjustable parameters. To make matters worse, in combustion chemistry attention must be given to the active free radicals responsible for explosions or active in the flame heart. Furthermore, other trace species are often found to govern certain behavior, as for pollutant emissions.

Specifically, consider control of pollutant sulfur emissions from stationary combustion systems. For pollutant control, it is necessary to obtain a rather detailed understanding of the mechanism responsible for the formation of sulfur species in flames. To study the mechanisms experimentally, a hydrogen-sulfide-doped methane–oxygen combustion system might be used.

Table 7.10 lists all of the species likely to be present; the reactions in which these species participate are listed in Table 7.11, along with rough estimates of the individual rate constants as compiled by Corley (1978) from theory or experimentation on each individual reaction. The 23 species lead to 23 ODEs, not all of which are independent; the interdependence could be used as a check on numerical computations. The experimental technique used often involves the molecular beam which studies collisions on a molecular scale, i.e. one reaction at a time and very few molecules. The rate constants listed span sixteen orders of magnitude and could be incorrect by several orders of magnitude, particularly if obtained from theory.

7.4.3 *Estimation by least squares*

We intend here only to outline certain aspects of the estimation problem for use in later sections. We follow the treatment of Bard (1974) which we recommend for background and more detail. Consider the exact mathematical model

$$y = f(t, \boldsymbol{\theta}), \tag{7.86}$$

that describes a relation between the set of variables $\boldsymbol{y}$ and the variable t with parameter set $\boldsymbol{\theta}$. The m variables $\boldsymbol{y} = [y_1, y_2, \ldots, y_m]^{\mathrm{T}}$ are the dependent variables we wish to predict; the variable t is the independent variable time on the basis of which we wish to make a prediction. These are related by the set of l parameters $\boldsymbol{\theta} = [\theta_1, \theta_2, \ldots, \theta_l]^{\mathrm{T}}$.

We may think of t as input to a system modelled by (7.86), which operates on t to produce output $\boldsymbol{y}$, as shown in Figure 7.12. In order to obtain information concerning the system in the box, we perform a series of experiments (numbered by μ) in which the value of t is varied. For the case in which (7.86) represents

TABLE 7.10
Postulated active species in the combustion of hydrogen-sulfide-doped methane flame

potential slow:	CH_4, O_2, H_2S, SO_2, H_2O, CO_2, CO
potential fast:	CH_5, CH_2, CH, H, OH, O, CH_3O, H_2, CH_2O, CHO, HO_2, COS, CS_2, CS, HS, SO, S

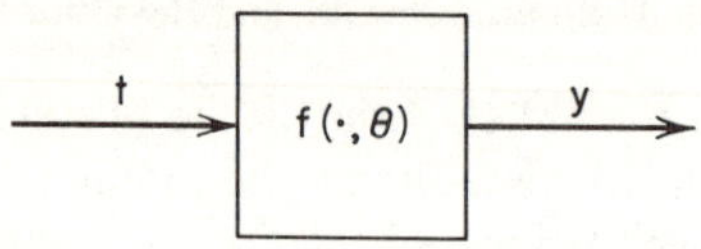

Fig. 7.12 Input–output system for parameter estimation.

the implicit solution to a dynamic model, that is to a differential equation, the experiments could merely constitute a number of consecutive $\boldsymbol{y}_\mu$ observations as time passes $\{t_\mu : \mu = 1, 2, \ldots, n\}$. The estimation problem consists of finding θ, defined as $\boldsymbol{\theta}^*$, that when used in (7.86) gives best (in some sense) agreement with the noisy experimental $\boldsymbol{y}$.

If we knew the true value of $\boldsymbol{\theta}$, defined as $\hat{\boldsymbol{\theta}}$, we could compute exactly the *error*, $\boldsymbol{\epsilon}_\mu$, in each experiment as

$$\boldsymbol{\epsilon}_\mu \equiv \boldsymbol{y}_\mu - \boldsymbol{f}(t_\mu, \hat{\boldsymbol{\theta}}), \tag{7.87}$$

where the subscript μ has been added to the variable vectors to denote their values for the μth experiment. Because $\hat{\boldsymbol{\theta}}$ is not known, however, we cannot determine $\boldsymbol{\epsilon}_\mu$; but we can replace $\hat{\boldsymbol{\theta}}$ by any other known value of $\boldsymbol{\theta}$ and compute the difference in (7.87). In particular, we can use our best estimate of $\hat{\boldsymbol{\theta}}$, namely $\boldsymbol{\theta}^*$:

$$\boldsymbol{e}_\mu \equiv \boldsymbol{y}_\mu - \boldsymbol{f}(t_\mu, \boldsymbol{\theta}^*), \tag{7.88}$$

where $\boldsymbol{e}_\mu$ is called the *residual*. An estimate of the experimental error in y is obtained if $\hat{\boldsymbol{\theta}}$ is approximated by $\boldsymbol{\theta}^*$; otherwise it indicates about how different $\boldsymbol{\theta}^*$ and $\hat{\boldsymbol{\theta}}$ are, within the limits of $\boldsymbol{y}$ measurement noise. We want to choose the $\boldsymbol{\theta}^*$ that minimizes some single numerical measure of the magnitude of the vector $\boldsymbol{e}_\mu = [e_{\mu 1}, e_{\mu 2}, \ldots, e_{\mu n}]^\mathrm{T}$.

One such measure is the sum of the squares of each element in the vector $\boldsymbol{e}_\mu$

$$\Phi(\boldsymbol{\theta}) = \sum_{a=1}^{m} \sum_{\mu=1}^{n} e_{\mu a}^2(\boldsymbol{\theta}), \tag{7.89}$$

where $\Phi(\boldsymbol{\theta})$ is the sum-of-squares objective function; the squared values of $e_{\mu a}$ assure cancellation of the terms if the summation does not occur. Now, one particular estimation problem can be stated precisely: find the $\boldsymbol{\theta}$ that minimizes $\Phi(\boldsymbol{\theta})$. This leads to solving

TABLE 7.11

Postulated reactions in the combustion of hydrogen-sulfide-doped methane

Reaction	Forward Rate* (moles/sec)
$CH_4 = CH_3 + H$	E −2
$CH_3O = CH_2O + H$	E 13
$CHO = CO + H$	E 12
$CO_2 = CO + O$	E --7
$H + O_2 = HO_2$	E 15
$OH + O = HO_2$	E 15
$H_2O = OH + H$	E −9
$CH_4 + CH_2 = CH_3 + CH_3$	E 11
$CH_4 + H = CH_3 + H_2$	E 10
$CH_4 + OH = CH_3 + H_2O$	E 12
$CH_4 + O = CH_3 + OH$	E 11
$CH_3 + OH = CH_2 + H_2O$	E 13
$CH_3 + O = CH_2O + H$	E 14
$CH_3 + O_2 = CH_3O + O$	E −7
$CH_2 + O = CHO + H$	E 13
$CH_2 + O_2 = CH_2O + O$	E 12
$CH + CO_2 = CHO + CO$	E 11
$CH + O = CO + H$	E 13
$CH + O_2 = CHO + O$	E 12
$CH_2O + OH = CHO + H_2O$	E 13
$CH_2O + O = CHO + OH$	E 13
$CHO + HO_2 = CH_2O + O_2$	E 14
$CHO + O = CO + OH$	E 13
$CO + OH = CO_2 + H$	E 10
$H + HO_2 = OH + OH$	E 14
$OH + OH = H_2O + O$	E 13
$OH + O + H + O_2$	E 13
$O + CS_2 = CS + SO$	E 13
$O + CS = CO + S$	E 14
$H_2S + H = SH + H_2$	E 12
$H_2S + OH = SH + H_2O$	E 12
$H_2S + O = OH + SH$	E 10
$SH + O = SO + H$	E 14
$SH + O_2 = SO + OH$	E 11
$SO + O_2 = SO_2 + O$	E 10

Table 7.11 (*cont.*)

Reaction	Forward Rate* (moles/sec)
$SO + OH = SO_2 + H$	E 14
$SO + O = SO_2$	E 17
$SH + H = H_2 + S$	E 13
$S + O_2 = SO + O$	E 12
$SH + SH = H_2S + S$	E 13
$H + SH = H_2 + S$	E 13
$S + H = SH$	E 15

* A very rough estimate from individual experiment or theory (Corley 1978).

$$\frac{\partial \Phi}{\partial \boldsymbol{\theta}} = 2 \sum_{a=1}^{m} \sum_{\mu=1}^{n} e_{\mu a} \frac{\partial e_{\mu a}}{\partial \boldsymbol{\theta}} = \boldsymbol{0}. \tag{7.90}$$

7.4.4 *Weighted least squares*

Because some observations may be more reliable (or more important in some sense, see §7.4.6) than others or because of scale differences among the $\boldsymbol{y}_\mu$, we might want to weight some terms in (7.89) more than others, as:

$$\Phi(\boldsymbol{\theta}) = \sum_{a=1}^{m} \sum_{\mu=1}^{n} b_{\mu a} e_{\mu a}^2(\boldsymbol{\theta}), \tag{7.91}$$

where $b_{\mu a}$ are the weights. We choose $b_{\mu a}$ small for the $y_{\mu a}$ which are relatively large, unreliable, or relatively unimportant; the converse holds for large $b_{\mu a}$. The more general weighting scheme which includes cross products can be found in Bard (1974).

7.4.5 *Multiple linear regression*

Consider the linear or linearized model equations

$$\boldsymbol{f}_\mu \equiv \boldsymbol{f}(t_\mu, \boldsymbol{\theta}) \approx \boldsymbol{B}_\mu(t_\mu)\boldsymbol{\theta}, \tag{7.92}$$

where $\boldsymbol{B}_\mu$ is a matrix function of t_μ, corresponding to the Jacobian of $\boldsymbol{f}$ with respect to $\boldsymbol{\theta}$,

$$\boldsymbol{B}_\mu(t_\mu) = \frac{\partial \boldsymbol{f}}{\partial \boldsymbol{\theta}}, \tag{7.93}$$

evaluated at a guessed or approximated value of θ. Adjoining the equations of (7.92) for all experiments, we have

$$\boldsymbol{F} = \boldsymbol{B}\boldsymbol{\theta},$$

where

$$\boldsymbol{F}^\top \equiv [\boldsymbol{f}_1^\top, \boldsymbol{f}_2^\top, \ldots, \boldsymbol{f}_n^\top], \text{ and } \boldsymbol{B}^\top \equiv [\boldsymbol{B}_1^\top(t_1), \boldsymbol{B}_2^\top(t_2), \ldots, \boldsymbol{B}_n^\top(t_n)].$$

The weighted least squares problem is to minimize

$$\Phi(\theta) = (Y - B\theta)^{\mathsf{T}} V^{-1}(Y - B\theta),$$

where $Y \equiv [y_1^{\mathsf{T}}, y_2^{\mathsf{T}}, \ldots, y_n^{\mathsf{T}}]^{\mathsf{T}}$, and the weighting matrix is V^{-1}. For the linear case and for the case of rich data with errors normally distributed, V^{-1} should correspond to the covariance matrix of the errors. It is well known that the optimal estimate is

$$\theta^* = (B^{\mathsf{T}} V^{-1} B)^{-1} B^{\mathsf{T}} V^{-1} Y \tag{7.94}$$

with a sampling variance

$$V_\theta = (B^{\mathsf{T}} V^{-1} B)^{-1}. \tag{7.95}$$

We want the sampling variance to be small and concentrated in the vicinity of the true values of the parameters. Note that many maximum likelihood problems with normal distributions have

$$V_\theta \approx H^{*-1},$$

where

$$H^* = \left[\frac{\partial^2 \Phi}{\partial \theta^2}\right]_{\theta=\theta^*} \tag{7.96}$$

so we can qualitatively see that we want the objective function to be very sensitive to the parameters in order to minimize V_θ.

7.4.6 *Special approaches*

We would now like to address the situation in which some parameters are much easier to estimate than others. The difficulties could arise because of critical data location (t_μ) or difficult-to-measure required observation variables (y). This problem could be the result of attempting to estimate parameters in a stiff differential model in which some parameters are primarily active only in a narrow time region and contribute only to certain variables.

For the linear or linearized model (7.92), the estimates are affected by two means, as shown in (7.95):

(i) The weight matrix V^{-1}. If this matrix corresponds to the inverse covariance matrix of errors, the more accurate the experiments, the longer will be measures of this matrix and the smaller V_θ. We would choose the weights so as to make certain experiments or certain variables more important, as shown in §7.4.4. These special experiments or variables could be selected on the basis of their accuracy or their sensitivity to certain parameters.

(ii) Data location (and richness) t_μ. Varying time is the only way to affect B for a given model. The choice of the a priori t_μ location is one of experimental design and could be based on minimizing any one of many measures of the V_θ size.

Effects (i) and (ii) are related in the sense that the decision as to the need of a data point in a certain location is similar to the decision of how to weight an existing data point that lies in a particular location.

One measure of the V_θ size was given by Box and Lucas (1959); they selected as their experimental design criterion minimizing $\det[\boldsymbol{V_\theta}]$. Their reason for this choice was that, if the estimates are approximately normally distributed they are proportional to the volume contained within any specific ellipsoidal probability contour for $\boldsymbol{\theta}$ about $\hat{\boldsymbol{\theta}}$ in parameter space.

Box and Lucas illustrate this with a serial chemical kinetics reaction sequence

$$\mathrm{A} \overset{\theta_1}{\rightarrow} \mathrm{B} \overset{\theta_2}{\rightarrow} \mathrm{C}.$$

The yield of intermediate product, y, is:

$$y = \frac{\theta_1}{\theta_1 - \theta_2} \left[\exp(-\theta_2 t) - \exp(-\theta_1 t)\right].$$

This reaches a maximum at

$$t_m = \frac{\ln(\theta_2/\theta_1)}{(\theta_2 - \theta_1)},$$

and exhibits a sharp initial transient if, for example, $k_1 \gg k_2$.

For a set of N experiments, the Jacobian

$$\boldsymbol{B}(t_\mu) = \frac{\partial \boldsymbol{y}}{\partial \boldsymbol{\theta}}$$

is a $2 \times N$ matrix. It is well known that the estimates obtained by minimizing the square of residuals have a covariance matrix which is approximated by $(\boldsymbol{B}^{\mathrm{T}}\boldsymbol{B})^{-1}\sigma^2$, where σ^2 is in general the (unknown) variance of the observations. In the Box and Lucas approach, the optimum times are selected so as to minimize the determinant $|(\boldsymbol{B}^{\mathrm{T}}\boldsymbol{B})^{-1}|$, which is the same as maximizing the determinant $|\boldsymbol{B}^{\mathrm{T}}\boldsymbol{B}|$. This criterion is essentially equivalent to minimizing the two-dimensional area of the confidence region for the parameters, if the experimental errors are approximately normally distributed and if the response relationship is approximately linear in the vicinity of the least-squares estimates. Since $\boldsymbol{y}$ is nonlinear, initial estimates of these parameters must be supplied.

For the particular case where the number of observations is equal to the number of parameters, we need only maximize the determinant $|\boldsymbol{B}|$. For our particular example, this involves maximizing a 2×2 determinant.

We performed this maximization using a nonlinear simplex algorithm by Nelder and Mead (1965). Table 7.12 shows the optimum data locations for several sets of rate constant values. As k_1 becomes much greater than k_2, the transients become sharper. In all cases, the optimum data-locations lie on either side of the maximum yield. Maximization of $|\boldsymbol{B}|$ seeks out times at which the elements of the Jacobian of $\boldsymbol{y}$ with respect to the parameters are large in absolute value (the sensitivity to the parameters is high at these points), but also as far as possible uncorrelated. We shall see this in the development of a sequential estimation procedure using singular-perturbation methods in §7.5.6.

From the entries in the last two columns of Table 7.12, we see that as the system becomes stiffer, the optimum data locations move away from the peak in either direction. Thus, not only does it become harder experimentally to track

TABLE 7.12
Optimum data locations

Case	k_1	k_2	Optimum locations t_1^*	t_2^*	t_1^*/t_{max}	t_2^*/t_{max}
1	2.0	1.0	0.3864	1.7126	0.5575	2.4708
2	10.0	1.0	0.0798	1.2983	0.3121	5.0754
3	15.0	1.0	0.0601	1.1012	0.3109	5.6939
4	25.0	1.0	0.0405	0.9045	0.3016	6.7453
5	35.0	1.0	0.0298	0.7983	0.2852	7.6323

earlier transients as stiffness increases because of the narrowing of the boundary-layer thickness, but there is an *additional* shortening of the relative location where data is most significant for the estimation, at least for this example.

For precisely estimating only a subset of parameters, Hunter, Hill, and Henson (1969) partition $\boldsymbol{\theta}$ into two parts:

$$\underset{(l \times 1)}{\boldsymbol{\theta}} = \begin{bmatrix} \boldsymbol{\theta}_1 \\ \\ \boldsymbol{\theta}_2 \end{bmatrix} \qquad \begin{matrix} (l_1 \times 1) \\ \\ (l_2 \times 1)\,. \end{matrix}$$

We wish to emphasize the estimation of $\boldsymbol{\theta}_1$. Suppose we also partition the $l \times l$ matrix $\boldsymbol{B}^{\mathrm{T}}\boldsymbol{B}$ so that

$$\boldsymbol{Y} = \boldsymbol{B}^{\mathrm{T}}\boldsymbol{B} = \begin{bmatrix} \boldsymbol{Y}_{11} & \boldsymbol{Y}_{12} \\ \\ \boldsymbol{Y}_{21} & \boldsymbol{Y}_{22} \end{bmatrix},$$

where $\boldsymbol{Y}_{11}$, $\boldsymbol{Y}_{12}$, $\boldsymbol{Y}_{21}$, and $\boldsymbol{Y}_{22}$ are $l_1 \times l_1$, $l_1 \times l_2$, $l_2 \times l_1$, and $l_2 \times l_2$ submatrices respectively. It can be shown (Hill and Hunter 1967) that in order to minimize the confidence region for $\boldsymbol{\theta}_1$ based on initial guess $\boldsymbol{\theta}(0)$, we must maximize the determinant

$$\Delta = |\boldsymbol{Y}_{11} - \boldsymbol{Y}_{12}\boldsymbol{Y}_{22}^{-1}\boldsymbol{Y}_{21}|$$

with respect to the independent variable, i.e. time.

Hunter et al. found for the

$$\mathrm{A} \xrightarrow{\theta_1} \mathrm{B} \xrightarrow{\theta_2} \mathrm{C}$$

reaction that an emphasis on the θ_1 rate constant moved optimal data location toward the origin. Thus again if we wish to weight sparse data in the boundary layer more heavily, this data would be best earlier in the boundary layer, compounding the experimental difficulty.

7.4.7 *Design-oriented estimation*

Before an estimation is performed, or experiments taken for this purpose, it is important to keep in mind the objective for the model. Most commonly, the parameters to be estimated correspond to physical or chemical constants; the constants seldom exist in nature but are devised from correlative models. These models themselves are not typically of interest but only as they help explain or predict some phenomenon. The predictions might be for the purpose of designing a process or a piece of equipment for which there is no data. If a stiff parameter does not contribute to this objective it does not need to be estimated; on the other hand, some parameters may have a very strong effect on, say, a prediction of interest, so that data requirements for it may take on additional importance.

Consider some single design parameter $\boldsymbol{P}$; let $\boldsymbol{P}^*$ be the predicted value of $\boldsymbol{P}$ using the parameter vector $\boldsymbol{\theta}^*$. Let the relation between $\boldsymbol{P}^*$ and $\boldsymbol{\theta}^*$ be:

$$\boldsymbol{P}^* = F(\boldsymbol{\theta}^*).$$

We assume that the prediction model is correct so that if we adjust the values of the parameters,

$$\boldsymbol{P} = \boldsymbol{F}(\boldsymbol{\theta}^* + \delta\boldsymbol{\theta}),$$

where

$$\boldsymbol{\theta}^* + \delta\boldsymbol{\theta} = \hat{\boldsymbol{\theta}}.$$

Thus

$$\boldsymbol{P} - \boldsymbol{P}^* = \boldsymbol{F}(\boldsymbol{\theta}^* + \delta\boldsymbol{\theta}) - \boldsymbol{F}(\boldsymbol{\theta}^*),$$

with expansion,

$$\boldsymbol{P} - \boldsymbol{P}^* \approx (\partial \boldsymbol{F}/\partial\boldsymbol{\theta})_{\boldsymbol{\theta}^*}\delta\boldsymbol{\theta}.$$

We can then speak of the covariance matrix for the predictor as

$$\begin{aligned} V_P &= \mathrm{E}[(\boldsymbol{P} - \boldsymbol{P}^*)(\boldsymbol{P} - \boldsymbol{P}^*)^{\mathsf{T}}] \\ &= (\partial F/\partial\boldsymbol{\theta})_{\boldsymbol{\theta}^*}\mathrm{E}[\delta\boldsymbol{\theta}\delta\boldsymbol{\theta}^{\mathsf{T}}](\partial F/\partial\boldsymbol{\theta})^{\mathsf{T}}_{\boldsymbol{\theta}^*}, \\ &= (\partial F/\partial\boldsymbol{\theta})_{\boldsymbol{\theta}^*} V_{\boldsymbol{\theta}}(\partial F/\partial\boldsymbol{\theta})_{\boldsymbol{\theta}^*}, \end{aligned}$$

where E[. . .] means 'the expected value of', $(\partial F/\partial\boldsymbol{\theta})_{\boldsymbol{\theta}^*}$ is a constant, and $V_{\boldsymbol{\theta}^*} \equiv \mathrm{E}[\delta\boldsymbol{\theta}\delta\boldsymbol{\theta}^{\mathsf{T}}]$.

Let us say our model is linear:

$$Y = B\boldsymbol{\theta}.$$

The weighted least-squares problem is then to minimize

$$\Phi(\boldsymbol{\theta}) = (Y - B\boldsymbol{\theta})^{\mathsf{T}}A(Y - B\boldsymbol{\theta}),$$

where $A \neq V^{-1}$. The normal equations lead to

$$\boldsymbol{\theta}^* = (B^{\mathsf{T}}AB)^{-1}B^{\mathsf{T}}AY,$$

or

$$\boldsymbol{\theta}^* \equiv RY,$$

where $\boldsymbol{R}$ is to be chosen to minimize some measure of V_P rather than $V_{\boldsymbol{\theta}}$. Note:

$$E[\boldsymbol{\theta}^*] = E[\boldsymbol{R}\boldsymbol{Y}] = \boldsymbol{R}E[Y] = \boldsymbol{R}\boldsymbol{B}\hat{\boldsymbol{\theta}},$$

hence for an unbiased estimate $\boldsymbol{RB} = \boldsymbol{I}$. Also,

$$\begin{aligned} \boldsymbol{V_\theta} &= E[(\boldsymbol{\theta}^* - \boldsymbol{\theta})(\boldsymbol{\theta}^* - \boldsymbol{\theta})^\mathsf{T}]. \\ &= \boldsymbol{R}\boldsymbol{V}\boldsymbol{R}^\mathsf{T}, \end{aligned}$$

so that we must minimize:

$$\boldsymbol{V_P} = (\partial \boldsymbol{F}/\partial \boldsymbol{\theta})\boldsymbol{R}\boldsymbol{V}\boldsymbol{R}^\mathsf{T}(\partial \boldsymbol{F}/\partial \boldsymbol{\theta})^\mathsf{T}.$$

The idea of a covariance matrix for prediction is mentioned by Bard (1974, p. 204). Peacock (1967) and Sealey (1970) consider the problem of the sensitivity in distillation-column design on the vapor–liquid–equilibrium parameters when a pinch point is present. Thus an arbitrary least-squares calculation of Henry's coefficient from a laboratory vapor-liquid-equilibrium bomb experiment may not be satisfactory for application to the distillation column design. This reasoning holds true for many other process design problems where the design is sensitive to certain operating conditions.

7.4.8 *Nonlinear confidence region*†

In addition to calculating the best parameter estimates $\boldsymbol{\theta}^*$, a nonlinear regression analysis also calculates the quality or statistical significance of these estimates. The usual way to measure the reliability of estimates is to state the $100(1 - \alpha')$ percent confidence limits on the estimates for some level of α', usually $\alpha' = 0.05$.

Unfortunately, the theory of confidence limits for parameter estimation involving nonlinear functions is not as well defined as the theory for the linear case. Consequently, three different measures of parameter reliability have been suggested.

The first is based on linear theory. That is, the nonlinear functions are assumed to behave as linear functions in the vicinity of the minimum $\boldsymbol{\theta}^*$. Then approximate or asymptotic standard deviations are calculated for each term as follows:

$$\mathrm{SD}_j = (x^{jj}s^2)^{1/2}, \qquad j = 1, \ldots, P, \tag{7.97}$$

where $s^2 = \mathrm{RSS}(\boldsymbol{\theta}^*)\ (N \quad P)$ is an estimate of the error variance under the assumption that the model is correct. $N - P$ is the number of degrees of freedom, and x^{jj} is the entry on the jth diagonal element of the matrix, $\boldsymbol{X}^{-1}$, where $\boldsymbol{X}$ is the $P \times P$ matrix of cross-products sums of $\boldsymbol{f}(\boldsymbol{\theta}^*, t_i)$ partial derivatives, evaluated at the minimum $\boldsymbol{\theta}^*$. Then the one-dimensional $100(1 - \alpha')$ percent confidence limits of the parameter θ_j^* are

$$\theta_j^* \pm t(\alpha', N-P)\mathrm{SD}_j, \tag{7.98}$$

where $t(\alpha, N-P)$ is Student's t distribution with $N-P$ degrees of freedom.

The one-dimensional confidence limit does not allow for correlations among the different parameters. Usually such correlations exist, particularly when more than one parameter is being estimated from one set of $(\boldsymbol{Y}_i, t_i)$ $i = 1, \ldots, N$ data.

† This material reprinted from "Pharmokinetic Studies in Evaluation of the Toxicological and Environmental Hazards of Chemicals," by Gehring, P. J., Watanabe, P. G., and Blau, G. E., in *Advances in Toxicology, Vol. I*, Mehlman, M. A., Shapiro, R. E., and H. Blumenthal, Eds., John Wiley (1975) with permission from Hemisphere Publishing Corporation.

The second measure of reliability takes this into account by calculating the $100(1-\alpha')$ percent ellipsoidal confidence region on $\boldsymbol{\theta}$ under the linear assumption from the relation.

$$(\boldsymbol{\theta}-\boldsymbol{\theta}^*)^{\mathrm{T}}X(\boldsymbol{\theta}-\boldsymbol{\theta}^*) \leqslant Ps^2\mathrm{F}(\alpha', P, N-P), \tag{7.99}$$

where $\mathrm{F}(\alpha', P, N-P)$ is the F distribution with P and $N-P$ degrees of freedom. This concept is best illustrated with an example.

Figure 7.13 presents sequential blood concentration-time data collected following the oral administration of a chemical (Metzler 1968). Assuming a one-compartment open model, the concentration of drug in the blood is given by

$$C(t)=\frac{D_0 k_a}{k_a-k_e}\;[\exp(-k_e t)-\exp(-k_a t)], \tag{7.100}$$

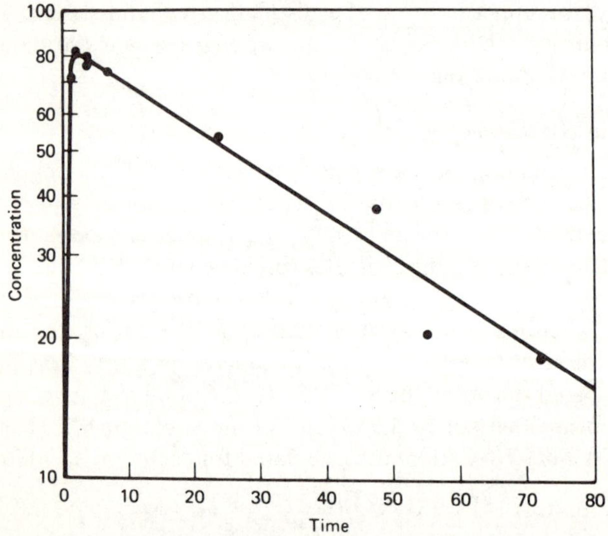

Fig. 7.13 Simulation of observed and calculated concentrations following oral administration of a chemical.

where D_0 is the dose absorbed divided by the apparent volume of distribution of the compartment (for simplicity we are assuming this volume is known). The optimal parameter estimates k_a^* and k_e^* characterizing the rates of absorption and elimination determined by a nonlinear regression analysis are $k_a^* = 1.89$ and $k_e^* = 0.0215$, with corresponding residual sum of squares $\mathrm{RSS}(\boldsymbol{\theta}^*) = 90.904$. A comparison between the observed data and these values calculated by the model is shown in Figure 7.13. Parameter space (k_a, k_e) is defined as all sets of points for which k_a and k_e are non-negative. Figure 7.13 is a plot of the 95 percent $(\alpha' = 0.05)$ and 66 percent $(\alpha' = 0.33)$ linearized confidence regions for k_a and k_e in parameter space. The 95 percent confidence region is the collection of points in parameter space defined by (7.99) such that in repeated runs or trials of the experiment, the true values of the parameters will lie within this confidence region in 95 percent of the cases. The standard deviations defined by (7.98) are simply the intersections of the coordinate axes with the ellipse defining the boundary of the 66 percent confidence region.

The problem of using standard deviations is apparent from Figure 7.14. If the parameters are substantially correlated, as is usually the case in fitting sums of experimental terms, the confidence ellipsoids are usually elongated. In practical terms this means that one may be able to change the parameter estimates substantially without changing the residual sum of squares to any extent. In other words, many different sets of parameters give equally good fits of the model to the data (see §7.5.4). Somewhat more conservative one-dimensional confidence limits than those of (7.98) are obtained by using support planes. These are simply tangent planes to ellipsoids parallel to the coordinate axes, as shown in Figure 7.14.

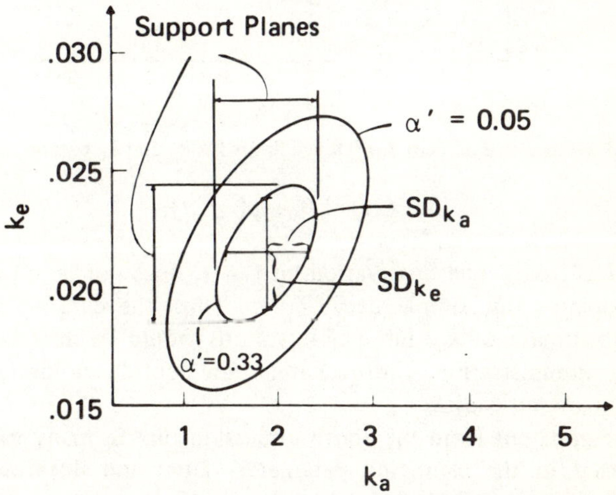

Fig. 7.14 Linear confidence regions for k_a and k_e parameter space.

The ellipsoidal confidence regions and associated one-dimensional confidence limits are a result of linearizing the functions about the minimizing point $\boldsymbol{\theta}^*$.

The third method for quantifying the reliability of the parameter estimate is to use a nonelliptical exact (that is, not linearizing) confidence contour defined by

$$\mathrm{RSS}(\boldsymbol{\theta}) = \mathrm{RSS}(\boldsymbol{\theta}^*)\left(1 + \frac{N}{N-P}\,\mathrm{F}(\alpha', P, N-P)\right). \qquad (7.101)$$

Although this contour is exact, the probability level is only approximately $(1-\alpha')$. This contour is difficult to compute and cannot be visualized when more than two parameters are involved. Despite these limitations, its importance can be demonstrated by our example. Figure 7.15 shows the 95 percent linearized confidence limits for k_a and k_e as well as the 95 percent exact confidence region plotted in parameter space. The exact confidence contour is shaped like a flask with an infinitely long neck parallel to the positive k_a axis. This means that at a level of confidence of 95 percent it is not possible to say whether k_a is less than infinity although it is greater than 1.0. This can be explained by examining the data and (7.100). As k_a increases,

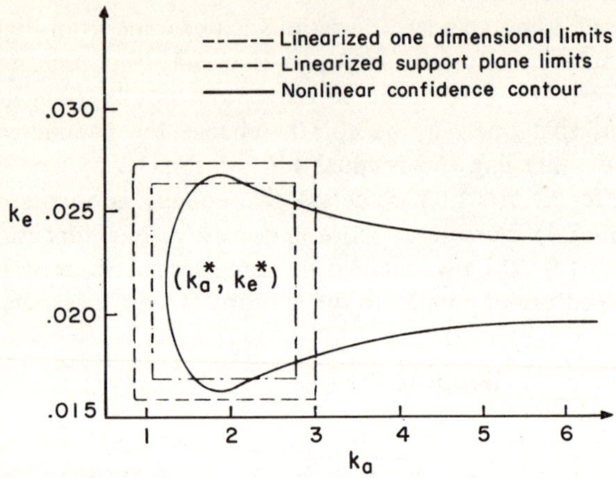

Fig. 7.15 Ninety-five percent confidence limits for k_a and k_e parameter space.

$$\lim_{k_a \to \infty} C(t) = D_0 \exp(-k_e t). \tag{7.102}$$

From Figure 7.13, only one observation, at $t = 1$, does not lie on the curve that closely approximates this simple decay curve. Thus, the data are inadequate to permit a good estimate of k_a; more observations would be necessary in the first hour following administration. Unfortunately, the confidence limits based on the linearized model are misleading.

It should be apparent from the above discussion that in many cases little faith should be placed in the estimated parameter values and linearized confidence regions. This is true even if the model is correct. If the correctness of the model is suspect, credibility in the resultant parameter values is further reduced. For example, fitting data that should be described by a two-compartment model with a one-compartment model cannot result in a meaningful parameter estimate even though the linearized region may be very small. These problems are even worse if the data points are not independent or are taken at improper intervals—that is, not designed properly. Therefore, it is imperative that nonlinear confidence regions be calculated as well as parameter estimates for alternative models, provided these models give almost equally good fits. Only in this manner is it possible to form some idea of the validity of the proposed model and parameter estimates from a purely statistical standpoint. Naturally, if extrastatistical arguments, for example, physiological ones, can be assimilated in discussion, the nature of the problem changes. Reilly and Blau (1974) discuss the various methods for calculating nonlinear confidence regions. A more extensive discussion of the problems of determining the precision of parameters may be found in the work of Westlake (1973) and Boxenbaum, Riegelmann, and Elshoff (1974).

7.4.9 *Stiff estimation packages*†

The earliest reported stiff estimation package was developed by Chandler, Hill, and Spivey (1972). It uses Treanor's algorithm (1966) to integrate numerically

† This section written by Venkateshwaran.

a stiff system. Treanor's method is adapted to the exponential nature of the solution of many stiff systems and fares much better than many explicit methods. However, it is not A-stable or stiffly stable which would make it truly well suited for stiff systems. This method requires smaller and smaller step sizes as the system becomes stiffer. A regression and experimental design package for nonlinear algebraic and ODE multi-response models (including stiff systems) is commerically available (Klaus and Rippin 1979). Clore and Chance (1978) modified a simulation package to allow parameter estimation in stiff systems.

We used a nonlinear weighted least-squares regression program from a statistical software package (BMDP 1981) for parameter estimation. This package is versatile, well designed and easy to use. The user can alter the convergence criteria and the maximum number of iterations permissible, transform variables and specify case weights. Plots of predicted and observed values of the dependent variables and their residuals can be readily obtained.

A set of subroutines is provided for use with dynamic models (Ralston 1979) that numerically integrates the model equation for a given set of parameters using a fourth-order Runge–Kutta method. We have replaced these subroutines with our own to successfully interface the regression program with LSODE (Hindmarsh 1981). The well organized and documented nature of the package permits this replacement in a relatively straightforward and simple manner. Several test cases in the report by Ralston et al. illustrate the handling of multiresponse situations.

7.4.10 *Comments on the singular-perturbation estimation approach*

Similar advantages in using the singular-perturbation theory for the numerical solution of stiff differential equations carry over to the stiff estimation problem. In fact, the advantages are magnified because the numerical solution must typically be solved many times during the course of an estimation procedure. Another advantage is that one is brought much closer to the nature of the estimation, whether or not the estimation is likely to succeed within given data location and accuracy. A natural process for increasing model detail sequentially is provided; for example, if the first-order approximations fail to produce estimates of additional parameters, the already found zeroth-order estimates are accepted.

The singular-perturbation approach offers yet another advantage over the use of the full stiff dynamic model. This advantage is related to stability with respect to initial guesses for the parameters and is discussed in §7.5.5.

Perturbation solution approximates an exact model equation. The error involved in this approximation could induce an error in the parameter estimates. Fortunately, for a stiff system, this induced error is smaller the stiffer the system.

7.5 Case study in stiff estimation: enzyme kinetics (Venkateshwaran and Aiken)

7.5.1 *Data*

The concentration-versus-time profile for the substrate and complex was simulated by solving (7.46) with LSODE. Simulated experimental error of various ranges was introduced into the data using a random number generator.

Table 7.13 lists the parameter values common to all the data sets. This table gives the exact values for the rate constants in (7.46) which we subsequently

TABLE 7.13
Parameter values common to all data sets simulated

Parameter	Exact value
k_1	1.0 E 08
k_{-1}	1.0 E 03
k_2	1.0 E 01
$s^*(0)$	1.0 E −3

attempt to estimate. Table 7.14a numbers the data sets and gives values of other experimental conditions and noise which may vary from run to run. The parameter ϵ is that used in (7.52), corresponding to initial concentration ratios between enzyme and substrate. Reference to data sets in subsequent subsections is consistent with this table. Table 7.14b gives the data location in time for data sets 1 (A, B, and C) and 2 (A, B, and C).

7.5.2 *Singular-perturbation method illustrated*

In this section we present numerical results that illustrate the use of singular perturbations as an alternative to the use of the full dynamic equations. The efficacy of this method is based on accuracy of estimates, quality of statistics, and computation time required.

The stiff enzyme system of interest involves rapidly changing transients in a boundary layer. Measurement of these transients presents problems peculiar to stiff systems (§7.4.1). It would be desirable to estimate all the parameters in the model using post boundary-layer data on the substrate alone. We have addressed this problem to the data requirement and data accuracy for reliable estimates of all parameters using the singular perturbation approach as well as the use of the full dynamic model.

The parameters appearing in a stiff dynamic model are grouped into stiff parameters and non-stiff parameters. The former are active mainly in the boundary layer and are difficult to estimate without data in the boundary layer while the latter are active primarily outside the boundary layer and are relatively easy to estimate there. In (7.46), k_1 and k_{-1} are the stiff parameters while k_2 is the non-stiff parameter. The initial guesses for the stiff parameters can adversely affect the quality of their estimates. This point, very critical for stiff estimation, is discussed in §7.5.5. A two-step sequence using the perturbation solution is developed which is not affected by initial guesses for the stiff parameters. Unless otherwise stated, the initial guesses for the parameters in all the runs is one order of magnitude below their exact values listed in Table 7.13.

It is theoretically possible to estimate all three rate constants using the uniform first-order perturbation approximation (7.61) and data on substrate concentrations alone, for some region of stiffness data-accuracy space. The last term on the

TABLE 7.14

(a)
Data sets simulated

Data Set No.	ϵ	Noise
1A	1.0 E –1	none
1B	1.0 E –1	up to 1%
1C	1.0 E –1	up to 5%
2A	1.0 E –3	none
2B	1.0 E –3	up to 5%
2C	1.0 E –3	up to 10%

(b)
Exact data locations

	Data set 1		Data set 2	
t^*	s^*	c^*	s^*	c^*
2.00 E –4	0.9837 E –5	0.1633 E –6	–	–
5.00 E –4	0.9689 E –5	0.3104 E 6	–	–
1.00 E –3	0.9756 E –5	0.4216 E –6	0.9996 E –5	0.4298 E –8
4.00 E –3	–	–	0.9995 E –5	0.4972 E –8
5.00 E –3	0.9494 E –5	0.4846 E –6	–	–
7.00 E –3	–	–	0.9995 E –5	0.4974 E –8
9.00 E –3	0.9475 E –5	0.4841 E –6	–	–
1.00 E –2	–	–	0.9995 E –5	0.4974 E –8
4.00 E –2	0.9329 E –5	0.4802 E –6	–	–
1.00 E –1	0.9051 E –5	0.4802 E –6	–	–
2.50 E –1	0.8375 E –5	0.4534 E –6	–	–
5.00 E –1	0.7316 E –5	0.4202 E –6	–	–
7.50 E –1	0.6343 E –5	0.3858 E –6	–	–
1.00 E +1	0.5457 E –5	0.3509 E –6	0.9945 E –5	0.4961 E –8
1.50 E +0	0.3948 E –5	0.2811 E –6	–	–
2.00 E +0	0.2774 E –5	0.2156 E –6	–	–
4.00 E +0	–	–	0.9797 E –5	0.4924 E –8
7.00 E +0	–	–	0.9650 E –5	0.4886 E –8
1.00 E +0	–	–	0.9504 E –5	0.4848 E –8
1.50 E +1	–	–	0.9263 E –5	0.4784 E –8
2.00 E +1	–	–	0.9026 E –5	0.4719 E –8
2.50 E +1	–	–	0.8792 E –5	0.4654 E –8
5.00 E +1	–	–	0.7670 E –5	0.4316 E –8
7.50 E +1	–	–	0.6635 E –5	0.3965 E –8
1.00 E +2	–	–	0.5689 E –5	0.3603 E –8
1.50 E +2	–	–	0.4071 E –5	0.2873 E –8
2.00 E +2	–	–	0.2812 E –5	0.2178 E –8

right-hand side of (7.61) is significant only in the boundary layer; this term is the one sensitive to the stiff parameters, so that early time data is essential to estimate all three parameters.

Table 7.15 summarizes the estimates using the full dynamic equation (7.61) and data on substrate concentration alone. Runs 1 and 2 in that table show that with exact substrate data alone and varying stiffness, it *is* possible to estimate all three parameters quite accurately. From run 3, we see that for $\epsilon = 0.1$ and up to 1 percent noise in the data, the estimates are reasonable. However, up to 5 percent noise in the data (run 4) gives poor estimates and standard deviations. At $\epsilon = 0.001$, even the slightest amount of noise proves fatal.

TABLE 7.15
Parameter estimates using (7.61) *and substrate concentration alone*

Run no.	Data set	Parameter	Estimate	σ
	2A	k_1	9.727 E+7	2.573 E+6
1	$\epsilon = 0.001$	k_{-1}	9.724 E+2	2.611 E+1
	exact	k_2	1.000 E+1	7.740 E+3
	1A	k_1	9.914 E+7	1.191 E+6
2	$\epsilon = 0.4$	k_{-1}	1.045 E+3	1.559 E+1
	exact	k_2	1.032 E+1	4.150 E−2
	1B	k_1	7.755 E+7	1.709 E+7
3	$\epsilon = 0.1$	k_{-1}	8.343 E+2	2.278 E+2
	1% noise	k_2	1.045 E+1	7.873 E−1
	1C	k_1	2.135 E+7	2.652 E+7
4	$\epsilon = 0.001$	k_{-1}	2.117 E+2	3.616 E+2
	5% noise	k_2	1.005 E+1	4.624 E+0

The disruptive effect of noise in the data has its roots in the nature of the inner solution. The zeroth-order inner solution for the substrate (Lin and Segel 1974) is

$$S_0(t, \epsilon) = 1.$$

At $\epsilon = 0.001$, for the rate constant value given by Table 7.15, changes in the substrate concentration in the boundary layer are of the order of only 0.005 percent. So any noise much greater than that is bound to be fatal. However, at $\epsilon = 0.1$, tolerance to noise in the data improves because changes in the substrate concentration are of the order of 0.3 percent. Therefore, reasonable estimates are obtained with up to 1 percent noise in the data. Up to 5 percent noise, however, leads to poor estimates even for this slight stiffness.

The complex concentration, unlike that of the substrate, changes rapidly in the boundary layer. So inclusion of the complex concentration also in the regression analysis reduces the sensitivity to noise in the data. The zeroth-order uniform approximation (7.60b) for the complex concentration (the stiff variable) is quite accurate even at $\epsilon = 0.1$ as seen from Figure 7.6b. We have used (7.60b) to predict the complex concentration.

Table 7.16 shows the dramatic improvement in the estimates obtained by the inclusion of the complex concentration. Run 5 shows that for $\epsilon = 0.1$ and up to 5 percent noise, the estimates are within 5 percent of the exact values. This contrasts sharply with run 4 in Table 7.15 where the estimates were about 50 percent in error for the same data set when only the substrate concentration was used. Run 6, however, shows that at $\epsilon = 0.001$, 5 percent noise is too much.

TABLE 7.16
Effect of inclusion of complex concentration

Run no.	Data set	Parameter	Estimate	σ
	2C	k_1	9.568 E+7	7.629 E+6
5	$\epsilon = 0.1$	k_{-1}	9.887 E+7	8.879 E+1
	5% noise	k_2	1.021 E+1	2.727 E−1
	2B	k_1	1.009 E+8	1.823 E+7
6	$\epsilon = 0.001$	k_{-1}	1.009 E+3	1.908 E+2
	5% noise	k_2	1.009 E+1	4.143 E−1

7.5.3 *Singular-perturbation versus dynamic-model approach*

The use of the full dynamic equations (7.46) to estimate the rate constants requires repeated integration of a stiff dynamic system. The use of singular-perturbation solutions (7.60b) and (7.61) requires the integration of only one non-stiff differential equation to calculate the zeroth-order term. So it is to be expected that the computation time for the latter approach would be greatly reduced.

For the same data set 2A, Table 7.17 compares the performance of the two approaches for identical initial guesses. The estimates obtained by using the perturbation equations (7.60b) and (7.61) together are as good as those obtained with the full dynamic equations (7.46). The residuals and standard deviations are almost identical. However, the computation time for the singular perturbation approach is about 33 percent of that required if the stiff dynamic model is used. For larger systems the saving in computation time would be even more significant.

Table 7.18 compares the two approaches using noisy data set 2C (up to 10 percent noise) and identical initial guesses. Again the estimates obtained with the singular-perturbation approach are as good as those obtained using the dynamic model, and the reduction in computation time is about 66 percent.

7.5.4 *Boundary layer information*

A characteristic of stiff systems is that the stiff variables change rapidly in the boundary layer while the non-stiff variables remain relatively constant there; outside

TABLE 7.17
Dynamic-model versus singular-perturbation approach

Data set		2A ($\epsilon = 0.001$, exact)	
Model equations		(7.46)	(7.60b), (7.61)
No. of iterations for convergence		20	20
Residuals at convergence		5.097 E –19	2.880 E –19
CPU time used (secs)		18.212	6.359
	k_1	9.930 E +7	9.927 E +7
Estimates	k_{-1}	9.929 E +2	9.983 E +2
	k_2	1.000 E +1	1.000 E +1
	k_1	1.064 E +4	9.304 E +3
σ	k_{-1}	1.117 E –1	9.746 E –2
	k_2	2.496 E –4	2.178 E –4

TABLE 7.18
Dynamic-model versus singular-perturbation approach

Data set		2C ($\epsilon = 0.001$, up to 10% noise)	
Model equations		(7.46)	(7.60b), (7.61)
No. of iterations per convergence		16	15
Residual at convergence		5.578 E –12	5.579 E –12
CPU time used (secs)		13.378	4.196
	k_1	1.026 E +8	1.026 E +8
Estimates	k_{-1}	1.025 E +3	1.026 E +3
	k_2	1.017 E +1	1.017 E +1
	k_1	3.758 E +7	3.767 E +7
σ	k_{-1}	3.932 E +2	3.941 E +2
	k_2	8.298 E –1	8.300 E –1

the boundary layer this is not true. It seems intuitively obvious that we would need points in the boundary layer to estimate accurately the stiff parameters. Intuition is well supported by our numerical experiments. The expression for the first-order outer singular-perturbation solution for the substrate, which we have not shown but which is given in Lin and Segel (1974; p. 315), does include all parameters. We have not been able to estimate them, however. There appears to be a fine balance among data accuracy, degree of stiffness (as it affects the perturbation-solution accuracy), and the robustness of the estimation procedure that determines estimability. A point in the boundary layer is necessary regardless of whether the dynamic-model or the singular-perturbation solution is used or whether exact or noisy data is used.

To illustrate, an attempt to estimate all three rate constants using the dynamic model and data set 2A ($\epsilon = 0.001$, no noise) with post-boundary-layer data only yielded results which are summarized in Table 7.19. The initial guesses are very close to the exact values, yet the estimates for the stiff parameters are poor. More important is the fact that the estimates are misleading because the statistics are reported to be good.

TABLE 7.19
Effect of using post-boundary-layer data

	Dynamic model	
	Data set 2A ($\epsilon = 0.001$, exact)	
k_1	Initial	0.90 E+8
	Estimate	5.31 E+7
	σ	2.98 E+5
k_{-1}	Initial	0.90 E+3
	Estimate	5.26 E+2
	σ	3.01 E+0
k_2	Initial	0.90 E+1
	Estimate	1.00 E+0
	σ	1.17 E−5

Thus the indication through this example is: a stiff system is sensitive to the stiff parameters only in the boundary layer. In practical terms, this means that outside the boundary layer one may be able to change the stiff parameters substantially without changing the residual sum-of-squares to any significant extent. Table 7.20 follows an attempt to estimate all three rate-constants using the perturbation equations (7.60b) and (7.61) and the same post-boundary-layer data-set 2A as Table 7.19. While the estimates of the stiff parameters change enormously, the residual sum-of-squares changes very little. The last entry in the table corresponds to the converged values of the parameters.

7.5.5 *Initial guesses for stiff systems*

As mentioned earlier, all the results from parameter estimation runs summarized in Table 7.15–7.20 were initiated with guesses for the parameters which were one order of magnitude lower than their exact values unless otherwise stated. Similar results were found for all the cases with even lower starting guesses. However, any attempt to initiate the iteration with guesses for the stiff parameters at an order of magnitude higher than their exact values either failed to converge in the

TABLE 7.20
Effect of using post-boundary-layer data

	Singular-perturbation model			
	Data set 2A ($\epsilon = 0.001$, exact)			
Iteration No.	RSS	k_1	k_{-1}	k_2
0	3.34266 E −11	1.0000 E +7	1.0000 E +2	1.0000 E +0
10	4.65494 E −20	5.6017 E +7	5.5604 E +2	1.0003 E +1
15	2.98500 E −20	4.5075 E +9	4.5538 E +4	1.0003 E +1
20	2.98037 E −20	5.3357 E +10	5.3915 E +5	1.0003 E +1
25	2.98027 E −20	1.2741 E +13	1.2874 E +8	1.0003 E +1
30	2.98007 E −20	2.3637 E +17	2.3880 E +12	1.0003 E +1
40	2.97487 E −20	1.3864 E +29	1.4010 E +14	1.0003 E +1

maximum number of iterations permitted or converged to misleading estimates as with the situation corresponding to post-boundary-layer data. The order of magnitude of the initial guess for the non-stiff rate constant k_2 creates no such problem. This is true for the singular-perturbation approach as well as for the dynamic model. For example, starting with the initial guesses 1.0 E 6 (two orders of magnitude too low), 4.0 E 1 (two orders of magnitude too low), and 1.0 E 2 (two orders of magnitude too high) for k_1, k_{-1}, and k_2, we were able to estimate the parameters to within 1 percent accuracy using the dynamic model as well as the singular-perturbation approach; the data set used was 2B ($\epsilon = 0.001$, up to 5 percent noise).

This behavior can be explained in the light of the previous section. A higher initial guess for the stiff parameters in practical terms means a stiffer system; consequently the boundary layer becomes narrower. So the initial guesses and the data set correspond to a stiffer system and post-boundary-layer data rather than to what is actually boundary layer and post-boundary-layer data in a less stiff system.

7.5.6 *Sequential estimation*

The singular-perturbation approach offers a basis for sequentially estimating parameters in a stiff system. The non-stiff parameters are first estimated using only the outer solution. Subsequently, the stiff parameters can be estimated using the uniform solution while keeping the nonstiff parameters fixed at their previously estimated values. After this stage we have very good estimates for all the parameters in the model. These estimates can now be refined by evaluating all the parameters at once in the model with each estimate previously determined. Such a sequential approach, estimating the non-stiff parameters first and subsequently the stiff parameters independently, does not suffer from the problems associated with initial guesses on the full problem which were described in §7.5.5.

TABLE 7.21
Michaelis–Menten parameter estimates

Parameter	Initial	Exact	Estimate	σ	Run
k_2	1.0+00	1.00+1	1.081+1	3.240+00	a
	1.0+02	1.00+1	1.081+1	3.242+00	b
K_M	1.0−02	1.01−5	1.135−5	5.292−06	a
	1.0+00	1.01−5	1.135−5	5.294−06	b

TABLE 7.22
Refined estimates

Parameter	k_1	k_{-1}	k_2
Initial	1.0090+08	1.0090+03	1.087+01
Estimate	1.0087+08	1.0088+03	1.007+01
σ	1.8224+07	1.9076+02	1.1427−01

We demonstrate this sequential estimation approach in the context of the enzyme kinetics model. The data set used was 2B in Table 7.14 ($\epsilon = 0.001$, up to 5 percent noise). The parameters k_2 and K_M in the Michaelis–Menten approximation (7.53) were first estimated using post boundary-layer data on the substrate concentration alone. The estimation proceeds uneventfully and provides very good estimates for the parameters even with rather arbitrary initial guesses. This is evident from the results summarized in Table 7.21. In the next stage of the estimation (7.60b) and (7.61) were used to predict the complex concentration and substrate concentration. The value of the non-stiff parameter k_2 was held constant at 10.1 while k_1 and k_{-1} were estimated (since $K_M = k_2/(k_{-1} + k_1)$, its estimate could be used to require only one independently guessed parameter—k_{-1} or k_1, but we did not do this). The estimates for k_1 and k_{-1} were identical to those given in Table 7.16 for the same data set (run 6). The estimates were identical regardless of whether the initial guesses for the stiff parameters k_1 and k_{-1} were higher or lower than their exact values. This clearly establishes, at least for this example, that the stiff parameters and non-stiff parameters are best estimated independently. The existing estimates for all the parameters provide excellent starting guesses for the refinement stage. In this stage all the parameters are estimated simultaneously with the available starting guesses. Table 7.22 summarizes the results.

8

THE FUTURE

8.1 The questionnaire

Eight 'stiff questions' were asked of each attendee following the Park City International Conference on Stiff Computation. A total 36 written responses were received, representing 92 percent of the attendence. The questions were intended to be important, general, and not of the type typically asked speakers during a conference presentation. The questions and quite candid responses follow:

QUESTION 1: *Should the term stiff be reserved for the initial-value ODE problem with sharp initial transients or could it be used in a wider sense (differential-algebraic, discontinuous, large sets, highly oscillatory, TPBVPs, PDE/BVPs with sharp transients, etc.)? Should another general term be defined? Should a number of terms, one for each specific type of difficulty, be defined? How important is terminology?*

Bickart: I tend to be against the coining of too many terms. I am quite content to have the term stiff mean, as I think it has come to mean, a dynamic system exhibiting both quickly and slowly changing modes in such a manner that a numerical solution process for its solution must be stable for all step sizes to facilitate efficiency in the solution process.

Byrne: The indicated definition is really appropriate *only* for linear, constant-coefficient, ordinary, initial-value problems (first-order equation or system of equations). Other definitions of stiffness are concerned with the size of the smallest time constant and the length of time of the phenomenon that causes the small time constant. For parabolic PDEs, stiffness can be attributed to a reactive source/sink term. I would not be uncomfortable with the use of stiffness for differential-algebraic systems (non-stiff integrator costs too much—a pragmatic definition of a stiff system in any case). Of course, terminology is used in the ODE community to identify a specific problem type and software requirements. I feel definition is important.

Cash: I think that the term stiff should only be used for initial value ODE problems at present. Certainly, it should not be used so widely for two-point boundary-value problems as it is. *When* we define more precisely what we mean by stiff I see no reason why the term could not be extended to certain problems in the other classes you mention (particularly to parabolic PDEs). I feel that the correct terminology is extremely important, particularly to guard against using stiff solvers as black boxes on the wrong sort of problem.

Cellier: . . . What do you need the terminology for? I am rather pragmatic. I feel that the term should help people in solving their problems. Therefore, if I use the term freely to denote almost any integration problem which is numerically difficult to solve, the only thing a potential user can know is the fact that he has a difficult problem to solve and may be forced to consult a specialist. This may be useful information, but it may then be more straightforward to call this term "difficult problem" rather than "stiff problem". If we restrict the term stiffness to mean only a particular class of difficult problems . . . this information is more useful, as the potential problem solver may then already know which integration technique to apply. Other classes of difficult problems (i.e. highly oscillatory problems) may also be defined, which again would tell the user which integration technique to apply. Ideally, an algorithm should be designed which can analyze a system description and find out for the user to which class of problem his system belongs. . .

Dove: I suspect that the term stiff will be used in a broader sense whether we like it or not. From the user's viewpoint this poses real risks that inappropriate software will be selected for particular problems. For this reason, there may be real advantages in developing a more descriptive terminology for the various types of difficulty. The objectives of such terminology would include alerting the user to the fact that there are various kinds of difficulty and helping to guide him or her towards appropriate software.

Enright: There are two characteristics of stiffness that are essential. The first is that the mathematical problem is well conditioned and stable in the sense that small perturbations in initial conditions or in the differential equation lead to small changes in the solution. The second characteristic is that when an attempt is made to integrate the problem with standard methods, a severe step-size restriction results from the constraints of numerical stability. Without both of these characteristics a problem should not be considered stiff. For example, highly oscillatory initial-value problems are not stiff even though it is frequently the case that the initial conditions contain no high-frequency components and the latter characteristic is present. The former characteristic is not present as a small change in the initial conditions or in the accuracy requirements could result in a very different solution and, in this case, the step size of standard methods would be determined by accuracy rather than numerical stability.

Finlayson: I believe that a two-point boundary-value problem, when treated as a PDE that gives a set of equations whose eigenvalues are widely separated (should be considered) a stiff problem. It might be useful to have a separate nomenclature, but such a problem qualifies as stiff according to criteria such as

$$\max_i |\lambda_i| / \min_i |\lambda_i|.$$

Perhaps there should be sub-classifications under stiff.

Gear: It should be reserved for systems . . . for which the stable time step of an explicit method is much smaller than that necessary to track the solution. This does not imply sharp initial transients, or vice versa. $y' = -10^6(y-1)$, $y(0) = 1$ is stiff but has no transient; $y' = -10^{-6}e^{-10^{-6}t} - e^{-t}$ has a transient, but is not stiff.

Hindmarsh: "Stiff" should be restricted to initial-value problems for ODEs, and possibly differential-algebraic systems, with one or more highly damped modes (and time constant small compared to the solution scale). Using "stiff" for other types of difficulties only causes confusion.

Krogh: The only other place where use of the term might be justified is for boundary-value problems with sharp transients, or which would have sharp transients with a small perturbation in the boundary conditions. For the other case, the words you use are just fine.

Mattheij and Söderlind: Today it is widely recognized that stiff IVPs call for methods with special stability properties, i.e. the stability set should contain most of the left half plane. It is also well known that methods aimed at nonstiff problems (e.g. implicit Adams methods) for stability reasons are very inefficient when applied to stiff problems, despite the fact that locally they are highly accurate. Thus stiffness is a conflict between stability and accuracy requirements that appears in certain problems. Needless to say, definitions of stiffness that cannot account for this simple observation are not conceptually correct. Unfortunately, the classical definitions of stiffness fail because they are based exclusively on mathematical properties of the differential equation and have thus caused severe misunderstandings. Obviously, a number of different terms, one for each specific type of difficulty, would be appropriate as long as the difficulties have different origins. Note, however, that the mentioned conflict between stability and accuracy requirements is very general indeed. Since it may well appear in differential-algebraic, discontinuous and highly oscillatory problems, there is no reason why the term stiffness should be not used in such cases. Even for certain classes of BVPs, a proper definition of stiffness caries over *ad verbatim*, although for general BVPs the notion of numerical stability seems far from being as well understood as in the IVP case. This is, however, probably due to the fact that numerical methods for BVPs in general are more complex than methods for IVPs. For this reason there is also a significant lag between the development of mathematical software for IVPs and BVPs.

O'Malley: The IVP people seem to have developed a special, nearly self-contained literature for ODEs which now includes a sequence of rapidly converging definitions of stiff (for their problems). The broader computing public, however, views stiff problems as tough ones whose analytical solutions involve intervals of rapid transition and whose numerical solutions can easily require (due to stability considerations) use of very small step sizes in regions where solutions are smooth. I prefer and would suggest using the term very loosely, recognizing that IVPs which are stiff "in the technical sense" are best understood. I would not, in any way, restrict or deter research on grounds of established terminology or turf. In particular, I would like to emphasize the point that stiff BVPs for ODEs and PDEs are in very critical need of analytical and numerical study.

Schiesser: The term stiff should be reserved for the initial-value problem in ODEs with sharp initial transients. The clearest definition of a stiff problem for me is one in which the product of the problem time scale and the largest eigenvalue is large (much greater than one). All of the other problems mentioned in the question are important, but I think they should somehow be described with other words and terminology so we know what problems we are discussing. Otherwise, as mentioned at the conference, any difficult problem is termed stiff.

Shampine: I believe it is important to develop a reasonable terminology. It impedes research and causes users of software to select the wrong tools if the same term is used for fundamentally different phenomena. For example, I believe that the widespread belief that algebraic-differential systems are virtually the same as differential systems has caused both the difficulties I mentioned. Also, I believe that highly oscillatory problems are fundamentally different from stiff problems.

Wendt: The term stiff should *not* be reserved for initial-value ODE problems. A wide sense, certainly, involving boundary-value problems, must be included. Stiffness arises from certain aspects of physical problems. Flames . . . should be considered stiff, especially since attempts to solve (these) equations led to identification of stiffness as a mathematical problem.

QUESTION 2: *Has there been an interaction in the past between the hardware developers and the software (and theory) developers? Has this interaction or would such an interaction have been an advantage to either group? Will this interaction be important in the future: why?*

Bickart: I believe such interaction has taken place quite naturally, but largely informally. I see no special need to force it as concerns the numerical solution of equations for dynamic systems. Auxiliary array processors as a common-place item are on the way. I do not see a special need for auxiliary processors designed to implement a specific solution process.

Byrne: IEEE floating point chip design used by INTEL and others is an example (Kahan) . . . the Gleneder Beach, Oregon, conferences sponsored by DOE, LLNL, and Los Alamos National Laboratories bring users, manufacturers, and code writers together to discuss issues related to large-scale computing—future and present. Of course it is important . . .

Cash: I believe that in the future there will be more interaction between hardware and software developers. About to appear is an important paper by Ower and Chershaw, who propose a new form of computer arithmetic to deal with error analysis. The interaction between software and theory developers is absolutely vital and of mutual benefit. On the one hand, we know that theory is unable to predict much of what happens in practice at present and software writers need to discover the way ahead by numerical experiment. . .

Cellier: Such an interaction has partly taken place—and is indeed useful. To state an example: array processors have been developed which are useful tools but incredibly difficult to program without appropriate interface to the software side. In numerical integration of differential equations, many subtasks can be formulated as vector operations (using the same operation on different data, that is: a SIMD structure). It may thus be advantageous to compute these vector operations in parallel. This is, however, only possible if

(1) the hardware specialist has designed his computer such that the parallel features can be addressed conveniently and efficiently, and
(2) the software designer makes use of these features by separating such vectorial operations in a modular way (e.g. by using LINPACK of which a version may be provided which makes use of the hardware vector operations).

This is only one possible application. In principle, this is a question of properly designing the interface. Some interfaces may have to do with other pieces of already existing or still-to-be generated software, others with pieces of hardware.

Enright: The evolution of vector machines is an example of a hardware development that can be exploited by developers of packages for stiff problems but I do not feel that our needs are particularly special. Hardware capable of performing standard linear algebra operations is all that would be required.

Gear: Excepting Kahan (University of California at Berkeley), no.

Gellinas: Such interaction is extremely important for the future in view of the overriding importance that will become attached to linear solvers for sparse matrix systems which will be encountered increasingly in ODE/PDE applications. Such interactions will be even more crucial as multiprocessor computers become widely used. Critical opportunities for great advances in applications will be lost if software development continues to languish or dies...

Hindmarsh: On the level of computer arithmetic, there has lately been much such interaction, and it benefits both groups greatly. On higher levels, notably in numerical PDEs, there has also been much mutually beneficial interaction, with improved parallel/pipeline computers and software for them as results. Stiff ODEs that arise from PDEs are a subset of that area.

Krogh: Little interaction in the past . . . IEEE Floating Point standard is an exception and that because of an exceptional man, W. Kahan.

Liniger: I do not think there has been much and I doubt whether in the ODE area this will become very useful because of the general-purpose nature of our algorithms.

Miranker: Yes (to all three questions). The development of hardware depends on problem types (data types and operations).

Pratt: My knowledge is limited to having to structure code so that a vector processor can be employed; for example, Gauss–Seidel iteration is better (faster) than

Jacobi iteration for solving systems of algebraic equations, but G–S is not vectorizable, while Jacobi iteration is, so that vectorized Jacobi is faster than G–S!

Schiesser: To the best of my knowledge, there has been no significant interaction between hardware developers and software developers with regard to stiff problems. Furthermore, I do not think there will be such interaction between the large mainframe manufacturers and people who work with stiff ODEs simply because the stiff ODE problem is not important enough commerically to warrant special computer designs. Perhaps the microcomputer manufacturers will find the problem important enough to justify special designs, but I am not sure small computers will ever be good for solving large, stiff ODE problems because they will be compute-bound; however, parallelism with small computers may be worth considering.

Shampine: I have not observed an interaction and I have no particular feelings about it. So far I have not noticed hardware designs calling for significantly different approaches. This could well change as parallelism is exploited, and then I would have to get involved. Even then I would expect to continue trying to make best use of the available hardware rather than influence its design.

QUESTION 3: *How automatic can and should a differential equation solver package be now and in the future? What advancements are seen in the near and far terms; how significant are they?*

Bickart: As automatic as possible. The solver built into the HP-34C is an example of what should be strived for in the more sophisticated software—almost complete tolerance of user ignorance. Only when software suspects it is not handling a problem properly should it demand that the user become less ignorant.

Brennan: Designers should be aware of the many hidden dangers in totally automating a differential equation solver—i.e. it may not be wise to implement software that requires little or no attention from the user. I currently maintain the integration package for a large trajectory simulation program (at the Aerospace Corporation). By a specialized input language, users simulate complex trajectories by specifying flight-dynamic and physical models with simple key words. Since the differential equations are automatically set up, the users may not even know how many differential equations are involved. The user is required to specify the type of integration method (Adams, R–K, or BDF; fixed or variable step) although this requirement could conceivably be automated, too. There are dangers in automating too many features since the typical user fully accepts whatever the computer spills out as the correct answer. . .

Byrne: How portable is the first issue. I can visualize an ODE solver that would: (1) select machine parameters, (2) select initial step, (3) advise progress, (4) switch methods (stiff, non-stiff), (5) pick step size dynamically, (6) report errors, (7) be graphics driven, (8) have graphics output, with zoom-in; 1–6 are here now, 7–8 are not quite here. How significant is it to look at gas kinetics with rates of 10^{10}? Pretty significant.

Cash: I think that an ODE solver should be as automatic as possible. There are many black-box users who are only concerned with obtaining a solution to their problem with a certain number of correct digits, and it is important that we should satisfy such people. There are numerous advances to be made in ODE software—Professor Gear's talk outlines many of these. Many of these advances will be very significant and in particular I feel control of global truncation error for stiff ODEs and automatic stiffness detection to be most pressing. Problems like the numerical solution of stochastic ODEs have not even got off the ground!

Cellier: The more that can be made automatic the better. Keep in mind that more and more problems are no longer solved by specialists in numerical mathematics, but rather by engineers out in the field. These people need as much help as possible to get their problems solved quickly and correctly. Please keep in mind the gullibility of these average users. They believe in a result as soon as no error message is printed out. Automation helps to make sure that correct answers are produced if at all. I know that this opinion is about 180° different from the average attitude of a mathematician. The reason is that a numerical mathematician is proud of solving a problem as efficiently as possible. Automation included in the algorithm obviously slows it down. However, for the average user, it is entirely immaterial if a solution of a problem requires 20 percent more computing time if in return he gets more robustness. Automation also helps in tuning.

Chen: From a user's point of view, the more automatic and flexible, the better. I had occasion to use LSODE for the first time recently. It is so well documented and easy to use that I was successful on the first attempt. . .

Churchill: Automatic solvers are of very little interest to me. They seldom are useful for real problems. Special purpose methods are almost always more effective. . .

Deuflhard: . . . Future advancements:

(1) large systems, possibly in the linear algebra part of stiff solvers,
(2) stiff/non-stiff automatic forth- and back-switching,
(3) differential-algebraic equations with stiff or non-stiff ODE part.

Devooght: My preference would be for a flexible approach where the user can decide what option to take. A code like LSODA which switches automatically between stiff and non-stiff methods is certainly useful but too much of a black box could be dangerous if one loses means to critically appraise the results.

Edsberg: According to my experience much of the software available in e.g. NAG, IMSL, and other libraries are hard to use for people from application areas. More robustness is needed and more needs to be done on the step-size regulation. Perhaps much software is designed for a too narrow problem class (from the mathematical point of view).

Enright: The package should match the user's requested accuracy to the accuracy of the numerical solution in a method-independent way. Although we do not have this yet, it should be possible in the near future. A package should also recognize when it is inappropriate and report this to a user. For example, a non-stiff code should report to a user when a problem is stiff.

Finlayson: . . . need low-accuracy ones for 2-D, 3-D elliptic PDEs. . .

Gear: Depends on applications. Very automatic for simple uses, but very complex problems or those solved repeatedly will require hand tailoring for efficiency.

Gellinas: Automatic solvers can and should compile and solve automatically broad classes of ODEs and PDEs by simple user inputs. These solvers should provide for dynamic scaling of incommensurate variables and for a choice of alternative error norms. The significance cannot be overstated because current practices of exerting large, frequently redundant efforts on dedicated computer programs for each new application are unacceptably wasteful of both computer and human resources.

Hindmarsh: Levels of automation are continually advancing with research and experience. Near-term advancements include automatic method selection and Jacobian analysis (we have a version of the first item). Far-term automation will allow a user to get solutions on specifying nothing but the problem itself, i.e. nothing about solution method, tolerances, etc. This will impact the casual users greatly by reducing their set-up time. But heavy users will still want to have their hands on their controls for optimal efficiency.

Hwang: A package of complete hands-off features sometimes may not be the best choice when computation speed is of prime concern. This is especially so when large-scale dynamic systems are being modelled. With a limited amount of user's effort, the solver may prove to be considerably more efficient. As long as user's interference is not annoying and tedious, a "semiautomatic" package is perfectly acceptable.

Krogh: I believe we have reached the point where the software should be completely automatic. But it should still allow the user to provide information about the characteristics of the problem if such information would be helpful. I believe there are still significant improvements in reliability to be made. Routines can be made more automatic and more flexible. Efficiency for general problems can probably be improved at least 30 percent, with much larger improvements for some problems.

Mattheij and Söderlind: There is always a risk that general-purpose solvers tend to be no-purpose solvers. Today there exist very sophisticated general-purpose codes, some of which suffer from being too integrated in the sense that generality prevents them from being efficient for certain applications. An advancement that may overcome this difficulty would be to develop highly modular ODE solvers so that suitable applications software can be obtained by the change of a driver subroutine or a reconfiguration of subroutines in the package. Another very important improvement over existing stiff solvers would be to have more built-in reliability checks, e.g. to monitor the stability of the differential equation (which may sometimes be quite different from that of the difference equation). In particular, robustness is important for codes used as black boxes.

O'Malley: One can always find bizarre problems and counterexamples to ordinary experience. Thus, the need for special-purpose codes will remain. The aim is to obtain a differential equation solver which can seldom be tricked. One needs tough problems to challenge all-purpose packages and to suggest improvements of them. Substantial confusion now occurs because naive users like me do not understand what the TOL setting means. Much progress has already been made, but user reluctance to blindly accept output from current codes is healthy and justified. There will certainly remain plenty of advantages for special codes for many kinds of restricted problems.

Petzold: With few exceptions a differential equation solver should be as automatic as possible. The reason for this is mainly to relieve the users of the solver from having to worry about how the solver works as opposed to how best to model their problems. There are now solvers which diagnose stiffness and even switch to the most efficient methods for a given problem, but there are other difficulties which are not diagnosed very well by the current codes. For example, it would be useful to know whether a stiff code is inefficient (or fails) because of an inaccurate Jacobian matrix or because of a poorly conditioned Jacobian or too stringent error tolerances or frequent discontinuities in some derivative of the solution. At present, the diagnostics which are provided by stiff solvers are nowhere near as good as for the non-stiff codes.

Seider and White III: . . . We envision a broadening in the scope of systems to be integrated, including systems with (1) discontinuities and (2) constraints on the variables. We also envision improvements that produce a smooth, differential result, as required by parameter identification algorithms that require derivatives without scatter. . .

Shampine: Wherever possible they should be made automatic. If additional information on the part of the user could affect decisions in an important way, the user should be able to influence the computation. I foresee codes which are rather successful regardless of the type—stiff or not. This will be a great convenience for users and may result in a net gain in efficiency. Perhaps more important is run-time monitoring of the computation to ascertain that the code being used was properly selected for the problem at hand and is being properly applied.

Watts: It is an important goal to strive for improving the robustness and capabilities of present-day ODE software. We should continue to provide packages which can automatically handle difficulties which arise frequently—of course, research and current state-of-the-art techniques may not be sufficiently well advanced to completely relieve the user of some burdens. Software which automatically detects and copes with stiffness, type-insensitive software, will be the most important new development. We will also see capabilities, highly oscillatory problems, differential–algebraic systems, and low-accuracy solution requirements (such as with real-time or tabular data computations).

Wendt: It seems to me that initial-value problems are being beaten to death. In many combustion-kinetics problems, the 1960s software of the NASA kinetics program is more effective than subsequent developments such as EPISODE. A moderate continuing effort to improve automatic solvers may be useful—but there are many more important problems involving stiffness in BVPs that still require significant effort.

QUESTION 4: *Is theory for the mathematics of computation complementing well software development needs of today and tomorrow? What are the major needs and what are the prospects for filling these?*

Bickart: To the first question, my answer is yes. To the second question, my answer is colored by my relationship to electronic-circuit analysis and design; we need an ever growing arsenal of tools for the analysis of large systems—where large is always getting larger in people's minds—of algebraic equations.

Byrne: Most theory supports linear ODEs or slowly varying nonlinear ODEs. We left them behind ten years ago. Need good work on error analysis, step selection, method switching. Prospects are slim; not many theoreticians have ever seen a real problem and universities will not support code work, in my opinion.

Cash: There has always been quite a wide gap between theory of computation and software development. It is important that they should come closer together but it is hard to see this in the near future.

Cellier: Principally, yes. I feel that a stronger interaction between numerical mathematicians and people from computer science (that is: software engineers) may be fruitful. However, even this interaction takes place more and more in that some people (like Alan Hindmarsh or Cleve Moler) are really both at the same time. . .

Churchill: Theory has never contributed signficiantly to numerical computation. Almost all advances have come from practitioners. I doubt that this will change.

Deuflhard: Theory is still open in extrapolation methods, implicit R–K methods, and stability. There is an unclear risk whether implicit R–K methods or stability will help software developments—at least unclear to me.

Gellinas: There is some complementation between theory and software in some areas where certain combinations of research talent have been assembled and when the right combinations of individuals decide to communicate closely and collaborate. There is a great need for supportive theory in PDE convergence and stability areas, particularly for nonlinear systems. Prospects for filling these needs in the short term are not too good: theorists frequently do not understand application needs and scientific practitioners frequently ignore theorists. Perhaps we can do better over the long term.

Hindmarsh: The majority of places where theory is done still have a low regard for software, and vice versa, but the exceptions are growing in number and impact. Increased communication between both types, and also between them and those doing applications, is needed. The biggest problem is for theoreticians to learn of and address the features of realistic application problems.

Krogh: When developing software, I have usually found that the theory people did not provide what was needed. What theory I needed, I needed to do for myself. There are of course a very few exceptions. Major needs (some hope of being filled) are:

(1) cheap and reliable global error estimation,
(2) good algorithms for boundary-value problems with very strong boundary layers,
(3) effective tests for precision requests that are too stringent for the arithmetic being used.

Liniger: The problem is that the theory usually limps behind the software development and by the time the theory points out new algorithms and strategies, the (software) adopted earlier based on heuristics has become so widely used that it becomes very hard to convince people that something new and perhaps better is around.

Mattheij and Söderlind: There is still a significant gap between theory and practice. In terms of stiff solvers, many codes contain "tuned" strategies and approaches that have not yet been justified theoretically. Thus the effects of variable steps on multistep integration procedures are not fully understood. Some of the major needs for theoretical improvement include: robustness, step-size strategies, global error estimation and automatic detection of stiffness. Large problems and differential–algebraic systems also need further investigation.

Miranker: Yes. A major need is to make the computer itself a scientific instrument. One computes but the operations which a computer executes are not known to the user . . . Numerical analysis is essentially independent of computers. This is a concealed but treacherous gap and moreover, it can now be closed.

O'Malley: I feel that current software for boundary-value problems is woefully inadequate. Adaptive mesh generation is essential. Nonlinear problems require much more practical experimentation and theory. In the singular perturbations context, careful numerical experimentation can motivate the necessary theory and vice versa.

Schiesser: My impression is that there is really very little useful exchange between the people developing theory and those developing codes. We enjoy being with each other and drinking coffee at the breaks between papers, but the useful exchange of information seems limited. I could not help but notice the number of theorists who left the room when an applications paper was presented, and visa versa.

Seider and White III: We need better theory for the stability analysis of nonlinear systems, for improving estimates of the global truncation error, and for adjusting the step-size and order of accuracy. The prospects of meeting these needs are excellent in the long term, but not promising in the immediate future.

Shampine: Many fundamental issues need attention. For example, variable-order codes are of the greatest importance but the theory describing them is, at best, fragmentary. I happen to be looking at the neglected areas of the effect of changing step size and of local error estimation—a lot needs to be done.

QUESTION 5: *What are a few of the most important application areas that demand far more improved stiff solution techniques? How do you interpret "most important"—by number of users, number of computer hours, the need for the solution (say for national defense)? Of these applications, how many are difficult because of the size or number of stiff sets to be solved?*

Bickart: Dynamic processes described by other than ordinary differential equations, such as Volterra integral equations, functional differential equations, etc. At the present these tend to take too much computer time for their solution.

Brennan: At the Aerospace Corporation, we want to solve a differential-algebraic system of nilpotency three, but there is no software or known algorithm which can solve such systems. By solving this system, we could generate feasible state-vectors during a trajectory from which a successful re-entry of the space shuttle could be initiated. We have modified the problem to reduce the nilpotency to two, so we can obtain a solution using a modification of GEAR's algorithm. Initializing the variables to be consistent is also a problem. Software capable of solving general DAE systems (and initializing the variables) would be very useful in all prescribed path control problems.

Byrne: (1) Chemical kinetics, (2) enzyme kinetics, (3) circuit design. Chemical kinetics is part of reactor design, process design, air pollution studies, reactive fluid flow—including combustion. I do not know how many users are concerned —several thousand, I would guess.

Cash: I feel that the numbers of users and the need for a solution are the most important reasons. However, if there are uses for an application area, I also feel that the contribution made to the theory is also of great importance when new techniques are developed. There is also a need to take a special look at large systems arising from PDEs where things like storage of the Jacobian, which is perfectly OK for small systems, may not be possible.

Cellier: In fact, most difficult "stiff" problems result from large systems. In particular, parabolic and/or hyperbolic PDEs translated into sets of ODEs by the method-of-lines approach are typical candidates of difficult problems. Parabolic PDEs result almost always in stiff sets of ODEs. Hyperbolic PDEs result in a different class in that some of the eigenvalues of the Jacobian tend to lie close to the imaginary axis and be complex. For this reason, the method-of-lines approach is known to be better applicable to parabolic PDEs than to hyperbolic PDEs. I suspect that this commonly stated sentence is not true. The method-of-lines could well be applied to hyperbolic problems if appropriate numerical integration techniques for this class of problems would be developed—and I do not see any reason why this should be impossible. Up to now, such algorithms do not exist or have at least not been implemented in a general-purpose package . . . Typical areas: reactor kinetics, helicopter simulation (stiffness of the blades), and many more. I do not think that the number of users is really the important issue. Number of computer hours is often a key point. However, for some of the problems there exist alternative solutions (e.g. a piece of hardware simulating the behavior in some sense). Some problems (like reactor kinetics) are key problems, because it is too dangerous to build these systems without knowing pretty well in advance what is going to happen.

Chua: These areas include the variable-step integration of differential–algebraic equations and the location of discontinuities. They are important because most large-scale practical problems can be modelled by a set of differential–algebraic equations containing frequent discontinuities.

Churchill: Split boundary-value problems, unstable behavior, multiple stationary states, recycle problems, partial differential equations, and integro-differential equations. Computational requirements and numbers of users. The inherent behavior, not the number of equations, is the primary source of difficulty.

Deuflhard: (1) Large chemical kinetics, (2) large circuit design, (3) inverse problems in chemistry and electronics, (4) many-component systems; most important: contribution to progress outside mathematics, mainly in sciences.

Devooght: My only experience with stiff problems concerns nuclear reactor kinetics. In this field the situation is rather good because *ad hoc* methods have been devised taking into account the specific form of the differential system. My secondary experience concerns the integration of the Liouville equation for the density matrix in quantum mechanics. In this case the eigenvalues are widespread but close to the imaginary axis and BDF methods are in this respect very deficient. The integration can be very time consuming and the use of a good program would help atomic physics a great deal.

Dew: There is still a long way to go before we get ODE integrations that can interface satisfactorily to PDE software based on the method of lines. The problem is selecting ODE integrations that can correctly match the stability of the PDE.

Edsberg: Chemical kinetics—in which case the kinetics is only part of a larger model including, e.g. flow, diffusion, thermal exchange, etc. I think model building with chemical kinetics is of great importance for environmental research, e.g. atmosphere chemistry, combustion, pollution processes, etc.

Enright: Solving large systems of loosely coupled stiff equations is an area where improvements are required. This is an important area since many problems arise in a variety of areas and they are now expensive to solve. It is very likely that the special structure they possess can be exploited.

Finlayson: . . . time-dependent finite-element codes—criterion is computer hours.

Gear: Highly structured large problems arising from PDEs, large systems from networks, VLSI modelling. (The basis is) computer time.

Gellinas: Oscillatory systems. I interpret "most important" in the context of breaking through important scientific problems which are presently unresolved because of inadequate numerical solution methods.

Hindmarsh:

(1) PDE-based problems, especially in 2-D and 3-D. Size is the main obstacle. Importance is due to number of problems, computer cost, and need.
(2) Systems with high-frequency oscillations where only an envelope is desired. Size usually not big. Importance due to numbers and cost.
(3) Systems with discontinuities (switches, etc.). Usually large size, but size is not the problem. Importance is due to numbers and cost.

Hwang: Computer simulation of complex chemical reaction processes (e.g. CH_4 oxidation), especially at the initial stage where sensitivity analysis of the mechanistic models with respect to initial reactant concentration and uncertainties in rate coefficients is desired. Typically, these models involve ~ 50 components and hundreds of parameters. The rate equations are very stiff (stiffness ratio could easily reach 10^8 or higher). Solutions to such problems are important for a number of reasons: energy conservation, environmental protection, new energy resources, etc.

Miranker: Chemical reactions, large circuits, satellites, weaponry.

O'Malley: Large-scale problems are important. Further, the use of aggregation methods and other hierarchical techniques which allow one to build increasingly complicated models of physical problems are essential. Chemical kinetics provide a very important applications area. One needs to figure out at what level detailed kinetics are necessary, as well as when experimental detail is needed. The meaning of various steady-state models also needs clarification, as well as their use in computational procedures.

Petzold: I think one of the most important application areas involves the solution of relatively small systems of equations with modest accuracy repeatedly and inexpensively. For example, when the operator-splitting technique is used for solving PDEs, there often results a system of ODEs which must be solved at any mesh point and at every time step. It is essential that this be done as quickly as possible, or else the computation becomes very expensive. This situation occurs in several diverse areas of application, including structural-deformation problem and combustion modelling.

Pratt: In my field, no one is solving finite-rate chemical kinetics in gas turbine, piston engine, or power generation furnace simulation models because GEARB is not fast enough. I do not (know of) anyone but myself (who) is attacking that problem seriously at the moment.

Schiesser: The recent developments in ODE integrators have had a major impact on the computer-based solution of PDEs, and I think this will continue. In fact, I will go out on a limb and say that the numerical method of lines has the potential for replacing most of the classical methods for solving PDEs. The importance of PDEs seems evident, and I think we will see a growing use of PDE applications in industry as well as in academic research.

Seider and White III: Improved integration methods are definitely needed for:

(1) Systems that pass through oscillatory regimes into the explosion mode, even for short periods of time; for example, in combustion and in limit cycles.
(2) Systems that involve combined integro-differential equations, as in heat transfer problems with radiation, conduction, and convection.
(3) Systems that are multidimensional; for example, 3-D natural convection.
(4) Systems involving determination of parameters to give a close fit to experimental data, especially when the integration results are very sensitive to small changes in the parameters or where discontinuities are encountered in the objective function with changes in parameters.

Shampine: I am concerned about the convenience and reliability of solution—all solutions would benefit from improvements in these areas. Comparison with the solution of nonstiff problems makes this clear although the latter would also benefit from research.

Thompson: Various simulation packages are starting to see use in the solution of really sophisticated problems, e.g. in the nuclear industry. As this trend continues, deficiencies in current methods (e.g. BDF) will become more apparent. As I see it, a crying need will surface for stiff solution techniques which better take into account the structure of large problems (e.g. sparsity, subsystems with different characteristics, partitioning, and global error control).

Watts: Chemical kinetics, elasticity/plasticity mechanical modelling, PDE modelling by ODEs, nuclear-reactor analysis. Each of the criteria—number of users, expense, and need for solution—defines a perfectly acceptable "most important" label. The PDE modelling by ODEs leads to the largest class of equations to be solved.

Wendt: Boundary-value problems based on occurrences in nature, computer hours and storage required, need; combustion, energy, catalysts, etc. Size is a most important factor in the above.

QUESTION 6: *Can you cite examples where the stiffness is a necessary part of the mode i.e. applications where the nature of the stiffness directly influences the objectives of the modeller?*

Byrne: Nonlinear models in chemical kinetics. If it is not stiff, it is not stable . . .

Cash: No!

Cellier: There exist some problems in the literature consisting of a fast and slow subsystem (especially in control literature). Although only the slow system response is important for the user, it is influenced by the fast modes as well. Some of the slow frequencies disappear if the fast subsystem is left out. . .

Chen: . . . A reactor with vastly different time scales linked in series is an example where stiffness is an inherent property of the entire system.

Chua: No.

Deuflhard: Parameter identification in chemical kinetics or electronics does require measurements in both the transient and the stationary phase of the process. Automatic simulation of mixed systems with fast and slow processes. Method of lines for parabolic PDEs.

Dove: There is large class of problems for which the ODEs of chemical kinetics must be integrated, often in conjunction with ODEs or PDEs representing, for example, transport processes in gases or gas–surface interactions. Examples abound in the chemical industry. In many cases, these systems of equations must be simultaneously solved at a very large number of points over a mesh in 3-D space. The main factor in the often very high cost of solving such problems is that of solving the large stiff systems of chemical kinetic equations. Thus there is great interest in simplifying such systems to their bare essentials, and in reducing the cost in other ways. Sensitivity analysis plays a very important role in such problems. There is, in my view, a potentially very large role in such areas as computer design—combustion, for example—of automobile engines.

Edsberg: Chemical kinetics. The use of e.g. the steady-state approximation in order to eliminate stiffness may be disastrous for following the correct solution trajectory, see e.g. TRITANA-8005, example 3, oxidation of propane, where iterations converge to wrong solution trajectory if the reduced problem is solved.

Enright: Stiffness is inherent in any model where transient behavior can affect the overall system behavior.

Gear: Some problems, e.g. relaxation oscillators, are limits of stiff systems as stiffness approaches infinity. Without knowledge of (the) stiff form, the behavior of the limit problem can not be determined.

Gellinas: Yes. Extreme, nonequilibrium behavior in most branches of the physical and chemical sciences (e.g. plasma physics, radiative systems, interactive fluids, combustion, and most "strongly driven" systems) has frequently remained out of the reach of modellers really understanding essential physical/chemical processes because sufficiently accurate, highly resolved numerical solutions of the defining PDEs/ODEs could not be attained.

Hindmarsh: Most chemical kinetics problems are necessarily stiff if the model is to be accurate. Steady-state assumptions are of unreliable accuracy and efficiency. We are at a point in the quality of stiff solvers where the modeller should *not* be influenced by the presence or absence of stiffness in forming the model—only by its accuracy. I have been solving some oil-shale particle models where *not* using ssa gave solutions over 400 times faster than the best one could do with the assumptions, and also solved the harder cases.

Liniger: Regions of rapid transitions in semiconductor equations, models in hyperbolic PDEs, chemical kinetics.

O'Malley: Power-system modellers cite an example where traditional reduced-order models lead to mechanical failure, but where including highly oscillatory transients circumvent the failure and explain the earlier difficulty.

Petzold: This is certainly the case for combustion modelling, and the modelling of control systems.

Pratt: If stiffness is a necessary part of the model—that is, if you wish to resolve the solution on the small time scales—then the problem is only wasted effort on computing the slow modes . . . is the problem then properly called "stiff"?

Schiesser: As the grid spacing in PDE solutions becomes smaller, the classical theory indicates that the resulting ODEs should be stiffer. However, we seem to have some evidence that with adaptive regridding, and with the use of higher-order approximations for the spatial derivatives, this may not be the case. I think this is still an unanswered question that warrants some research. Also, the effect of nonlinearities on the apparent stiffness of ODEs is an important, unanswered question. We find computer run times to be extremely sensitive to small variations in nonlinearities.

Seider and White III: Most systems stiffen as they stabilize or approach a steady state. In many cases the model cannot be easily simplified as stiffness sets in, and stiffness is a necessary part of the model. In some cases, it is possible to simplify the model as the system stiffens, without loss of accuracy, to reduce $|\lambda|_{max}$ and stiffness (see, for example, the fluidized-bed reactor model in our paper). Limit cycles, with regular oscillations over large time segments, are examples of systems that cannot be easily simplified when stiffness sets in. These systems stiffen periodically and model reduction would not apply at all times following the first onset of stiffness. In these cases, it seems impractical to periodically alter between models.

Thompson: Examples abound. Those with which I am most familiar involve simulation of nuclear steam systems in which the treatment of kinetics is often of paramount importance.

Wendt: Detailed combustion kinetics. Remove the factors causing stiffness and you have thrown out the baby with the bathwater.

QUESTION 7: *Do you see an advantage to considering the interactions of model identification, parameter estimation, and solution? Be as specific as you can and please cite any literature in this area of which you are aware.*

Byrne: Suppose a reaction mechanism for a system is postulated for a general temperature range. Laboratory and literature data can then be used to develop a preliminary set of parameters for a generalized Arrhenius reaction form. The kinetics model can be solved (stiff ODEs) to obtain a set of computed data. The computed data is then matched against the observed data for one or more final (product) species. The parameters for the reactions can then be adjusted by an algorithm to get new rates. The system can be solved again, etc. until the computed and observed data match up. If the match-up is awful, then so might be the proposed model. I suppose the variance-covariance matrix could provide some idea of goodness. Usually not much helps unless the fit is right on.

Cellier: It is indeed realized meanwhile that a pure simulation tool is not that useful, as often no appropriate models are known (particularly in soft sciences—so called "ill-defined systems"). Only very recently some attempts have been made to automate the model building and model validation business. The research is here, however still far from an answer. A good review may be: Vansteenkiste, G. C., and J. Spriet, "Computer Assisted Modelling of Ill-defined Systems" in *Progress in Modelling and Simulation*, F. E. Cellier, ed., Academic Press (1982). In fact, the first half of this book deals with modern issues of computer-assistance in modelling. The second half deals with modern issues of computer simulation.

Churchill: These interactions should be considered. However, parameter estimation without consideration of the uncertainty of the input is idle. See Churchill, "The Interpretation and Use of Rate Data," Hemisphere, Washington, D.C. (1979).

Deuflhard: Numerical parameter estimation techniques should also be able to monitor the model in terms of sensitivities of the model parameters. (See forthcoming proceedings volume: P. Deuflhard, E. Hairer (eds): *Numerical Treatment of Inverse Problems in Differential and Integral Equations.*

Dove: The use of sensitivity analysis, which deals with the interaction of the form of the model, the magnitudes of the parameters, and the effects of these factors on the solution, is absolutely crucial to problems in which chemical kinetics are important. There is a real need for professionally developed software packages in this area. We have had to write our own, so far! Leads to much of the literature in this area can be located under the names of H. Rabitz and K. E. Shuler. The topic is also important for the "inverse problem" where one is given data on $y(t)$ and needs to find information about the underlying ODEs. For example, given experimental data on a chemical kinetics problem, what do those data tell us about the form of the model and about its parameters? There is already some software to deal with this, e.g. Curtis's CHEKMAT, but in my view a lot more work is needed. For example, very often the data available do not enable a unique determination of all of the parameters—or even of any of them. Nevertheless, one would like to have a way of deciding, and expressing, the constraints which those data place on the possible values of the parameters.

Edsberg: Of course it is important to work interactively with a model, not only to solve one initial-value problem but to have also a qualitative idea of how the model behaves with respect to structure, parameters, initial values, etc. I (almost) agree with Aris in his book "Introduction to the Analysis of Chemical Reactors," Prentice Hall, 1965, pp. 325: "It cannot be too strongly emphasized that it is folly of the first magnitude to approach the computer without first having as good a feel as possible for the structure of the problem."

Finlayson: This is a very important area in engineering. Rarely do we know the parameters well enough to justify excellent mathematics solutions, but we would like to be close.

Hindmarsh: I have no experience with model/parameter identification problems. But at present, some interaction seems wise. Solution accuracy requests should be as loose as possible when large model adjustments are being made. Methods and codes for calculating sensitivities with respect to parameters are useful here, and this is an active area.

Liniger: Yes, I think these questions are very important in simulation, e.g. in chemical kinetics it is very hard to evaluate a model, even qualitatively, if one has little or no information about the rate constants. The solution may lie in a completely different ball park if those constants are off.

O'Malley: Certainly. Not enough talented mathematicians and engineers have faced the tough problem of parameter estimation. This intermediate link is critical to any such effort.

Petzold: I think there is definitely an advantage to considering these interactions. For example, parameter estimation often involves the solution of several problems which differ from each other very little. This structure can sometimes be taken advantage of to speed the whole process. Parameter-estimation procedures could also benefit from having solvers which produce solutions that have a smooth (or nearly so) behavior with respect to changes in initial conditions and other parameters.

Schiesser: Parameter estimation and model identification of PDE systems will become increasingly important in industrial applications and will serve as an important link between theory and experiments. Efficient methods for the repetitive solution of large sets of ODEs should therefore be pursued.

Seider and White III: Yes, George Byrne answered my question (Seider) by stating that discontinuities in the results of the numerical integrator cause problems for nonlinear programming algorithms that approximate a Hessian matrix; for example, using Powell's method. This problem needs to be resolved. However, in addition, methods to avoid a complete integration for each set of parameters should be examined.

Shampine: My own research has not been directed at such issues although I think about them from time to time. I do not see how one can avoid considering the interactions.

Wendt: Yes—again with detailed kinetics. Sensitivity analysis and parameter estimation are most important.

QUESTION 8: *What will be the major advancements in simulation of stiff systems in the near and far terms? Are these of fundamental primary interest or are they secondary improvements?*

Bickart: In the analysis of very, very large sets of equations—really, the systems they describe—mixed mode analysis and the concept of latency have come to the fore as a means of reducing the problem size to manageable proportions.

Byrne: Automatic stiffness/non-stiffness detection and method switching with dynamic memory allocation—prototypes available now will become standard.

Cellier: . . . Certainly, parallel processing may bring some key impulses into the game. Some problems may become solvable which are currently simply too expensive to solve. Whether the algorithms themselves may still improve very much, I do not know . . . Certainly, the software will—and has to—improve. In particular, the modularity of code has to be improved, and the interfaces have to be better defined (e.g. with respect to the data involved—large systems mostly require a large amount of data to be entered; for this purpose, an appropriate interface to a data management system should be defined.

Churchill: I expect new special-purpose algorithms will be developed. They may in some cases involve general principles.

Edsberg: Easier-to-use software for nonexperts—primary interest; automatic scaling of ODEs—secondary interest; sensitivity analysis.

Enright: I see three distinct developments which are of fundamental importance:

(1) the development of more effective techniques which can exploit special structure such as only a few transients or weak coupling between subsystems;
(2) the acceptance by software developers of a uniform interpretation of accuracy. This will enable users to view packages more as black boxes;
(3) the acceptance and wide distribution and publication of new improved packages as they become available. (This can best be accomplished through meetings such as this one.)

Gear: Better techniques for handling the linear algebra in very large problems. Techniques for decoupling subsystems.

Gellinas: Those associated with solving extremely stiff PDEs with newly emerging adaptive mesh techniques—both near and far term. We are now seeing fundamental advances in this area which will require extremely robust ODE (and linear system) solvers.

Hindmarsh:

(1) Better automation.
(2) Larger computer capacity and higher speeds will help a lot even with present methods.
(3) Advances in sparse nonlinear algebraic system methods, with and without consideration of special machine architecture, will contribute greatly.
(4) Better interfaces with users, via discipline-dependent simulation software, will be of secondary importance.

Krogh: Better reliability and efficiency. Primary verus secondary is hard to answer, but I would probably choose the latter. But the improvements to be had are well worth doing.

Mattheij and Söderlind: . . . the gap between theory and software will decrease. Such advancements are important although they may not have a large influence on the software design. The importance is mainly from the robustness point of view—a more rigorous theoretical framework is needed to sustain black-box use of the software, especially in nonlinear applications.

Miranker: . . . ill-conditioned problems generally (PDEs, integral equations, turning points, highly oscillatory) . . .

O'Malley: I believe extension of related research to partial differential equations and boundary-value problems will require fundamental new concepts.

Petzold: Some of the areas where we can expect (or hope) to see major advancements in the near future are:

(1) solution of differential–algebraic systems, and more generally, constrained differential systems;
(2) solution of highly oscillatory ODEs;
(3) global error estimation for stiff ODEs;
(4) inexpensive solution of small systems which must be solved repeatedly;
(5) better diagnostic messages for stiff solvers.

(Of course, not all of these problems are stiff, depending upon your definition.) These problems are of primary interest. It would require quite a few pages to list problems of secondary importance.

Pratt: Special-purpose or hardwired computers, and rewriting or reselecting algorithms to take advantage of computer architecture; in short, doing for stiff systems what has been done for Fourier transforms in signal-processing hardware!

Schiesser: . . . the development of integrators which can handle a changing number of ODEs during the solution to accommodate adaptive regridding in the solution of PDEs.

Thompson: The incorporation of stiff system methodology into widely available simulation packages, adequate implementation of root-finding techniques, valid automatic stiffness detection and method switching, and global error estimation, is of fundamental and primary importance. Potential advances in partitioning and subsystem identification are at least of secondary importance. Also of crucial importance are the needed improvements in documentation and the development of reliable, user-friendly software.

Wendt: I think that current technology on rapid Poisson-solvers and techniques for handling large systems involving huge matrices will evolve from the classified literature and become very useful for stiff problems. I do not know much about them, but I hear—through the grapevine—that very powerful computational techniques have been developed by DOD and could be very useful in many other applications, such as stiff boundary-value problems. These would involve radical changes in how stiff problems are attacked and solved.

8.1.1 *Comments on questionnaire* (*Aiken*)

Table 8.1 contains the number of responses to each question; generally, the most answered questions were answered in greatest detail. The most popular—and most emotional—questions were numbers 1 and 3. Opinions on question 1, concerning use of the word "stiff" fell basically into two categories: practitioners who want a free hand with the usage, and developers/theorists who want a precise, restricted usage. The almost universal answer to question 3 was that automatic packages should be automatic to the limit, with options for control; significant advances (in addition to ultimate robustness) are seen for special problem types.

Questions 2, 4, 5, and 8 drew about the same number of responses. From question 2, we would conclude that there has not been much interaction between

TABLE 8.1
Number of responses to stiff questions

Question No.	No. of responses
1	35
2	28
3	36
4	27
5	30
6	21
7	18
8	27

hardware people and software people, but the usefulness of such interaction is definitely recognized. It is clearly realized (question 4) that theory lags well behind practice in stiff computation (unlike in many physical sciences). Question 5 resulted in a variety of most important application areas being mentioned, but particularly those involving partial differential equations and large numbers of equations; this was echoed in answer to question 8 on what the major advances will be.

Questions 6 and 7 were not answered nearly as often as the others, and this could be significant. Generally, few respondees could cite cases where stiffness was necessary to the model (question 6). The idea of "system analysis" for stiff computation (§2.1) was definitely regarded as important, as was the parameter estimation problem, but these notions were considered new (question 7).

A ninth question was asked concerning the usefulness of the Park City meeting. The consensus was that it was a very good idea to bring together individuals with widely differing backgrounds who share a common problem area. This would support the response sparsity to questions 6 and 7.

8.2 Panel discussion at International Conference on Stiff Computation, Park City, Utah†

Shampine: We have heard during these talks so far quite a lot of computation described as stiff . . . anyone have opinions?

Gear: Yes . . . the purpose of having a classification and giving names to classes is to group them into moderate size groups; if you say a problem belongs to class x, then you can say it can be solved by certain methods . . . it is madness to classify BVPs that require totally different techniques from IVPs as being stiff . . . BVPs are unstable as well as stable . . . in both directions, while stiff problems are strongly stable in one direction.

† Moderator, Shampine; panelists: Byrne, Dahlquist, Gear, Hindmarsh.

Miranker: This guy was walking down a street in New York and passed this store in which he saw a sign that said "Stiff Differential Equation Solver." He quickly ran into the store and said "Can you solve my stiff differential equation?" The storekeeper said, 'I don't know what you are talking about, I make signs!"

Shampine: Many people think that our problems will be mopped up if we just build a bigger computer . . . or many more little tiny computers . . . what implications (from next generation computers) does this have on software and algorithmic developments?

Hindmarsh: There are going to be some *minor* changes as a result of new architecture, but I do not think they are going to be profound as far as the stiff computation problem is concerned. We are going to see higher speeds, certainly, and we are going to see far greater use of parallelism, and we will have to try to make use of that new architectural environment, but the obstacle we refer to as stiffness will still be around . . . the most that will happen, I think, is that the boundary between mildly stiff problems and the stiff problem will move a bit. But the *really* stiff problems will be around.

Enright: What about the emphasis of different codes? For example, extrapolation codes become more competitive as we get into parallel architecture. . .

Hindmarsh: They may very well . . . What I am saying will not happen if we will fall back on non-stiff methods and do everything explicitly.

Pratt: . . . I have been trying for some years to solve large-scale systems . . . what I see happening is as we get bigger and bigger computers we can do things faster and faster, but we will always want to solve bigger problems and we need sophistication at the low end of the scale. . .

Byrne: . . . there are two obvious directions that computing is moving . . . the day is not far off when . . . every engineering professional will have on his desk a mini which is linked to a supercomputer mainframe . . . this is mind-boggling . . . the mainframe is probably going to be a big parallel processor . . . we see in the literature snake algorithms . . . you put something on your mini and it snakes out for unused mini-computers in the network . . . you have all this computing power. What is the impact on stiff computation? There will be a change obviously in the algorithms. The linear algebra will be changed dramatically . . . we will see much more detailed modelling. . .

Shampine: Many people feel (the stiff problem) is rather mopped up . . . I think we need to have a period of true confession . . . what areas do we not know what we are doing or do not know how to do anything at all . . . this is an opportunity for you to say where you think things are going . . . (for example) Runge-Kutta methods for stiff problems, we are just beginning to explore them . . . doing an integration and locating where some function had a zero, our theoretical and practical understanding of that problem is almost zero. There are terrible difficulties in stating theoretically what it means to locate a root of a function when that function is being defined by the noisy process of solving a differential equation . . . it is easy to write down something that sort of works . . . there is one example of a true confesion. . .

Petzold: . . . I do not think anyone really has a good understanding of nonlinear differential-algebraic equations, in general, and there is this whole question of turning points. . .

Cash: Some problems we can get nowhere with . . . a boat connected to the shore and the tension in the chain is some random variable, stochastic differential equations. . .

Shampine: Stability analyses have been tilled since the earlier days . . . Some of us think this has very little to do with practice, but I have a feeling at the current time stability analysis is in a real state of ferment and that the theory is moving closer to the practice. . .

Dahlquist: I am glad to hear you say that. Well, of course we try. All theory must be based on simplifications . . . see how much one can do with simple model problems . . . I think all progress must be based on this—that one makes a theory or makes software for some standard situations . . . Henrici hoped when he wrote his book that the theory or the practice of differential equations could be handled just by the assumption that the step size times the Lipschitz constant should be chosen sufficiently small. Then it became more and more obvious that there were practical problems for which this was not the case. My own experience at the time was, at the beginning of the fifties, I encountered *one* problem where this difficulty existed. I tried to ask a lot of people from different application disciplines if they recognized this kind of situation and everybody said, "No"! Around 1960, things became completely different and everyone became aware that the world was full of stiff problems. Even if a theoretician tries to get a good formulation of the real problems, it is sometimes really difficult to get them so well formulated that it can be a good starting point for a theoretical analysis . . . to return to differential-algebraic equation analysis . . . it is still a little uncertain how many real nasty situations are left that require a general theory . . . a lot of practical situations particularly in classical mechanics are differential equations with inequality constraints . . . think of all the problems with Couloub's Law of friction . . . or two-body collisions, it is an inequality constraint that they are not allowed to penetrate . . . perhaps you should widen the discussion of differential-algebraic systems also to cover the case of inequalities . . . I would really like to hear about other applications (of differential equations with inequality constraints). . .

Shampine: My impression is we have no satisfactory way of solving delay- or difference-differential equations. . .

Bickart: Some of the codes work well . . . as soon as you run into nonconstant delays you run into problems.

Krogh: I do not know what the theoretical problems are with handling discontinuities, but I have not seen any practical difficulties. I have had features for handling discontinuities in my codes since 1969. . .

Shampine: That is because you have infrequent discontinuities. . .

Krogh: The difficulty is in the restarting?

Shampine: The order plummets. It has a disastrous effect if it happens often enough.

Shampine: The inverse problem of parameter estimation is certainly a very difficult problem. . .

Byrne: The difficulties are . . . the kinetics themselves are wicked . . . you start off with so-called literature rate coefficients which for the particular system at hand are not really right so you are really off the mark for a starting point . . . generally you have interval constraints . . . we need a cheap nonlinear constrained least-squares package . . . the problem is quite typically badly scaled. . .

Churchill: It is an impossible problem. There is a non-uniqueness that will never be solved by any package . . . you never will have good enough information so that you can trust what you get. . .

Dove: . . . virtually no kinetic experiment gives you information on one rate constant only . . . there is information on several constants and that information has to be unscrambled. . .

Pratt: . . . we must not forget the third leg of this thing and that is sensitivity analysis . . . without it you are groping. . .

Seider: We have had a great deal of success in tuning the parameters in (models of) non-ideal solutions with Powell's nonlinear programming algorithm written in 1977; it is very good for highly nonlinear functions as well as highly nonlinear constraints. . .

Byrne: It does not work out for us . . . the line search is too expensive.

Shampine: When we actually get down to solving a problem we have to choose a solver to do it . . . we have actually to pick one . . . how do we decide?

Enright: . . . we can certainly identify some codes that are not competitive . . . testing will allow us to choose between two algorithms that have a very similar design structure, but to compare two codes with very different structure I think is almost impossible unless we restrict the class of problems to a very small class that exhibits a special structure. . .

Pratt: . . . How do we chose between a BMW and a Vespa?

Byrne: I think we all have built-in prejudices . . . if you use a code and you are not comfortable with it, you are not going to use it again.

Enright: . . . the problem here is that codes are not trying to do the same things. Twenty years ago they were. Fixed step size (codes) we could compare . . . now codes determine accuracy requirements in a completely method-dependent way . . .

8.3 Stiff computation: where to go? (Cellier)

8.3.1 *Introduction*

It is the aim of this discussion to list some of the more severe shortcomings of available stiff computation codes as they have been collected through discussions with the end users of such software. These complaints, however, are formulated in terms that are understandable to the producers of integration algorithms. Some suggestions are given on how these deficiencies may be overcome in the future, and where open research fields can still be found.

8.3.2 *Discontinuity handling: where to go?*

The initial-value problem (1.1) and (1.2) may contain an $f(x, y)$ that is discontinuous in x. Typical examples of such discontinuities are (i) in mechanical engineering: friction phenomena, (ii) in electrical engineering: diodes, thyristors, any combined analog and digital circuitry, (iii) in chemical engineering: charging and discharging of batch reactors.

We can distinguish between two different types of discontinuities:

(i) Discontinuities of which we know the time when they are expected to happen. As a typical example of this type of discontinuity, consider when output voltage switches from negative to positive, or vice versa, a discontinuity takes place. The time instances when these discontinuities are expected to take place are precisely known beforehand. These discontinuities are in simulation usually referred to as "time events." Time events are scheduled events in the sense that the time instances when the discontinuity is to take place may be collected into a "calendar of events."

(ii) Discontinuities of which we do not beforehand know the time when they are expected to happen. Instead, we know the condition under which the discontinuity takes place, e.g. a state variable x_i crossing a prescribed level into positive direction. As a typical example of this type of discontinuity, we mention the output voltage of a rectifier circuit. Here we do not know when the output voltage is going to have a discontinuity (in its first time derivative). We just know that each time the input voltage crosses though zero in either direction, the output voltage has a break. These discontinuities are in simulation usually referred to as "state events." For state events, no scheduling is possible. Instead, we need a mechanism to describe the condition under which the event is going to happen. This is usually referred to as "state condition."

It makes sense to distinguish between these two types of discontinuities for two reasons:

(i) The user of the code requires two different mechanisms to describe the discontinuities to the program.

(ii) The algorithm requires two different mechanisms to handle them. Time events may be handled simply by inspection of the calendar of events for the next scheduled event time. When this moment approaches, the step size must be reduced to hit the event time precisely, or the algorithm must interpolate back to the just-passed event time, depending on the algorithm in use. State events, on the other hand, must be handled by either iteration or interpolation (depending on the algorithm).

Unfortunately, until very recently none of the existing codes for numerical integration provided for appropriate discontinuity handling. When I mentioned this problem to a prominent numerical mathematician during the Urbana ACM/SIGNUM Conference on Numerical Ordinary Differential Equations in 1979, he answered that this was a problem which should be left to the end user to be optimally adapted to his personal needs and should not be of concern to the numerical mathematician. I do not agree for several reasons:

(i) The numerical mathematician usually thinks that given a particular algorithm everything is said and clear. However, this is not the case for reasons of communication problems. The end user in general is not able to understand an integration algorithm well enough to be able to code it into a properly executing program by himself. This holds equally true for discontinuity handling. Concerning the integration algorithm, the attitude originally was also that coding of the program be basically the user's responsibility, while it has meanwhile been realized that precut standardized codes are superior in many respects, even though there are still some people around to call this the "vacuum cleaner" approach. I suggest that the handling of discontinuities is a more recent problem, and that in the long run standardized codes shall replace home-tailored software here also.

(ii) If you look into the work done by Hull and Enright at Toronto, it should become clear that one has no chance generally to compare integration algorithms. What one can do, however, is to compare the aptitude of different integration codes to solve a particular application problem. Hopefully, one will be lucky enough to be able to extend the experiences gained from entire sets of problems to a classification of applications. It should then become clearer which algorithm is best for any particular practical application without being forced to try them all. Results indicate that the step from the integration *algorithm*, once given, to the integration *code* is very large indeed. In fact, more computation time is used by the "dirty" overhead surroundings than by the integration algorithm itself. Again, this applies to the problem of discontinuity handling as well.

(iii) It is not usually economical to develop larger pieces of software for each applica-separately. On the contrary, we should concentrate on determining what large classes of problems have in common, extract this information, and provide for a program which handles all that in a standardized manner. This is time-efficient, cost-efficient, and much safer, as the amount of software left to be

> coded by the end user is minimized while obviously much more care and expertise can flow into the design and production of the standardized part of the software. It is here, where simulation software comes into the game. Definitely, discontinuity handling belongs to the part of the code which should be standardized.

To conclude these considerations, let me propose some very concrete steps toward (in my view) a satisfactory solution. We need a standardized user interface for ODE solvers (Hindmarsh 1978). It is evident that I cannot know in advance which integration code will be optimally suited for my particular application under all circumstances. Thus, it should be such that I can code my application software entirely independently of the ODE solver. Ideally seen, I would like to be able to replace any ODE solver by any other ODE solver simply by replacing one subroutine name by another subroutine name without being forced to change a bit of my application program. In fact, I would want to maintain a library of integration codes to have a "remedy against all diseases." An additional advantage of this approach is that I may easily update my ODE solver library when new releases become available as there are no side effects to be expected from simply replacing the old subroutine by its modified version. However, this also means that I should not be forced to change a single line within the ODE solver for any particular application. It is now very easily shown that, when I try to graft the discontinuity handling upon the integration code (which I have done in my simulation software GASP-V—Cellier (1978, 1979)—for precisely the aforementioned reasons), I obtain inefficient code. For this reason, although I fully support Hindmarsh's idea concerning a standardized interface for ODE solvers, I do not agree to his standard proposal. I feel strongly about the need for enhancing the standard proposal by adding to it an appropriate description of conditional termination criteria (discontinuity functions). In addition, the standard also should be expanded in another sense. If I want to maintain a library of ODE solvers, I shall most certainly need a linear system solver (LSS) in the majority of them. If this (lower end) interface is not standardized, too, I have to maintain any number of basically identical LSS in my ODE library. For this reason, it would make sense to standardize this lower end of the ODE code, also. Different LSS may then reside in a possibly separate library (e.g. general LSS and sparse LSS) out of which I may select whatever seems appropriate to go with whatever ODE solver I want to use. It ought to be mentioned that Hindmarsh has introduced a root solver into one of his programs, LSODAR, and uses sparse matrix techniques for linear systems solving in another program (LSODES) (see §4.3.4). However, there currently exists no version which combines root finding with sparse linear system solving. I would strongly suggest a version which combines all these features at the expense of a somewhat reduced efficiency.

A basic problem here is that the producers of the integration algorithms are not necessarily equally well trained in computer science. However, the above-mentioned considerations are basically those of a computer scientist, and it is sometimes hard to convince algorithm developers of the fruitfulness of such considerations.

When I asked Wanner after his presentation of a newly developed A-contractive algorithm (that is, in terms of Dahlquist, a G-stable algorithm, if I understood Wanner correctly), during the Rutishauser Symposium in Zurich (Wanner 1980c), whether he had already produced any executable code for his algorithm, he looked at me rather puzzled and answered that his work was purely theoretical, and that the job of coding would definitely have to be someone else's task. (For reasons of fairness it should be mentioned that Wanner has personally produced several integration codes for some other algorithms.)

I can assure you that the end user is certainly *not* willing to try his hands at such an adventure because he is (1) not able to do it, and (2) not able to judge beforehand whether this new algorithm will get him anywhere or not. It has already too often happened (also to myself) that a good-looking new algorithm turned out to be a complete failure after it had been coded into a program. Once I tried to code a DIRK algorithm: the effort spent on coding this algorithm was tremendous, it never worked satisfactorily, and it never became clear afterwards whether the problem was really with the algorithm or just with the code!

Even Henrici (who is known for his sympathy for practical solutions to real-world problems as opposed to ivory-tower solutions to green-grass problems), when I asked him a couple of years ago whether he would agree to join the consenting committee for my PhD thesis which I was to write on "combined continuous/discrete system simulation" (Cellier 1979), answered that he did not know anything about simulation but that he was certainly willing to learn something about it. I then told him that this was beside the point in that I was sure that he knew a lot about simulation. The problem was simply that he did not know that he knew something about it. This indeed is a severe problem in that it indicates that the average mathematician, even being a specialist in numerical integration, does not scan the literature for articles on simulation, even though these articles could be as important to his work as any contribution on numerical integration. I was enchanted to learn during the Park City meeting that several of the participating mathematicians had very much adopted the engineering viewpoint in contrast to the view of a pure theoretician (see §2.2 and §4.2).

Coming back to my former issue on discontinuity handling: What has happened since the Urbana meeting? I was glad to realize that there were at least a few mathematicians who took my comments seriously enough to think of some remedies:

(i) For nonstiff problems, the discontinuity-handling problem had already been solved prior to the Urbana meeting by people like Pritsker (1974) and myself (Cellier 1979) in a fairly general way. Applying an R-K algorithm, we just have to make sure that the discontinuity takes place at the end of an integration step. Time events are handled simply by reducing the step size if the next event is shortly ahead. State events are handled by iterating back to the unknown event time by any available method. Pritsker uses bisection, whereas I resorted to inverse Hermite interpolation. For obvious numerical reasons, it is important *not* to code the discontinuity itself in the ODE set but only the state conditon (e.g. by means of conditional termination criteria). The discontinuity itself is then expressed by context switching during the execution of the event (after execution of the event, another set of ODEs becomes active). A somewhat more mathematical view of this procedure can be found from Mannshardt (1978).

(ii) For stiff problems, one usually wants to apply multistep integration, for which the step-size adjustment at least creates a certain amount of overhead. The now common approach is to maintain two independent clocks, the *external* simulation clock and the *internal* integration clock. The integration proceeds with optimized step size and order, whereas the synchronization with the simulation clock is done by interpolating back using the Nordsieck vector. This methodology also may be applied to discontinuity handling if the state conditions are formulated as an adjoint set of discontinuity functions with the meaning that the simulation run terminates at either the final time or when the first of these functions crosses through zero in either direction, whatever comes first. According to my knowledge, we owe this formulation to (Carver 1977).

(iii) Meanwhile, two of the available GEAR codes, one by Carver, Stewart, Blair, and Selander (1978) and the other by Kahaner (1979), have been upgraded to contain a discontinuity handling mechanism. The Kahaner implementation differs from the Hindmarsh implementation in that the former DIFSUB subroutine has been modularized into about twenty smaller subroutines which are now much more understandable and which avoid all those "dirty" GOTO statements pointing backward in code.

(iv) I might want to suggest the introduction of an additional flag to determine whether all crossings are to be detected or whether only positive or only negative crossings are important. This is not really essential, but it is useful in many applications to formulate hysteresis effects, e.g. a heating system is switched on when the temperature falls below 18 degree centigrade, while it is switched off only after the temperature has reached 21 degrees centigrade as too frequent switching may damage the switching mechanism. By use of the above mentioned flag, a lot of unnecessary context switching can be avoided which makes the user programs execute more efficiently.

(v) Unfortunately, the aforementioned mechanism does not yet solve all the problems we would like it to solve. This is because after a discontinuity has taken place, the integration needs to be restarted. In the case of the GEAR codes, this means that the algorithm has to start again at an order of one. If discontinuities occur at frequent rates, the integration algorithm needs to be restarted again and again. One easily ends up having an extremely inefficient implementation of the trapezoidal rule, as higher orders get no chance to build up.

(vi) It is quite frequent in engineering applications that the accuracy requirements are not very severe. Then, an efficient low-order algorithm may do a better job on the problem.

(vii) Point (vi) was realized by Deuflhard who developed a new low-order code for stiff integration which looks very promising (Deuflhard, Bader, and Nowak 1980). For such an algorithm, the discontinuity handling also becomes easier.

(viii) If higher accuracy requirements are important, a higher-order algorithm has its advantages. A typical engineering application for this would be the simulation of a combined analog and digital circuitry containing memory elements (flip flops). In such applications, it is extremely important to know whether spikes are around, and what the maximum trasient voltages in some parts of the analog circuit are. The answer to these questions is very sensitive to parameter variations and also to event timing. For this reason, the system usually must be simulated with a relative accuracy of from 10^{-6} to 10^{-8}.

(ix) After I mentioned this problem to Gear, he realized that his algorithm could be substantially improved if the warming-up period could be made more efficient, that is, if one could avoid having to restart at one one after discontinuities. He developed in the meantime R–K starters for GEAR codes (Gear 1980d). Although I have not yet found the time to implement such an algorithm personally, I am fully convinced that this amendment will make the code substantially faster when frequent discontinuities occur.

The question of discontinuity handling is by no means settled. To illustrate, let me discuss the following application problem which I offered some time ago as a new bench mark problem for simulation software (Cellier 1979).

Given: a set of domino stones from a domino game (usually 55, but any number will do). Place these stones in a distance *d* from each other. If the first of these stones is pushed, all stones will fall. The question to be answered is, at which distance *d* between two consecutive stones is the chain velocity maximized.

This problem is of a sufficiently "green-grass" nature to refresh the heart of even a mathematician. However, the problem is by no means academic as precisely the same simulation problem arises in many practical applications, e.g. the heating of steel ingots in a steel soaking pit and slabbing mill, or chemical batch reactions with charging and discharging of batch reactors, or the traffic flow though an intersection where each car may be modelled by a set of discontinuous ODEs whereby new cars may enter the considered area at any time while old cars disappear from the region after they have stayed in the system for some time.

What is common to all these applications? Obviously, whenever a discontinuity arises, the entire structure of the problem may change, and even the number of ODEs varies with time. We call these problems "variable structure problems." Each domino stone in my benchmark problem has to obey Newton's law, being represented by a second-order system or by a set of two first-order ODEs. Taking the 55 stones of the game altogether, we obtain a 110th-order model. However, it makes little sense to program the model in this way as only a very small number of stones are moving simultaneously. Some stones may have fallen while others remain untouched. Moreover, the physical law governing the motion of any stone in the system is the same. It therefore makes much more sense to code an entity to represent any "model" stone and allow new stones of the prescribed type to be generated at event times while others may be destroyed at event times. It is evident that there will exist an interaction (possibly even continuous) between falling stones which has to be taken into account. Any ODE solver as they are currently marketed is theoretically able to handle this situation as long as the code provides for appropriate discontinuity handling mechanisms in the previously discussed sense. However, the portion of the program which remains to be user-coded will still be substantial. We feel that, again, the user should be relieved of that part of the coding which is common to all the aforementioned examples.

Software for this type of application is currently under development at our group. This software will consist of a FORTRAN coded subroutine package GASP-VI (Rimvall and Cellier 1982), an extension to the existing package GASP-V (Cellier and Blitz 1976), together with a PSCAL coded preprocessor (front end) COSY (Cellier and Bongulielmi 1979; Cellier and Rimvall 1981) to make the user interface a little more convenient and less error prone.

8.3.3 *Automated partitioning*

It is quite common to many applications (e.g. in chemical engineering) that some portions of the model are considerably faster than others (e.g. a chemically reacting system consisting of fast and slow reactions). It seem intriguing to try to reduce the overhead involved in the numerical integration of the slow subsystem by splitting the system into fast and slow portions and by using different step sizes and possibly different algorithms for the two of them (Palusinski and Wait, 1977). This works quite well for some applicatons, e.g. for self-tuning regulators with a fast inner loop and a considerably slower outer adaptation loop. We tried this out by use of the SIMNON software (Elmquist 1976, 1977).

However, this partitioning scheme is not always easily done. It requires expertise to master such software. Moreover, it is not guaranteed that such a partitioning scheme even exists for a particular stiff system. In fact, if the system is nonlinear, the eigenvalues of the Jacobian may move around freely with time, and it may well happen that some modes of the system are fast during some period of time and slow during other periods. Carver came up with a brilliant idea (because extremely simple) for automated partitioning, which he presented during the International Conference on Simulation at Interlaken (Carver and MacEwen 1980). His method requires one single additional parameter to be user-tuned, a parameter which even has some physical meaning assigned to it and which makes the adjustment reasonably easy. Except for this parameter, the partitioning is fully automated and even adaptive in that it may vary with time.

8.4 Using advanced scientific computers to solve stiff systems (Lomax)

8.4.1 *Introduction*

In some instances stiffness tends to prevent our finding any solution to a problem; in others it tends to make the computation of a solution not impossible but costly. It is the latter form with which the author has had experience. Most of this experience has been in that part of computational fluid dynamics (CFD) which relates to the finite-difference computation of the PDEs governing compressible fluid flow. These equations can be stiff, and in this article our discussion of CFD problems is based on this form of stiffness.

There are at least two ways to combat stiffness. One is to design a better computer, the other, to design a better algorithm. These two are not necessarily mutually exclusive and that fact motivates this article. The following material contains a definition of stiffness, where it enters CFD formulations, what methods are used to overcome it, what one can expect in advanced computers, and how this is all related.

We start with the assumption that some physical event has been modelled with sufficient accuracy by a coupled set of ODEs. Just how that happens in the case of fluid flows will be discussed briefly later. These, usually nonlinear, set of equations expressed in (1.1), are rewritten as

$$\frac{\mathrm{d}y}{\mathrm{d}t} = f(y, t), \qquad a \leqslant t \leqslant b. \tag{8.1}$$

In order to be precise, we proceed to locally linearize the right-hand side of (8.1) about some reference state $\boldsymbol{p}$ which is a vector similar to $\boldsymbol{y}$. We start with the Taylor series expansion

$$f(y, t) = f(p, t) + \left(\frac{\partial f}{\partial y}(p, t)\right)(y - p) + \tfrac{1}{2}\left(\frac{\partial^2 f}{\partial y^2}(p, t)\right)(y - p)^2 + \ldots, \tag{8.2}$$

where the index on $(y - p)$ denotes the tensor power, which is the foundation of Newton's method for the iterative solution of nonlinear equations. For simplicity, at this point we drop the dependence of f on t, since its exclusion does not affect the concept being developed. By introducing some more notation and substituting (8.2) into (8.1), we can write

$$\begin{aligned} A(p) &\equiv \frac{\partial f}{\partial y}(p), \\ F(p) &\equiv A(p)p - f(p), \\ \frac{\mathrm{d}y}{\mathrm{d}t} &= A(p)y - F(p) + O[(y - p)^2]. \end{aligned} \tag{8.3}$$

From now on we neglect terms $\sim O[(y - p)^2]$, assuming that we are only interested in the stiffness of our system when the solution $\boldsymbol{y}$ is close to the reference $\boldsymbol{p}$. In practical problems, of course, this reference would vary widely throughout a computational path.

The eigensystem of the matrix A is defined as

$$X^{-1} A X = \Lambda, \tag{8.4}$$

where $\Lambda = \operatorname{diag}[\lambda_1, \lambda_2, \ldots, \lambda_M]$, with

$$|\lambda_1| \leqslant |\lambda_2| \leqslant \ldots |\lambda_m| \leqslant \ldots \leqslant |\lambda_M|. \tag{9.5}$$

Assume that the eigensystem of A is not defective. Then each λ_m of A has a corresponding eigenvector $\boldsymbol{x}_m$ which appears as a column in X. With this we can express the solution of (8.3) as

$$y = \underbrace{\sum_{m=1}^{M} c_m e^{\lambda_m t} x_m}_{\text{transient}} + \underbrace{X \Lambda^{-1} X^{-1} F}_{\text{steady-state}}. \tag{8.6}$$

It is now necessary to introduce the fact that we are seeking to find a numerical solution of (8.3) by means of some time-marching method. This solution will, of course, be an approximation of (8.6), but in particular it will be a different approximation for each of the eigenvectors $\boldsymbol{x}_m$. In general the eigenvectors associated with the small $|\lambda_m|$ will be well resolved and those associated with the large $|\lambda_m|$

will be resolved less accurately. The whole concept of stiffness in CFD arises because it is very often the case that we do not need to resolve the eigenvectors associated with the large $|\lambda_m|$ in the transient solution although we still need them in the particular or steady-state solution. For this reason it is convenient to subdivide the transient solution part given in (8.6) further into two parts. Since we have ordered our eigenvalues by their magnitude, this can easily be done as follows:

$$\text{Transient Solution} = \underbrace{\sum_{m=1}^{P-1} c_m e^{\lambda_m t} x_m}_{\text{Driving}} + \underbrace{\sum_{m=P}^{M} c_m e^{\lambda_m t} x_m}_{\text{Parasitic}}. \tag{8.7}$$

This concept is crucial to our discussion. Rephrased it states that we can separate our eigenvalue spectrum into two groups: one $(\lambda_1, \ldots, \lambda_{P-1})$, called the driving eigenvalues (our choice of a time-marching method must accurately approximate the eigenvectors associated with these), and the other $(\lambda_P, \ldots, \lambda_m)$, called the parasitic eigenvalues (no accuracy whatsoever is required for the eigenvectors associated with these).

Each of the following definitions refers to the current reference state p.

(i) A coupled set of ODEs is said to be inherently stable if

$$\text{Re}(\lambda_m) \leqslant 0 : m = 1, 2, \ldots, M$$

(ii) An inherently stable set of ODEs typically is part of a stiff problem here if

$$|\lambda_P| \ll |\lambda_M|.$$

(iii) We define the ratio (compare with the stiffness ratio of (7.47))

$$\text{Cr} = |\lambda_M| / |\lambda_P|,$$

and then define the categories

Mildly-stiff	$\text{Cr} < 10^2$
Strongly-stiff	$10^3 < \text{Cr} < 10^5$
Extremely-stiff	$10^6 < \text{Cr} < 10^8$
Pathologically-stiff	$10^9 < \text{Cr}$

(iv) For relaxation methods $P = 1$ and the transient is identical to the error. By our definitions all relaxation methods are at least mildly stiff.

These are merely some working definitions that have proven to be of qualitative utility with CFD. For a general definition of stiffness see §1.2. It should be mentioned that the gaps in the category definitions are intentional because the bounds are arbitrary.

It is extremely important to notice that these definitions make no distinction between real, complex, and imaginary eigenvalues. In fact, in convection-dominated fluid-flow calculations the eigenvalues are often very close to the imaginary axis.

8.4.2 *The effect of mesh choice*

It is impossible to review here all possible causes of stiffness in fluid dynamic computations. In fact, we limit ourselves for the most part to example situations that have occurred in the numerical solution of the compressible Navier–Stokes equations for high Reynolds' number flows. Many of these flows are characterized by a few regions having high spatial gradients of the dependent variables and a domain that is otherwise characterized by relatively low gradient phenomena. As a result it is quite common to cluster mesh points in certain regions and spread them out otherwise. Examples of where this clustering might occur are at a shock wave, near an airfoil leading edge, and in a boundary layer. Very often the adaptation of the mesh to the geometry and physics of the problem is carried out by introducing a generalized coordinate system which transforms a highly distorted adaptive grid in physical space to a uniform, equispaced grid in the computational domain. Typically the space derivatives are approximated by central differences in the computational plane, which can be expressed as (8.6).

It is easy to show for simple model problems that the details of the procedure just discussed strongly effect the eigensystem of the resulting A matrix. In fact, it has been shown for model problems, and verified by experience for practical problems, that the stiffness parameter Cr is strongly attached to the minimum space interval in the mesh. If we let this minimum interval be Δ then

$$\mathrm{Cr} \sim 1/\Delta^2 \quad \text{for dissipating terms, and}$$

$$\mathrm{Cr} \sim 1/\Delta \quad \text{for periodic convecting terms.}$$

This leads to the rather annoying fact that the more we try to increase the resolution of our spatial gradients, the stiffer our equations tend to become. Typical CFD problems without chemistry vary between being mildly and strongly stiff, and are greatly effected by the resolution of a boundary layer.

8.4.3 *A matrix for two and three dimensions*

Before we can make any practical comments about the effect of computer architecture on the computation of stiff finite-difference CFD-problems, we need a rudimentary briefing on the underlying matrix structure of these problems when they are cast into ODE by first approximating all of the spatial derivatives.

Recall that a generalized coordinate system has been used to transform a distorted physical grid structure to a uniform mesh in a simple box-like computational domain. Suppose we use three-point central-difference approximations for all of the first and second derivatives, ignoring for this exercise all cross derivatives. The format of the resulting **A** matrix is shown in Figure 8.1 for a three-dimensional mesh having three points in the y and z directions. The lines in the schematic represent diagonals along which small block matrices are entered. These blocks are 4×4 for a two-dimensional case, and 5×5 for the three-dimensional one shown. Whether or not the elements in the matrices represented by Figure 8.1 are ever actually calculated (and very often, of course, they are not) it is, nevertheless, the eigenvalues of these matrices that determine the stiffness of the fluid flow computation.

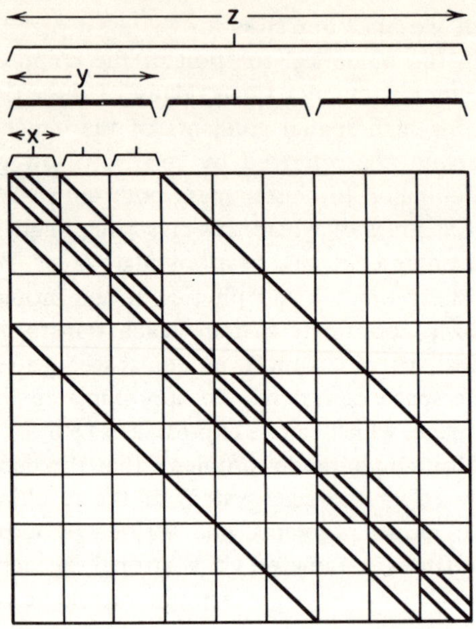

Fig. 8.1. Schematic of an ODE matrix formed by central differencing the space derivatives.

8.4.4 *Comments on methods for computing stiff problems*

The use of a time-marching method on (8.3) converts the differential system to a coupled set of difference equations and the analytic solution of these equations depends upon the roots (which we refer to as the σ-roots) of another characteristic polynomial closely related to that for the differential system. To be brief let us use a constant time increment, h, and set $t = nh$. Then it is well known that any numerical method that produces just one σ for each λ produces a solution which has the same form as (8.6) except that

$$e^{\lambda_m t} \rightarrow [\sigma(\lambda_m h)]^n. \tag{8.8}$$

The explicit fourth-order Runge-Kutta method (R-K4) and the implicit trapezoidal method produce

$$\sigma(\lambda h) = 1 + \lambda h + \tfrac{1}{2}\lambda^2 h^2 + \tfrac{1}{6}\lambda^3 h^3 + \tfrac{1}{24}\lambda^4 h^4, \tag{8.9}$$

$$\sigma(\lambda h) = (1 + \tfrac{1}{2}\lambda h)/(1 - \tfrac{1}{2}\lambda h), \tag{8.10}$$

respectively.

One can easily show using (8.9) that R-K4 is stable for any inherently stable A if the step size is chosen such that $|\lambda_M| < 2.6$. It is possible, therefore, if one has the computer resources, to use it even on strongly stiff systems. This may not be as foolish as it first seems, as we shall see.

8.4.5 *Multigrid methods*

The A matrices formed by space differencing the equations governing fluid flows have certain special properties related to the fact that a spatial continuity exists in the data structure. This is reflected in the relation between the magnitude of the eigenvalues and the spatial frequencies of the eigenvectors. If properly designed, relaxation methods will efficiently remove the high-frequency error from the solution vector. The remaining error and the residual can be interpolated to a coarser grid. The resulting A matrix on the coarser grid will have a reduced Cr and the process can be repeated. The method works very well for certain elliptic systems but can be quite subtle for hyperbolic ones.

8.4.6 *Specially designed methods and preconditioning*

It is quite easy to construct a predictor–corrector sequence that is first-order time-accurate and will give any polynomial relation between σ and λh for powers of λh greater than one. If one knows that the eigenvalues of A are clustered in a given region (for example, spread along a portion of the negative real axis), one can design highly stable methods for these special cases; see Stetter (1968) and Lomax (1968).

If one is interested in finding a steady-state solution, it is a well known practice to precondition the A matrix by multiplication with some nonsingular matrix which has the effect of lowering Cr for the matrix product. In the simplest cases the preconditioner is diagonal or block diagonal and amounts to a form of scaling. This is often used in CFD applications and can be combined with any of the other methods mentioned.

8.4.7 *Implicit methods*

By far the most popular numerical technique to combat stiffness is the use of implicit techniques. This is just as true in the field of computational fluid dynamics as it is elsewhere. In the analysis of explicit methods the matrix A is used to establish concept but the evaluation of its elements is seldom carried out in a code that actually makes use of the method. This is not true in the case of implicit methods. The implementation of implicit methods requires: (i) the calculation and use of the elements of A; (ii) the solution of simultaneous equations. In CFD applications for one space dimension, A is typically a 3×3 block tridiagonal matrix. For such cases the trapezoidal method (also known as the Crank–Nicolson method) is relatively easy to implement, and it is well known to be stable for all inherently stable ODEs. This follows, for model problems, from (8.10).

In CFD applications to two and three space dimensions, however, the size (order) of A quickly becomes very large. For a 50×50 point mesh, with four dependent variables at every point, the order is $(50 \times 50 \times 4)^2 = 10^8$. Admittedly A is very sparse and the calculation of its elements is actually quite feasible in practice. However, using it in the process of a Gaussian elimination would require a forward-sweep temporary-storage area proportional to the area between the central and outermost diagonal in Figure 8.1. For our example this would be $(4 \times 50)(4 \times 50 \times 50) = 2 \times 10^6$ words.

8.4.8 *Comments on advanced scientific computer architecture*

In the next several sections an attempt is made to address the minimum information required to make sense in evaluating the effect of computer architecture on constructing codes and algorithms used to solve stiff CFD systems. We try to speak in generalities, although to clarify the meaning we also refer to specific computers. The main part of the author's personal experience was on the Illiac IV which no longer exists (see the article by Lomax and Pulliam in Rodrigue 1982). Nevertheless, the experience gained there has several implications that apply to the more recent classes of computers such as the Cray 1, Cyber 205, etc., which we refer to as "advanced computers." These implications are what will be considered.

8.4.9 *The arithmetic processors*

Users of advanced computers can expect their arithmetic commands to be routed to one of two arithmetic units: the scalar unit, which executes operations one at a time in a random address sequence, and a vector unit, which executes commands in a locked-step "vector" sequence, which could make use of pipeline or parallel construction or both. The word *vector* is used to refer to a string of words stored with some uniform skip distance. In some machines the storage must be contiguous, that is, the skip distance must be one. The vector unit executes a series of arithmetic commands much faster than the scalar unit; for example, from about five to ten times faster on a Cray 1, and from about ten to sixty times faster on the Cyber 205. The spread depends on the word length, number of pipes, etc. These times are attained for vector lengths of 64 words (or multiples thereof) for the Cray 1, and 2000 words for the Cyber 205. The point is, in any case, that the identification and use of vector strings in one's algorithm is essential if the advanced performance in the advanced computers is to be realized.

8.4.10 *Memory storage allocation*

This aspect of computer configuration is a volatile item at the present time. Increased availability of memory at reduced cost is announced at frequent intervals, so what is installed at a computer site now will probably be changed in a year or two. Nevertheless, it is an important item for consideration because three-dimensional CFD problems seem never to have enough memory. On the Cray 1 computer at the Ames Research Center at the present writing the user has three forms of memory at his disposal for constructing a code and choosing an algorithm (magnetic tape back-up is not considered viable for this purpose). One of these is composed of one million, 64-bit randomly addressable words (RAM). Another is referred to as a "solid state disk" (SSD), and is composed of 16 million, 64-bit words randomly addressable in blocks of 512 words each. This memory has a very high transfer rate and can be, and often is, packed into 32-bit words randomly addressable in blocks of 1024 each. The third is a hierarchy of standard disks with relatively slow transfer rates. Good programming strategies move the blocks from the SSD to RAM and back, RAM being used as a buffer area in which the algebraic part of an algorithm takes place. The details of mapping the storage areas on and coding the transfers between these areas is referred to as data management. Other Cray 1 computer sites may be configured differently, and different computers, e.g. the Cyber 205, will have quite different arrangements.

In the Cray 1 example given above, the optimization of the data management system for massive databases is the responsibility of the user and it must include: (i) identification and storage of vector lengths in RAM; (ii) management of vector storage on the SDD; (iii) transfer of vectors between RAM and SDD; (iv) avoidance of "bank conflicts" on vector interchanges. Unfortunately, each site will probably require its own data management optimization for problems of the type we are considering, very little of which may be portable. Data management strategies and capabilities can play an important role in choice of algorithm for stiff problems in general. This is true for stiff CFD problems, but because of the very large memories that are becoming available, to a much lesser extent than might be expected. This is discussed in the next section.

8.4.11 *Effect of architecture on brute force approach*

The vector requirements of advanced computers actually encourage the use of explicit methods to solve mildly stiff and even strongly stiff problems. The principal reasons are:

(i) Explicit methods are easy to code and, therefore, are reliable and easy to maintain and modify. This can save calendar time.
(ii) In two- and three-dimensional CFD problems the vector lengths can be made equal to the total number of points in the mesh. This can be used to good advantage on those machines that require very long vectors for very fast speeds.
(iii) The number of arithmetic operations per mesh point for explicit methods can be less than one fifth of that required for implicit methods.

One of the most popular codes used to solve the inviscid, compressible, Navier-Stokes equations (the so-called Euler equations) uses a version of the R–K4 method which produces the characteristic polynomial signature given by (8.9), see Jameson, Schmidt, and Turkel (1981). The code has been highly vectorized to get maximum speeds out of advanced computers. The problems for which it is being used are stiff enough (for some choices of mesh) to be run more efficiently with implicit methods if the vector capability were absent. Yet for the reasons listed above the explicit approach may be the more economical on an advanced computer.

8.4.12 *Effect of architecture on multigrid methods*

The use of multigrid methods in advanced computers emphasizes data management problems involving vector identifications and transfers. Very little has been published in this area with regard to the Euler equations, but for a publication on highly parallel multigrid solvers for elliptic equations see Gannon and Van Rosendale (1982); also Brandt (1981) and Ni (1981). Look for much more in the Euler area in the future.

8.4.13 *Effect of architecture on specially designed methods and preconditioning*

The historical development of the choice of a Runge-Kutta sequence came from attempts to find a specially designed explicit method that would be stable for combinations of λ_m, some near the negative real axis and others near the imaginary axis. Conventional first- and second-order Runge–Kutta methods do not have this property. At one time a method due to Stetter was used, but eventually it was discovered that a version of the R-K4 method is near optimum. Higher-order Runge–Kutta methods have been tried but they are not significantly better. Work continues in this area.

The checker board and zebra relaxation methods (see Stuben and Trottenberg 1981) are techniques for combating stiffness in relaxation. They have received great popularity because of their easy adpatability to advanced computers that require long vectors.

A form of preconditioning is used in many codes attacking the Euler equations. It amounts to a scaling that is often described by the phrase "keeping the local Courant number close to one." The technique is very compatible with advanced computer hardware and can sometimes reduce stiffness in the Euler codes by a factor of three to five.

8.4.14 *Effect of architecture on implicit methods*

The computer architecture can play a very important role in the choice, design, and detailed coding of an implicit method in CFD problems having more than one space dimension. We saw in §4.6 that direct Gaussian elimination is generally not practical for these problems, especially in three dimensions. At present this is a memory limitation, but it may well be impractical even if this restriction were removed. The usual alternative, which is very widely used, is to factor the implicit operator, represented by the matrix shown in Figure 8.1, into the product of three implicit operators, each of which requires only a block tridiagonal solver in one of the space directions, Briley and McDonald (1975), Beam and Warming (1976), and Pulliam and Steger (1980). The matrix form of each of these operators is represented by the nine central diagonal blocks in the figure. Notice that these blocks are uncoupled from one another so each can be solved separately and in any order.

The factoring strategy mentioned above reduces the implicit part of a three-dimensional finite difference calculation to three one-dimensional sweeps through a box-like computational mesh. This in turn immediately suggests that the vector length needed for the efficient use of advanced computers can be the number of dependent variables in y in an entire plane of the mesh. In this way the algorithm combines each plane with its neighbors just as points are combined in a problem having only one dimension. This is possible because during each one-dimensional sweep the algorithm along every line is independent of the algorithm along every other line so that all lines can be computed at the same time. Furthermore, in many CFD problems the algorithms are not only the same but they are identical to one that can be designed for efficiency without regard to vectorization.

The concept just described, combined with various clever forms of machine-dependent data-management structures, has been used to construct programs for strongly stiff ODEs resulting form the spatial discretization of the compressible Navier–Stokes equations on the Illiac IV, Cray 1, and Cyber 205 computers. In each case the basic algorithm and resulting numerical solution was the same.

However, the data management parts of the codes were drastically different due to the differences in computer architecture.

8.4.15 *Concluding remarks*

Stiffness in CFD problems arising from the numerical solution of the compressible Navier–Stokes equations is related to the mesh intervals used for the space differences in the physical domain; the finer the intervals, the stiffer the problem. Stiffness is expressed in terms of the magnitudes of the eigenvalues occurring in the locally time-linearized ODE matrix. The ratio of the maximum to the minimum magnitude can be very large and can occur between eigenvalues located very near the imaginary axis or along the negative real axis. Characteristically a solution of one of these CFD problems in three dimensions requires a very large amount of computer memory to get good resolution and a very fast computer running a sophisticated algorithm to combat the stiffness.

Advanced scientific computers such as the Cray 1 and the CDC 205 can match these requirements under certain conditions. One condition is that the data must be mapped onto, and moved between, the various hierarchies of memory available at a given computer site, so that the speed of the computation is not seriously impaired. This is an annoying problem that is not only computer-dependent but even site-dependent. Fortunately, it diminishes as the amount of rapid-access memory increases, and technology is providing this trend. Another condition is that the algorithm be such that it will realize the vector capability that is built into all advanced computers proposed to date. Most conventional explicit time-marching methods can be made to do this very well. Space-factored implicit methods in three dimensions can also be vectorized rather easily. In the latter case, most implicit concepts based on experience with serial computers are equally useful on advanced vector computers.

APPENDIX I
LISTING OF SOME METHOD AND THEORY WORK PERTAINING TO STIFFNESS (1974–1982)

Subject listing in alphabetical order (note various stability types are listed by heading under Stability).

Averaging

Miranker (1982); Hoppensteadt (1979); Miranker (1978); Persek and Hoppensteadt (1978); Hoppensteadt and Miranker (1977); Coleman et al. (1977); Liniger (1976).

Blended

Cash (1981, 1982); Skeel and Kong (1977).

Block

Bond and Cash (1979).

Collocation

Billingsley (1982); Muelthei (1982); Burka (1982); Finlayson (1982); Dias et al. (1982); Chin et al. (1979); Guertin et al. (1977); Varma et al. (1976); Wouk (1976); Bushard (1976); Shah and Paraskos (1975); Serth (1975); Carey and Finlayson (1975); Michelsen and Villadsen (1974); Karanth and Hughes (1974); Scholander and Svedberg (1974); Brunner (1974).

Contractive

Liniger (1982); Soderlind (1981); Mingyou (1981); Ruehli et al. (1980); Odeh and Liniger (1980); Dahlquist and Jeltsch (1979); Nevanlinna and Liniger (1979); Dahlquist (1979); Nevanlinna and Liniger (1978).

Composite Multistep

Tendler et al. (1978); Lucey (1975); Bickart (1974).

Cyclic

Tisher and Sacks-Davis (1981); Gupta (1979); Tendler et al. (1978); Mihelcic (1978); Michelcic (1977); Cash (1977).

Defect Correction

Ueberhuber (1979); Frank and Ueberhuber (1977).

Delay

Bickart (1982); Cryer (1974).

Discontinuous

Chua and Dew (1982); Struwe (1981); Tuttle (1981); O'Malley and Flaherty (1980); Halin (1979); Srivastava et al. (1979); De Doncker (1978); Mannshart (1978); De Hoog and Weiss (1977); Carver (1977); Halin (1976); Luke et al. (1976); Luke et al. (1975); Macccomack and Paullay (1974); Hay et al. (1974).

Explicit

Fatunla (1980); Alfeld (1979); Kubicek and Visnak (1974).

Exponential Fitting

Cash (1981); Iserles (1981); Cash (1981); Raptis (1980); Rentrop (1980); DeGroen and Hamker (1979); Miller (1979); Iserles (1979); Strehel and Peper (1979); Miranker (1978); Norsett (1978); Iserles (1978); Rosenbaum (1978); Iserles (1977); Murphy (1977); Gear (1977); Jackson (1976); Ehle (1975); Sarkany and Liniger (1974); Jackson and Kenue (1974); Chu and Berman (1974); Makela et al. (1977); Chun-Yat (1974); Meister (1974).

Extrapolation

Deuflhard (1982); Duff and Nowak (1982); Stetter and Weinmueller (1981); Deuflhard (1980); Deuflhard (1979); Hoppensteadt and Miranker (1979); Lawrence and Dorning (1977); Gladwell (1976); Cash (1976); Lindberg (1974).

Functional Fitting

Iserles (1977).

Hybrid Methods

Fatunla (1982); Jain and Jain (1981).

Implicit Euler

Frank and Ueberhuber (1977).

Linear Multistep

Nolting (1982); Liniger (1981); Van Veldhuizen (1981); Butcher (1981); Fatunla (1980); Sinha (1980); Gear (1980); Warming and Beam (1979); Soderling (1979); Dahlquist (1979); Alfeld (1979); Grigorieff and Schroll (1978); Dahlquist (1978); Jeltsch (1978); Varah (1978); Nevanlinna (1977); DeHoog and Weiss (1977); Jeltsch (1977); Kong (1977); Prothero (1976); Bryne and Hindmarsh (1975); Liniger (1975); Gupta and Wallace (1975); Williams and DeHoog (1974).

Multiderivative

Twizell and Khaliq (1982); Cahs (1981); Burrage (1980); Jeltsch (1979); Jeltsch (1977); Brown (1977); Jeltsch (1976); Fuchs (1976); Gennin (1974).

Nonlinear Multistep

Lee and Preiser (1978); Lee (1974).

One-step

Van Brokhoven (1980); Cash (1978); Mannshart (1978); Watanabe (1978); Cash (1975); Prothero and Robinson (1974); Van Veldhuizen (1974); Brunner (1974); Norsett (1974).

One-leg

Liniger (1982); Dahlquist (1981); Sand (1981); Dahlquist et al (1980); Brown (1979); Nevanlinna and Liniger (1979); Dahlquist (1979); Liniger (1979); Dahlquist (1978).

Order Stars

Hairer (1979); Wanner et al. (1978).

Oscillatory Equations

Miranker (1982); Fatunla (1982); Gear and Gallivan (1981); Jain and Jain (1982); Petzold (1981); Cash (1981); Gear (1980); Gallivan (1980); Gourlay (1980); Ruehhli et al. (1980); Brown (1980); Fatunla (1980); Rinzel and Miller (1980); Warming and Beam (1979); Hoppensteadt (1979); Hoppensteadt and Miranker (1979); Bui (1979); Kreiss (1979); Auslander and Miranker (1979); Jain and Jain (1979); Ruehli et al. (1978); Miranker and Veldhuizen (1978); Miranker (1978); Jeltsch (1978); Hoppensteadt and Miranker (1977); Skelboe (1977); Smith (1977); Amdursky and Ziv (1977); Lambert and Watson (1976); Miranker and Wahba (1976); Fatunla (1976); Gupta (1976); Snider and Fleming (1974).

Quadrature

Iserles (1981).

Runge-Kutta

Implicit

Burrage (1982); Day and Murthy (1982); Cash (1982); Hairer and Wanner (1981); Zlatev (1981); Hundsdorfer and Spijker (1981); Stetter and Weinmueller (1981); Mingyou (1981); Cash (1981); Burrage (1979); Van der Houwen (1979); Bui and Bui (1979); Burrage and Butcher (1979); Crouzeix (1979); Scherer (1979); Dahlquist and Jetsch (1979); Eitelberg (1979); Curtis (1979); Butcher (1979); Varah (1979); Freidli (1978); Burrage (1978); Palusinski (1978); Iserles (1978); Alexander (1977); Bickart (1977); Butcher (1976); Fuchs (1976); Cash (1975); Ehle (1975); Ehle and Lawson (1975).

Semi-explicit

Cooper (1979).

Semi-implicit

Cash (1982); Prokopakis and Seider (1981); Weimer and Boness (1979); Kaps (1979); Clough (1979); Cash (1979); Bui (1979); Kaps and Rentrop (1979); Burrage (1978); Lapidus et al. (1978); Freidli (1978); Cash (1976); Michelsen (1976).

Singly implicit

Burrage (1980); Strehel and Weiner (1982).

Second Derivative

Kovvali and Gupta (1982); Sacks-Davis and Shampine (1981); Sacks-Davis (1980); Enright (1978); Gupta (1978); Sacks-Davis (1977); Kennealy and Moore (1977); Hill (1976); Brown (1976); Kubicek and Visnak (1974); Enright (1974).

Second-order Equations

Addison (1980); Odeh and Liniger (1980); Heinrich and Zienkiewicz (1979); Van der Houwen (1979); Hairer (1979); Gear (1978); Jensen (1976); Liniger and Gagnebin (1974).

Separably Stiff

Lambert (1981).

Singular Perturbation

Mattheij (1982); Mattheij and O'Malley (1982); Kreiss and Kreiss (1981); Petzold (1981); Soderlind and Dahlquist (1981); Smith (1981); Sanchez-Palencia and Lobo-Hidalgo (1980); DeVooght and Dahlquist et al. (1980); Mund (1980); Kahil and Kokotovic (1980); Barton (1980); O'Malley and Flaherty (1980); Brandt (1979); Auslander and Miranker (1979); De Groen and Hamker (1979); Heinrich and Zienkiewicz (1979); Michell and Christie (1979); Hsiao and Jordan (1979); Mattheij (1979); Miller (1979); Flaherty and O'Malley (1979); Bourgeat and Tapiero (1979); Aundrus (1979); Hoppensteadt and Miranker (1979); Come (1979); Hoppensteadt (1979); Kreiss (1979); Miranker (1978); Emel'yanov (1978); Persek and Hoppensteadt (1978); David (1977); Flaherty and O'Malley (1977); Robertson (1975); Aiken and Lapidus (1975); Flaherty and O'Malley (1975); Dontchev (1974); Kreiss (1974); Aiken and Lapidus (1974).

Splines

Muelthei (1982); Rentrop (1980); Hill (1976).

Stability

General

Butcher (1981); Lambert (1980); Brown (1979); Jury (1978); Dahlquist (1978); Glaser (1978); Bickart and Jury (1978); Jury (1977); Dahlquist (1976).

A-stable

Crisci and Russo (1982); Muelthei (1982); Bui and Poon (1981); Zlatev (1981); Iserles (1981); Odeh and Liniger (1980); Tadeusiewicz (1980); Van Brokhoven (1980); Wanner (1980); Galantai (1980); Burrage and Butcher (1979); Bui (1979); Warming and Beam (1979); Cooper and Sayfy (1979); Scraton (1979); Kaps (1979); Tripathy and Rao (1978); Bickart and Jury (1978); Dahlquist (1978); Iserles (1978); Jeltsch (1978); Wanner et al. (1978); Wantanabe (1978); Lee and Preiser (1978); Trigiante (1977); Brown (1977); Butcher (1977); Jeltsch (1977); Lombardi (1977); Fuchs (1976); Liniger (1976); Chipman (1976); Wanner (1976); Jackson (1976); Jeltsch (1976); Cash (1976); Cash (1975); Ehle and Lawson (1975); Butcher (1975); Ehle (1975); Prothero and Robinson (1974); Williams and DeHoog (1974); Marzulli (1974); Brandon (1974); Brayton (1974); Liniger and Gagnebin (1974); Norsett (1974); Axelsson (1974); Genin (1974); Nevanlinna and Sipila (1974).

A (α) Stable

Shampine (1982); Galantai (1980); Kaps (1979); Bickart and Jury (1978); Grigorieff and Schroll (1978); Mihelcic (1978); Jeltsch (1977); Liniger (1975).

A (α, r) Stable

Nolting (1982).

A(0) Stable

Rodabaugh and Thompson (1979); Freidli and Jetsch (1978); Rodabaugh and Thompson (1979); Freidli and Jetsch (1978); Jeltsch (1976); Liniger (1975).

A_0 Stable

Feinberg (1982); Rodabaugh and Thompson (1979); Jeltsch (1976);

A_n Stable

Zlatev (1981).

Algebraically Stable

Burrage (1982); Hairer and Wanner (1982); Burrage (1978);

Almost A Stable

Mihelcic (1977).

Asymptotically Stable

Krogh (1981).

B Stable

Hundsdorfer (1981); Burrage and Butcher (1979); Crouzeix (1979); Scherer (1979); Jeltsch (1979); Dahlquist and Jeltsch (1979).

D Stable

Veldhuizen (1981); Verwer (1982).

F Stable

Bickart (1982).

G Stable

Dahlquist (1978); Nevanlinna (1976).

I Stable

Wanner et al. (1978).

L Stable

Scraton (1981); Day (1980); Cash (1980); Lambert (1990); Bui (1979); Scraton (1979); Bui (1979); Eitelberg (1979); Fatunla (1978); Bui (1977); Trigiante (1977).

L(α) Stable

Cash (1980).

L(0) Stable

Gourlay (1980).

L_n Stable

Zlatev (1981).

Nonlinear Stability

Dahlquist (1982); Soderlind (1981); Burrage and Butcher (1980); Burrage (1980); Brown (1979); Burrage and Butcher (1979); Dahlquist (1978); Burrage (1978); Cooper and Whitworth (1978); Rodabaugh and Thompson (1978); Trigiante (1977); Liniger (1977); Nevanlinna (1977); Wanner (1976); Liniger and Odeh (1976); Dahlquist (1975); Butcher (1975); Lambert (1974); Brandon (1974).

P Stable

Costabile and Costabile (1982); Bickart (1982); Fatunla (1982); Jain and Jain (1981); Cash (1981); Lambert and Watson (1976).

S Stable

Day and Murthy (1982); Day (1980); Alexander (1977); Verwer (1977); Prothero and Robinson (1974).

Stiffly A Stable

Ehle and Lawson (1975).

Stiffly Stable

Nolting (1982); Watkins (1981); Jeltsch (1979); Watanabe (1978); Albrecht (1978); Jain and Srivastava (1978); Tendler et al. (1978); Varah (1978); Jeltsch (1977); Rao and Iyengar (1976); Jeltsch (1976); Gupta and Wallace (1975); Prothero and Robinson (1974); Jensen (1974); Bickart and Rubin (1974).

Strong A Stable

Watanabe (1978).

Strongly Stablc

Struwe (1981); Taubert (1976); Gear and Watanabe (1974); Lee (1974).

Strong Stiffly Stable

Watanabe (1978).

Zero Stability

Zlatev (1978);

Taylor Series

Corliss and Chang (1982).

Turning Points

Ponisch and Schwetlick (1981); Moore and Spence (1980); Miranker and Morreeuw (1974).

Two-step

Iserles (1981); Dahlquist et al. (1980); Odeh and Liniger (1980).

Unstable Equations

Aiken (1982); Hoppensteadt et al. (1981); Mazurkin (1980); Kreiss (1979); Serth (1975); Lindberg (1974).

REFERENCES

Ackerberg, R. C., "Boundary Layer Separation at a Free Streamline: Part 2. Numerical Results." *J. Fluid Mech.*, **46**, 727 (1971).

Ackerberg, R. C. and J. H. Pyillips, "A Numerical Method for Highly Accelerated Laminar Boundary-Layer Flows," *SIAM J. Numer. Anal.*, **10**, 147 (1973).

Adachi, H., Basco, N., and D. G. L. Hames, "Mutual Interactions of the Methyl and Methylperoxyl Radicals Studied by Flash Protolysis and Kinetic Spectroscopy," *Int. J. Chem. Kinetics,* **12**, 949 (1980).

Addison, C. A., "Numerical Methods for a Class of Second-Order ODEs Arising in Structural Dynamics," Ph.D. Thesis, Dept. Comp. Sci., Rpt. 147, University of Toronto, Canada (1980).

Aguilar, F. and S. Thompson, "Non-Equilibrium Flashing Model for Rapid Pressure Transients," *ASME J. Heat Trans*, to appear.

Ahmad, S. Y. and M. B. Carver, "A System for the Automated Solution of Sets of Implicitly Coupled Partial Differential Equations." *5th Canadian Congr. Applied Mechanics,* Fredericton, Canada, p. 570 (1975).

Aiken, R. C., "Numerical Integration of General Stiff Initial-Value Problems by Singular Perturbation Methods," *AICHE-GVC Mtg.*, Munich, Germany (1974a).

Aiken, R. C., "Numerical Integration of Stiff Chemical Systems: The Pseudo-Steady-State Approximation," Jamshidi, M. and M. H. Nehrir, Eds., *Proc. of the First Iranian Conf. on Electrical Eng.*, Shiraz, Iran (1974b).

Aiken, R. C., "A Criterion for the Isothermal First Explosion Limit," *Combustion and Flame,* **47**, 281 (1982a).

Aiken, R. C., "Approximate Solutions to Explosion Kinetics," *AICHE J.*, **28**, 639 (1982b).

Aiken, R. C., "Selectivity in Gas Purification: Film Theory, Semi-Infinite Medium," *Chem. Eng. Sci.*, **37**, 1031 (1982c).

Aiken, R. C. and L. Lapidus, "An Effective Numerical Integration Method for Typical Stiff Systems," *AICHE J.*, **20**, 368 (1974).

Aiken, R. C. and L. Lapidus, "Problem Approximation For Stiff Ordinary Differential Equations," *AICHE J.*, **21**, 627 (1975a).

Aiken, R. C. and L. Lapidus, "Pseudo-Steady-State Approximation for the Numerical Integration of Stiff Kinetic Systems," *AICHE J.*, **21**, 817 (1975b).

Aiken, R. C. and V. Srinivasan, "Parameter Estimation in Stiff Ordinary Differential Equations," *Fourth International Conference on Mathematical Modelling,* 15–17 August, Swiss Federal Institute of Technology, Zurich, Switzerland (1983a).

Aiken, R. C. and V Srinivasan, "Parameter Estimation in Enzyme Kinetic Models," to appear in *Math. Biosci.* (1983b).

Albrecht, P., "On the Order of Composite Multistep Methods for Ordinary Differential Equations," *Numer. Math* (1978).

Alexander, R., "Diagonally Implicit Runge–Kutta Methods for Stiff ODEs," *Siam J. Numer. Anal.*, **14**, 1006 (1977).

Alfeld, P., "An Improved Version of the Reduction to Scalar CDS Method for the

Numerical Solution of Separably Stiff Initial Value Problems," *Math. Comp.*, **33**, 535 (1979a).

Alfeld, P., "Inverse Linear Multistep Methods for the Numerical Solution of Initial Value Problems of Ordinary Differential Equations." *Math. Comp.*, **33**, 111 (1979b).

Alfeld, P., "An Improved Version of the Reduction to Scalar CDS Method for the Numerical Solution of Separably Stiff Initial Value Problems," *Math. Comp.*, **33**, 535 (1979c).

Alfeld, P., "A Special Class of Explicit Linear Multistep Methods as Basic Methods for the Correction in the Dominant Space Technique," *Math. Comp.*, **33**, 1195 (1979d).

Alfeld, P., "A Method of Skipping the Transient Phase in the Solution of Separably Stiff Ordinary Differential Equations," *Math. Comp.*, **35**, 1173 (1980).

Alfeld, P. and J. D. Lambert, "Correction of the Dominant Space: A Numerical Technique for a Certain Class of Stiff Initial Value Problems," *Math. Comp.*, **31**, 922 (1977).

Allen, R. H. and Pottle, C., "Stable Integration Methods for Electronic Circuit Analysis With Widely Separated Time Constants," *Sixth Annual Allerton Conference on Circuit and System Theory,* Trick, T. and R. T. Chien, Ed., 311 (1966).

Alvarado, F. L., "Parallel Solution of Transient Problems by Trapezoidal Integration," *IEEE Trans. Power Appar. Sys. Pas-98,* **3**, 1080 (1979).

Amdurksy, V. and A. Ziv, "On the Numerical Solution of Stiff Linear Systems of the Oscillatory Type," *SIAM J. Ap. Math.*, **33**, 593 (1977).

Amundson, N. R., "Some Further Observations on Tubular Reactor Stability," *Can. J. Chem. Eng.*, **43**, 49 (1965).

Anderson, L. R., "Order Reduction in the Numerical Simulation of Time-Varying-Two-Time-Scale," *Control Conf., Pt. I,* San Francisco, California (1980).

Anderssen, R. S. and P. Bloomfield, "Numerical Differentiation Procedures for Non-Exact Data," *Numer. Math.*, **22**, 157 (1974).

Andrus, J. F., "Numerical Solution of Systems of Ordinary Differential Equations Separated into Subsystems," *SIAM J. Numer. Anal.*, **16**, 605 (1979).

Arnout, G. and H. J. De Man, "The Use of Threshold Functions and Boolean-Controlled Network Elements for Macromodelling of LSI Circuits," *IEEE J. Solid-State Circuits,* 13 (1978).

Aronowitz, D., Nategeli, D. W., and I. Glassman, "Kinetics of the Pyrolysis of Methanol," *J. Phys. Chem.*, **81**, 2555 (1977).

Atherton, R. W., Schainker, R. B., and E. R. Ducot, *AICHE J.*, **21**, 441 (1975).

Atkinson, R., Carter, W. P. L., Darnall, K. R., Winer, A. M. and J. M. Pitts, Jr., "A Smog Chamber and Modelling Study of the Gas Phase NOX-Air Photooxidation of Toluene and the Cresols," *Int. J. Chem. Kinetics,* **12**, 779 (1980).

Auslander, L. and W. L. Miranker, "Algebraic Methods in the Study of Stiff Differential Equations," *Numerical Analysis of Singular Perturbation Problem,* Hemker, P. W. and J. J. H. Miller, Eds., Academic Press, New York (1979).

Axelsson, O., "A Class of A-Stable Methods," *BIT,* **9**, 185 (1969).

Axelsson, O., "On the Efficiency of a Class of A-Stable Methods," *BIT,* **14**, 279 (1974).

BMDP Statistical Software, Department of Biomathematics, University of California, Los Angeles, University of California Press, 305 (1981).

Babcock, P. D., Stutzman, L. F. and T. G. Koup, "General Equation Modelling Systems (GEMS)—A Mathematical Simulation Language," *Simulation,* 36, 41 (1981).

Bader, G. and P. Deuflhard, "A Semi-Implicit Mid-Point Rule for Stiff Systems of Ordinary Differential Equations," *Inst. Appl. Math.*, Rpt. 114 (1981).

Bader, G., Nowak, U., and P. Deuflhard, "An Advanced Simulation Package for Very Large Chemical Reaction Systems," Univ. Heidelberg, SFB123: Tech Rpt., Feb. (1982a).

Bader, G., Nowak, U. and P. Deuflhard, "An Advanced Simulation Package for Large Chemical Reaction Systems," Proc. Int. Conf. Stiff Comp., Park City, Utah (1982b).

Baker, P. W., "The Solution of Differential Equations on Short-Word-Length Computing Devices," *IEEE Trans. Comp.*, **C-28**, 205 (1979).

Baldwin, A. C., Barker, J. R., Golden, D. M. and D. G. Hendry, "Photochemical Smog: Rate Parameter Estimates and Computer Simulations," *J. Phys. Chem.*, **81**, 25 (1977).

Ballard, D. M. and C. B. Brosilow, "Dynamic Simulation of Multicomponent Distillation Columns," paper 42a presented at the 71st Annual Meeting of *AICHE*, Miami, Florida (1978).

Bank, R. E., Sherman, A. H., and A. Weiser, "On the Regularity of Local Mesh Refinement," *Proc. 10th IMACS World Congress On System Simulation and Scientific Computing,* August 8–13, Montreal, Canada, p. 61 (1982).

Bardiaux, M., Devooght, J., Dugnoille, P., Goldwasser, M., Joly, J., Mund, E., Saragossi, I., and A. Siebertz, "Cassandre: A 2D Multigroup Diffusion Benchmark Code for fast Reactors Power Excursions Analysis," *Proc. of the ENS/ANS Topical Meeting on Nuclear Power Reactor Safety,* Brussels (1978).

Bar-Eli, K. and S. Haddad, "The Belousov-Zhabotinskii Reaction. Comparison of Experiments with Calculations," *J. Phys. Chem.*, **83**, 2944 (1979a).

Bar-Eli, K. and S. Haddad, "Effect of Oxygen on the Belousov-Zhabotinskii Oscillating Reaction," *J. Phys. Chem.*, **83**, 2952 (1979b).

Barker, J. R., "Infrared Multiphoton Decompositon: A Comparison of Approximate Models and Exact Solutions of The Energy-Grained Master Equation," *J. Chem. Phys.*, **72**, 3686 (1980).

Barney, J. R. and A. I. Johnson, "Numerical Solution of Large Systems of Stiff Ordinary Differential Equations in a Modular Simulation Framework," Rpt., Faculty of Engineering Science, University of Western Ontario, Ontario, Canada (1974).

Bartine, D., Oblow, E. M., and F. R. Mynatt, *Nuc. Sci. Eng.*, **55**, 147 (1974).

Barwell, V. K., "Numerical Solution of Differential-Difference Equations," Rpt. CS-76-04, Dept. of Computer Science, University of Waterloo, Waterloo, Ontario, Canada (1976).

Beam, R. M., and R. F. Warming, "An Implicit Finite-Difference Algorithm for Hyperbolic Systems in Conservation-Law Form," *J. Comput. Phys.*, **22**, 87 (1976).

Bell, G. and S. Glasstone, "Nuclear Reactor Theory," Van Nostrand Reinhold Co. (1970).

Bell, J. B., Shubin, G. R., and J. M. Solomon, "Fully Implicit Shock Tracking," *J. Comp. Phys.*, **48**, 223 (1982).

Bellman, R., *Introduction to Matrix Analysis,* McGraw-Hill, New York (1960).

Bellman, R., *Perturbation Techniques in Mathematics, Physics, and Engineering,* Holt, Rinehart, and Winston, New York (1964).

Ben-Shaul, A., Feiiks, S., and O. Kafri, "Time Evolution of the Pulsed HF Chemical Laser System I. Kinetic Modelling—Rotational Nonequilibrium," *J. Chem. Phys.*, **36**, 291 (1979).

Bennett, C. H., "Molecular Dynamics and Transition State Theory: The Simulation of Infrequent Events," Algorithms for Chemical Computations, Rpt., T. J. Watson Research Center, IBM, Yorktown Heights, New York (1977).

Bernard-Weil, E., Duvelleroy, M. and Y. Cherruault, "Computer Analog Simulation of a Model for the Regulation of Ago-Antagonistic Couples," *Int. J. Bio-Medical Computing,* **9**, 191 (1978).

Bettis, D. G., "Recent Developments with Implicit Runge–Kutta Algorithms," *Proc. ACM Annual Conf.*, San Diego (1974).

Bickart, T. A., "An Efficient Solution Process for Implicit Runge–Kutta Methods," *SIAM J. Numer. Anal.*, **14**, 1022 (1977).

Bickart, T. A., "Stiff Differential-Difference Equation Solver: Algorithm and Program," Technical Report, TR81-7, Electrical and Computer Eng. Dept., Syracuse Univ., Syracuse, New York (1981).

Bickart, T. A., "P-Stable and P(α, β)-Stable Integration-Interpolation Methods in the Solution of Retarded Differential-Difference Equations," *Proc. Int. Conf. Stiff Comp.*, Park City, Utah (1982a).

Bickart, T. A., "F-Stable and F(α, β)-Stable Integration/Interpolation Methods in the Solution of Retarded Differential-Difference Equations," *J. Comp. Appl. Math.*, **8**, 279 (1982b).

Bickart, T. A. and E. I. Jury, "Arithmetic Tests for A-Stability, A(α)-Stability, and Stiff-Stability," *BIT*, **18**, 9 (1978).

Bickart, T. A. and Z. Picel, "High Order Stiffly Stable Composite Multi-Step Methods for Numerical Integration of Stiff Differential Equations," *BIT*, **13**, 272 (1973).

Bickart, T. A. and W. B. Rubin, "Composite Multistep Methods and Stiff Stability in Stiff Differential Systems," R. A. Willoughby, Ed., Plenum Press, New York (1974).

Bickart, T. A., Burgess, D. A., and H. M. Sloate, "High Order A-Stable Composite Multistep Methods for Numerical Integration of Stiff Differential Equations," *Proc. Ninth Annual Allerton Conference on Circuit and System Theory*, University of Illinois, Urbana, Illinois, 465 (1971).

Billingsley, D. S., "Analytic and Numeric Solution of 'Stiff' Multiphase Convective Diffusion Equations with Variable Coefficients," *Ind. Eng. Chem. Fund.*, **21**, 272 (1982).

Birkhoff, G., "Numerical Solution of Non-linear Differential Equations." *Preoc. Symp. of the Math. Research Center*, U.S. Army, University of Wisconsin, Madison (1966).

Birkhoff, G. and R. S. Varga, "Discretization Errors for Well-Set Cauchy Problems," *J. Math & Phys.*, **44**, 1 (1965).

Birnbaum, K. and L. Lapidus, "Studies in Approximation Methods II.," *Chem. Eng. Sci.*, **33**, 427 (1978).

Bittker, D. A. and V. J. Scullin, "General Chemical Kinetics Computer Program for Static and Flow Reactions, NASA TN D-6586 (1972).

Balkemore, J. E. and W. H. Corcoran, "Validity of the Steady-State Approximation Applied to the Pyrolysis of n-Butane," *IEC Proc. Des. Deve.*, **8**, 2 (1969).

Belnski, T., *Atomkernenergie*, **31**, 232 (1978).

Blenski, T., Gadomski, A., and J. Mika, "Higher Order Prompt-Jump Approximation in Reactor Kinetics," *Nuclear Sci. & Eng.*, **66**, 66 (1978).

Bloch, R., Ingram, D., Sweeney, G. D., and C. J. Dickinson, "MAC DOPE: A Simulation of Drug Disposition in the Human Body: Mathematical Considerations," *J. Theor. Biol.*, **87**, 211 (1980).

Blottner, F. G., "Variable Grid Scheme for Discontinuous Grid Spacing and Derivatives," *Comp. & Fluids*, **8**, 421 (1980).

Bodenstein, M. and H. Lutkemeyer, "Die Photochemische Bildung von Bromwasserstoff, und die Bildungsgeschwindigkeit der Brommolekel aus den Atomen," *Z. Phys. Chem.*, **114**, 208 (1925).

Boisvert, R. F., and R. A. Sweet, "A Survey of Mathematical Software for Elliptic Boundary Value Problems," *Proc. 10th IMACS World Congress on System Simulation and Scientific Computing*, Montreal, Canada, p. 449 (1982).

Bond, J. E., and J. R. Cash, "A Block Method for the Numerical Integration of Stiff Systems of Ordinary Differential Equations," *BIT*, **19**, 429 (1979).

Boness, R., "A Two-Stage Semi-Implicit Runge–Kutta Process for Numerical Analysis of Nonlinear Dynamic Networks," *Wiss. Z. Tech.*, Univ. Dresden, Germany, **28**, 389 (1979).

Boni, A. A. and R. C. Penner, *Comb. & Sci. Tech.*, **15**, 99 (1977).

Boston, J. F., Britt, H. I., Jirapongphan, S., and V. B. Shah, "An Advanced System For the Simulation of Batch Distillation Operations," *Foundations of Computer Aided Chemical Process Design Vol. II,* Mah, R. S. H. and W. D. Seider, Eds., *AICHE* (1981).

Boston, J. F. and S. L. Sullivan, Jr., "A New Class of Solution Methods for Multicomponent, Multistage, Separation Process," *Can. Jour. Chem. Eng.,* **52**, 1 (1974).

Borgonovi, G. M., Hammelman, J. E., and C. L. Miller, "Dynamic Process Model of a Plutonium Oxalate Precipitator," *Nuclear Tech.,* **49**, 214 (1980).

Bottenheim, J. W. and O. P. Strausz, "Gas-Phase Chemistry of Clean Air at 55 Degrees N Latitude," *Environ. Sci. & Tech.,* **14**, 709 (1980).

Boudry, M. R. and J. P. Stagg, "The Kinetic Behavior of Mobile Ions in the Al-SI102-SI System," *J. Appl. Phys.,* **50**, 942 (1979).

Bourgeat, A. and R. Tapiero, "Singular Perturbations and Boundary Layer Correctors for a System of Two Quasi-Linear Parabolic Evolution Equations with Two Parameters," *Numerical Analysis of Singular Perturbation Problems,* Hemker, P. W. and J. J. H. Miller, Eds., Academic Press (1979).

Bowen, J. R., Acrivos, A., and A. K. Oppenheim, "Singular Perturbation Refinement to Quasi-Steady State Approximation in Chemical Kinetics," *Chem. Eng. Sci.,* **18**, 177 (1963).

Box, G. E. P. and H. L. Lucas, "Design of Experiments in Non-linear Situations," *Biometrica,* **46**, 77 (1959).

Boxenbaum, H. G., Riegelmann, S. and S. M. Elshoff, *Stat. Est. Pharmacokinet,* **2(2)**, 123 (1974).

Boy, G., "A Numerical Method for the Convective and Diffusive Process Simulation," *Proc. 2nd World Congress of Chemical Engineering,* **5**, 327 (1981).

Brackbill, J. U. and J. S. Saltzman, "Adaptive Zoning for Singular Problems in Two Dimensions," *J. Comp. Phy.,* **46**, 342 (1982).

Braekhus, J. and J. O. Aasen, "Experiments with Direct Integration Algorithms for Ordinary Differential Equations in Structural Dynamic," *Computers & Structures,* **13**, 91 (1981).

Brandon, D. M., Jr., "A New Single-Step Implicit Integration Algorithm With A-Stability and Improved Accuracy," *Simulation,* **23**, 17 (1974).

Brandon, D. M., Jr., "Improvements in a Single-Step Integration Algorithm," *Simulation,* **28**, 1 (1979).

Brandt, A., "Multi-Level Adaptive Techniques (MLAT) for Singular-Perturbation Problems," *Numerical Analysis in Singular Perturbation Problems,* Hemker, P. W. and J. J. H. Miller, Eds., Academic Press, New York (1979).

Brandt, A., "Multigrid Solutions to Steady-State Compressible Navier–Stokes Equations," *Proc. 5th Int. Symp. Computing Meth. in Applied Science and Engineering,* Glowinski, R. and J. L. Lions, Eds. (1981).

Brayton, R. K. and C. C. Conley, "Some Results on the Stability and Instability of the Backward Differentiation Methods With Non-Uniform Time Steps," Rpt. RC 7027, IBM Research Center, Yorktown Heights, New York (1978).

Brayton, R. K. and C. H. Tong, "Stability of Dynamic Systems: A Constructive Approach," Rpt. RC 7027, IBM Research Center, Yorktown Heights, New York (1978).

Brent, R. P., "An Algorithm With Guaranteed Convergence for Finding a Zero of a Function," *Computer J.,* **14(4)**, 422 (1971).

Briley, W. R. and H. McDonald, "Solution of the Three-Dimensional Compressible Navier–Stokes Equations by an Implicit Technique," *Proc. of the Fourth International Conference on Numerical Methods in Fluid Dynamics, Lecture Notes in Physics, Vol. 35,* Springer-Verlag, Berlin (1975).

Broughton, J. Q. and F. F. Abraham, "A Comparison of the FCC (111) and (100) Crystal-Melt Interfaces by Molecular Dynamics Simulation," *Chem. Phys. Let,* **71**, 456 (1980).

Brown, P. S., Jr., "More Efficient Numerical Solution of Diurnal Kinetics Problems," *J. Comp. Phys.,* **34**, 111 (1980).

Brown, R. L., "Numerical Integration of Linearized Stiff Ordinary Differential Equations," *Numerical Methods for Differential Systems,* Lapidus, L. and W. E. Schiesser, Eds., Academic Press, 39 (1976).

Brown, R. L., "Some Characteristics of Implicit Multistep Multi-Derivative Integration Formulas," *SIAM J. Numer. Anal.,* **14**, 982 (1977).

Brown, R. L., "Error Behavior of Multistep Methods Applied to Unstable Differential Systems," *Appl. Math. Modelling,* **2**, 115 (1978a).

Brown, R. L., "Evaluation of Ordinary Differential Equation Software," *BIT,* **18**, 103 (1978b).

Brown, R. L., "Stability Analysis of Nonlinear Differential Sequences Generated Numerically," *Comp. & Math. Appl.,* **5**, 187 (1979a).

Brown, R. L., "Stability of Sequences Generated by Nonlinear Differential Systems," *Math. Comp.,* **33**, 637 (1979b).

Brown, R. L. and C. W. Gear, "Documentation for DFASUB—A Program for the Solution of Simultaneous Implicit Differential and Nonlinear Equations," Dept. Computer Sci., Rpt. UIUCDCS-R-73-575, Univ. Illinois (1973).

Brummer, J. G. and R. J. Field, "Kinetics and Mechanism of the Oxidation of Ferrous Ion by Iodate Ion in Strong Perchloric Acid, Aqueous Media," *J. Phys. Chem,* **83**, 2328 (1979).

Brunner, H. "Recursive Collocation for the Numerical Solution of Stiff Ordinary Differential Equations," *Math. Comp.,* **28**, 475 (1974).

Bui, T. D., "On an L-Stable Method for Stiff Differential Equations," *Inf. Process. Lett.,* Netherlands, **6**, 158 (1977).

Bui, T. D., "A Note on the Rosenbrock Procedure," *Math. Comp.,* **33**, 971 (1979a).

Bui, T. D., "Some A-Stable and L-Stable Methods for the Numerical Integration of Stiff Ordinary Differential Equations," *J. Assn. for Computing Machinery,* **26**, 483 (1979b).

Bui, T. D., "Solving Stiff Differential Equations for Simulations," *Math. & Comp. Sim.,* **23**, 149 (1981).

Bui, T. D. and T. R. Bui, "Numerical Methods for Extremely Stiff Systems of Ordinary Differential Equations," *Appl. Math. Modelling,* **3**, 355 (1979).

Bui, T. D. and S. W. H. Poon, "On the Computational Aspects of Rosenbrock Procedures with Built-In Error Estimates for Stiff Systems," *BIT,* **21**, 168 (1981).

Bulirsch, R. and H. W. Branca, "Computation of Real-Time Control in Aerospace Applications by Multiple Shooting Procedures: Extended Abstract," *Stiff Differential Systems,* R. A. Willoughby, Ed., Plenum Press, New York (1974).

Bulirsch, R. and J. Stoer, *Num. Math.,* **8**, 1 (1966).

Burka, M. K., "Solution of Stiff Ordinary Differential Equations by Decomposition and Orthogonal Collocation," *AICHE J.* **28**, 11 (1982).

Burrage, K., "A Special Family of Runge–Kutta Methods for Solving Stiff Differential Equations," *BIT,* **18**, 22 (1978a).

Burrage, K., "High Order Algebraically Stable Runge–Kutta Methods," *BIT*, **18**, 373 (1978b).

Burrage, K., "Stability and Efficiency Properties of Implicit Runge–Kutta Methods," Ph.D. Thesis, Dept. of Mathematics, Auckland University (1978c).

Burrage, K., "Non-Linear Stability of Multivalue Multiderivative Methods," *BIT,* **20**, 316 (1980a).

Burrage, K., "An Implementation of Singly-Implicit Runge–Kutta Methods," *BIT,* **20**, 326 (1980b).

Burrage, K., "Efficiently Implementable Algebraically Stable Runge–Kutta Methods," *SIAM J. Numer. Anal.,* **19**, 245 (1982).

Burrage, K., "Order Properties of Multivalue Methods," in preparation (1983).

Burrage, K. and J. C. Butcher, "Stability Criteria for Implicit Runge–Kutta Methods," *SIAM J. Numer. Anal.*, **16**, 46 (1979).

Burrage, K. and J. C. Butcher, "Non-Linear Stability of a General Class of Differential Equation Methods," *BIT*, **20**, 185 (1980).

Burrage, K., Butcher, J. C., and F. Chipman, "An Implementation of Singly-Implicit Runge–Kutta Methods," *BIT*, **20**, 326 (1980).

Bushard, L. B., "Subroutine Packages for the Numerical Solution of Stiff Ordinary Differential Equations," Alliance Research Center Rpt., Babcock and Wilcox (1976).

Butcher, J. C., "Coefficients for the Study of Runge–Kutta Integration Processes," *J. Austral. Math. Soc.*, **3**, 185 (1963).

Butcher, J. C., "Implicit Runge–Kutta Processes," *Math. Comp.*, **18**, 50 (1964).

Butcher, J. C., "On the Attainable Order of Runge–Kutta Methods," *Math. Comp.*, **19**, 408 (1965).

Butcher, J. C., "On the Convergence of Numerical Solutions to Ordinary Differential Equations," *Math. Comp.*, **20**, 1 (1966).

Butcher, J. C., "The Order of Numerical Methods for Ordinary Differential Equations," *Math. Comp.*, **27**, 793 (1973).

Butcher, J. C., "A Stability Property of Implicit Runge–Kutta Methods," *BIT*, **15**, 358 (1975).

Butcher, J. C., "Implicit Runge–Kutta and Related Methods," *Modern Numerical Methods for Ordinary Differential Equations*, Hall, G. and J. M. Watt, Eds., Clarendon Press, Oxford 136 (1976a).

Butcher, J. C., "On the Implementation of Implicit Runge–Kutta Methods," *BIT*, **16**, 237 (1976b).

Butcher, J. C., "On A-Stable Implicit Runge–Kutta Methods," *BIT*, **17**, 375 (1977).

Butcher, J. C., "Some Implementation Schemes for Implicit Runge–Kutta Methods," *Numer. Anal. Proc.*, Dundee 12 (1979a).

Butcher, J. C., "A Transformed Implicit Runge–Kutta Method," *J. ACM.*, **26**, 731 (1979b).

Butcher, J. C., "Stability Properties for a General Class of Methods for Ordinary Differential Equations," *SIAM J. Numer. Anal.*, **18**, 37 (1981a).

Butcher, J. C., "A Generalization of Singly-Implicit Methods," *BIT*, **21**, 175 (1981b).

Butcher, J. C., Burrage, K., and F. Chipman, "An Implementation of Singly-Implicit Runge–Kutta Methods," Research Rpt. No. 149, University of Auckland (1979a).

Butcher, J. C., Burrage, K., and F. Chipman, "STRIDE: Stable Runge–Kutta Integrator for Differential Equations," Rpt. No. 150, Dept of Computer Sci., Auckland University (1979b).

Bykov, A. A., "Stable Numerical Method of Solution of Boundary Problems for a System of Linear Ordinary Differential Equations," *Dokl. Akad. Nauk SSSR*, **251**, 1040 (1980).

Bykov, A. A. and A. S. Il'Inskii, "Solution of Boundary Value Problems for Linear Systems of Ordinary Differential Equations by the Method of Directed Orthogonalization," *USSR Comput. Math. Phys.*, **19**, 74 (1979).

Byrne, G. D., "A Report on the ODE Workshop, Held at San Antonio, Texas, 1976," *ACM-SIGNUM Newsl.*, **11(1)**, 27 (1976).

Byrne, G. D., "Some Software for Stiff Systems of Differential Equations," *Proc. IMACS(AICA) Int. Symp. on Simulation Software & Numerical Methods for Diff. Equa.*, Blacksburg, Virginia (1977).

Byrne, G. D., "The ODE Solver EPISODE, Its Variants, and Their Use," *Proc. ANS Topical Meeting on Computational Methods in Nuclear Engineering*, Williamsburg, Virginia (1979).

Byrne, G. D., "Software for Differential Systems and Applications Involving Macroscopic Kinetics," *Comp. & Chem.*, **5**, 151 (1981a).

Bryne, G. D., "Some Software for Solving Ordinary Differential Equations," *Foundations of Computer-Aided Chemical Process Design, Vol. I,* Mah, R. S. H. and W. D. Seider, Eds., *AICHE* (1981b).

Byrne, G. D. and A. C. Hindmarsh, "A Polyalgorithm for the Numerical Solution of Ordinary Differential Equations," *ACM Trans. on Mat. Software*, **1**, 71 (1975).

Byrne, G. D. and A. C. Hindmarsh, "EPISODEB: An Experimental Package for the Integration of Systems of Ordinary Differential Equations with Banded Jacobians," UCID-30132, Lawrence Livermore Lab., Livermore, California (1976).

Byrne, G. D. and A. C. Hindmarsh, "Nonsymmetric Conjugate Gradient and Chemical Kinetics-Transport," Tech. Mem. No. 77-1, Num. Mat. Group, User Systems Div., Comp. Dept., Lawrence Livermore Lab., Livermore, California (1977).

Byrne, G. D., Hindmarsh, A. L., Jackson, K. R., and H. G. Brown, "A Comparison of Two ODE Codes: GEAR and EPISODE," *Comp. & Chem. Eng.*, **1**, 133 (1977).

Byrne, G. D., Gear, C. W., Hindmarsh, A. C., Hull, T. E., Krogh, F. T., and L. F. Shampine, "Panel Discussion of Quality Software for ODEs," *Numerical Methods for Differential Systems,* Lapidus, L. and W. E. Schiesser, Eds., Academic Press, 267 (1976).

Cahill, L. W., "A Stable Numerical Integration Scheme for Nonlinear Systems," *7th Hawaii Int. Conf. on Syst. Sciences,* Honolulu, Hawaii (1974).

Caillaud, J. B. and Padmanabhan, L., "An Improved Semi-Implicit Runge–Kutta Method for Stiff Systems," *Chem. Eng. J.*, **2**, 227 (1971).

Calahan, D. A., "A Stable, Accurate Method of Numerical Integration for Nonlinear Circuits," *Proc. IEEE,* **56**, 744 (1968).

Carey, G. F. and B. A. Finlayson, "Orthogonal Collocation on Finite Elements," *Chem. Eng. Sci.*, **30** (1975).

Carmichael, G. R., Kitada, T. and L. K. Peters, "Application of a Galerkin Finite Element Method to Atmospheric Transport Problems," *Comp. & Fluids,* **8**, 155 (1980).

Carter, W. P. L., Lloyd, A. C., Sprung, J. L. and J. N. Pitts, Jr., "Computer Modelling of Smog Chamber Data: Progress in Validation of a Detailed Mechanism for the Photooxidation of Propene and *n*-Butane in Photochemical Smog," *Inter. J. Chem.*, **11**, 45 (1979).

Carver, M. B., "FORSIM: A FORTRAN-Oriented simulation Package for the Automatic Solution of Partial and Ordinary Differential Equation Systems," Atomic Energy of Canada Rpt. AECL-4608, 1973.

Carver, M. B., "The Choice of Algorithms in Automated Method of Lines Solution of Partial Differential Equations," *Numerical Methods for Differential Systems,* L. Lapidus and W. E. Schiesser, Eds., Academic Press, p. 243 (1976).

Carver, M. B., "Efficient Handling of Discontinuous and Time Delays in Ordinary Differential Equations," *Proc. Int. Conf. Simulation,* Montreux, Switzerland (1977).

Carver, M. B., "FORSIM VI: A Program Package for the Automated Solution of Arbitrarily Defined Differential Equations," *Comp. Phys. Commun.*, **17**, 239 (1979).

Carver, M. B., "Pseudo Characteristic Method of Lines Solution of the Conservation Equations," *J. Comp. Phys.*, **35**, 57 (1980).

Carver, M. B., "The Role of General Purpose Software in the Simulation of Nuclear Components," Invited Paper ANS-NRC-EPRI Conference on Simulation Methods for Nuclear Power Systems, Tucson, Arizona (1981).

Carver, M. B. and A. P. Baudouin, "Solution of Reactor Kinetics Problems Using

Sparse Matrix Techniques in an ODE Integrator for Stiff Equations," *AECL-5177 Simulation Symposium on Reactor Dynamics and Plant Control* (1976).

Carver, M. B. and A. W. Boyd, "A Program Package Using Stiff, Sparse, Integration Methods for the Automatic Solution of Mass Action Kinetics Equations," *Int. J. Chem. Kinetics,* **11**, 1097 (1979).

Carver, M. B. and H. W. Hinds, "The Method of Lines and Advection Equation," *Simulation,* **31b**, no. 2, 56.

Carver, M. B. and S. R. MacEwen, "Automatic Partitioning in Ordinary Differential Equation Integration," *Prog. in Modelling and Simulation,* Cellier, F. E., Ed., Academic Press, London (1982).

Carver, M. B. and W. E. Schiesser, "Biased Upwind Difference Approximations for First-Order Hyperbolic (Convective) Partial Differential Equations," paper presented at the 73rd Annual Meeting of the American Institute of Chemical Engineers, November (1980).

Carver, M. B., Liu, J. L. and D. G. Stewart, "Assessment of Available Integration Algorithms for Initial Value Ordinary Differential Equations: Selection for the CRNL Subroutine Library and Simulation Packages," Report AECL-6220, Atomic Energy Canada Ltd., Chalk River, Ontario, p. 92 (1979).

Carver, M. B., Stewart, D. G., Blair, J. M., and W. N. Selander, "The FORSIM VI Simulation Package for the Automated Solution of Arbitrarily Defined Partial and/or Ordinary Differential Equation Systems," Rpt. No. AECL-5821. Available from: Atomic Energy of Canada Ltd., Chalk River Nuclear Laboratories, Chalk River, Ontario KOJ 1JO, p. 154.

Cash, J. R., "The Numerical Solution of Systems of Stiff Ordinary Differential Equations," *Proc. Camb. Phil. Soc.,* **76**, 443 (1974).

Cash, J. R., "A Class of Implicit Runge–Kutta Methods for the Numerical Integration of Stiff Ordinary Differential Equations," *J. Assoc. Comp. Math.,* **22**, 504 (1975a).

Cash, J. R., "The Numerical Solution of Linear Stiff Boundary Value Problems," *Comput. J.,* **18**, 371 (1975b).

Cash, J. R., "Semi-Implicit Runge–Kutta Procedures with Error Estimates for the Numerical Integration of Stiff Systems of Ordinary Differential Equations," *J. Assoc. Comp. Math,* **23**, 455 (1976).

Cash, J. R., "On a Class of Cyclic Methods for the Numerical Integration of Stiff Systems of ODEs," *BIT,* **17**, 270 (1977).

Cash, J. R., "High Order Methods for the Numerical Integration of Ordinary Differential Equations," *Numer. Math.,* **30**, 385 (1978).

Cash, J. R., "Diagonally Implicit Runge–Kutta Formulae With Error Estimates," *J. IMA,* **24**, 293 (1979).

Cash, J. R., "On the Integration of Stiff Systems of ODEs Using Extended Backward Differentiation Formulae," *Numer. Math.,* **34**, 235 (1980a).

Cash, J. R., "A Semi-Explicit Runge–Kutta Formula for the Numerical Integration of Stiff Systems of ODEs," *Chem. Eng. J.,* **20**, 219 (1980b).

Cash, J. R., "A Note on the Exponential Fitting of Blended, Extended Linear Multistep Methods," *BIT,* **21**, 450 (1981a).

Cash, J. R., "On the Design of Higher Order Exponentially Fitted Formulae for the Numerical Integration of Stiff Systems," *Numer. Math.,* **36**, 253 (1981b).

Cash, J. R., "High Order P-Stable Formulae for the Numerical Integration of Periodic Initial Value Problems," *Numer. Math.,* **37**, 355 (1981c).

Cash, J. R., "Second Derviative Extended Backward Differentiation Formulas for the Numerical Integration of Stiff Systems," *SIAM J. Numer. Anal.,* **18**, 21 (1981d).

Cash, J. R., "On the Exponential Fitting of Composite, Multiderivative Multistep Methods," *SIAM J. Numer. Anal.,* **18**, 808 (1981e).

Cash, J. R., "Block Runge–Kutta Methods for the Numerical Integration of Initial Value Problems in ODEs," to appear.

Cash, J. R. and C. B. Liem, "On the Computational Aspects of Semi-Implicit Runge–Kutta Methods," *Comp. J.*, **21**, 363 (1978).

Cash, J. R. and C. B. Liem, "On the Design of a Variable Order, Variable Step Diagonally Implicit Runge–Kutta Algorithm," *J. Inst. Maths, Applics.*, **26**, 87 (1980).

Cash, J. R. and A. Singhal, "Mono-Implicit Runge–Kutta Formulae for the Numerical Integration of Stiff differential systems," to appear.

Cavendish, J. C. and S. H. Oh, "A Computationally Efficient Galerkin Technique for Approximating Transient Diffusion-Reaction Equations in Composite Media," *Chem. Eng. J.*, **17**, 41 (1979).

Cellier, F. E., "The GASP-V Users' Manual," Available from the author, (1978).

Cellier, F. E., "Combined Continuous/Discrete System Simulation by Use of Digital Computers: Techniques and Tools," Ph.D. Thesis, ETH Zurich, Switerland (1979).

Cellier, F. E., "Stiff Computation: Where to Go," *Proc. Int. Conf. Stiff. Comp.*, Park City, Utah (1982).

Cellier, F. E. and A. E. Blitz, "GASP-V: A Universal Simulation Package," *Proc. 8th AICA Congress on Simulation of Systems,* Delft, The Netherlands, North-Holland Pub. Co., 391 (1976).

Cellier, F. E. and A. P. Bongulielmi, "The COSY Simulation Language," *Proc. 9th IMACS Congress on Simulation of Systems,* Sorrento, Italy, North-Holland Pub. Co., Dekker, L., Savastano, G., and G. C. Vansteenkiste, Eds., 271 (1979).

Cellier, F. E. Rimvall, M., and A. P. Bongulielmi, "Discrete Processes in COSY," *Proc. Int. Workshop on Modelling and Simulation Methodology,* Cosenze, Italy (1981).

Chamber, T., "Some Numerical Techniques for Use in a Large Digital Simulator," *Proceed. of UKSC Conf. on Computer Simulation,* Chester, England (1978).

Chamorro, R. S., "Fast Transient Contingency Evaluation in Power Systems," *IEEE Trans. Power Appar. & Syst.*, **100**, 1796 (1981).

Chan, Y. N. I., Birnbaum, I. and L. Lapidus, "Solution of Stiff Differential Equations and the Use of Imbedding Techniques," *Ind. Eng. Chem. Fundam.*, **17**, 133 (1978).

Chance, B., "The Kinetics of the Enzyme-Substrate Compound of Peroxidase," *J. Biol. Chem.*, **151**, 553 (1943).

Chance, B., "A Simple Relationship for Calculations of the 'on' Velocity Constant in Enzyme Reactions," *Arch. Biochem. Biophys.*, **71**, 130 (1957).

Chance, E. M., Curtis, A. R., Jones, I. P., and C. R. Kirby, "FACSIMILE: A Computer Program for Flow and Chemistry Simulation and General Initial Value Problems," Harwell Rpt. AERE-R8775, Oxfordshire, England (1977).

Chandler, J. P., Hill, D. E., and H. O. Spivey, "A Program for Efficient Integration of Rate Equations and Least-Squares Fitting of Chemical Reaction Data," *Comp. Biomed.*, **5**, 515 (1972).

Chang, J. S., Hindmarsh, A. C., and N. K. Madsen, "Simulation of Chemical Kinetics Transport in the Stratosphere," *Stiff Differential Systems,* R. A. Willoughby, Ed., Plenum Press, New York (1974).

Chang, J. S., Wuebbles, D. J., and W. H. Duewer, "Sensitivity to Parameter Uncertainties for Ozone Reduction from Chlorofluoromethanes," LLNL Rpt. UCRL-77432 (1976).

Chatterjee, A. and J. Magee, "A Spur Unfolding Model for the Radiolysis of Water," *Rad. Phys, & Chem.*, **15** (1980).

Chen, F. T., Douglas, J. M. and C. Georgakis, "Approximate Evaluation of Process Time Constants and Dominant Variables by Quasidiagonalization," *Ind. Eng. Chem. Fundam.*, **18**, 321 (1979).

Chen, J. C. and S. W. Churchill, "Radiant Heat Transfer in Packed Beds," *AICHE J.*, **9**, 35 (1963).

Chen, J. C. and S. W. Churchill, "A Theoretical Model for Stable Combustion Inside a Refractory Tube," *Combustion and Flame,* **18**, 27 (1972).

Chen, K. L. and W. E. Schiesser, "Some Experiences in the Selection of Integrators for Large-Scale ODE Problems in Chemical Engineering," *Proc. Int. Conf. Stiff Comp.,* Park City, Utah (1982).

Chen, K. L. and W. E. Schiesser, "Upwind Approximations in the Numerical Methods of Lines. Integration of Hyperbolic Partial Differential Equations," Lehigh University Rpt. DSS2, Bethlehem, Pennsylvania (1980).

Chen, Y. M. and D. T. S. Lee, "Numerical Methods for Solving Differential Equations with Inadequate Data," *J. Comp. Phys.,* **42**, 238 (1981).

Chin, R. C. Y. and R. L. Braun, "Numerical Solution of Chemically Reacting Flows in Porous Media," *J. Comp. Phys.,* **34**, 74 (1980).

Chin, R. C. Y., Hedstrom, G. W., and K. E. Karlsson, "A Simplified Galerkin Method for Hyperbolic Equations," *Math. Comp.,* **33**, 647 (1979).

Chipman, F. H., "A-Stable Runge–Kutta Methods," *BIT,* **11**, 384 (1971).

Chipman, F. H., "A Note on Implicit A-Stable R–K Methods with Parameters," *BIT,* **16**, 223 (1976).

Cho, Y. C., "Rigorous Solutions for Sound Radiation from Circular Ducts with Hyperbolic Horns or Infinite Plane Baffle," *J. Sound & Vibration,* **69**, 405 (1980).

Cho, Y. S. and B. Joseph, "Heterogeneous Model for Moving-Bed Coal Gasification Reactors," *Ind. Eng. Chem. Process Des. Dev.,* **20**, 314 (1981).

Choi, B. and S. W. Churchill, "A Model for Combustion of Gaseous and Liquid Fuels in a Refractory Tube," *17th Symp. Comb.,* The Combustion Inst., 915 (1979).

Chorin, A. J., "Numerical Methods for Use in Combustion Modelling," *Comp. Methods in Applied Sci. Eng.,* R. Glowinski and J. L. Lions, Eds., North-Holland Publishing Co., Inria p. 229 (1980).

Chow, J. H, Avramovic, B., Kokotovic, P. V., and J. R. Winkelman, "Singular Perturbations, Coherency and Aggregation of Dynamic Systems," Rpt., Electric Utility Systems Engineering Dept., General Electric Co., Schenectady, New York (1981).

Christiansen, J. P. and N. K. Winsor, "A Numerical Model for Laser Targets," *J. Comp. Phys.,* **35**, 291 (1980).

Chu, S. C. and M. Berman, "An Exponential Method for the Solution of Systems of Ordinary Differential Equations," *Comm. ACM,* **17**, 699 (1974).

Chua, T. S. and P. M. Dew, "The Design of A Variable-Step Integrator for the Simulation of Gas Transmission Networks," Rpt. No. 157, Dept. Computer Studies, University of Leeds, Leeds, England (1981).

Chua, T. S. and P. M. Dew, "Numerical Integration of Stiff Differential–Algebraic Equations with Severe Discontinuities," *Proc. Int. Conf. Stiff Comp.,* Park City, Utah (1982).

Chun-Yat-Lau, T., "A Class of Approximations to the Exponential Function for the Numerical Solution of Stiff Differential Equations," Univ. Waterloo, Ontario Canada, Ph.D. Thesis (1974).

Churchill, S. W., "Bounded and Patched Solutions for Boundary Value Problems," *AICHE J.,* **11**, 431 (1965).

Churchill, S. W., "Stiffness in Heat Transfer," *Proc. Int. Conf. Stiff Comp.,* Park City, Utah (1982).

Churchill, S. W. and J. P. Gupta, "Approximations for Conduction with Freezing and Melting," *Int. J. Heat Mass Transfer,* **20**, 1251 (1977).

Churchill, S. W., and H. Ozoe, "Correlations for Laminar Forced Convection with Uniform Heating Flow Over a Plate and in Developing and Fully Developed Flow in a Tube," *J. Heat Transfer,* **95c**, 78 (1973a).

Churchill, S. W. and H. Ozoe, "Correlations for Laminar Forced Convection in

Flow Over an Isothermal Flat Plate and In Developing and Fully Developed Flow in an Isothermal Tube," *J. Heat Transfer,* **95C**, 416 (1973b).

Churchill, S. W. and Usagi, R., "A General Expression for the Correlation of Rates of Transfer and Other Phenomena," *AICHE J.,* **18**, 1121 (1972).

Clasen, R. J., Garfinkel, D., Shapiro, N.Z. and G. C. Roman, "A Method for Solving Certain Stiff Differential Equations," *SIAM Appl. Math.,* **34**, 732 (1978).

Clore, G. M. and E. M. Chance, "The Mechanisms of Reaction of Fully Reduced Membrane-Bound Cytochrome Oxidase with Oxygen at 176K," *Biochem. J.,* **173**, 799 (1978).

Clough, R. W., "The Finite Element Method After Twenty-Five Years, A Personal View," Report, University of California, Berkeley, California (1978).

Clouston, J. G., Gilbert, C. P., and P. Miskelly, "An Autocatatlytic Model for Bacterial Spore Germination," *J. Theor. Bio.,* **76**, 387 (1979).

Coleman, T. G., Mesick, H. C., and R. L. Darby, "Numerical Integration, A Method for Improving Solution Stability in Models of the Circulation," *Ann. Biomed. Eng.,* **5**, 322 (1977).

Collatz, L., "The Numerical Treatment of Differential Equations," Springer-Verlag, Berlin, 323 (1960).

Collins, R. J., "Bandwidth Reduction by Automatic Renumbering," *Int. J. Num. Methods Eng.,* **6**, 345 (1973).

Come, G. M., "Mechanistic Modelling of Homogeneous Reactors: A Numerical Method," *Comp. & Chem. Eng.,* **3**, 603 (1979).

Cooper, G. J., "The Order of Convergence of General Linear Methods for Ordinary Differential Equations," *SIAM J. Numer. Anal.,* **15**, 643 (1978).

Cooper, G. J. and J. C. Butcher, "An Iteration Scheme for Runge–Kutta Methods," to appear *IMA J. Numer. Anal.* (1983).

Cooper, G. J. and A. Sayfy, "Semiexplicit A-Stable Runge–Kutta Methods," *Math. Comp.,* **33**, 541 (1979).

Cooper, G. J. and F. C. P. Whitworth, "Liapunov Functions and Error Bounds for Approximate Solutions of Ordinary Differential Equations," *Numer. Math.,* **30**, 411 (1978).

Coppel, W. A., "Stability and Asymptotic Behavior of Differential Equations," D.C Heath, Boston, Mass. (1965).

Corbo, V. amd L. Lapidus, "An Invariant Imbedding Difference Equation Algorithm, *Chem. Eng. J.,* **3b**, 35 (1972).

Corley, T. L. and C. T. Bowman, *Pulverized Coal Combustion, Pollutant Formation and Control,* EPA Project Decade Monograph, Chpt. 5, to be published by U.S. Gov't. Printing Office (1983).

Corliss, G. and Y. F. Chang, "Solving Ordinary Differential Equations Using Taylor Series," *ACM Trans. Math. Software,* **8**, 114 (1982).

Costabile, F. and C. Costabile, "Two-Step Fourth Order P-Stable Methods For Second Order Differential Equations," *BIT,* **22**, 384 (1982).

Covello, R. J. and A. B. White, "Application of Stiff Methods to Analytical Assessment Problems," *AFWL-AFOSR Workshop,* p. 250 (1977).

Creighton, J., "The Role of Numerical Stability in the Computer Solution of Chemical Chain Reactions," *Proc. Summer Simulation Conf.,* Boston (1971).

Crisci, M. R. and E. Russo, "A-Stability of a Class of Methods for the Numerical Integration of Certain Linear Systems of Ordinary Differential Equations," Math. Comp. **38**, 431 (1982).

Crosbie, R. E., "Interactive and Real-Time Simulation," *Progress In Modelling and Simulation,* F. E. Cellier, Ed., Academic Press, p. 393 (1982).

Crouzeix, M. "Sur l'Approximation des Equations Differentielles Operationelles Lineaires par des Methodes de Runge–Kutta," Ph.D. Thesis, University of Paris VI, Paris, France (1975).

Crouzeix, M. "On B-Stability of the Methods of Runge–Kutta," *Numer. Math.,* **32**, 75 (1979).

Crow, J. F. and M. Kimura, *An Introduction to Population Genetics Theory,* Harper Row, New York (1970).

Cryer, C. W., "A New Class of Highly Stable Methods: A_0-Stable Methods," *BIT,* **13**, 153 (1973).

Cryer, C. W., "Highly Stable Multistep Methods for the Retarded Differential Equations," *SIAM J. Numer. Anal.,* **11**, 788 (1974).

Cukier, R. I., Levine, H. B. and K. E. Shuler, "Nonlinear Sensitivity Analysis of Multiparameter Systems," *J. Comp. Phys.,* **26**, 1 (1978).

Curtis, A. R., "Solution of Large, Stiff Initial Value Problems—The State of the Art," *Comp. Sci. & Syst. Div.,* AERE Harwell, Oxfordshire, 1 (1978a).

Curtis, A. R., "The FACSIMILE Numerical Integrator for Stiff Initial Value Problems," *Computational Techniques for Ord. Dif. Equations, Proceed. of Conf.,* Manchester, Eng., London, England, Academic Press (1978b).

Curtis, A. R., "The Implementation of Implicit Runge–Kutta Formulae for the Solution of Initial Value Problems," *J. Inst. Math. & Appl.,* **23**, 339 (1979).

Curtis, A. R., "A Guide to the Theory of Gear Method for the Numerical Solution of Systems of Stiff Ordinary Differential Equations," Rpt. NPL REP DNACS 29-38, Nat. Phys. Lab., Teddington, Eng. p. 13 (1980).

Curtis, A. R., Powell, M. J. D., and J. K. Reid, "On the Estimation of Sparse Jacobian Matrices," *J. Inst. Math. Appl.,* **13**, 117 (1974).

Curtiss, C. F. and J. O. Hirschfelder, "Integration of Stiff Equations," *Proc. Nat. Acad. Sci.,* **38**, 235 (1952).

Dahlquist, G., "A Special Stability Problem for Linear Multistep Methods," *BIT,* **8**, 27 (1963).

Dahlquist, G., "Problems Related to the Numerical Treatment of Stiff Differential Equations," *International Computing Symp.,* Gunther, A., Levrat, B., and H. Lipps, Eds., Amer. Elsevier, New York, p. 307 (1974a).

Dahlquist, G., "The Sets of Smooth Solutions of Differential and Difference Equations," *Stiff Differential Systems,* R. A. Willoughby, Ed., Plenum Press, New York (1974b).

Dahlquist, G., "On Stability and Error Analysis for Stiff Nonlinear Problems," Royal Inst. Tech., Stockholm, Sweden (1975a).

Dahlquist, G., "Recent Work on Stiff Differential Equations," *Czechoslavakian-Swedish Symp. on Appl. Math.,* (1975b).

Dahlquist, G., "Error Analysis for a Class of Methods for Stiff Nonlinear Initial Value Problems," Num. Anal. Dundee, Lecture Notes in Math 506, Springer-Verlag, Berlin, 60 (1975c).

Dahlquist, "On the Relationship of G-Stability to Other Stability Concepts for Linear Multistep Methods," *Topics In Numerical Analysis III,* J. Miller, Ed., Academic Press, London, England, 349 (1977).

Dahlquist, G., "On Accuracy and Unconditional Stability of Linear Multistep Methods for Second-Order Differential Equations," *BIT,* **18**, 133 (1978a).

Dahlquist, G., "G-Stability is Equivalent to A-Stability," *BIT,* **18**, 384 (1978b).

Dahlquist, G., "Positive Functions and Some Applications to Stability Questions for Numerical Methods," *Recent Adv. in Num. Anal.,* Deboor, C. and G. Golub, Eds., Academic Press, 1 (1978c).

Dahlquist, G., "Some Properties of Linear Multistep and One-Leg Methods for ODEs," *1979 SIGNUM Meeting on ODEs,* R. D. Skeel, Ed., also available as Rpt. TRITA-NA-7904 (1979a).

Dahlquist, G., "Some Contractivity Questions for One-Leg and Linear Multistep Methods," Royal Inst. Tech., Stockholm, Sweden (1979b).

Dahlquist, G., "On the Control of Global Error in Stiff Initial Value Problems," Dept. Num. Anal., Royal Inst. of Tech., Stockholm, Sweden (1981a).

Dahlquist, G., "On the Local and Global Errors of One-Leg Methods," Rpt. TRITA-

NA-8110, Dept. of Numerical Analysis and Computing Science, Royal Inst. of Tech., Stockholm, Sweden (1981b).

Dahlquist, G., "On Matrix Majorants and Minorants With Applications to Differential Equations," accepted for publication in *Lin. Alg. Appl.*, (1983a).

Dahlquist, G., "On One-Leg Multistep methods," submitted to *SIAM J. Num. Anal.* (1983b).

Dahlquist, G., "On Transformations of Graded Matrices, With Applications to Stiff ODEs," to appear as a TRITA-NA-Rpt. (1983c).

Dahlquist, G. and R. Jeltsch, "Genralized Disks of Contractivity for Explicit and Implicit Runge–Kutta Methods," Royal Inst. Tech., Stockholm, Sweden (1979).

Dahlquist, G. and Mao Zu-Fan, "Partitioning a Stiff Initial Value Problem by Scaling Technique," Royal Institute of Technology, Stockholm, Sweden (1982).

Dahlquist, G. and G. Soderlind, "Some Problems Related to Stiff Non-Linear Differential Systems," *Computing Methods in Applied Sciences and Engineering V,* Glowinski, R. and J. L. Lions, Eds., North-Holland Pub. Co., 57 (1982).

Dahlquist, G., Edsberg, L., Skollermo, G., and G. Soderlind, "Are the Numerical Methods and Software Satisfactory for Chemical Kinetics?" Rpt. TRITA-NA-8005, Royal Institute of Technology, Stockholm, Sweden (1980).

Dahlquist, G., Edsberg, K., Skollermo, G., and G. Soderlind, "Are the Numerical Methods and Software Satisfactory for Chemical Kinetics?" Springer Lecture Notes in Mathematics, **968**, 149 (1982).

Dahlquist, G., Liniger, W., and O. Nevanlinna, "Stability of Two-Step Methods for Variable Integration Steps," Research Report, RC 8494 (§36967) IBM (1980).

Dahlquist, G., Liniger, W., and O. Nevanlinna, "Stability of Two-Step Methods for Variable Integration Steps," to appear in *SIAM J. Numer. Anal.* (1983).

Dahlquist, G., Mingyou, H., and R. Leveque, "On the Uniform Boundedness of a Family of Matrices and the Application to One-Leg and Linear Multistep Methods," Royal Inst. Tech., Stockholm, Sweden (1981).

Datta, N. K. and H. R. Martens, "Automatic Step Size Selection in the Numerical Solution of Mixed Differential–Algebraic Equations," *Modelling and Simulation,* **5**, 1035 (1974).

Davey, A., "On the Numerical Solution of Difficult Boundary-Value Problems," *J. Comp. Phys.,* **35**, 36 (1980).

David, C. W., "Direct Integration of Chemical Kinetic Rate Equations," *Comp. & Chem.,* **1**, 167 (1977).

Davison, E. J. and W. G. Cross, "An Algorithm for the Simulation of Very Large Dynamic Composition Systems," *Modelling and Simulation,* **7**, 26 (1976).

Day, J. D., "On the Internal S-Stability of Rosenbrock Methods," *Inf. Process. Lett.,* **11**, 27 (1980a).

Day, J. D., "Comments On: T. D. Bui 'On an L-Stable Method for Stiff Differential Equations,' " *Inf. Process. Lett.,* **11**, 31 (1980b).

Day, J. D. and D. N. P. Murthy, "Two Classes of Internally S-Stable Generalized Runge–Kutta Processes Which Remain Consistent with an Inaccurate Jacobian," *Math. Comp.,* **39**, 491 (1982).

De Doncker, E., "An Adaptive Extrapolation Algorithm for Automatic Integration," *SIGNUM Newsl.,* **13**, 12 (1978).

De Groat, J. J. and M. J. Abbett, "A Computation of One-Dimensional Combustion of Methane," *AIAA J.* **3**, 381 (1965).

De Groen, P. P. N. and P. W. Hemker, "Error Bounds for Exponentially Fitted Galerkin Methods Applied to Stiff Two-Point Boundary Value Problems," *Numerical Analysis of Singular Perturbation Problems,* Hemker, P. W. and J. J. H. Miller, Eds., Academic Press, p. 217 (1974).

De Hoog, F. R. and R. Weiss, "The Application of Linear Multistep Methods to Singular Initial Value Problems," *Math Comp.,* **31**, 676 (1977).

Dekker, K., "Algebraic Stability of General Linear Methods," Rpt. No. 25, Dept. Computer Sci., Auckland University (1981).

Demiralp, M. and H. Rabitz, *J. Chem. Phys.*, **74**, 3362 (1981).

Descloux, J., "On the Numerical Integration of the Heat Equation," *Numer. Math.* **15**, 371 (1970).

De Silva, B. M. E. and G. C. Grant, "Comparison of Some Optimal Control Methods for the Design of Turbine Blades," *J. Mech. Design,* **100**, 173 (1978).

Desoer, C. A. and H. Haneda, "The Measure of a Matrix as a Tool to Analyse Computer Algorithms for Circuit Analysis," *IEEE Trans.*, **CT-19**, 480 (1972).

Deuflhard, P., 'A Study of Extrapolation Methods Based on Multistep Schemes Without Parasitic Solutions," *J. Appl. Math. Phys. (ZAMP),* **30**, 17 (1979).

Deuflhard, P., "Order and Stepsize Control in Extrapolation Methods," Univ. Heidelberg, SFB 112: Tech. Rpt. 93 (1980a).

Deuflhard, P., "Kepler Discretization in Regular Celestial Mechanics," *Cel. Mech.*, **21**, 213 (1980b).

Deuflhard, P., "Recent Progress in Extrapolation," *SIAM 30th Anniversary Meeting* Standford, California, July (1982).

Deuflhard, P., Bader, G., and V. Nowak, "LARKIN—A Software Package for the Numerical Solution of Large Systems Arising in Chemical Reaction Kinetics," *Modelling of Chemical Reaction Systems,* Ebert, K. H., Deuflhard, P., and W. Jaeger, Eds., Springer, Berlin (1981).

Devooght, J., "Quasistatic Solutions of Reactor Kinetics," *Anal. Nuclear Energy,* **7**, 47 (1979a).

Devooght, J., *N. S. E.,* **67**, 147 (1979b).

Devooght, J., *Anal. Nucl. Energy,* **7**, 47 (1980).

Devooght, J., "An Overview of Stiffness Problems in Nuclear Reactor Kinetics," *Proc. Int. Conf. Stiff Comp.,* Park City, Utah (1982).

Devooght, J. and E. Mund, "A-Stable Algorithms for Neutron Kinetics," *Proc. NEACRP/CSNI Specialist Meeting on New Developments in 3D-Neutron Kinetics and Review of Kinetics Benchmark Calculations,* T. U. Munchen-Garching (1975).

Devooght, J. and E. Mund, "Generalized Quasi-Static Method for Nuclear Reactor Space-Time Kinetics," *Nuclear Science & Eng.,* **76**, 10 (1980).

Devooght, J. and E. Mund, "International Topical Meeting," *Advances in Mathematical Methods for the Solution of Nuclear Engineering Problems,* **2**, 103 (1981).

Devooght, J., Mund, E., et al., "Developments of New Numerical Methods for Space-Dependent Nuclear Reactor Dynamics," EEC-DGIII-E2/ULB Contract, Vols. 1–3 (1977).

Dew, P. M. and J. E. Walsh, "Library Routines for Solving Parabolic Equations," *Trans. Math. Software,* **7(3)**, 295 (1981).

Dew, P. M. and M. R. West, "Estimating and Controlling the Global Error in Gear's Method," *BIT,* **19**, 135 (1979).

Dey, S. K., "Numerical Solution of Nonlinear Finite Difference Equations of Flow Problems by Perturbed Functional Iterations," *Computational Fluid Dynamics* W. Kollmann, Ed., Hemisphere Pub. Co., New York (1980a).

Dey, S. K., "Analysis and Applications of a Perturbed Iterative Scheme to Nonlinear Discretized Flow Models," *Proc. Int. Cong. Num. Methods Eng.* Dunod, Paris, France (1980b).

Dey, S. K., "Numerical Solution of Stiff Nonlinear Systems by Perturbed Functionals," *Proc. Int. Conf. Stiff Comp.,* Park City, Utah (1982).

Dias, M. Q., Lopes, J. C., Costa, C. A., and A. E. Rodrigues, "Dynamics of Fixed Bed Adsorbers," *Proc. Int. Conf. Stiff Comp.,* Park City, Utah (1982).

Dickinson, R. P. and Gelinas, R. J., "GENKIN-I, A General Atmospheric Code for Atmospheric Applications," Lawrence Livermore Lab. Rpt. UCID-16577 (1974).

Dickinson, R. P. and R. J. Gelinas, "Sensitivity Analysis of Ordinary Differential Equation Systems—A Direct Method (Atmospheric Chemical Kinetics Application)," *J. Comput. Phys.*, **21**, 123 (1976a).

Dickinson, R. P., Jr. and R. J. Gelinas, "SETKIN: A Chemical Kinetics Preprocessor Code," *Numerical Methods for Differential Systems,* Lapidus, L. and W. E. Schiesser, Eds., Academic Press, p. 167 (1976b).

Dietze, S. and A. Reibiger, "Stability and Stiff Differential Equations (Circuit Computer-Aided Analysis and Numerical Integration)," *Nachrichtentech Elektron.*, **28**, 82 (1978).

Dill, C. and C. W. Gear, "A Graphical Search for Stiffly Stable Methods for Ordinary Differential Equations," *J. Assoc. Comput. Math.*, **18**, 75 (1971).

Dimitrov, V. I., "Some Mathematical Modelling Problems in Chemical Kinetics of High-Speed Processes", *Archivum Termodynamiki i Spalania,* **6b**, No. 1, 117 (1975).

Dissinger, G. R. and W. E. Schiesser, "Experiences in Heuristic Techniques for Evaluation and Processing the Jacobian Matrix of an ODE System." *Proc. 10th IMACS World Congress on System Simulation and Scientific Computing,* August 8–13, Montreal, Canada, p. 443 (1982).

Distefano, G. P., "Mathematical Modelling and Numerical Integration of Multicomponent Batch Distillation Equations," *AICHE J.*, **14**, 190 (1968).

Divakaruni, S. M., Long, A. B., and A. F. Armor, "The EPRI Modular Modelling System for Power Plant Transient Simulation," presented at 1982 Summer Comp. Simulation Conf., July 19–21, Denver, Colorado, p. 363 (1982).

Dommel, H. W. and W. S. Meyer, "Computation of Electromagnetic Transients," *Proceed. of the IEEE,* **62**, 983 (1974).

Dongarra, J. J., Bunch, J. R., Moler, C. B., and G. W. Stewart, "LINPACK User's Guide," *SIAM*, Philadelphia, Pennsylvania (1979).

Dontchev, A. L., "Sensitivity Analysis of Linear Infinite-Dimensional Optimal Control Systems Under Changes of System Order," *Control & Cybern.*, **3**, 21 (1974).

Dorning, J. and G. Spiga, *Trans. Am. Nucl. Soc.*, **28**, 761 (1978).

Dougherty, E. P. and H. Rabitz, "Computational Kinetics and Sensitivity Analysis of Hydrogen-Oxygen Combustion," *J. Chem. Phys.*, **72**, 6571 (1980).

Dougherty, E. P., Hwang, J. and H. Rabitz, "Further Developments and Applications of the Green Function Method of Sensitivity Analysis in Chemical Kinetics," *J. Chem. Phys.*, **71**, 1794 (1979).

Dove, J. E. and S. Raynor, "An Ab Initio Calculation of the Rate of Vibrational Relaxation and Thermal Dissociation of Hydrogen by Helium at High Temperatures," *J. Phys. Chem.*, **83**, 127 (1979).

Dove, J. E. and S. Raynor, "A Master Equation Study of the Rate and Mechanism of Vibrational Relaxation and Dissociation of Molecular Hydrogen by Helium," *Proc. Int. Conf. Stiff Comp.*, Park City, Utah (1982).

Dove, J. E. and H. Teitelbaum, "The Vibrational Relaxation of H2. II. Measured Density Changes; Master Equation Study of the Mechanism of Relaxation," *Chem. Phys.*, **40**, 87 (1979).

Dove J. E., Scroggie, B. J., and H. Semerjian, "Velocity Deficits and Detonability Limits of Hydrogen-Oxygen Detonations," *Acta Astrn.*, **1**, 345 (1974).

Duff, I. S. and V. Nowak, "On Sparse Matrix Techniques in a Stiff Integrator of Extrapolation Type," Univ. Heidelberg SFB 123: Techn. Rep. (1982).

Dwyer, H. A., Kee, R. J., and B. R. Sanders, "An Adaptive Grid Method for Problems in Fluid Dynamics and Heat Transfer," AIAA 79-1464 (1979).

Earles, J. W., Wagner, S. G., and R. G. Foster, "FRST-A Fast-Running Space-Time Kinetics Code for Use in a Nuclear Power Plant Simulation," presented at 1982 Summer Comp. Simulation Conf., July 19–21, Denver, Colorado, p. 571 (1982).

Ebert, K. H., Ederet, H. J., and G. Isbard, *Angew. Chem.*, **19**, 333 (1980).

Ebert, K. H., Deuflhard, P., and W. Jaeger, Eds., *Modelling of Chemical Reaction Systems, Springer Series Chem. Hys.*, **18** (1981).

Eccher, J., "Report from Symposium on Applied Mathematics," *Program and Abstracts*, April (1975).

Edelson, D., "On the Solution of Differential Equations Arising in Chemical Kinetics," *J. Comp. Phys.*, **11**, 455 (1973).

Edelson, D., "A Simulation Lanuage and Compiler to Aid Computer Solution of Chemical Kinetic Problems," *Comp. & Chem.*, **1**, 29 (1976b).

Edelson, D., "Computational Techniques for the Study of Reaction Processes," *Comp. & Chem.*, **5**, 149 (1981).

Edelson, D. and D. L. Allara, "Parameterization of Complex Reaction Systems: Model Fitting versus Fundamental Kinetics," *AICHE J.*, **19**, 638 (1973).

Edleson, D. and D. L. Allara, *Int. J. Chem. Kinetics*, **12**, 605 (1980).

Edelson, D. and N. L. Schryer, "Modelling Chemically Reaction Flow Systems —I. A Comparison of Finite Difference and Finite Element Methods for One-Dimensional Reactive Diffusion," *Comp. & Chem.*, **2**, 71 (1978).

Edsberg, L., "Integration Package for Chemical Kinetics," *Stiff Differential Systems*, R. A. Willoughby, Ed., Plenum Press, (1974).

Edsberg, L., "Numerical Methods for Mass Action Kinetics," *Numerical Methods for Differential Systems*, Lapidus, L. and W. E. Schiesser, Eds., Academic Press, 181 (1976).

Edsberg, L., "Chemical Kinetics—An Overview from the Point of View of Numerical Analysis and Software Implementation," *Proc. Int. Conf. Stiff. Comp.*, Park City, Utah (1982).

Ehle, B. L., "On Pade Approximations to the Exponential Function and A-Stable Methods for the Numerical Solution of Initial Value Problems," University of Waterloo, Dept. Applied Analysis and Computer Sci., Research Rpt. No. CSRR 2010 (1969).

Ehle, B. L., "Two-Parameter, Arbitrary Order, Exponential Approximations for Stiff Equations," *Math. Comp.*, **29**, 501 (1975).

Ehle, B. L. and J. D. Lawson, "Generalized Runge–Kutta Processes for Stiff Initial-Value Problems," *J. Inst. Math. & Appl.*, **16**, 11 (1975).

Eigenberger, G. and J. B. Butt, "A Modified Crank-Nicolson Technique with Non-Equidistant Space Steps," *Chem. Eng. Sci.*, **31**, 681 (1976).

Eisenstat, S. C., Gursky, M. C., Schultz, M. H., and A. H. Sherman, "Yale Sparse Matrix Package: I. The Symmetric Codes," Research Rpt. No. 112, Dept. Computer Sci., Yale University (1977a).

Eisenstat, S. C., Gursky, M. C., Schultz, M. H., and A. H Sherman, "Yale Sparse Matrix Package: II. The Nonsymmetric Codes," Research Rpt. No. 114, Dept. Computer Sci., Yale University (1977b).

Eisenstat, S. C., Gursky, M. C., Schultz, M. H., and A. H. Sherman, "Yale Sparse Matrix Package I: The Symmetric Codes," *Int. J. Numer. Method, Eng.*, **18**, 1145 (1982).

Eitelberg, E., "Numerical Simulation of Stiff Systems with a Diagonal Splitting Method," *Mat. & Comput. Simulation*, **21**, 109 (1979).

Eitelberg, E., "Modular Simulation of Large Stiff Systems," *Progress in Modelling and Simulations*, F. E. Cellier, Ed., Academic Press, p. 281 (1982).

El-Genk, M. S. and A. W. Cronenberg, "On the Thermal Stability of a Frozen Crust in Forced Flow on an Insulated Finite Wall," *Int. J. Heat Mass Transfer*, **22**, 1719 (1979).

El-Zorkany, H. I., "Hybrid Computer Soluton of PDEs Using Laplace-Modified Galerkin Approximation," *Math. & Comp. Simul.*, **23**, 304 (1981).

Ellison, D., "Efficient Automatic Integration of Ordinary Differential Equations with Discontinuities," *Math. Comp. Simul.*, **23**, 12 (1981).

Elmqvist, H., "SIMNON—An Interactive Simulation Program for Nonlinear Systems User's Manual," Rpt. No. 7502, Dept. Auto. Control, Lund Institute of Tech., Lund, Sweden (1975).

Elmqvist, H., "SIMNON—An Interactive Simulation Program for Nonlinear Systems," *Proc. Int. Conf. SIMULATION '77,* Montreux, Switzerland, M. Hamza, Ed., Acta Press, 85 (1977).

Emel Yanov, K. V., "A Different Scheme for an Ordinary Differential Equation with a Small Parameter," *Zh. Vychisl. Mat. & Mat. Fiz,* **18**, 1146 (1978).

Enright, W. H., "Optimal Second Derivative Methods for Stiff Systems," *Stiff Differential Systems,* R. A. Willoughby, Ed., Plenum Press, p. 95 (1974a).

Enright, W. H., "Second Deriviative Multistep Methods for Stiff Ordinary Differential Equations," *SIAM J. Numer. Anal.,* **11**, 321 (1974b).

Enright. W. H., "Improving the Efficiency of a Matrix Operations in the Numerical Solution of Stiff Ordinary Differential Equations," *ACM Trans. On Math. Software,* **4**, 127 (1978a).

Enright, W. H., "The Efficient Solution of Linear Constant-Coefficient Systems of Differential Equations," *Simulation,* **30**, 129 (1978b).

Enright, W. H., "On the Efficient and Reliable Numerical Solution of Large Linear Systems of ODEs," *IEEE Trans Aut. Cont. AC-24,* 905 (1979).

Enright, W. H., "On the Efficient Time Integration of Systems of Second-Order Equations Arising in Structural Dynamics," *Int. J. Numerical Methods Eng.,* **16**, 13 (1980).

Enright, W. H., "The Use of Partitioning in Stiff Solvers, Numerical Methods for Solving Stiff Initial Value Problems," *Proc. Oberwolfach,* **28.6**, 4.7. (1981).

Enright, W. H., "Pitfalls in the Comparison of Numerical Methods for Stiff Ordinary Differential Equations," *Proc. Int. Conf. Stiff Comp.,* Park City, Utah (1982).

Enright, W. H. and T. E. Hull, "Comparing Numerical Methods for the Solution of Stiff Systems of ODEs Arising in Chemistry," *Numerical Methods for Differential Systems,* Lapidus, L. and W. E. Schiesser, Eds., Academic Press, p. 45 (1976).

Enright. W. H. and M. S. Kamel, "Automatic Partitioning of Stiff Systems and Exploiting the Resulting Structure," *ACM Trans. on Math. Software,* **5**, 374 (1979).

Enright, W. H. and M. S. Kamel, "On Selecting a Low-Order Model Using the Dominant Mode Concept," *IEEE Trans. Aut. Cont. AC-25,* 976 (1980).

Enright, W. H., Hull, T. E., and B. Lindberg, "Comparing Numerical Methods for Stiff Systems of ODEs," *BIT,* **15**, 10 (1975).

Fagley, J. C. and B. Carnahan, "Comparison of Integration and Equipment-Grouping Strategies in Dynamic Chemical Plant Simulation," presented at 1982 Summer Comp. Simulation Conf., July 19–21, Denver, Colorado, p. 543 (1982).

Fan, L., Tojo, K., and L. T. Fan, "Cyclic Dynamics of a Nonisothermal Controlled Release," *Math. Biosci.,* **45**, 69 (1979).

Fan, L. T., Tojo, K., and C. C. Chang, "Modelling of Shallow Fluidized Bed Combustion of Coal Particles," *Ind. Eng. Chem. Process Des. Dev.,* **18**, 333 (1979).

Farrow, L. A. and D. Edelson, "The Steady-State Approximation: Fact or Fiction?" *Int. J. Chem. Kinetics,* **1**, 787 (1974).

Farrow, L. A. and T. E. Graedel, "Steady State Approximations and Urban Atmospheric Chemistry," *J. Phys. Chem.,* **81**, 2480 (1977).

Fatunla, S. O., "A New Algorithm for Numerical Solution of Ordinary Differential Equations," *Comp. Math. Appl.,* **2**, 247 (1976).

Fatunla, S. O., "An Implicit Two-Point Numerical Integration Formula for Linear and Non-Linear Stiff Systems of Ordinary Differential Equations," *Math. Comp.,* **32**, 1 (1978).

Fatunla, S. O., "Numerical Integrators for Stiff and Highly Oscillatory Differential Equations," *Math. of Computation,* **34**, 373 (1980).

Fatunla, S. O., "P-Stable Hybrid Schemes for Initial Value Problems," *Proc. Int. Conf. Stiff Comp.*, Park City, Utah (1982).

Fehlberg, E., "New High Order Runge–Kutta Formulas With Step-Size Control for Systems of First- and Second-Order Differential Equations," *Z. Angew, Math. Mech.*, **44**, 17 (1964).

Feinberg, M., "Reaction Network Structure, Multiple Steady States, and Sustained Composition Oscillations: A Review of Some results," *Modelling of Chemical Reactions Systems,* Ebert, K. H., Deuflhard, P., and W. Jager, Eds., Springer Berlin, Heidelberg (1981).

Feinberg, R. B., "A_0-Stable Formulas of Adams Type," *SIAM J. Numer. Anal.*, **19**, 259 (1982).

Fendell, F. E., "Ignition and Extinction in Combustion of Initially Unmixed Reactants," *J. Fluid Mech.*, **21**, 281 (1965).

Ferguson, W. E., Jr., "A Singularly Perturbed Linear Two-Point Boundary Value Problem," Doctoral Dissertation, California Institute of Technology, Pasadena, California (1975).

Field, R. J. and R. M. Noyes, "Oscillations in Chemical Systems. IV. Limit Cycle Behavior in a Model of a Real Chemical Reaction," *J. Chem. Phys.*, **60** (1974).

Field, M. A., Gill, D. W., Morgan, B. B., and P. G. W. Hawksley, "Combustion of Pulverized Coal," *BCURA Leatherhead,* Cherey & Sons Ltd., Bamburg, England (1967).

Finlayson, B. A., "The Method of Weighted Residuals and Variational Principles," Academic Press (1972).

Finlayson, B. A., "Nonlinear Analysis in Chemical Engineering, McGraw-Hill (1980).

Finlayson, B. A., "Solution of Stiff Equations Resulting from Partial Differential Equations," *Proc. Int. Conf. Stiff Comp.*, Park City, Utah (1982a).

Finlayson, B. A., "Methods for Solving Partial Differential Equations in Chemical Engineering," Paper Presented at Joint *AICHE-CIESC* meeting, Beijing, China (1982b).

Finlayson, B. A. and R. W. Nelson, "A Preliminary Investigation into the Theory and Techniques of Modelling the Natural Moisture Movement in Unsaturated Sediments," Rpt., BCSR-40, Boeing Computer Services, Richland (1977).

Flaherty, J. E. and R. E. O'Malley, Jr., "The Numerical Solution of Boundary Value Problems for Stiff Differential Equations," Rpt. TR-9, University of Arizona (1979).

Flaherty, J. E. and R. E. O'Malley, Jr., "On the Numerical Integration of Two-Point Boundary Value Problems for Stiff Systems of Ordinary Differential Equations," *Boundary and Interior Layers—Computational and Asymptotic Methods,* J. J. H. Miller, Ed., Boole Press, Dublin 93 (1980).

Flaherty, J. E. and R. E. O'Malley, Jr., "The Numerical Solution of Boundary Value Problems for Stiff Differential Equations," *Math. Comp.*, **31**, 66 (1977).

Flaherty, J. E. and R. E. O'Malley, "The Numerical Solution of Singularly-Perturbed Initial Value Problems," *Numerical Analysis of Singular Perturbation Problems,* Hemker, P. W. and J. J. H. Miller, Eds., Academic Press (1979).

Forsythe, G. E., Malcolm, M. A., and C. B. Moler, *Computer Methods for Mathematical Computations,* Prentice-Hall, 129 (1977).

Frank, L., "On Asymptotic Behavior of High Oscillating Solutions," *Uspechi Math. Nauk XXV,* **5**, 155 (1970).

Frank, R. and C. W. Ueberhuber, "Iterated Defect Correction for the Efficient Solution of Stiff Systems of Ordinary Differential Equations," *BIT,* **17**, 46 (1977).

Frank, R. and Ueberhuber, C. W., "Iterated Defect Correction for Differential Equations, Part 1: Theoretical Results," *Computing,* **20**, 207 (1978).

Franke, R., "An Efficient Method for Solving Stiff Transient Field Problems Arising from FEM Formulations," *Comput. & Math. with Appl.*, **6**, 15 (1980).

Franklin, R. D., "Kinetic Modelling of Chemical Laser Systems," AFWL–AFOSR Workshop, p. 135 (1977).

Franks, R. G. E., "DYFLO Update: DYFLO2," presented at 1982 Summer Comp. Simulation Conf., July 19–21, Denver, Colorado, p. 507 (1982).

Franzone, P. C., Stefanelli, M. and C. Viganotti, "A Distributed Model of Iron Kinetics for Clinical Assessment of Normal-Abnormal Erythropoietic Activity," *IEEE Trans. on Biomed. Eng.*, **26**, 586 (1979).

Friedli, A., "Generalized Runge–Kutta Procedure for Solution of Stiff Differential Equations," *Proc. of Conf. on Numerical Treatment of Diff. Equ.*, Germany, p. 35 (1978).

Friedli, A. and R. Jeltsch, "An Algebraic Test for A_0–Stability," *BIT*, **18**, 402 (1978).

Froment, G. F., *Ind. Eng. Chem.*, **59(2)**, 18 (1967).

Fuchs, F., "A-Stability of Runge–Kutta Methods with Single and Multiple Nodes," *Computing*, **16**, 39 (1976).

Gaffney, P. W., "A Survey of Fortran Subroutines Suitable for Solving Stiff Oscillatory Ordinary Differential Equations," Tech. Memo 134, Oak Ridge (1982a).

Gaffney, P. W., "Using the Method of Lines Technique to Solve Initial Boundary Value Partial Differential Equations," *Proc. 10th IMACS World Congress on System Simulation and Scientific Computing*, August 8–13, Montreal, Canada, p. 433 (1982b).

Galantal, A., "New Stability Property Concerning Stiff Methods," *Numerical Methods*, Keszthely, Hungary (1980).

Gallivan, K. A., "Detection and Integration of Oscillatory Differential Equations with Initial Stepsize, Order and Method Selection," Dept. Comp. Sci., Univ. Illinois (1980).

Gallun, S. E. and C. D. Holland, "GEAR's Procedure for the Simultaneous Solution of Differential and Algebraic Equations with Application to Unsteady State Distillation Problems," *Comp. Chem. Eng.*, **6**, 231 (1982).

Gambart-Ducros, D. and G. Maral, "An Efficient Algorithm Reduces Computing Time in Solving a System of Stiff Ordinary Differential Equations," *IEEE Trans. on Circuits and Systems*, **27**, 747 (1980).

Gannon, D. and J. Van Rosendale, "Highly Parallel Multigrid Solvers for Elliptic PDEs: An Experimental Analysis," NASA Langley Research Center, ICASE Rpt. No. 82-36 (1982).

Gardiner, W. C., Jr., "Derivation of Elementary Reaction Rate Constants by Means of Computer Modelling," *J. Phys. Chem.*, **83**, 37 (1979).

Garfinkel, D., "Numerical Integration of Enzyme Reactions Systems," personal communication (1973).

Garfinekl, D., Marbach, C. B., and N. Z. Shapiro, "Stiff Differential Equations," *Ann. Rev. Biophys. Bioeng.*, **6**, 525 (1977).

Gary, J., "The Method of Lines Applied to a Simple Hyperbolic Equations," *J. Comp. Phys.*, **22**, 131 (1976).

Gawronski, W., "The Effect of System Parameter Variations on Natural Frequencies," *Computers & Structures*, **5**, 31 (1975).

Gear, C. W., *Numerical Initial Value Problems in ODEs*, Prentice-Hall, Englewood Cliffs, New Jersey (1971a).

Gear, C. W., "The Automatic Integration of Ordinary Differential Equations," *Comm. ACM*, **14**, 176 (1971b).

Gear, C. W., "Simultaneous Numerical Solution of Differential-Algebraic Equations," *IEEE Trans. Circuit Theory*, **CT-18**, 89 (1971c).

Gear, C. W., "What Do We Need In Programming Languages for Mathematical Software?", Dept. Computer Sci. Rpt. UIUCDCS-R-74-652, Univ. Illinois (1974).

Gear, C. W., "Future Developments in Stiff Integration Techniques—Stability of Methods that Do Not Use an Exact Jacobian," AFWL-AFOSR Workshop (1977a).

Gear, C. W., "ODE Techniques for Partial Differential Equations," *Formulations and Computational Algorithms in Finite Element Analysis,* Bathe, K. T., Oden, J. T., and W. Wunderlich, Eds., 691 (1977b).

Gear C. W., "Simulation: Conflicts Between Real-Time and Software," *Mathematical Software III,* J. R. Rice, Ed., Academic Press, New York 121 (1977c).

Gear, C. W., "The Stability of Numerical Methods for Second Order Ordinary Differential Equations," *SIAM J. Numer. Anal.,* **15**, 188 (1978).

Gear, C. W., "Method of Initial Stepsize Selection in Multistep ODE Solvers," Rpt. R-80-1006 (1980a).

Gear, C. W., "Numerical Software: Science or Alchemy?," *Advances in Computers,* **19**, 229 (1980b).

Gear, C. W., "Numerical Solution of Ordinary Differential Equations: Is There Anything Left to Do?," *SIAM Review,* **23**, 13 (1980c).

Gear, C. W., "Runge–Kutta Starters for Multistep Methods," *ACM Trans. on Mathematical Software,* **6**, no. 3, 263 (1980d).

Gear, C. W., "The Automatic Treatment of Stiff and of Oscillatory Equations," *Proc. Bielefield Symp. Num. Tech. Comp. Chem.,* Germany (1980e).

Gear, C. W., "Automatic Multirate Methods for Ordinary Differential Equations," *Proc. IFIP,* **80**, North-Holland Pub. Co., 717 (1980f).

Gear, C. W., "Stiff Systems: What Do We Have and What Do We Need?" *Proc. Int. Conf. Stiff Comp.,* Park City, Utah (1982).

Gear, C. W. and K. A. Gallivan, "Automatic Methods for Highly Oscillatory Ordinary Differential Equations," *Proc. Biennial Dundee Conf. Num. Anal.,* Scotland (1981).

Gear, C. W. and O. Osterby, "Solving Ordinary Differential Equations With Discontinuities," Dept. Computer Sci. Rpt. UIUCDCS-R-81-1064, Univ. Illinois (1981).

Gear, C. W. and K. W. Tu, "The Effect of Variable Mesh Size on the Stability of Multistep Methods," *SIAM J. Numer. Anal.,* **11**, 1025 (1974).

Gear, C. W., Tu, K. W., and D. S. Watanabe, "The Stability of Automatic Programs for Numerical Problems," *Stiff Differential Systems,* R. A. Willoughby, Ed., Plenum Press, New York (1974).

Gear, C. W. and D. S. Watanabe, "Stability and Convergence of Variable Order Multistep Methods," *SIAM J. Numer. Anal.,* **11**, 1044 (1974).

Gear, C. W., Hsu, H. H., and L. Petzold, "Differential Algebraic Equation Revisited," *Proc. Workshop Num. Meth. Solving Stiff Initial Value Problems,* Germany (1981).

Gear, C. W. and Y. Saad, "Iterative Solution of Linear Equations in ODE Codes," Dept. Computer Sci. Rpt. UIUCDCS-R-81-1054, University of Illinois (1981).

Gehring, P. J. and G. E. Blau, "Mechanisms of Carcinogenesis: Dose Response," *J. Envirn. Patn. Tox.,* **1**, 163 (1977).

Gelinas, R. J., "Stiff Systems of Kinetic Equations—A Practitioner's View," *J. Comp. Phys.,* **9**, 222 (1972).

Gelinas, R. J. and P. D. Skewes-Cox, "Tropospheric Photochemical Mechanisms," *J. Phys. Chem,* **81**, 2468 (1977).

Gelinas, R. J., "Diurnal Kinetic Modelling," LLNL jRpt. UCRL-75373, 1974, pub. *Proc. Intern. Conf. Structure, Composition and General Circulation of the Upper and Lower Atmospheres and Possible Anthropogenic Perturbations, Melbourne, Australia,* Derco, N. J. and E. H. Trublar, Eds., Atmospheric Environment Service, Downsview, Ontario (1974).

Gelinas, R. J., Doss, S. K., Vajk, J. P., Djomehri, J., and K. Miller, "Moving Finite

Elements in Two Dimensions," *Proc. IMACS 10th World Congress,* Montreal, **1**, 58 (1982).

Genin, Y., "An Algebraic Approach to A-Stable Linear Multistep-Multiderivative Integration Formulas," *BIT,* **14**, 137 (1977).

Gerlich, D., Nowotny, U., Schlier, Ch., and E. Teloy, "Complex Formation in Proton-D2 Collisions," *Chem. Physics,* **47**, 245 (1980).

Gersting, J. M., Jr., "Numerical Methods for Eigensystems: The Orr-Sommerfeld Problem as an Initial Value Problem," *Comput. & Math. with Appl.* **6**, 167 (1980).

Giddings, J. C. and J. O. Hirschfelder, *Sixth Symposium on Combustion,* Reinhold Pub. Corp., New York, 199 (1975).

Girijashankar, P. V., Ketrick, D. L., Keepin, W. N., and O. A. Palusinski, "A Technique for Efficient Simulation of Large Dynamic System," *Joint Automatic Control Conf.,* **1**, WP9-A(1980a).

Girijashankar, P. V., Hetrick, D. L., Keepin, W. N., and O. A. Palusinski, "An Efficient Simulation Technique for PWR Systems," *Trans. ANS,* **34**, 321 (1980b).

Girijashankar, P. V., Hetrick, D. L., Keepin, W. N., and O. A. Palusinksi, "An Efficient Technique for the Simulation of Large Systems," *Advances in Mathematical Methods for the Solution of Nuclear Engineering Problems, ANS-ENS Int. Topical Meeting,* Munich, Germany (1981).

Girijashankar, P. V., Hetrick, D. L., Keepin, W. N., Paluskinski, O. A., and P. G. Bailey, "Multirate Integration Algorithms Applied to Nuclear Power System Dynamics," *Proc. Int. Conf. Stiff Comp.,* Park City, Utah (1982).

Gladwell, I., "Extrapolation Methods for Stiff Systems and a Comparison of Methods for Stiff Problems," *Modern Numer. Methods for Ordinary Differential Equations,* Hall, G. and J. M. Watt, Eds., Clarendon Press, Oxford, p. 164 (1976).

Gladwell, I., "Initial Value Routines in the NAG Library," *ACM Trans. Math. Software,* **5**, 386 (1980).

Glaser, J., "On the Problem of Stability in the Numerical Treatment of Differential Equations," *Computing,* **19**, 221 (1978).

Goldstein, R. and L. M. Shotkin, *N. S. E.,* **38**, 94 (1969).

Goldstein, R. and L. M. Shotkin, *N. S. E.,* **40**, 498 (1970).

Gopfert, W., "Optimizing Discrete Step Width for Implicit Integration Processes of Variable Step Width and Order," *NSTG-Fachber,* **51**, 53 Germany (1975).

Gordon, S. and B. McBride, "Computer Program for Calculation of Complex Chemical Equilibrium Compositions," NASA SP-273 (1971).

Gottwald, "KISS—A Digital Simulation System for Coupled Chemical Reactions," *Simulation,* **37**, 169 (1981).

Gottwald, B. A. and G. Wanner, "A Reliable Rosenbrock Integration for Stiff Differential Equations," *Computing,* **26**, 355 (1981).

Gottwald, B. A. and G. Wanner, "Comparison of Numerical Methods for Stiff Differential Equations in Biology and Chemistry," to appear in *Simulation* (1982).

Gourlay, A. R., "The Extrapolation of First Order Methods for Parabolic Partial Differential Equations II," *SIAM J. Numer. Anal.,* **17**, 641 (1980).

Gourlay, A. R. and H. D. D. Watson, "An Implementation of Gear's Algorithm for CSMP III," *Stiff Differential Systems,* R. A. Willoughby, Ed., Plenum Press, New York (1974).

Graedel, T. E., Farrow, L. A., and T. A. Weber, "The Influence of Aerosols on the Chemistry of the Troposhere," *Bell Lab,* Murray Hill, New Jersey (1974).

Gray, B. F., "The Unified Theory of Explosions with Fuel Consumption," *Combustion Studies,* 17th Combustion Symp. (1972).

Greenberg, H. J. and A. G. Konheim, "Linear and Nonlinear Methods in Pattern Classification," *IBM J. Res. Dev.,* **8**, 299 (1964).

Greene, A. E. and C. A. Brau, "Theoretical Studies of UV–Preionized Transverse Discharge KRF and ARF Lasers," *J. Quantum Electronics,* **14**, 951 (1978).

Gresho, P. M., Lee, R. L., and R. L. Sani, *Recent Advances in Numerical Methods in Fluids, Vol. 1,* Chpt. 2, Taylor, C., and K. Morgan, Eds., Pineridge Press Ltd., Swansea, U.K. (1980).

Griffiths, D. F., "Towards Time-Stepping Algorithms for Convective-Diffusion," Numerical Analysis of Singular Perturbation Problems, Hemker, P. W. amd J. J. H. Miller, Eds., Academic Press, p. 200, New York (1979).

Grigorieff, R. and J. Schroll, "Ueber A(α)-stabile Verfahren hoher Konsistenzordnung," Rpt. No. 34, Fachbereich Mathematik, technische Universitat Berlin (1977).

Grigorieff, R. D. and J. Schroll, "On A (α)-Stable Methods with High Order of Consistency," *Computing,* **20**, 343 (1978).

Grober, H., Erk, S. and R. Grigull, *Fundamentals of Heat Transfer,* 3rd Ed., Transl. J. R. Moszynski, New York (1961).

Gross, G., "An Efficient Algorithm for Simulation of Transients in Large Power Systems," *IEEE Trans. Circuits & Syst. Cas.,* **23**, 791 (1976).

Gross, G. and A. R. Bergen, "A Class of New Multistep Integration Algorithms for the Computation of Power System Dynamic Response," *IEEE Trans. on Power Apparatus & Sys.,* **96**, 293 (1977).

Guderley, K. G., "A Unified View of Some Methods for Stiff Two-Point Boundary Value Problems," *SIAM Review,* **17**, 416 (1975).

Guertin, E. W., Sorensen, J. P., and W. E. Stewart, "Exponential Collocation of Stiff Reactor Models," *Computers and Chem. Eng.,* **1**, 197 (1977).

Gupta, G. K., "Some New High-Order Multistep Formulas for Solving Stiff Equations," *Math. Comp.,* **30**, 417 (1976).

Gupta, G. K., "Implementing Second-Derivative Multistep Methods Using the Nordsieck Polynomial Representation," *Math. Comp.,* **32**, 13 (1978).

Gupta, G. K., "A Polynomial Representation of Hybrid Methods for Solving Ordinary Differential Equations," *Math. Comp.,* **33**, 1251 (1979a).

Gupta, G. K., "DSTIFF: A Set of Subroutines for Solving Stiff Equations, User's Guide Part 1," Dept. Computer Sci. Rpt. 3, Monash Univ., Clayton, Victoria, Australia (1979b).

Gupta, G. K., "A Note About Overhead Costs in ODE Solvers," *ACM Trans. on Math. Software,* **6**, 319 (1980).

Gupta, G. K., "Description and Evaluation of a Stiff ODE Code DSTIFF3," Dept. Computer Sci. Rpt., Monash Univ., Clayton, Victoria, Australia (1981).

Gupta, G. K., "Description and Evaluation of a Stiff ODE Code DSTIFF," *Proc. Int. Conf. Stiff Comp.,* Park City, Utah (1982).

Gupta, G. K. and C. S. Wallace, "Some New Multistep Methods for Solving Ordinary Differential Equations," *Math. Comp.,* **29**, 489 (1975).

Gupta, G. K. and C. S. Wallace, "A New Step-Size Changing Technique for Multistep Methods," *Math. Comp.,* **33**, 125 (1979).

Gustafson, R. R., Sleicher, C. A., McKean, W. T., and B. A. Finlayson, "A Theoretical Model for the Kraft Pulping Process," submitted to *Ind. Eng. Chem. Proc. Des. Dev.* (1982).

Hachtel, G. and M. Mach, "A Pseudodynamic Method for Solving Nonlinear Algebraic Equations," *Stiff Differential Systems,* R. A. Willoughby, Ed., Plenum Press, New York (1974).

Hachtel, G. D., Brayton, R. K., and F. G. Gustavson, "The Sparse Tableau Approach to Network Analysis and Design," *IEEE Trans. on Circuit Theory,* **18** (no. 1), 101 (1971).

Hahn, W. A., "Pollutant Formation of Flat Laminar Opposed Jet Diffusion Flames,"

Ph.D. Dissertation, Univ. of Arizona, Tucson, University Microfilms, Ann Arbot, Michigan (1979).

Hahn, W. A. and J. O. L. Wendt, "Integration of Stiff, Boundary Valued ODEs in Combustion Problems with Detailed Kinetics, Convection and Diffusion," *Int. Conf. Stiff Comp.*, Park City, Utah (1982).

Haile, J. M. and H. W. Graben, "Molecular Dynamics Simulations Extended to Various Ensembles, I. Equilibrium Properties in the Isoenthalpic-Isobaric Ensemble," *J. Chem. Phys.*, **73**, 2414 (1980).

Hairer, E., "On the Order of Iterated Defect Correction, An Algebraic Proof," *Numer. Math.*, **29**, 409 (1978).

Hairer, E., "Unconditionally Stable Methods for Second Order Differential Equations," *Numer. Math.*, **32**, 373 (1979).

Hairer, E., "Highest Possible Order of Algebraically Stable Diagonally Implicit Runge–Kutta Methods," *BIT*, **20**, 254 (1980a).

Hairer, E., "Unconditionally Stable Explicit Methods for Parabolic Equations," *Numer. Math.*, **35**, 57 (1980b).

Hairer, E., "Constructive Characterization of A-Stable Approximations to Exp(z), and Its Connection With Algebraically Stable Runge–Kutta Methods," *Num. Math.*, **39**, 247 (1982).

Hairer, E., Bader, G., and C. H. Lubich, "On the Stability of Semi-Implicit Methods for Ordinary Differential Equations," *BIT*, **22**, 211 (1982).

Hairer, E. and G. Wanner, "On the Butcher Group and General Multi-Value Methods," *Computing*, **13**, 15 (1974).

Hairer, E. and G. Wanner, "Algebraically Stable and Implementable Runge–Kutta Methods of High Order," *SIAM J. Numer. Anal.*, **18**, 1098 (1981)

Hajj, I. N. and S. Skelboe, "Time-Domain Analysis of Nonlinear Systems with Finite Numer of Continuous Derivatives," *Trans. Circ. Syst. Cas-26*, 297 (1979).

Halin, H. J., "Integration of Ordinary Differential Equations Containing Discontinuities," *Proc. of Comp. Simu. Conf.*, Wash. D.C. (1976).

Halin, H. J., "Integration Across Discontinuities in Ordinary Differential Equations Using Power Series," *Simulation*, **32**, 33 (1979).

Halin, H. J. and H. Benz, "PSCSP-A Continuous-System Simulation Program Based Upon Semianalytical Methods," *1982 Summer Comp. Simulation Conf.*, July 19–21, Denver, Colorado, 8 (1982).

Han, S. M., Wu, S. T., and Y. Nakagawa, "Numerical Study of an Explosion in a Non-Homogeneous Medium with and Without Magnetic Fields," *Comp. & Fluids*, **7**, 97 (1979).

Hansen, K. B., Livbjerg, H., and J. Villadsen, "Dynamic Modelling of Catalytic Fixed-Bed Reactors," Rpt., Institutet far Kemiteknik, Danmarks Tekniske Hoyskole, Lyngby, Denmark (1974).

Hansen, K. B., Wilbrandt, R., and P. Pagsberg, "Electro-Optical Multichannel Spectrometer for Transient Resonance Raman and Absorpotion Spectroscopy," *Rev. Sci. Instrum.*, **50**, 1532 (1979).

Harris, W. A., Jr., "Singularly Perturbed Boundary Value Problems revisited," *Lecture Notes in Math.*, **312**, 54, Springer-Verlag, Berlin (1973).

Hautman, D. J., Santoro, R. J., Dryer, F. L., and I. Glassman, "An Overall and Detailed Kinetic Study of the Pyrolysis of Propane," *Int. J. Chem. Kinetics*, **13**, 149 (1981).

Hay, J. L., Crosbie, R. E., and R. I. Chaplin, "Integration Routine for Systems with Discontinuities," *Comp. J.*, **17**, 275 (1974).

Hedstrom, G., Rodrigue, G., Berger, M., and J. Oliger, "Adaptive Mesh Refinement for 1-Dimensional Gas Dynamics," *Proc. 10th IMACS World Congress on System Simulation and Scientific Computing*, August 8–13, Montreal, Canada, p. 55 (1982).

Heineken, F. G., Tsuchiya, H. M., and R. Aris, "On the Mathematical Status of the Pseudo-Steady State Hypothesis of Biochemical Kinetics," *Math. Biosci.*, **1**, 95–113 (1967).

Heinrich, J. C. and O. C. Zienkiewicz, "Solution of Nonlinear Second Order Differential Equations With Significant First Derivatives by a Petrov-Galerkin Finite Element Method," *Numerical Analysis of Singular Perturbation Problems,* Hemker, P. W. and J. J. H. Miller, Eds., Academic Press, p. 251 (1979).

Heinzinger, K., Riede, W. O., Schaefer, L., and G. I. Szasz, "Molecular Dynamics Simulations of Liquids with Ionic Interactions," *Computer Modelling of Matter,* ACS (1978).

Hendry, W. L. and G. I. Bell, *N. S. E.,* **35**, 240 (1969).

Hennart, J. P., *Math. Comp.,* **31**, 24 (1977a).

Hennart, J. P., *N. S. E.,* **64**, 875 (1977b).

Hennart, J. P. and H. Gourgeon, Springer Notes to appear 1982.

Henri, V., *Compt. Rend,* **135**, 916 (1902).

Henrici, P., *Discrete Variable Methods in Ordinary Differential Equations,* J. Wiley & Sons, New York (1962).

Henry, A., "Nuclear Reactor Analysis," The MIT Press (1975).

Henry, A., *N. S. E.,* **20**, 338 (1974).

Herriot, G. E., Eckert, R. C., and L. F. Albright, "Kinetics of Propane Pyrolysis," *AICHE J.,* **18**, 84 (1972).

Hersh, R. and T. Kato, "High-Accuracy Stable Difference Schemes for Well-Posed Initial-Value Problems," *SIAM J. Numer. Anal.,* **16**, 670 (1979).

Hetrick, D. L., Girijashankar, P. V., and O. A. Palusinski, "Solution Methods for Simulation of Nuclear Power Systems," EPRI NP-1928, Project 1381, Univ. of Arizona, Tucson, Arizona (1981).

Heydweiller, J. C. and R. F. Sincovec, "A Stable Difference Scheme for the Solution of Hyperbolic Equations Using the Method of Lines," *J. Comp. Phys.,* **22**, 377 (1976).

Hiebert, K. L., "An Evaluation of Mathematical Software That Solves Systems of Nonlinear Equations," *ACM Trans. Math. Software,* **8**, 5 (1982).

Hiebert, K. L., and L. F. Shampine, "Implicitly Defined Output Points for Solutions of ODEs," Sandia National Lab. Rpt. SAND80-0180 (1980).

Hiestand, J. W. and A. R. George, "Generalized Steady-State Method for Stiff Equations," *AIAA J.,* **14**, 1153 (1976).

Hill, D. R., "Second Derivative Multistep Formulas Based on G-Splines," *Numerical Methods for Differential Systems,* Lapidus, L., and W. E. Schiesser, Eds., Academic Press, New York (1976).

Hill, W. J. and W. G. Hunter, "Model Building," Rpt. No. 42, Depts. Chemical Engineering and Statistics, Univ. of Wisconsin (1967).

Hindmarsh, A. C., "GEAR: Ordinary Differential Equation Solver," Lawrence Livermore Rpt. UCID-30001, Rev. 3, December (1974).

Hindmarsh, A. C., "Note on the Startup of ODE Solvers," *SIGNUM Newsl.,* **11**, 12 (1976a).

Hindmarsh, A. C., "Preliminary Documentation of GEARIB: Solution of Implicit Systems of Ordinary Differential Equations with Banded Jacobian, LLNL Rpt. UCID-30130 (1976b).

Hindmarsh, A. C., "Preliminary Documentation of GEARBI: Solution of ODE Systems with Block-Iterative Treatment of the Jacobian," LLNL Rpt. UCID-30149 (1976c).

Hindmarsh, A. C., "The LLL Family of Ordinary Differential Equation Solvers," AFWL-AFOSR Workshop, p. 57 (1977a).

Hindmarsh, A. C., "GEARB: Solution of Ordinary Differential Equations Having Banded Jacobian," Lawrence Livermore Report UCID-30059, Rev. 2, June, Livermore, California (1977b).

Hindmarsh, A. C., "On the Use of Rank-One Updates in the Solution of Stiff Systems of Ordinary Differential Equations," *SIGNUM Newsl.,* **11**, 23 (1977c).

Hindmarsh, A. C., "A Tentative User Interface Standard for ODEPACK," Rpt. No. UCID-L7954, Lawrence Livermore Lab., Livermore, California (1978).

Hindmarsh, A. C., "A User Interface Standard for ODE Solvers," SIGNUM Meeting on Numerical ODEs, Champaign, Illinois (1979a).

Hindmarsh, A. C., "A Collection of Software for Ordinary Differential Equations," LLL Preprint UCRL-82091 (1979b).

Hindmarsh, A. C., "On Numerical Methods for Stiff Differential Equations—Getting the Power to the People," Lawrence Livermore Lab. Rpt. UCRL-83259, Livermore, California (1979c).

Hindmarsh, A. C., "Method-of-Lines Considerations in ODE Software Design," SIGNUM Meeting on Numerical ODEs, Champaign, Illinois (1979d).

Hindmarsh, A. C., "LSODE and LSODI, Two Initial Value Ordinary Differential Equation Solvers," *ACM SIGNUM Newsl.*, **15**, 10 (1980a)

Hindmarsh, A. C., "On Numerical Methods for Stiff Differential Equations—Getting the Power to the People," *COMPCON, VLSI: New Architectural Horizons,* San Francisco, California (1980b).

Hindmarsh, A. C., "ODE Solvers for Use with the Method of Lines," *Advances in Computer Methods for Partial Differential Equations – IV,* IMACS (1981).

Hindmarsh, A. C., "Stiff System Problems and Solutions at LLNL," *Proc. Int. Conf. Stiff Comp.,* Park City, Utah (1982a).

Hindmarsh, A. C., "Toward A Systematized Collection of ODE Solvers," *Proc. 10th IMACS World Congress on System Simulation and Scientific Computing,* August 8–13, Montreal, Canada, p. 427 (1982b).

Hindmarsh, A. C., "ODEPACK: A Systematized Collection of ODE Solvers," *Numerical Methods For Scientific Computation,* R. S. Stepleman, Ed., North Holland Pub. Co. (1983).

Hindmarsh, A. C. and G. D. Byrne, "Applications of EPISODE: An Experimental Package for the Integration of Systems of Ordinary Differential Equations," *Numer. Methods for Differential Systems,* Lapidus, L. and W. E. Schiesser, Eds., Academic Press, p. 147 (1976a).

Hindmarsh, A. C. and G. D. Byrne, "A Proposed ODEPACK Calling Sequence," LLNL Rpt. UCID-30134 (1976b).

Hindmarsh, A. C. and G. D. Byrne, "EPISODE: An Effective Package for the Integration of Systems of Ordinary Differential Equations," UCID-30112, Rev. 1, Lawrence Livermore Lab. (1977).

Hindmarsh, A. C., Sloan, L. J., and P. F. Dubois, "DEC/SOL: Solution of Dense Systems of Linear Algebraic Equations," UCID-30137, Rev. 1 (1978).

Hines, J. D., "CW CO_2 Electric Discharge Laser Modelling and Comparison with Experiment," AFWL-AFOSR Workshop, p. 207 (1977).

Hiromi, K., *Kinetics of Fast Enzyme Reactions,* Halsted Press (1979).

Hiromi, K., personal communication (1982).

Hirschfelder, J. O., "Pseudo-Stationary State Approximation In Chemical Kinetics," *J. Chem. Phys.,* **26**, 271 (1957).

Hofer, E., "A Partially Implicit Method for Large Stiff Systems of ODEs With Only Few Equations Introducing Small Time Constants," *SIAM J. Numer. Anal.,* **13**, 654 (1976).

Hofer, E. and W. Werner, CONF-730414, USAEC, 73 (1973).

Holmes, C. P., "Solution of the CO_2–N_2 Vibrational Kinetic Rate Equations in a Multizone Configuration," AFWL-AFOSR Workshop, p. 165 (1977).

Hoppensteadt, F. C., "Properties of Solutions of Ordinary Differential Equations With Small Parameters," *Comm. Pure Appl. Math.,* **24**, 807 (1971).

Hoppensteadt, F. C., "Numerical Solutions of Singular Perturbation Problems," Sonderforschungsbereich, 123 (1979).

Hoppensteadt, F. C. and W. L. Miranker, "Multitime Methods for Systems of Difference Equations," *Studies in Appl. Math.,* **56**, 273 (1977).

Hoppensteadt, F. C. and W. L. Miranker, "An Extrapolation Method for the

Numerical Solution of Stiff Differential Equations," Research Report RC 7697 Math. (1979).

Hoppensteadt, F. C. and W. L. Miranker, "Differential Equations Having Rapidly Changing Solutions: Analytic Methods for Weakly Nonlinear Systems," *J. Differential Equations,* **22**, 237 (1976).

Hoppensteadt, F. C., Alfeld, P., and R. Aiken, "Numerical Treatment of Rapid Chemical Kinetics by Perturbation and Projection Methods," *Modelling of Chemical Reaction Systems,* Ebert, K. H., Deuflhard, P., and W. Jaeger, Eds., Springer, Berlin (1981).

Houback, N. and P. G. Thomsen, "SPARKS, a FORTRAN Subroutine for the Solution of Large Systems of Stiff ODEs With Sparse Jacobians," NI-79-02, Inst. for Numer. Anal., Technical University of Denmark, Lyngby (1979).

Hsiao, G. C. and K. E. Jordan, "Solutions to the Difference Equations of Singular Perturbation Problems," *Numerical Analysis of Singular Perturbation Problems,* Hemker, P. W., and J. J. H. Miller, Eds., Academic Press, New York (1979).

Hu, S. S. and W. E. Schiesser, "An Adaptive Grid in the Numerical Method of Lines," *Proc. 4th IMACS Symposium on Computer Methods for Partial Differential Equations*, Lehigh University, Bethlehem, Pennsylvania, 305 (1981).

Hull, T. E., "The Validation and Comparison of Programs for Stiff Systems," *Stiff Differential Systems,* R. A. Willoughby, Ed., Plenum Press, New York (1974).

Hull, T. E., "Comparison of Algorithms for Initial Value Problems," *Computational Techniques for Ord. Dif. Equations, Proceed. of Conf.,* Manchester, England (1980).

Humpage, W. D., Wong, K. P., and Y. W. Lee, "Numerical Integration Algorithms in Power-System Dynamic Analysis," *Proc. Inst. Electr. Eng.,* **121**, 467 (1974).

Hunding, A., "Dissipative Structures in Reaction-Diffusion Systems: Numerical Determination of Bifurcations in the Sphere," *J. Chem. Phys.,* **72**, 5241 (1980).

Hundsdorfer, W. H., "Nonlinear Stability Analysis, for a Simple Rosenbrock Method," Rpt. 81–31, Institute of Applied Mathematics and Computer Science, University of Leiden, The Netherlands (1981).

Hundsdorfer, W. H. and M. N. Spijker, "A Note on B-Stability of Runge–Kutta Methods," *Numer. Math.,* **36**, 319 (1981).

Hunter, W. G., Hill, W. J., and T. L. Henson, "Designing experiments for Precise Estimation of All or Some of the Constants in a Mechanistic Model, *Canadian J. Chem. Engr.* **47**, 76 (1969).

Hwang, J.-T., *Proc. Natl. Sci. Counc. B,* Rep. of China, **6**, 20 (1982a).

Hwang, J.-T., *Proc. Natl. Sci. Counc. B,* Rep. of China, **6**, 37 (1982b).

Hwang, J.-T., "Nonlinear Sensitivity Analysis in Chemical Kinetics—The Scaled Green's Function Method," *Proc. Int. Conf. Stiff Comp.,* Park City, Utah (1982c).

Hwang, J.-T., "Sensitivity Analysis in Chemical Kinetics by the Method of Polynomial Approximations," submitted to *Int. J. Chem. Kinetics* (1983).

Hwang, J.-T., "A Computational Algorithm for the Scaled Green's Function Method in Chemical Kinetics," *Proc. Natl. Sci. Counc.,* in press (a).

Hwang, J.-T., "On the Proper Usage of Sensitivities of Chemical Kinetics Models to the Uncertainties in Rate Coefficients," *Proc. Natl. Sci. Counc.,* in press (b).

Hwang, J.-T., Dougherty, E. P. and H. Rabitz, "The Green's Function Method of Sensitivity Analysis in Chemical Kinetics," *J. Chem. Phys.,* **69**, 5180 (1978).

Hwang, J.-T. and H. Rabitz, *J. Chem. Phys.,* **70**, 4609 (1979).

Hwang, J.-T. and K. H. Wang, "On Nonlinear Sensitivity Analysis and Logarithmic Solutions of Chemical Rate Equations," Chinese Chemical Society Conference (1980).

Hwang, J.-T., and Y.-S. Chang, *Proc. Natl. Sc. Counc. B,* **6**, 308 (1982).
Hyka, J., "Methode zur Ermittlung von Daten fuer die Dimensionievung von Mehrstoffrektiti Kationsanlagen aus Messungon Aneiner Laboratoriumsdestillatimskolonne," Ph.D. Thesis, Eidgenoessiche Technische Hochschule, Zuerich (1977).
Hylton, B., "Dynamic Simulation of Process Systems Using Continuous System Modelling Program (CSMP)," Presented at AICHE Met., Orlando, Florida (1982).
Hyman, J. M., "MOLID: A General Purpose Subroutine Package for the Numerical Solution of Partial Differential Equations," Rpt., Los Alamos Scientific Laboratory (1975).
Hyman, T. M., "The Method of Lines Solution of Partial Differential Equations," Rpt. §COO-3077-139, ERDA Math. and Comp. Lab., New York University (1976).
Isaacson, E., "Mountain Winds Revisited," *SIAM J. Numer. Anal.,* **18**, 277 (1981).
Isbarn, G, private communication (1981).
Iserles, A., "Functional Fitting—New Family of Schemes for Integration of Stiff ODE," *Math. Comp.,* **32**, 19 (1977).
Iserles, A., "On the A-Stability of Implicit Runge–Kutta Processes," *BIT,* **18**, 157 (1978).
Iserles, A., "On the Generalized Pade Approximations to the Exponential Function," *SIAM J. Numer. Anal.,* **16**, 631 (1979).
Iserles, A., 'Quadrature Methods for Stiff Ordinary Differential Systems," *Math. Comput.,* **36**, 171 (1981a).
Iserles, A., "Rational Interpolation to Exp(-X) with Application to certain Stiff Systems," *SIAM J. Numer. Anal.,* **18**, 1 (1981b).
Iserles, A., "Two-Step Numerical Methods for Parabolic Differential Equations," *BIT,* **21**, 80 (1981c).
Ishimura, Y., Nozaki, M., and O. Hayaishi, "The Oxygenated Form of L-Tryptophan 2, C-Dioxygenase as Reaction Intermediate," *J. Biol. Chem.,* **245**, 3593 (1970).
Izumi, M. and T. Noda, *N. S. E.,* **41**, 299 (1970).
Jackson, K. R. and R. Sacks-Davis, "An Alterantive Implementation of Variable Step-Size Multistep Formulas for Stiff ODEs," *ACM Trans. On Math. Software,* **6**, 3 (1980).
Jackson, L. W., "The A-Stability of a Family of Fourth Order Methods," *BIT,* **16**, 383 (1976).
Jackson, L. W. and S. K. Kenue, "A Fourth Order Exponentially Fitted Method," *SIAM J. Numer. Anal.,* **11**, 965 (1974).
Jackson, R. C., "Kinetic Simulation of Anticancer Drug Interaction," *Int. J. Bio-Medical Computing,* **11**, 197 (1980).
Jacobs, D., *The State of the Art in Numerical Analysis,* Academic Press, New York (1977).
Jain, M. K. and R. K. Jain, "P-Stable Methods for Periodic Initial Value Problems of Second Order Differential Equations," *BIT,* **19**, 347 (1979).
Jain, M. K. and R. K. Jain, "Hybrid Numerical Methods for Periodic Initial Value Problems Involving Second-Order Differential Equations," *Appl. Math. Modelling,* **5**, 53 (1981).
Jameson, A., Schmidt, W., and E. Turkel, "Numerical Solutions of the Euler Equations by Finite Volume Methods Using Runge–Kutta Time Stepping Schemes," AIAA paper 81-1259, AIAA 14th Fluid and Plasma Dynamics Conference, Palo Alto (1981).
Janac, K., "CSDT Algorithm for 3-Dimensional Parabolic Equations with O(R Squared) Truncation Error," *Proc. of Summer Computer Simulation Conf.,* Los Angeles, California (1978).
Jeltsch, J., "Stiff Stability of Multistep Multiderivative Methods," *SIAM J. Numer. Anal.,* **14**, 760 (1977).

Jeltsch, R., "A Necessary Condition for A-Stability of Multistep Multiderviative Methods," *Math. Comp.*, **30**, 739 (1976a).

Jeltsch, R., "Note on A-Stability on Multistep Multiderivative Methods," *BIT*, **16**, 74 (1976b).

Jeltsch, R., "Stiff Stability and its Relation to A_0 and A(0) Stability," *SIAM J. Numer. Anal.*, **13**, 8 (1976c).

Jeltsch, R., "Multistep Methods Using Higher Derivatives and Damping at Infinite," *Math. Comp*, **31**, 124 (1977).

Jeltsch, R., "Stability on the Imaginary Axis and A-Stability of Linear Multistep Methods," *BIT*, **18**, 170 (1978).

Jeltsch, R., "A_0-Stability and Stiff Stability of Brown's Multistep Multiderivative Methods," *Numer. Math.*, **32**, 167 (1979a).

Jeltsch, R., "Corrigendum: Stiff Stability of Multistep Multiderivative Methods," *SIAM J. Numer. Anal.*, **16**, 339 (1979b).

Jeltsch, R. and O. Nevanlinna, "Lower Bounds for the Accuracy of Linear Multistep Methods," Rpt. No. 6 Inst. for Geometry and Practical Mathematics, Technical University Aachen, Aachen, Germany (1980).

Jensen, P. S., "Transient Analysis of Structures by Stiffly Stable Methods," *Comput. & Struc.*, **4**, 615 (1974).

Jensen, P. S., "Stiffly Stable Methods for Undamped Second Order Equations of Motion," *SIAM J. Numer. Anal.*, **13**, 549 (1976).

Jensen, O. K., "Numerical Modelling With a Moving Coordinate System: Application to Flow Through Porous Media," Ph.D. Thesis, University of Washington (1980a).

Jensen, O. K., "An Automatic Timestep Selection Scheme for Reservoir Simulation," Soc. Petroleum Eng. paper 9373, Dallas, Texas (1980b).

Jensen, O. K. and B. A. Finlayson, "Adv. Water Resources," **3**, 9 (1980).

Johnson, A. I. and J. R. Barney, "Numerical Solution of Large Systems of Stiff Ordinary Differential Equations in a Modular Simulation Framework," *Numerical Methods for Differential Systems*, Lapidus, L. and W. E. Schiesser, Eds., Academic Press, 97 (1976).

Johnson, H. S., *Gas Phase Reaction Rate Theory*, The Ronald Press, New York (1966).

Jones, R. E., "SLATEC Common Mathematical Library Error Handling Package," Sandia National Laboratories Rpt. SAND78-1189 (1978).

Jury, E. I., "Stability Tests for One-, Two- and Multi-Dimensional Linear Systems," *Proc. IEEE*, **124**, 1237 (1977).

Jury, E. I., "Stability of Multidimensional Scalar and Matrix Polynomials," *Proc. IEEE*, **66**, 1018 (1978).

Kahaner, D., "A New Implementation of the Gear Algorithm for Stiff Systems," Unpublished private communication. For further detail contact: Dr Mac Hyman, University of California, Los Alamos Scientific Research Laboratory, Contract W-7405-ENG-36, P. O. Box 1663, Los Alamos, New Mexico (1979).

Kamen, E. W., "On the Relationship Between Zero Criteria for Two-Variable Polynomials and Asymptotic Stability of Delay Differential Equations," *IEEE Trans. Auto. Control*, **AC-25**, 983 (1980).

Kansa, E. J., "An Algorithm for Multidimensional Combusting Flow Problems," *J. Comp. Phys*, **42**, 152 (1981).

Kao, Y. K., "A Simple Start-Up Policy for a Catalytic Reactor With Catalyst Deactivation," *Chem. Eng*, **20**, 237 (1980).

Kaps, P., "Generalized Runge–Kutta Methods of Order Four With Stepsize Control for Stiff Ordinary Differential Equations," *Numer. Math.*, **33**, 55 (1979).

Kaps, P. and P. Rentrop, "Generalized Runge–Kutta Methods of Order Four With Stepsize Control for Stiff Ordinary Differential Equations," *Numer. Math.*, **33**, 55 (1979).

Kaps, P. and G. Wanner, "A Study of Rosenbrock Type Methods of High Order," *Numer. Math.,* **38**, 279 (1981).

Karanth, N. G. and R. Hughes, "Simulation of an Adiabatic Packed Bed Reactor," *Chem. Eng. Sci.,* **29**, 197 (1974).

Karasalo, I., "On Smooth Solutions to Stiff Nonlinear Analytic Differential Systems," Dept. Infor. Processing Computer Sci., Royal Inst. Tech., Stockholm, Sweden, TRITA-NA-7501 (1975a).

Karasalo, I., "Minimum Norm Solutions of Single Stiff Linear Analytic Differential Equations," *J. Math. Anal. Appl.,* **51**, 516 (1975b).

Karasalo, I. and J. Kurylo, "On Solving the Stiff ODEs of the Kinetics of Chemically Reacting Gas Flow," Lawrence Berkeley Lab. Rpt. Berkeley, California (1979).

Karba, R., Bremsak, F., Kozjek, J., Mrhar, A. and D. Matko, "Hybrid Simulation in Drug Pharmacokinetics," *Proc. 9th ACS Congress,* Sorrento, Italy (1979).

Karetkina, N. V., "An Unconditionally Stable Difference Scheme for Parabolic Equations Containing First Derivatives," *U.S.S.R. Computational Mathematics and Mathematical Physics,* **20**, 257 (1980).

Karplus, M., Gelin, B. R., and J. A. McCannon, "Internal Dynamics of Proteins," *Biophy. J.* (1980).

Karplus, W. J., "Software for Distributed System Simulation," *Progress in Modelling and Simulations,* F. E. Cellier, Ed., Academic Press (1982).

Kato, T., *Perturbation Theory for Linear Operators,* Springer, 244 (1966).

Katz, I. N., "Optimally Stable Parallel Predictors for Adams-Moulton Correctors," *Comput. & Math. With Appl.,* **3**, 217 (1977).

Kayihan, F., "An Iterative Approach for the Solution of a Class of Stiff ODE Models of Reacting Polydispersed Particles," ACS Symposium Sevier No. 124 (1980).

Kee, R. J., Miller, J. A., and T. H. Jefferson, "CHEMKIN: A General-Purpose, Problem-Independent, Transportable, FORTRAN Chemical Kinetics Code Package," Tech. Rpt. SAND 80-8003, Sandia National Laboratories, Livermore, California (1980).

Keneshea, T. J., "A Technique for Solving The General Reaction-Rate Equations in the Atmosphere," U.S. Air Force Cambridge Research Lab, Environmental Research Paper, No. 263 (1967).

Kennealy, J. P. and W. M. Moore, "A Numerical Method for Chemical Kinetics Modelling Based on the Taylor Series Expansion," *J. Phys. Chem.,* **81**, 2413 (1977).

Kerber, R. L., Hough, J. J. T., and W. Jaul, "Modelling of Botanical Nonequilibrium Effects in Pulsed HF Chain Reaction Bases," *Proc. of the AFWL/AFOSR Workshop On the Numerical Solution of Stiff Systems of Ordinary Differential Equations,* August (1977).

Khalil, H. K. and P. V. Kokotovic, "Decentralized Stabilization of Systems with Slow and Fast Modes," *Large Scale Systems,* **1**, 141 (1980).

Kinderlerer, J. and S. Ainsworth, "A Computer Program to Derive the Rate Equations of Enzyme Catalysed Reactions With Up to Ten Enzyne-Containing Intermediates in the Reaction Mechanism," *Int. J. Bio-Medical Computing,* **7**, 1 (1976).

Klinzing, G. E., "Mass Transfer with Nth Order Chemical Reaction Around Spheres in the Presence of Surfactants," *Chem. Eng. Sci.,* **35**, 1667 (1980).

Klopfenstein, R. W., "Numerical Differentiation Formulas for Stiff Systems of Ordinary Differential Equations," *RCA Review,* **32**, 447 (1971).

Knuth, D. E., "The Art of Computer Programming, Vol. III Sorting and Searching," Addison-Wesley (1973).

Koda, M., Dogru, A. H., and J. H. Seinfeld, *J. Comp. Phys.,* **30**, 259 (1979).

Koda, M., McRae, J., and J. H. Seinfeld, "Automatic Sensitivity Analysis of Kinetic Mechanisms," *Inter. J. Chem. Kinetics,* **11**, 427 (1979).

Kogan, B. Y., Zykov, V. S., and A. A. Petrov, "Hybrid Computer Simulation of Stimulative Media," *Proc. 9th IMACS Congress,* Sorrento, Italy (1979).

Kondo, H., Nakatani, H., Matsuno, R. and K. Hiromi, "Product Distribution in Amylase-Catalyzed Hydrolsis of Amylase," *J. Biochem,* **87**, 1053 (1980).

Kong, A. K., "A Search for Better Linear Multistep Methods for Stiff Problems," Ph.D. Thesis, Univ. Illinois (1977).

Korn, G. A. and J. V. Wait, *Digital Continuous System Simulation,* Prentice Hall (1977).

Kovvali, S. and G. K. Gupta, "Polynomial Formulation of Second Derivative Multistep Methods," *Math. Comp.,* **38**, 447 (1982).

Krazinski, J. L., Bucklus, R. O., and H. Krier, "Coal Dust Flames: A Review and Development of a Model for Flame Propagation," *Prog. Energy Combust. Sci.,* **5**, 31 (1979).

Kreiss, B. and H. Kreiss, "Numerical Methods for Singular Perturbation Problems," *SIAM J. Numer. Anal.,* **18**, 262 (1981).

Kreiss, H. O., "Numerical Solution of Singular Perturbation Problems," *Stiff Differential Systems,* R. A. Willoughby, Ed., Plenum Press, New York (1974).

Kreiss, H. O., "Difference Methods for Stiff Ordinary Differential Equations," *SIAM J. Numer. Anal.,* **15**, 21 (1978).

Kreiss, H. O., "Problems with Different Time Scales for Ordinary Differential Equations," *SIAM J. Numer. Anal.,* **16**, 980 (1979).

Krishnamurthey, L. and F. A. Williams, "A Flame Sheet in the Stagnation-Point Boundary Layer of a Condensed Fuel," *Acta astronautica,* **1**, 711 (1974).

Krishnan, K. N. and V. M. K. Sastri, "Numerical Solution of Thermal Entry Length Problems with Variable Viscosities and Viscous Dissipation," *Waerme Stof.,* **11**, 73 (1978).

Krogh, F. T., "The Numerical Integration of Stiff Differential Equations," TRW Rpt. No. 99900-6573-R000, TRW systems, Redondo Beach, California (1968).

Krogh, F. T., "On Testing a Subroutine for the Numerical Integration of Ordinary Differential Equations," *J. ACM,* **20**, 545 (1973).

Krogh, F. T. and K. Stewart, "Asymptotic (h goes to infinity) Absolute Stability for BDFs Applied to Stiff Equations," Submitted to *ACM Trans. Math. Software.*

Krogh, F. T. and K. Stewart, "STRUT," in preparation (1983).

Krogh, F. T., "VODQ/SVDQ/DVDQ, Variable Order Integrators for the Numerical Solution of Ordinary Differential Equations," *Tech. Brief NPO-11643,* Jet Propulsion Laboratory, California Institute of Technology, Pasadena, California (1970).

Krogh, F. T., "Changing Stepsize in the Integration of Differential Equations Using Modified Divided Differences," *Proc. Conf. on the Numerical Solution of Ordinary Differential Equations, 1972,* Lecture Notes in Math., **362**, Springer-Verlag, New York, 22 (1974).

Krogh, F. T., "Recurrence Relations for Computing with Modified Divided Differences," *Math. Comput.,* **33**, 1256 (1979).

Krogh, F. T., "Recurrence Relations for Computing With Modified Divided Differences," *Math. Comp.,* **33**, 1265 (1979).

Krogh, F. T., "Recurrence Relations for Computing With Modified Divided Differences," *Math. Comp.,* **35**, 1445 (1980).

Krogh, F. T. and K. Stewart, "Asymptotic (H to Infinity) Stability for BDFs Applied to Stiff Equations," Jet Propulsion Lab., Comp. Memo. 478, Pasadena, California (1981a).

Krogh, F. T. and K. Stewart, "Implementation of Variable-Step BDF Methods for Stiff ODEs," *Numerical Methods for Solving Stiff Initial Value Problems,"* Dahlquist, G. and R. Jeltsch, Eds., Proc. Oberwolfach Meeting, Germany (1981b).

Kruger, P. B. and R. M. Mayer, "Calculation of the Nucleation and Growth of Defect Clusters," *Comp. Phys. Comm.,* **18,** 385 (1979).

Kubicek, M. and K. Visnak, "A Nonlinear Explicit Algorithm for Efficient Integration of Stiff Systems of Ordinary Differential Equations," *Chem. Eng. Commun.,* **1,** 291 (1974).

Kuchel, P. W. and K. Dalziel, "Kinetics of Co-Operative Ligand Binding," *J. Theor. Biol.,* **85,** 497 (1980).

Kuhlman, M. R., Fox, D. L., and H. E. Jeffries, "The Effect of CO on Sulfate Aerosol Formation," *Atmospheric Environment,* **12,** 2415 (1978).

Kurtz, L. A., Smith, R. E., Parks, C. L., and L. R. Boney, "A Comparison of the Method of Lines to Finite Difference Techniques in Solving Time-Dependent Partial Differential Equations," *Comp. & Fluids,* **8,** 49 (1978).

Lahey, R. T., Jr., Cheng, L. Y., Drew, D. A., and J. E. Flaherty, "The Effect of Virtual Mass on the Numerical Stability of Acceleration Two-Phase Flows," *Int. J. Multiphase Flow,* **6,** 281 (1980).

Lambert, J. D., *Computational Methods in Ordinary Differential Equations,* Wiley (1973).

Lambert, J. D., "Two Unconventional Classes of Methods for Stiff Systems," *Stiff Differential Systems,* R. A. Willoughby, Ed., Plenum Press, New York (1974).

Lambert, J. D., "The Initial Value Problem for Ordinary Differential Equations," *The State of the Art in Numerical Analysis,* D. Jacobs, Ed., Academic Press (1977).

Lambert, J. D., "Stiffness," *Computational Techniques for Ord. Dif. Equations, Proceed. of Conf.,* Manchester, England (1978).

Lambert J. D., "Safe Point Methods for Separably Stiff Systems of Ordinary Differential Equations," *SIAM J. Numer. Anal.,* **18,** 83 (1981).

Lambert, J. D. and J. A. Watson, "Symmetric Multistep Methods for Periodic Initial Value Problems," *J. Inst. Math. Appl.,* **18,** 189 (1976).

Lapidus, L. and R. C. Aiken, "Numerical Integration of Stiff Dynamic Systems," *Trans ASME, Series G,* **96,** 5 (1974).

Lapidus, L. and J. H. Seindfeld, *Numerical Solution of Ordinary Differential Equations,* Academic Press (1971).

Lapidus, L., Aiken, R. C., and Y. A. Liu, "The Occurrence and Numerical Solution of Physical and Chemical Systems Having Widely Varying Time Constants," *Stiff Differential Systems,* Plenum Press, 1987 (1974).

Larkin, B. K. and S. W. Churchill, "Heat Transfer by Radiation Through Porous Insulations," *AICHE J.,* **92,** 6 (1960).

Lastman, G. J., Wentzell, R. A., and A. C. Hindmarsh, "Numerical Solution of a Bubble Cavitation Problem," *J. Comp. Phys.,* **28,** 56 (1978).

Lawrence, R. D. and J. J. Dorning, *Trans. Am. Nucl. Soc.,* **24,** 199 (1976).

Lawrence, R. D. and J. J. Dorning, "Smoothing and Extrapolation Techniques for Reactor Kinetics," *Nucl. Science & Eng.,* **64,** 492 (1977).

Lawson, IC. L., "SFTRAN3, Programmer's Reference Manual," JPL Document No. 1846-98, Revision A(1981).

Lawton, S. A., Richards, J. B., Newman, L. A., Specht, L., and T. A. DeTemple, "The High-Pressure Neutral Infrared Xenon Laser," *J. Appl. Phys.,* **50,** 3888 (1979).

Layokun, S. K. and D. H. Slater, "Mechanism and Kinetics of Propane Pyrolysis," *Ind. Eng. Chem. Process Des. Dev.,* **18,** 232 (1979).

Leaf, G. K. and M. Minkoff, "Preliminary Documentation for DISPL2," Argonne National Laboratory, Argonne, Illinois (1981).

Leaf, G. K., Minkoff, M., Byrne, G. D., Sorensen, D., Bleakney, T., and J. Saltzman, "DISPL: A Software Package for One- and Two-Spatially Dimensional Kinetics-Diffusion Problems," Argonne National Laboratory Rpt. ANL-77-12, Rev. 1 (1978).

Lee, D., "Nonlinear Multistep Methods for Solving Initial Value Problems in Ordinary Differential Equations," Polytech. Inst. Ph.D. Thesis, Brooklyn, New York (1974).

Lee, D. and S. Preiser, "A Class of Nonlinear Multistep A-Stable Numerical Methods for Solving Stiff Differential Equations," *Comp. & Math. Appl.,* **4**, 43 (1978).

Levethal, S. H., "An Operator Compact Implicit Method of Exponential Type," *J. Comp. Phys.,* **46**, 138 (1982).

LeVeque, R., Dahlquist, G., and D. Andree, "Computations Related to G-Stability of Linear Multistep Methods, Rpt. STAN-79-738, Computer Sci. Dept., Stanford University (1979).

Lewins, J., "Nuclear Reactor Kinetics and Control," Pergamon Press (1978).

Lewis, J. G., "The Biggs-Poole-Stockmeyer and Gibbs-King Algorithms for Reordering Sparse Matrices," *ACM Trans. on Mathematical Software,* **8**, no. 2, 190 (1982).

Li, B., Yang, W. H., and X. Ye, "An Effective Method for a System of Ordinary Differential Equations With Right Hand Side Functions Containing Discontinuities," *Proc. 10th IMACS World Congress on System Simulation and Scientific Computing,* August 8–13, Montreal, Canada, p. 6 (1982).

Lin, C. C. and L. A. Segel, *Mathematics Applied to Deterministic Problems in the Natural Sciences,* MacMillan Pub. Co., Inc., New York 303 (1974).

Lindberg, B., "On Smoothing for the Trapezoidal Rule, an Analytical Study of Some Representative Test Examples, Rpt. TRITA-NA-7131 (1971a).

Lindberg, B., *BIT,* **11**, 29 (1971b).

Lindberg, B., "A Stiff System Package Based on the Implicit Midpoint Method With Smoothing and Extrapolation," *Stiff Differential Systems,* R. A. Willoughby, Ed., Plenum Press, New York (1974a).

Lindberg, B., "On a Dangerous Property of Methods for Stiff Differential Equations," *BIT,* **14**, 430 (1974b).

Lindberg, B., "Characterization of Optimal Stepsize Sequences for Methods for Stiff Differential Equations," *SIAM J. Numer. Anal.,* **14**, 859 (1977).

Liniger, W., "Zur Stabilitat der Numerischen Integrationsmethoden fur Differentialgleichungen, Faculte des Sciences de l'Universite de Lausanne (1957).

Liniger, W., "Connections Between Accuracy and Stability Properties of Linear Multistep Formulas," *Comm. ACM,* **18**, 53 (1975).

Liniger, W., "High-Order A-Stable Averaging Algorithms for Stiff Differential Systems," *Numerical Methods for Differential Systems,* Lapidus, L. and W. E. Schiesser, Eds., Academic Press, p. 1 (1976).

Liniger, W., "On Stability and Accuracy of Numerical Integration Methods for Stiff Differential Equations," AFWL-AFOSR Workshop, p. 36 (1977a).

Liniger, W., "Stability and Error Bounds for Multistep Solutions of Nonlinear Differential Equations," *Proc. ISCAS, IEEE Int. Symp. on Circuits and Systems,* Phoenix, Arizona (1977b).

Liniger, W., "Multistep and One-Leg Methods for Implicit Mixed Differential Algebraic Systems," *IEEE Trans. on Circuits and Systems,* **CAS-26**, 755 (1979).

Liniger, W., 'A-Contractivity of K-Step Methods with Variable Steps," Report RC-9281, IBM, Oberwolfach Meeting, Germany, July (1981).

Liniger, W., "A-Contractivity of Second-Order Multistep Formulas with Variable Steps," Report RC-9281, IBM (1982a).

Liniger, W., "Contractivity of Multistep and One-Leg Methods with Variable Steps," *Proc. Int. Conf. Stiff Comp.,* Park City, Utah (1982b).

Liniger, W., "A-Contractivity of Second-Order Multistep Formulas With Variable Steps," to appear in *SIAM J. Numer. Anal.* (1983).

Liniger, W. and T. Gagnebin, "Construction of a Family of Second Order A-Stable K-Step Formulas Depending on the Maximum Number, 2K-2, of Parameters,"

Stiff Differential Systems, R. A. Willoughby, Ed., Plenum Press, New York (1974).

Liniger, W. and F. Odeh, "On Liapunov Stability of Stiff Nonlinear Multistep Difference Equations," Research Rpt. RC 5900 (1976).

Liniger, W. and F. Odeh, "Non-Linear Fixed-*h* Stability of Linear Multistep Formulas," *J. Math. Anal. Applic.,* **61**, 691 (1977).

Liniger, W. and R. A. Willoughby, "Efficient Integration Methods for Stiff Systems of Ordinary Differential Equations," *SIAM J. Numer. Anal.,* **7**, 47 (1970).

Loeb, A. M. and W. E. Schiesser, "Stiffness and Accuracy in the Method of Lines Integration of Partial Differential Equations," *Stiff Differential Systems,* R. A. Willoughby, Ed., Plenum Press, p. 229 (1974).

Lomax, H., "On the Construction of Highly Stable, Explicit Numerical Methods for Integrating Coupled Ordinary Differential Equations with Parasitic Eigenvalues," NASA Tech. Note D-4547 (1968).

Lomax, H., NASA Ames, private communications (1983).

Lomax, H. and H. E. Bailey, "A Critical Analysis of Various Numerical Integration Methods for Computing the Flow of a Gas in Chemical Non-Equilibrum," NASA Tech. Note D-4109 (1967).

Lomax, H. and T. H. Pulliam, "A Fully Implicit, Factored Code for Computing Three-Dimensional Flows on the ILLIAC IV," *Parallel Computations,* G. Rodrigue, Ed., Academic Press (1982).

Lombardi, G., "A-Stable Predictor Corrector Methods for Numerical Solution of Differential Systems," *Calcolo.,* **14**, 369 (1977).

Loomis, R. S., Rabbinge, R., and E. Ng, "Explanatory Models in Crop Physiology," *Ann. Rev. Plant Physiol.,* **30**, 339 (1979).

Lowe, A., Wall, T. F., and I. M. Stewart, "Combustion Kinetics in the Modelling of Large Pulverized Fuel Furnaces: A Numerical Experiment in Sensitivity," *AICHE J.,* **23**, 440 (1977).

Lucey, J. M., "A Fifth Order Composite Multistep Method for Solving Certain Stiff Ordinary Differential Equations," Avail. from Univ. Microfilms, Ann Arbor Mich., Order No. 75-26280 (1975).

Luke, Y. L., Fair, W., and J. Wimp, "Predictor-Corrector Formulas Based on Rational Interpolants," *Comp. & Math.,* **1**, 3 (1975).

Lundstrom, E. A., "Integration of CO Laser Equations Using a Modified Gear Program," AFWL-AFOSR Workshop, p. 150 (1977).

Luss, D. and N. R. Amundson, "Stability of Batch Catalytic Fluidized Beds," *AICHE J.,* **14**, 211 (1968).

MacCormack, R. W. and A. J. Paullay, "The Influence of the Computational Mesh on the Accuracy for Initial Value Problems with Discontinuous or Non-unique Solutions," *Comp. & Fluids,* **2**, 339 (1974).

MacCracken, C., "DOT-CIAP Final Rpt.," LLNL Rpt. UCRL-51336 (1975).

MacCracken, M. C., Wuebbles, D. J., Walton, J. J., Duewer, W. H., and K. E. Grant, "Livermore Regional Air Quality Model: 1. Concept and Development," *J. Appl. Meterology,* **17**, 254 (1978).

Machura, M. and R. A. Sweet, "A Survey of Software for Partial Differential Equations," *ACM Trans. on Math. Software,* **6**, 461 (1980).

MacMillan, D. B., "Asymptotic Methods for Systems of Differential Equations in Which Some Variables Have Very Short Response Times," *SIAM J. Appl. Math.,* **16**, 704 (1968).

Madala, R. V. and S. A. Piacsek, "A Semi-Implicit Numerical Model for Baroclinic Oceans," *J. Comp. Phys.,* **23**, 167 (1977).

Madsen, N. K. and R. F. Sincovec, "General Software for Partial Differential Equations," *Numerical Methods for Differential Systems,* Lapidus, L., and W. E. Schiesser, Eds., Academic Press, p. 229 (1978).

Madsen, N. K. and R. F. Sincovec, "Algorithm 540. PDECOL, General Collocation Software for Partial Differential Equations," *ACM-Trans. Math. Software,* **5**, 326 (1979).

Magee, J. and A. Chatterjee, Lawrence Berkeley Lab., private communications (1982).

Mah, R. S. H., Michaelson, S., and R. W. H. Sargent, "Dynamic Behavior of Multicomponent Multistage Systems. Numerical Methods for the Solution," *Chem. Eng. Sci.,* **17**, 619 (1962).

Makela, M., Nevanlinna, O., and A. H. Sipila, "Exponentially Fitted Multistep Methods by Generalized Hermite-Birkoff Interpolation," *BIT,* **14**, 437 (1974).

Mannshart, R., "One-Step Methods for Any Order for Ordinary Differential Equations with Discontinuous Right-Hand Sides," *Numer. Math.,* **31**, 131 (1978).

Mattheij, R. M. M., "On Approximating Smooth Solutions of Linear Singularly Perturbed ODE," *Numerical Analysis of Singular Perturbation Problems,* Hemker, P. W., and J. J. H. Miller, Eds., Academic Press, New York p. 457 (1979).

Mattheij, R. M. M., "Estimates for the Errors in the Solutions of Linear Boundary Value Problems, Due to Perturbations," *Computing,* **27**, 229 (1981a).

Mattheij, R. M. M., "Accurate Estimates for the Fundamental Solutions of Discrete Boundary Value Problems," Rpt. 8112, Mathematisch Instituut Katholieke Universiteit, Nijmegen, The Netherlands (1981b).

Mattheij, R. M. M., "Riccati Type Transforms and Decoupling of Singularly Perturbed ODE," *Proc. Int. Conf. Stiff Comp.,* Park City Utah (1982).

Mattheij, R. M. M. and R. E. O'Malley, Jr., "Decoupling of Boundary Value Problems for Two-Time-Scale System," *Proc. Int. Conf. Stiff Comp.,* Park City, Utah (1982).

Mattheij, R. M. M. and G. W. M. Staarink, "A Method for Solving General Linear Boundary Value Problems," Rpt. 8029, Mathematisch Instituut, Katholieke Universiteit, Nijmegen, The Netherlands (1980).

Mattheij, R. M. M. and G. W. M. Staarink, "An Efficient Algorithm for Solving General Linear Two-Point Boundary Value Problems," Mathematisch Instituut, Katholieke Universiteit, Nijmegen, The Netherlands (1982).

Mazhukin, V. I., Uglov, A. A., and B. N. Chetverushkin, "Numerical Investigation of the Problem of Laser Breakdown of a Dense Gas," *U.S.S.R. Computational Mathematics and Mathematical Physics,* **20**, 182 (1980).

McCammon, J. A., Wolynes, P. G., and M. Karplus, "Picosecond Dynamics of Tyrosine Side Chains in Proteins," *ACS,* **18**, 927 (1979).

McDonald, H., "Combustion Modelling in Two and Three Dimensions—Some Numerical Considerations," *Prog. Energy Combust. Sci.,* **5**, 97 (1979).

McKee, S., "Cyclic Multistep Methods for Solving Volterra Integro-Differential Equations," *SIAM J. Numer. Anal.,* **16**, 106 (1979).

McKenna, J. and N. L. Schryer, "Analysis of Field-Aided Charge-Coupled Device Transfer," *The Bell System Tech. J.,* **54**, 667 (1975).

McRae, G. J., Goodin, W. R., and J. H. Seinfeld, "Numerical Solution of the Atmospheric Diffusion Equation for Chemically Reacting Flows," *J. Comp. Phys.,* **45**, 1 (1982).

Meister, G., "On the Integration of Systems of First Order Differential Equations with Exponentially Fitted Numerical Methods," *Computing,* **13**, 327 (1974).

Melgaard, D. K. and R. F. Sincovec, "Algorithm 565, PDETWO/PSETM/GEARB: Solution of Systems of Two-Dimensional Nonlinear Partial Differential Equations," *ACM Trans. Math. Software,* **7**, (1), 126 (1981a).

Melgaard, D. K. and R. F. Sincovec, "General Software for Two-Dimensional Nonlinear Partial Differential Equations," *ACM Trans. on Math. Software,* **7**, 106 (1981b).

Metzler, C. M., "A Brief Introduction to Nonlinear Least Squares Estimation," *Proc. 5th Natl. Meeting,* Pharmaceut. Sci., Washington DC (1968).

Michaelis, M. and M. L. Menten, *Biochem. Z.,* **49**, 333 (1913).

Michelsen, M. L., "An Efficient General Purpose Method for the Integration of Stiff Ordinary Differential Equations," *AICHE J.*, **22**, 594 (1976a).

Michelsen, M. L., "Semi-Implicit Runge–Kutta Methods for Stiff Systems Program Description and Application Examples," Institute for Kemiteknik (1976b).

Michelsen, M. L. and J. Villadsen, "The Graetz Problem with Axial Heat Conduction," *Int. J. Heat Mass Transfer,* **17**, 1391 (1974).

Mihelcic, M., "Almost A-Stable Donelson-Hansen Cyclic Methods for the Numerical Integration of Stiff Systems of Differential Equations," *Angew. Inf.*, **19**, 299 (1977).

Mihelcic, M. "A(Alpha) Stable Cyclic Composite Multi-Step Methods of Order 5" *Computing,* **20**, 267 (1978).

Mika, J., Rpt. INR 700/XXI (1966).

Mika, J., *Nukleonika,* **19**, 1013 (1975).

Mika, J., *Nukleonika,* **21**, 573 (1976).

Mika, J., *Int. Rpt. INR,* Swierk (1979).

Miller, C., Meakin, P., Franks, R. G. E., and J. P. Jesson, "The Fluorocarbon-Oxone Theory—V. One-Dimensinal Modelling of the Atmosphere: The Base Case," *Atmospheric Environment,* **12**, 2481 (1978).

Miller, J. H., "Topics in Numerical Analysis III," *Proc. Royal Irish Acad. Conf. on Num. Anal.*, Academic Press, London (1976).

Miller, J. H., "On the Convergence, Uniformly in E, of Difference Schemes for a Two Point Boundary Singular Perturbation Problem," *Numerical Analysis of Singular Perturbation Problems,* Hemker, P. W. and J. J. H. Miller, Eds., Academic Press, New York, p. 467 (1979).

Miller, R. E. and S. E. Stein, "Liquid-Phase Pyrolysis of 1, 2-Diphenylethane," *J. Phys. Chem.*, **85**, 580 (1981).

Milne, W. E. and R. R. Reynolds, "Stability of a Numerical Solution of Differential Equations—Part II.," *J. ACM,* **7**, 46 (1960).

Milonni, P. W., "Computer Codes for Predicting Molecular Laser Performance: An Overview," AFWL-AFOSR Workshop, p. 122 (1977).

Milstein, J. and H. J. Bremermann, "Parameter Identification of the Calvin Photosynthesis Cycle," *J. Math. Biology,* **7**, 99 (1979).

Mingyou, H., "On the Contractivity Region of Runge–Kutta Methods," Rpt. Trita-Na-8106, Royal Inst. Tech., Stockholm, Sweden (1981).

Miranker, W. L., "Numerical Methods of Boundary-Layer Type For Stiff Systems of Differential Equations," IBM Research Rpt. RC 3881 (1972).

Miranker, W. L. "The Computational Theory of Stiff Differential Equations," Publication Mathematique D'Orsay No. 219-7667 (1978).

Miranker, W. L., *Numerical Methods for Stiff Equations,* D. Reidel Pub. Co., Boston (1981).

Miranker, W. L., "An Overview of the Highly Oscillatory Initial Value Problem," *Proc. Int. Conf. Stiff Comp.*, Park City, Utah (1982).

Miranker, W. L. and F. Hoppensteadt, "Numerical Methods for Stiff Systems of Differential Equations Related With Transistors, Tunnel Diodes, etc.," *Lecture Notes in Computer Science,* **10**, Springer-Verlag, 416 (1973).

Miranker, W. L. and J. P. Morreeuw, "Semianalytic Numerical Studies of Turning Points Arising in Stiff Boundary Value Problems," *Math. Comp.*, **28**, 1017 (1974).

Miranker, W. L. and M. Van Veldhuizen, "The Method of Envelopes," *Math. Comp.*, **32**, 453 (1978).

Miranker, W. L. and G. Wahba, "An Averaging Method for the Stiff Highly Oscillatory Problem," *Math. Comp.*, **30**, 383 (1976).

Mitchell, A. R. and I. Christie, "Finite Difference-Finite Element Methods at the Parabolic-Hyperbolic Interface," *Numerical Analysis of Singular Pertrubation Problems,* Hemker, P. W. and J. J. H. Miller, Eds., Academic Press, New York, P. 339 (1979).

Mitchell, E. E. L., "Advanced Continuous Simulation Language (ACSL): An Update,"

Proc. 10th IMACS World Congress on System Simulation and Scientific Computing, August 8–13, Montreal, Canada, 462 (1982).

Moller, O., "Quasi Double-Precision Floating-Point Addition," *BIT,* **5**, 37 (1965).

Monat, J. P., Hanson, R. K., and C. H. Kruger, *Combust. Sci. Tech.,* **16**, 21 (1977).

Moore, E., "Differential–Difference Equations," Univ. Waterloo, Ontario, Canada, Ph.D. Thesis (1974).

Moore, G. and A. Spence, "The Calculation of Turning Points of Nonlinear Equations," *SIAM J. Numer. Anal,,* **17**, 567 (1980).

Moore, W. M. and T. K. Eccles, "Developmental Studies of the Hydrogen-Nitrogen-Oxygen Systems," Rpt. No. AFCRL-Tr-73-0747, Utah State University, Logan. Utah (1974).

Morris, D. B., Hindmarsh, A. C., and P. F. Dubois, "GEARV and GEARST: Vectorized Ordinary Differential Equation Solvers for the 7600 and Star Computers," UCID-30119, Rev. 1, Lawrence Liv. Lab. (1977).

Moyle, M. P., Churchill, S. W., and R. B. Morrison, "Detonation Characteristics of Hydrogen-Oxygen Mixtures," *AICHE J.,* **92**, 6 (1960).

Multhei, H. N., "A-stabile Kollakationsverfahren mit mechrfachen Knoten," *Computing,* **29**, 51 (1982).

Murphy, W. D., "Hermite Interpolation and A-Stable Methods for Stiff Ordinary Differential Equations," *Appl. Math. Comp.,* **3**, 103 (1977).

Nakhle, M. and Roux, P., "Integration of Nonlinear Systems of Stiff Differential Equations With Discontinuities," *Proc. 10th IMACS World Congress on System Simulation and Scientific Computing,* August 8–13, Montreal, Canada, p. 13 (1982).

Nandakumar, K. and R. P. Andres, "A New Composition Method for Dynamic Simulation of Large Scale Systems," *Proc. Summer Computer Simulation Conf.,* Los Angeles, California (1978).

Nedoma, J., "The Solution of Parabolic Models by Finite Element Space and A-Stable Time Discretization," *Equadiff IV,* Prague, Czechoslovakia (1979).

Nelder, J. A. and R. Mead, "A Simplex Method for Function Minimization," *Comp. J.,* 7, 308 (1965).

Neundorf, W., "Iterative Block Methods for the Hybrid Computer Solution of the Method of Lines," *Math. & Comput. Simulation,* **23**, 142 (1981).

Nevanlinna, O., "On Error Bounds for G-Stable Methods," *BIT,* **16**, 79 (1976).

Nevanlinna, O., "On the Numrical Integration of Nonlinear Initial Value Problems by Linear Multistep Methods," *BIT,* **17**, 58 (1977).

Nevanlinna, O., "Stability of Variable Step Integration, Rpt. No. 32, Mathematics Dept., University of Oulu, Finland (1979).

Nevanlinna, O. and W. Liniger, "Contractive Methods for Stiff Ordinary Differential Equations," Rpt. RC 7122, IBM Research Center, Yorktown Heights, New York (1977).

Nevanlinna, O. and W. Liniger, "Contractive Methods for Stiff Differential Equations, Part I," *BIT,* **18**, 457 (1978).

Nevanlinna, O. and W. Liniger, "Contractive Methods for Stiff Differential Equations," *BIT,* **19**, 53 (1979).

Nevanlinna, O. and F. Odeh, "Multiplier Techniques for Linear Multistep Methods, 2 *Numer. Funct. Anal. and Optimiz.,* **3**, 377 (1981).

Nevanlinna, O. and A. H. Sipila, "A Nonexistence Theorem for Explicit A-Stable Methods," *Math. Comp.,* **28**, 1053 (1974).

Ni, R.-H., "A Multiple Gridscheme for Solving the Euler Equations," AIAA Paper 81-1025 (1981).

Nicholls, J. A., "Stabilization of Gaseous Detonation Waves With Emphasis on the Ignition Delay Zone," Ph.D. Thesis, University of Michigan, Ann Arbor, Michigan (1960).

Nickerson, G. R., Frey, H. M., and D. E. Coats, "Generalized Kinetic Analysis Program," Dynamic Science, Inc. (1971).

Nilsen, R. N. and W. J. Karplus, "Continuous-System Simulation Languages: A State-Of-The-Art Survey," *Annales de L'Association Internationale Pour Le Calcul Analogique,* 1 (1974).

de Nobrega, J. A. W., *N. S. A.,* **46**, 366 (1971).

Nolting, D. J., "Linear Multistep Methods With Near-Optimal Stability," Ph.D. Dissertation, Univ. of Missouri at Columbia (1979).

Nolting, D. J. and D. J. Rodabaugh, "Stiffly Stable Linear Multistep Methods," *Proc. Int. Conf. Stiff Comp.,* Park City, Utah (1982).

Norsett, S. P., "Semi Explicit Runge–Kutta Methods," Mathematics and Computing Rpt. No. 6/74, University of Trodheim (1974a).

Norsett, S. P., "One-Step Methods of Hermite Type for Numerical Integration of Stiff Systems," *BIT,* **14**, 63 (1974b).

Norsett, S. P., "C-polynomials for Rational Approximation to the Exponential Function," *Numer. Math.,* **25**, 39 (1975).

Norsett, S. P., "Runge–Kutta Methods with a Multiple Real Eigenvalue Only," *BIT,* **16**, 388 (1976).

Norsett, S. P., "Restricted Pade Approximations to the Exponential Function," *SIAM J. Numer. Anal.,* **15**, 1008 (1978).

Norsett, S. P. and G. Wanner, "Perturbed Collocation and Runge–Kutta Methods," *Numer. Math.,* **38**, 193 (1981).

Norsett, S. P. and A. Wolfbrandt, "Order Conditions for Rosenbrock Type Methods," *Numer. Math.,* **28**, 193 (1977).

Norsett, S. P. and A. Wolfbrandt, "Order Conditions for Rosenbrock Type Methods," *Numer. Math.,* **32**, 1 (1979).

Noyes, R. M., "Generalized Kinetics of Chemical Change: Some Conditions for the Validity of the Steady State Approximation," *Suppl. Prog. of Theo. Physics,* **64**, 295 (1978).

Nuutinen, E. M., Peuhkurinen, K. J., Pietilainen, E. P., Hiltunen, J. K., and E. I. Hassinen, "Elimination and Replenishment of Tricarboxylic Acid-Cycle Intermediates in Myocardium," *Biochem. J.,* **194**, 867 (1981).

Ockendon, J. R., "A Survey of Oxford Industry Problems," *Computational Techniques for Ord. Dif. Equations, Proc. of Conf.,* Manchester Eng.: 163 (1980).

Odeh, F. and Liniger, W., "On A-Stability of Second-Order Two-Step Methods for Uniform and Variable Steps," *Proc. IEEE Intl. Conf. Circuits and Computers,* **1**, 123 (1980).

Oesterhelt, G., "Multistep Methods for the Numerical Integration of Systems of Differential Equations with Widely Separated Time Constants," *Computing,* **13**, 279 (1974).

Oey, L., "Implicit Schemes for Differential Equations," *J. Comp. Phy.,* **45**, 443 (1982).

Ojika, T., Nishikawa, Y., and M. Okudaira, "A Time-Decomposition Algorithm for a Stiff Linear Two-Point Boundary-Value Problem and Its Application to a Nonlinear Optimal Control Problem," *J. Optimization Theory & Appl.,* **27**, 231 (1979).

O'Malley, R. E., Jr., "Boundary Value Problems for Linear Systems of Ordinary Differential Equations Involving Many Small Parameters," *J. Math. Mech.,* **18**, 835 (1969).

O'Malley, R. E., Jr., "A Nonlinear Singular Perturbation Problem Arising in the Study of Chemical Flow Reactors," *J. Inst. Math. Appl.,* **6**, 12 (1970).

O'Malley, R. E., Jr., "Singular Perturbation f the Time-Invariant Linear State Regulator Problem," *J. Diff. Equations,* **12**, 117 (1972).

O'Malley, R. E., Jr., *Introduction to Singular Perturbations,* Academic Press, New York (1974).

O'Malley, R. E., Jr., and L. R. Anderson, "Singular Perturbations, Order Reduction, and Decoupling of Large Scale Systems," *Numerical Analysis of Singular Perturbation Problems,* Hemker, P. W. and J. J. H. Miller, Eds., Academic Press, p. 317 (1979).

O'Malley, R. E., Jr., and L. R. Anderson, "Time-scale Decoupling and Order Reduction for Linear Time-Varying Systems," *Optimal Control Applications and Methods,* **3** (1982).

O'Malley, R. E. and J. E. Flaherty, "Analytical and Numerical Methods for Nonlinear Singular Singularly-Perturbed Initial Value Problems," *SIAM J. Appl. Math.,* **38**, 225 (1980).

Ong, B. G. and D. M. Mason, "Numerical Solution of Equations Arising in Energy Transport in Non-Equilibrium Chemically Reacting Systems," *Chem. Eng. Sci.,* **31**, 661 (1976).

Oran, E. S. and J. P. Boris, "Detailed Modelling of Combustion Systems," *Prog. Energy Combust. Sci.,* **7**, 1 (1981).

Otey, G. R., "Numerical Methods for Solving Reaction-Diffusion Problems," Sand 78-8025 (1978).

Ott, K. and D. A. Meneley, *N. S. E.,* **36**, 402 (1969).

Ozoe, H., Sato, N., and S. W. Churchill, "The Effect of Various Parameters on Thermoacoustic Convection, Part I – Vertical One-Dimensional Flow," *Chem. Eng. Commun.,* **5**, 203 (1980).

Ozoe, H., Yamamoto, K., and S. W. Churchill, "Three-Dimensional Numerical Analysis of Natural Convection in an Inclined Channel With a Square Cross Section," *AICHE J.,* **25**, 709 (1979).

Painter, J. F., "Solving The Navier–Stokes Equations with LSODE and the Method of Lines," LLNL Rpt. UCID-19262 (1981).

Palusinski, O. A., "Simulation Methods for Combined Linear and Nonlinear Systems," *Simulation,* **30**, 85 (1978).

Palusinski, O. A. and J. V. Wait, "Simulation of Partitioned Linear/Nonlinear Systems," *Proc. Int. Conf. SIMULATION '77,* M. Hamza, Ed., Montreux, Switzerland, 134 (1977).

Palusinski, O. A. and J. V. Wait, "Numerical Techniques for Partitioned Dynamic System Simulation," *Proc. Summer Computer Simulation Conf.,* Los Angeles, California (1978).

Park, K. C., "An Improved Stiffly Stable Method for Direct Integration of Nonlinear Structural Dynamic Equations," *Trans. ASME Ser. E,* **42**, 464 (1975).

Patterson, W. R. and D. L. Cresswell, "A Simple Method for the Calculation of Effectiveness Factors," *Chem. Eng. Sci.,* **26**, 605 (1971).

Patterson, G. K. and R. B. Rozsa, "DYNSYL: A General-Purpose Dynamic Simulator for Chemical Processes," *Comp. Chem. Eng.,* **4**, 1 (1980).

Peaceman, D. W., "Fundamentals of Numerical Reservoir Simulations," Elsevier Scientific Pub. Co., New York (1977).

Perelson, A. S., "A Model for the Reversible Binding of Bivalent Antigen to Cells," Elsevier North Holland, Inc., 147 (1979).

Perelson, A. S. and C. DeLisi, "Receptor Clustering on a Cell Surface. I. Theory of Receptor Cross-Linking by Ligands Bearing Two Chemically Identical Functional Groups," *Math. Biosci.,* **48**, 71 (1980).

Persek, S. C. and F. C. Hoppensteadt, "Integrated Averaging Methods for Systems of Ordinary Differential Equations with a Small Parameter," *Comm. Pure Appl. Math.,* **31**, 123 (1978).

Pert, G. J., "Algorithms for the Calculation of Ionization in Rapidly Changing Plasmas," *J. Comput. Phys.,* **27**, 241 (1978).

Peters, N., "An Asymptotic Analysis of Nitric Oxide Formation in Turbulent Diffusion Flames," *Comb. Sci. and Tech.,* **19**, 39 (1978).

Petzold, L. R., "An Efficient Numerical Method for Highly Oscillatory Ordinary

Differential Equation," Dept. Computer Sci. Rpt. UIUCDCS-R-78-933, Univ. Illinois, Ph.D. Thesis (1978).

Petzold, L. R., "Automatic Selection of Methods for Solving Stiff and Nonstiff Systems or Ordinary Differential Equations," Sand 80-8230 (1980).

Petzold, L. R., "An Efficient Numerical Method for Highly Oscillatory Ordinary Differential Equations," *SIAM J. Numer. Anal.,* **18**, 455 (1981a).

Petzold, L. R., "Differential-Algebraic Equations Are Not ODEs," Sandia National Lab.-Livermore, Rpt. Sand81-8668 (1981b).

Petzold, L. R., "A Description of DASSL: A Differential-Algebraic System Solver," *Proc. 10th IMACS World Congress On System Simulation and Scientific Computing,* August 8–13, Montreal, Canada, p. 430 (1982a).

Petzold, L. R., "Automatic Selection of Methods for Solving Stiff and Non-Stiff Systems of Ordinary Differential Equations," to appear SIAM J. Sci. Stat. Comp. (1982b).

Petzold, L. R., private communication (1982c).

Piotrowski, O., "Stability, Consistency and Convergence of Variable K-step Methods for Numerical Integration of Large Systems of Ordinary Differential Equations," *Proc. Conf. Dundee,* Lect. Notes in Math. No. 109, Springer-Verlag, Berlin 221 (1969).

Pirkle, J. C., Poehler, T. O., and V. G. Sigillito, "Numerical Procedure for a Pulsed DF-CO_2 Transfer Laser," *J. Comp. Phys.,* **15**, 293 (1974).

Pitts, J. N., Jr., and B. J. Finlayson, "Mechanisms of Photochemical Air Pollution," *Angewandte Chemie, Int. Ed. in Eng.,* **14**, 1 (1975).

Plesser, T., Wurster, B., and B. Hess, "Determination of the Kinetic Constants of Glucosephosphate Isomerase by Non-Linear Optimization," *Eur. J. Biochem.,* **98**, 93 (1979).

Ploch, P. W., "KAPP-VI: Digital Code for the Simulation of Primary System Transients in Pressurized Water Reactors," NPGD-12-1102640-02, Babcock and Wilcox, Lynchburg, Virginia (1980).

Ponisch, G. and H. Schwetlich, "Computing Turning Points of Curves Implicitly Defined by Nonlinear Equations Depending on a Parameter," *Computing,* **26**, 107 (1981).

Ponzo, P. J. and N. Wax, "Relaxation Oscillations, Parasitics, and Singular Perturbations," *IEEE Trans. Circuit Theory,* **CT-19**, 623 (1972).

Porching, T., *SIAM J. Appl. Math..,* **16**, 301 (1968).

Porter, B., "Design of Stabilizing Feedback Controllers for a Class of Multivariable Linear Systems with Slow and Fast Modes," *Int. J. Control,* **23**, 49 (1976).

Powles, J. G., Evans, W. A. B., McGrath, E., Gubbins, K. E., and S. Murad, "A Computer Simulation for a Simple Model of Liquid Hydrogen Chloride," *Molecular Physics,* **38**, 893 (1979).

Prasad, S. S. and W. T. Huntress, Jr., "A Model for Gas Phase Chemistry in Interstellar Clouds: I. The Basic Model, Library of Chemical Reactions, and Chemistry Among C, N, and O Compounds," *The Astrophysical J. Supply Series,* **43**, 1 (1980).

Pratt, D. T., "Calculation of Chemically Reacting Flows with Complex Chemistry," *Studies in Convection, Vol. II,* B. F. Launder, Ed., Academic Press (1976).

Pratt, D. T., "New Computational Algorithms for Chemical Kinetics," National Bureau of Standards Special Publ. 561, 1265, October (1979a).

Pratt, D. T., "Coalescence/Dispersion Modelling of Turbulent Combustion," *AIAA J. Energy,* **3**, 177 (1979b).

Preussner, P. R. and K. P. Brand, "Application of a Semi-Implicit Euler Method to Mass Action Kinetics," *Chem. Eng. Sci.,* **36**, 1653 (1981).

Pritsker, A. A. B., "The GASP-VI Simulation Language," John Wiley (1974).

Prokopakis, G. J., and W. D. Seider, "Adaptive Semi-Implied Runge–Kutta Method for Solution of Stiff Ordinary Differential Equations," *Ind. Eng. Chem.,* **20**, 255 (1981).

Prokopakis, G. J. and W. D. Seider, "Dynamic Simulation of Distillation Towers," Submitted to AICHE J. (1982).

Prokopakis, G. J., Ross, B. A., and W. D. Seider, "Azeotropic Distillation Towers with Two Liquid Phases," *Foundations of Computer-aided Chemical Process Design, Vol. II,* Mah, R. S. H. and W. D. Seider, Eds., *AICHE* (1981).

Prothero, A., "Introduction to Stiff Problems," Chapter 9, *Modern Numerical Methods for Ordinary Differential Equations,* Hall, G., and J. M. Watt, Eds., Clarendon Press, Oxford, 123 (1976a).

Prothero, A., "Multistep Methods for Stiff Problems," *Modern Numerical Methods for Ordinary Differential Equations,* Hall, G., and J. M. Watt, Eds., Clarendon Press, Oxford 152 (1976b).

Prothero, A. and A. Robinson, "On the Stability and Accuracy of One-Step Methods for Solving Stiff Systems of Ordinary Differential Equations," *Math. Comp.,* **28**, 145 (1974).

Pulliam, T. H., and J. L. Steger, "On Implicit Finite Difference Simulations of Three-Dimensional Flow," *AIAA J.,* **18**, 159 (1980).

Purcell, S. P., "Special Problem Assignment," ChE 591, Univ. of Utah (1979).

Rabbat, N. B. G., Sangiovanni-Vincente, A. L. and H. Y. Hsieh, "A Multilevel Newton Algorithm with Macromodelling and Latency for the Analysis of Large-Scale Nonlinear Circuits in the Time Domain," *IEEE Trans. Circ. Sys. Cas-26,* 733 (1979).

Rabitz, H., "Chemical Sensitivity Analysis Theory with Applications to Molecular Dynamics and Kinetics," *Comp. & Chem.,* **5**, 167 (1981).

Radhakrishnan, K., private communication (1982).

Rai, M. M. and D. A. Anderson, "Numerical Grid Techniques," *NASA Conf. Publ.,* **2166**, 409 (1980).

Ralston, M. L., Jennrich, R. I., Sampson, P. F., and F. K. Uno, "Fitting Models Defined by Differential Equations," BMDP Tech. Rpt. No. 70 (1980).

Ramshaw, J. D., "Partial Chemical Equilibrium in Fluid Dynamics," *Phys. Fluids,* **23**, 675 (1980).

Ranson, V., "RELAP4/MOD5: A Computer Program for Transient Thermal-Hydraulic Analysis of Nuclear Reactor and Related Systems," ANCR-Nureg-1335, Idaho National Engineering Laboratory, Idaho falls, Idaho (1976).

Rao, C. P. and R. K. Iyengar, "High Order Numerical Differentiation Type Formulas with an Off-Step Point for Stiff Ordinary Differential Equations," *J. Inst. Math. & Appl.,* **17**, 281 (1976).

Raptis, A. D., "Exponential-Fitting Methods for the Numerical Integration of the Fourth-Order Differential Equation Y to the IV + F Times $Y = G$," *Computing,* **24**, 241 (1980).

Ray, W. H., "The Quasi-Steady-State Approximation in Continuous Stirred Tank Reactors," *Can. J. Chem. Eng.,* **47**, 503 (1969).

Ray, W. H. and M. R. Ajinkya, "Experiences in Simulating Axially Dispersed Tubular Reactors," *Chem. Eng. Comm.,* **1**, 103 (1973).

Reid, J. K., "Sparse Matrix Routines," For further detail contact: Dr J. K. Reid, Atomic Energy Research Establishment (AERE), Harwell, U.K.

Reily, N., "Surface Reactions in Similar Boundary Layers," *Quart. J. Mech. Appl. Math.,* **25**, 273 (1972).

Reilly, P. M. and Blau, G. E., "The Use of Statistical Methods to Build Mathematical Models of Chemical Reacting Systems," *Can. J. Chem. Eng.,* **52**, 289 (1974).

Rentrop, P., "An Algorithm for the Computation of the Exponential Spline," *Num. Math., Handbook Ser.,* **35**, 81 (1980).

Rice, J. R., "Machine and Compiler Effects of the Performance of Elliptic PDEW Software," *Proc. 10th IMACS World Congress on System Simulation and Scientific Computing,* August 8–13, Montreal, Canada, p. 446 (1982).

Rice, J. R., *Numerical Methods, Software and Analysis,* McGraw-Hill Book Company, New York (1983).

Ridler, P. F., "CAKE: A Computer Program for the Solution of Kinetic Rate Equations," *J. Phys. Chem.,* **81**, 2419 (1977).

Rimvall, M. and F. E. Cellier, "The GASP-VI Simulation Package for Process-Oriented Combined Continuous and Discrete System Simulation," *Proc. 10th IMACS Congress on Simulation of Systems,* Montreal, Canada (1982).

Rinzel, J. and R. N. Miller, "Numerical Calculation of Stable and Unstable Periodic Solutions to the Hodgkin-Huxley Equations," *Math. Biosci.,* **49**, 27 (1980).

Roberts, S. A., "Numerical Modelling of a Chemical Plasma II. PLASKEN: A Program to Predict the Variation with Time of the Number Densitites of Chemical Species Within a Plasma," *Computer Phys. Commun.,* **18**, 363 (1979).

Roberts, S. M., "Simultaneous and Sequential Solution of a Stiff Differential Equation System," *J. Optimiz. Theory & Appl.,* **25**, 207 (1978).

Robertson, H. H., "Numerical Integration of Systems of Stiff Ordinary Differential Equations with Special Structures," *J. Inst. Math. & Appl.,* **18**, 249 (1976).

Rodabaugh, D. J. and S. Thompson, "Adams-Type Methods with Increased Ranges of Stability," *Com. & Math.,* **4**, 349 (1978).

Rodabaugh, D. J. and S. Thompson, "Low-Order A(Zero)-Stable Adams-Type Correctors," *J. Comp. Appl. Math.,* **5**, 225 (1979).

Rodrigues, A. E., and E. C. Beira, "Sorption Processes in a Perfectly Mixed Reactor: Influence of Nonlinear Equilibrium Isotherm and External Mass Transfer Resistance," *AICHE J.,* **25**, 416 (1979).

Rosenbaum, J. S., "Conservation Properties of Numerical Integration Methods for Systems of Ordinary Differential Equations," *J. Comp. Phys.,* **20**, 259 (1976).

Rosenbaum, J. S., "Implicit and Explicit Exponential Adams Methods for Real Time Simulations," *Proc. IMACS(AICA) Int. Symp. Simulation Software and Numerical Methods for Dif. Equa.,* Blacksburg, Virginia (1977).

Rosenbrock, H. H., "Some General Implicit Processes for the Numerical Solution of Differential Equations," *Comp. J.,* **5**, 329 (1963).

Ross, R. T., "Steady State of Stiff Linear Kinetic Systems by a Markov Matrix Methods," *J. Comp. Phys.,* **24**, 70 (1977).

Rossky, P. J. and M. Karplus, "Solvation: A Molecular Dynamics Study of a Dipeptide in Water," *J. ACS,* **101**, 1913 (1979).

Rossky, P. J., Karplus, M., and A. Rahman, "A Model for the Simulation of an Aqueous Dipeptide Solution," *Biopolymers,* **18**, 825 (1979).

Roth, M. G., "Difference Methods for Stiff Delay Differential Equations," Dept. Comp. Sci., Rpt. UIUCDCS-R-80-1012 Ph.D. Thesis, Univ. Illinois (1980).

Ruehli, A. E., Brennan, P. A. and W. Liniger, "Control of Numerical Stability and Damping in Oscillatory Differential Equations," *Proc. IEEE Int. Conf. Circuits and Computers,* **1**, 111 (1980).

Sacks-Davis, R., "Error Estimates for a Stiff Differential Equation Procedure," *Math. Comps.,* **31**, 939 (1977a).

Sacks-Davis, R., "Solution of Stiff Ordinary Differential Equations by a Second Derivative Method," *SIAM J. Numer. Anal.,* **14**, 1088 (1977b).

Sacks-Davis, R., "Fixed Leading Coefficient Implementation of SD-Formulas for Stiff ODEs," *ACM Trans. on Mathematical Software,* **6**, 540 (1980).

Sacks-Davis, R. and L. F Shampine, "A Type Insensitive ODE Code Based on Second Derivative Formulas," *Comp. & Math. Appl.* **7**, 487 (1981).

Samarskii, A. A., "Introduction to the Theory of Difference Schemes," *Nauka,* Moskow (1971).

Sanchez-Palencia, E. and M. Lobo-Hidalgo, "Perturbation of Spectral Properties for a Class of Stiff Problems," *Computing Methods in Appl. Sci. and Eng.,* Glowsinski, R. and J. L. Lions, Eds., North-Holland Publ. Co., Inria (1980).

Sand, J., "On One Leg and Linear Multistep Formulas with Variable Step Sizes," Rpt. Royal Inst. Tech., Stockholm, Sweden (1981).

Sand, J., "On Contractive One-Leg Methods," Rpt. TRITA-NA-8218 (1982a).

Sand, J., "Stability and Boundedness Results for Variable-Step Variable Formula Methods," Rpt. TRITA-NA-8219 (1982b).

Sand, J. and O. Osterby, "Regions of Absolute Stability," Rpt. DAIMI PB-102," Comp. Sci. Dept., Aarhus University, Denmark (1979).

Sandberg, I. W., "Two Theorems on the Accuracy of Numerical Solutions of Ordinary Differential Equations," *Bell Syst. Tech. T.,* **46,** 1243 (1967).

Sandusky, H., Bellan, J., and T. J. Ohlemiller, "Linear Finite Element Numerical Technique for Combustion Problems Requiring Variable Step Size," *Comb. & Flame,* **36,** 193 (1979).

Sanz-Serna, J. M., "Some Aspects of the Boundary Locus Method," *BIT,* **20,** 97 (1980).

Sarkany, E. F. and W. Liniger, "Exponential Fitting of Matricial Multistep Methods for Ordinary Differential Equations," *Math. Computation,* **28,** 1035 (1974).

Scaccia, C. and L. A. Kennedy, "Calculating Two-Dimensional Chemically Reacting Flows," *AIAA,* **12,** 1268 (1974).

Schaumburg, K., Wasniewski, J., and Z. Zlatev, "The Use of Sparse Matrix Technique in the Numerical Integration of Stiff Systems of Linear Ordinary Differential Equations," *Computers & Chemistry,* **4,** 1 (1980).

Scherer, R., "On the Numerical Treatment of Differential Equation Systems Having Appreciable Stiffness," *Z. Angew. Math. & Mech.,* **56,** 317 (1976).

Scherer, R., "A Necessary Condition for B-Stability," *BIT,* **19,** 111 (1979).

Schiesser, W. E., "DSS/2 (Differential System Simulator, Version 2), An Introduction to the Numerical Method of Lines Integration of Partial Differential Equations," Lehigh University, Bethlehem, Pennsylvania (1977).

Schiesser, W. E., "An Introduction to the Use of Mathematical Software for Ordinary and Partial Differential Equations; Guidelines for the Programming of a Numerical Method of Lines Code," Rpt. DSS/2 No. 8, Lehigh University, Bethlehem, Pennsylvania (1983).

Schneider, D. R., Amundson, N. R., and R. Aris, "On a Mechanism for Autocatalysis," *Chem. Eng. Sci.,* **27,** 895 (1972).

Schneider, D. R., Aris, R., and N. R. Amundson, "An Analysis of Chemical Reactor Stability and Sensitivity-XIV: The Effect of the Steady State Hypothesis," *Chem. Eng. Sci.,* **28,** 885 (1973).

Scholander, P. and G. Svedberg, "Solution of a Stiff Boundary Value Problem by Orthogonal Collocation," Project Assignment, Numerical Model Anal. in Chem. Eng., Stockholm, Sweden (1975).

Schonauer, W., Raith K., and G. Glotz, "Ein Selbststeuerndes Lösungsverfahren für Grobe Systeme von Gewöhnlichen Differentialgleichungen, wie sie zum Beispiel aus der Linienmethode Resultieren," *ZMM,* **59,** T79 (1979).

Schryer, N. L., "The State of the Art in the Numerical Solution of Time-Varying Partial Differential Equations," *J. Phys. Chem.,* **81,** 2335 (1977).

Scraton, R. E., "Some New Methods for Stiff Differential Equations," *J. Comp. Math.,* **7,** 55 (1979).

Scraton, R. E., "Some L-Stable Methods for Stiff Differential Equations," *J. Comp. Math.,* **9,** 81 (1981).

Seager, M. K. and S. Balsdon, "LSODIS, A Sparse Implicit ODE Solver," *Proc. 10th IMACS World Congress on System Simulation and Scientific Computing,* August 8–13, Montreal, Canada, p. 437 (1982).

Sebastian, D. J. G., Noble, R. G., Thambynayagam, R. K. M., and R. K. Wood, "DPS-A Unique Tool for Process Simulation," Rpt., University of Alberta, Edmonton, Alberta (1980).

Seider, W. D., "Integration of Stiff Ordinary Differential Equations," Lecture Notes AIChE Cont. Ed., Course on Math. Models for Process Design and Analysts (1980).

Seider, W. D., White, III, C. W., and G. J. Prokopakis, "Integration of Stiff Ordinary Differential Equations In Chemical Process Analysis," *Proc. Int. Conf. Stiff. Comp.,* Park City, Utah (1982).

Seinfeld, J. H., Lapidus, L., and M. Hwang, "Review of Numerical Integration Techniques for Stiff Ordinary Differential Equations," *Ind. Eng. Chem. Fund.,* **9**, 266 (1970).

Semenov, N. N., "Some Problems in Chemical Kinetics, Vol. 1" Pergamon, New York, 47 (1959).

Serth, R. W., "Solution of Stiff Boundary Value Problems by Orthogonal Collocation," *Inter. J. Numer. Methods in Eng.,* **648** (1975).

Shah, Y. T. and J. A. Paraskos, "Approximate Solutions to Single and Two Phase Axial Dispersion Problems in Isothermal Tubular Flow Reactors," *Chem. Eng. Sci.,* **30**, 465 (1975).

Shampine, L. F., "The User's View of Solving Stiff Differential Equations," AFWL-AFOSR Workshop, p. 6 (1977a).

Shampine, L. F., "Stiffness and Nonstiff Differential Equation Solvers, II: Detecting Stiffness with Runge-Kutta Methods," *ACM Trans. on Math. Software,* **3**, 44 (1977b).

Shampine, L. F., "Limiting Precision in Differential Equation Solvers. II: Sources of Trouble and Starting a Code," *Math. Comp.,* **32**, 1115 (1978a).

Shampine, L. F., "Stability Properties of Adams Codes," *ACM Trans. on Math. Software,* **4**, 323 (1978b).

Shampine, L. F., "Better Software of ODEs," *Comp. & Math. Appl.,* **5**, 157 (1979a).

Shampine, L. F., "Evaluation of Implicit Formulas for the Solution of ODEs," *BIT,* **19**, 495 (1979b).

Shampine, L. F., "What Everyone Solving Differential Equations Numerically Should Know," SAND 78-0315, Sandia Corporation, Albuquerque, New Mexico (1979c).

Shampine, L. F., "Global Error Estimation for Stiff ODEs, Sandia National Laboratories-Albuquerque Rpt. SAND 79-1587 (1979d).

Shampine, L. F., "Type-Insensitive ODE Codes Based on Implicit A-Stable Formulas," Rpt. SAND 79-2444, Sandia Lab (1980a).

Shampine, L. F., "Implementation of Rosenbrock Methods," Rpt. No. SAND 80-2367J, Sandia National Laboratories, Albuquerque, New Mexico (1980b).

Shampine. L. F., "Lipschitz Constants and Robust ODE Codes," *Computational Methods in Nonlinear Mechanics,* North Holland, 427 (1980c).

Shampine. L. F., "What Everyone Solving Differential Equations Numerically Shown Know," *Computational Techniques for Ord. Dif. Equations, Proc. Conf.,* Manchester, England (1980d).

Shampine, L. F., "Evaluation of a Test Set for Stiff ODE," *ACM Trans. Math. Software,* **7**, 409 (1981a).

Shampine, L. F., "Stiffness and The Automatic Selection of ODE Codes," Sandia Rpt. SAND 81-0525, Albuquerque, New Mexico (1981b).

Shampine, L. F., "Type-Insensitive ODE Codes Based on Implicit A-Stable Formulas," *Math. Comput.,* **36**, 499 (1981c).

Shampine, L. F., "Implementation of Rosenbrock Methods," Rpt. SAND 80-2367 J. Sandia Lab J (1980); *ACM Trans. Math. Soft.,* **8**, 93 (1982a).

Shampine, L. F., "Type-Insensitive ODE Codes Based on Implicit A(Alpha)-Stable Formulas," *Math. Comp.,* **39**, 109 (1982b).

Shampine. L. F., "What is Stiffness?" *Proc. Int. Conf. Stiff Comp.,* Park City, Utah (1982c).

Shampine, L. F. and C. W. Gear, "A User's View of Solving Stiff Ordinary Differential Equations," *SIAM Review,* **21**, (1979).

Shampine, L. F. and M. K. Gordon, *Solution of Ordinary Differential Equations—The Initial Value Problem,* W. H. Freeman and Co., San Francisco, California (1975).

Shampine, L. F. and K. L. Hiebert, "Detecting Stiffness with the Fehlberg Formulas," **41** (1976).

Shampine. L. F. and H. A. Watts, "Algorithm 504 GERK: Global Error Estimation for Ordinary Differential Equations," *ACM Trans. Math. Software,* **2(2)**, 200 (1978).

Shampine, L. F., Gordon, M. K., and J. A. Wisniewski, "Variable Order Runge–Kutta Codes," *Computational Techniques for Ord. Dif. Equations, Proc. Conf.,* **83**, 101, Manchester, England (1980).

Shampine, L. F., Watts, H. A., and S. M. Davenport, "Solving Non-Stiff Ordinary Differential Equations—The State of the Art," SAND 75-0182, Sandia Corp., Albuquerque, New Mexico (1975).

Sherman, A. H. and A. C. Hindmarsh, "GEARS: A Package for the Solution of Sparse, Stiff Ordinary Differential Equations," *Electrical Power Problems: The Mathematical Challenge, SIAM* (1980).

Shimizu, H., Yoshihara, M., and K. Watanabe, "An Iterative Method for Computing Time-Responses of Stiff and Nonlinear Systems," *Trans. Soc. Instrum. & Control Eng.,* **16**, 858 (1980).

Simonian, S. S., "Inverse Problems in Structural Dynamics-II, Applications," *Int. J. Numer. Methods in Eng.,* **17**, 367.

Sincovec, R. F. and N. K. Madsen, "Software for Nonlinear Partial Differential Equations," *ACM Trans. on Mathematical Software,* **1**, 232 (1975a).

Sincovec, R. F. and N. Madsen, "Algorithm 494 PDEONE. Solutions of Systems of Partial Differential Equations," *ACM Trans. Math. Software,* **1(3)**, 262 (1975).

Sincovec, R. F., Erisman, A. M., Yip, E. L., and M. A. Epton, "Analysis of Systems Using Numerical Algorithms," *IEEE Trans. on Auto Control,* **AC-26**, 139 (1981).

Sinha, A. S. C., "High-Order Stiffly Stable Methods for Analysis of Nonlinear Circuits and Systems," *International J. Syst. Sci.,* **11**, 1001 (1980).

Skeel, R. D., "An Incomplete List of Better Fortran Codes for IVPs in ODEs," *SIGNUM Newsl.,* **14**, 22 (1979).

Skeel, R. D. and A. K. Kong, Rpt. UIUCDCS-R-76-800, Dept. Computer Sci., Univ. Illinois at Urbana-Champaign (1976).

Skeel, R. D. and A. K. Kong, "Blended Linear Multistep Methods," *ACM Trans. Math. Software,* **3**, 326 (1977).

Skelboe, S., "The Control of Order and Steplength for Backward Differentiation Methods," *BIT,* **17**, 91 (1977).

Sloat, H. M. and T. A. Bickart, "A-Stable Composite Multistep Methods," *J. ACA,* **20**, 7 (1973).

Smith, B. T., Boyle, J. M., Dongarra, J. J., Garbow, B. S., Ikebe, Y., Klema, V. C., and C. B. Moler, "Matrix Eigensystem Routines—EISPACK Guide," Springer-Verlag, New York, 2nd Ed. (1976).

Smith, B. T., Boyle, J. M., Garbow, B. S., Ikebe, Y., Klema, V. C., and C. B. Moler, "Matrix Eigensystem Routines-EISPACK Guide," *Lecture Notes Comp. Sic.,* **6(2)**, Springer-Verlag, New York (1976).

Smith, D. R., "On Boundary Layer Approximations in the Numerical Solution of Certain Stiff Boundary Value Problems," *SIAM J. Numer. Anal.,* **18**, 377 (1981).

Smith, H. V., "A Method for the Integration of Oscillatory Functions," *BIT,* **17**, 388 (1977).

Smith, R. M., "Finite Element Solutions of the Energy Equation at High Peclet Number," *Comp. & Fluids,* **8**, 335 (1980).

Snider, A. D., "Error Analysis for a Stiff System Procedure," *Math. Comp.,* **39**, 216 (1976).

Snider, A. D. and G. C. Fleming, "Approximation by Aliasing with Application to 'Certaine' Stiff Differential Equations," *Math. Computation,* **28**, 126 (1974).

Snow, R. H., "A Chemical Kinetics Program for Homogeneous and Free-Radical Systems of Reactions," *J. Phys. Chem.,* **70**, 2780 (1966).

Söderlind, G., "Some Stability Properties of Linear Multistep Compound Discretizations of Partitioned Differential Systems," Royal Inst. Tech., Stockholm, Sweden (1979).

Söderlind, G., "DASP3–A Program for the Numerical Integration of Partitioned Stiff ODEs and Differential–Algebraic System," Tech. Rpt., TRITA-Na-8008, Royal Inst. Tech., Stockholm, Sweden (1980).

Söderlind, G., "Multistep G-Contractivity with Applications to Slowly Varying Linear Systems," Rpt. TRITA-Na-8107, Royal Inst. Tech., Stockholm, Sweden (1981a).

Söderlind, G., "On the Efficient Solution of Nonlinear Equations in the Numerical Solution of Stiff Differential Equations," Rpt. TRITA-Na-8114, Royal Inst. Tech., Stockholm, Sweden (1981b).

Söderlind, G., "Theoretical and Computational Aspects on Partitioning in the Numerical Solution of Stiff Differential Systems," Rpt. TRITA-Na-8115, Royal Inst. Tech., Stockholm, Sweden (1981c).

Söderlind, G. and G. Dahlquist, "Error Propagation in Stiff Differential Systems of Singular Perturbation Type," Rpt. TRITA-Na-8108 (1981).

Sommerfeld, J. T. and R. C. Schuker, "Acceleration of the Numerical Integration of Stable Differential Equations," Paper Presented at the 78th Natl. *AICHE* Meeting, Salt Lake City, Utah (1974).

Soresen, J. P. and W. E. Stewart, "Structural Analysis of Multicomponent Reaction Models: Part II, Formulation of Mass Balances and Thermodynamic Constraints," *AICHE J.,* **26**, 98 (1980).

Spellmann, J. W. and A. C. Hindmarsh, "GEARS: Solution of Ordinary Differential Equations Having a Sparse Jacobian Matrix," LLNL Rpt. UCID-30116 (1975).

Srivastava, B. N., Werle, M. J., and R. T. Davis, "A Finite Difference Technique Involving Discontinuous Derivatives," *Comp. & Fluids,* **7**, 69 (1979).

Stabler, R. N. and J. P. Chesick, "A Program System for Computer Integration of Multistep Reaction Rate Equations Using the Gear Integration Method," *Int. J. Chem. Kin.,* **10**, 461 (1978).

Stacey, W., "Space-time Nuclear Reactor Kinetics," Academic Press, New York (1969).

Starner, J. W., "A Numerical Algorithm for the Solution of Implicit Algebraic-Differential Systems of Equations," Ph.D. Thesis, Univ. N. M., Albuquerque, New Mexico (1976).

Starner, J. W., Jr., "A Numerical Algorithm for the Solution of Over-Determined Non-Square Implicit Algebraic-Differential Systems of Equations," AFWL-AFOSR Workshop, p. 298 (1977).

Stayton, M. M. and H. J. Fromm, "A Computer Analysis of the Validity of the Integrated Michaelis-Menten Equation," *J. Theor. Biol.,* **78**, 309 (1979).

Steihaug, T. and A. Wolfbrandt, "An Attempt to Avoid Exact Jacobian and Nonlinear Equations in the Numerical Solution of Stiff Differential Equations," *Math. Comp.,* **33**, 521 (1979).

Stetter, H. J., "Improved Absolute Stability of Predictor-Corrector Schemes," *Computing,* **3**, 286 (1968).

Stetter, H. J., "Analysis of Discretization Methods for Ordinary Differential Equations," Springer-Verlag, Berlin-Heidelberg-New York (1973).

Stetter, H. J., "Economical Global Error Estimation," *Stiff Differential Systems,* R. A. Willoughby, Ed., Plenum Press, New York (1974).

Stetter, H. J., "Towards a Theory for Discretizations of Stiff Differential Systems," *Lecture Notes Math.,* **506**, 190 (1975).

Stetter, H. J., "The Defect Correction Principle and Discretization Methods," *Numer. Math.,* **29**, 425 (1978).

Stetter, H. J., "Global Error Estimation in Adams PC Codes," *ACM Trans. Math. Software,* **5**, 415 (1979).

Stetter, H. J. and E. Weinmuller, "On the Error Control in ODE Solvers with Local Extrapolation," *Computing,* **27**, 169 (1981).

Stewart, H. B., "Calculation of Transient Boiling Flow in Channels," *J. Comp. Phys.,* **30**, 61 (1979).

Strang, G., "Trigonometric Polynomials and Differencing Methods of Maximum Accuracy," *J. Math. Phys,* **41** 147 (1962).

Streett, W. B. and D. J. Tildesley, "Multiple Time Step Methods and an Improved Potential Function for Molecular Dynamics Simulations of Molecular Liquid," ACS 0-8412-0463-47-086-144, 13 (1978).

Strehmel, K., "Multistep Methods for Systems of First Order Differential Equations with Order Differential Equations with Exponential Adaption," *Computing,* **17**, 247 (1976).

Strehmel, K., "A New Algorithm for the Numerical Solution of Ordinary Stiff Differential Equations of the N-th Order," *SIGNUM Newsl.,* **12**, 20 (1977).

Strehmel, K. and C. Peper, "Numerical Solution of Initial Value Problems for Ordinary Differential Equations of *N*-th Order with One-Step Methods," *Computing,* **22**, 125 (1979).

Strehmel, R. and R. Weiner, "Behandlung Steifer Anfangswertprobleme gewohnlicher Differentialgleichungen mit Adaptiven Runge–Kutta Methoden," *Computing,* **29**, 153 (1982).

Strikwerda, J. C., "Initial Boundary Value Problems for the Method of Lines," *J. Comp. Phys.,* **34**, 94 (1980).

Strom, T., "On Logarithmic Norms," *SIAM J. Num. Anal.,* **12**, 741 (1975).

Struwe, M., "On the Solvability of Ordinary Differential Equations with Only Partially Continuous Right-Hand Side By Difference Methods," *Computing,* **27**, 165 (1981).

Stuben, K. and U. Trottenberg, "Multigrid Methods: Fundamental Algorithms, Model Problem Analysis and Applications," *Lecture Notes in Mathematics, Multigrid Methods*, Dold, A. and B. Eckmann, Eds., Springer-Verlag, 177 (1981).

Sundaram, K. M. and G. F. Froment, "Modelling of Thermal Cracking Kinetics, 3. Radical Mechanisms for the Pyrolysis of Simple Paraffins, and their Mixtures," *Numerical Methods for Differential Systems*, Lapidus, L. and W. E. Schiesser, Eds., Academic Press (1976).

Sundaram, K. M. and G. F. Froment, "Modelling of Thermal Cracking Kinetics, 3. Radical Mechanisms for the Pyrolysis of Simple Paraffins, and their Mixtures," *IEC Fund.,* **17**, 3 (1978).

Sundaresan, S. and N. R. Amundson, "Diffusion and Reaction in a Stagnant Boundary Layer About a Carbon Particle. 5. Pseudo-Steady-State Structure and Parameter Sensitivity," *Ind. Eng. Chem. Fundam.,* **19**, 344 (1980).

Suuberg, E. M., Ph.D. Thesis, Dept. Chem. Eng., MIT, Cambridge, Massachusetts (1977).

Szeverenyl, Z., Budo-Zahonyi, E., and L. I. Simandi, "Autocatalytic Reduction of Pyridinecobaloxime (III) by Molecular Hydrogen," *J. Coord. Chem.,* **10**, 41 (1980).

Tadeusiewicz, M., "A-Stable Algorithm for Solving Dynamic Networks," *Anch. Elektrotech.,* **29**, 689 (1980).

Taubert, K., "An Extension of the Theory Given by G. Dahlquist," *Computing,* **17**, 177 (1976).

Tendler, J. M., "STINT: Stiff (Differential Equations) Integrator," *ACM Trans. Math. Software,* **4**, 399 (1978).

Tendler, J. M., Bickart, T. A., and Z. Picel, "A Stiffly Stable Integration Process Using Cyclic Composite Methods," *ACM Trans. On Mathematical Software,* **4**, 339 (1978).

Thambynayagam, R. K. M., Wood, R. K., and P. Winter, "DPS-An Engineer's Tool for Dynamic Process Analysis," *Chem. Eng.* (1981).

Thompson, S., "A Comparison of Available Software for the Numerical Solution of Stiff Ordinary Differential Equations," NPGD-TM-368, Babcock & Wilcox, Nuclear Power Generation Division, Lynchburg, Virginia (1977).

Thompson, S. and D. J. Rodabaugh, "Corrector Methods With Increased Ranges of Stability," *Comp. Math. With App.*, **3**, 197 (1977).

Thompson, S. and P. G. Tuttle, "DSTPGT, Automatic Software for the Numerical Solution of Differential Equations," NPGD-TM-521, Babcock & Wilcox, Lynchburg, Virginia (1980).

Thompson, S. and P. G. Tuttle, "The Evolution of an ODE Solver in an Industrial Environment," *Proc. 10th IMACS World Congress on System Simulation,* Montreal, Canada (1982a).

Thompson, S. and P. G. Tuttle, "The Solution of Several Representative Stiff Problems in an Industrial Environment: The Evolution of an ODE Solver," *Proc. Int. Conf. Stiff Comp.*, Park City, Utah (1982b).

Tikhonov, A. N., "Systems of Differential Equations Containing Small Parameters in the Derivatives," Eng. Trans. A. Muzyka *Math. Sboruitk.*, **31**, 73 (1952).

Tischer, P. and R. Sacks-Davis, "A New Class of Cyclic Multistep Formulae for Stiff Systems," Tech. Rpt. No. 15, Monash Univ., Dept. Computer Science, Clayton, Victoria, Australia (1981).

Treanor, C. E., "A Method for the Numerical Integration of Coupled First-Order Differential Equations With Greatly Different Time Constants," *Math. Comp.*, **20**, No. 93, 39 (1966).

Trigiante, D., "Asymptotic Stability and Discretization on an Infinite Interval," *Computing*, **18**, 117 (1977).

Tripathy, S. C. and N. D. Rao, "Comparison of Stability Properties of Numerical Integration Methods for Switching Surges," *IEEE Trans. Power Appar. & Syst.*, **PAS-97**, 2138 (1978).

Turnage, J. C., *N. S. E.*, **51**, 67 (1973).

Turner, J. S., "From Microphysics to Macrochemistry Via Discrete Simulations," *ACS,* **18** (1978).

Tuttle, P. G., "The Numerical Solution of Ordinary Differential Equations with a Removable Singularity," Ph.D Thesis, School of Engineering and Applied Science, Univ. Virginia (1981).

Tuttle, P. G. and L. L. Barinks, "Scientific Computing Practices of the Nuclear Power Generation Division of Babcock and Wilcox," *Proc. Topical Meeting On Computational Methods in Nuclear Engineering, Vol. 2,* Williamsburg, Virginia, 5b.1 (1979).

Twizell, E. H. and A. Q. M. Khaliq, "One-Step Multiderivative Methods for First Order Ordinary Differential Equations," *BIT,* **21**, 518 (1981).

Tyreus, B. D., Luyben, W. L., and W. E. Schiesser, "Stiffness in Distillation Models and the Use of an Implicit Integration Method to Reduce Computation Times," *Ind. Eng. Chem., Procss Des. Dev.*, **14**, 427 (1975).

Tyson, T. J., "An Implicit Integration Method for Chemical Kinetics," Paper 9840-60002-R000, TRW Space Tech. Lab., Redondo Beach, California (1964).

Tzafestas, S. G., "Hybrid Simulation of Distributed Parameter Systems: The State of the Art," *Progress in Modelling and Simulation,* F. E. Cellier, Ed., Academic Press (1982).

Ueberhuber, C. W., "Implementation of Defect Correction Methods for Stiff Differential Equations," *Computing,* **23**, 205 (1979).

Uhlen, M., "KINRATE and KINBOX, Two Program Packages for Interactive Simulation of Chemical Systems," Dept. Info. Processing Computer Science, Royal Inst. Tech., Stockholm, Sweden (1981).

Uhlen, M. and L. Edsberg, "KINRATE and KINBOX, Two Program Packages for Interactive Simulation of Chemical Systems," TRITA-NA-7912, Dept.

Num. Anal. and Comp. Sci., The Royal Institute of Technology, Stockholm, Sweden (1979).

Unno, H. and I. Inoue, "Size Reduction of Bubbles by Orifice Mixer," *Chem. Eng. Sci.,* **35**, 1571 (1980).

Van Brokhoven, W. M. G., "Efficient Higher Order Implicit One-Step Methods for Integration of Stiff Differential Equations," *BIT,* **20**, 34 (1980).

Van Der Houwen, P. J., "Explicit and Semi-Implicit Runge–Kutta Formulas for the Integration of Stiff Equations," Rpt. TW132, Mathematisch Centrum, Amsterdam (1972a).

Van Der Houwen, P. J., "One Step Methods With Adaptive Stability Functions for the Integration of Differential Equations," *Lecture Notes in Math.,* **333**, Springer-Verlag, Berlin-Heidelberg-New York (1972b).

Van Der Houwen, P. J., "Stabilized Runge–Kutta Methods for Second Order Differential Equations without First Derivatives," *SIAM J. Numer. Anal.,* **16**, 523 (1979).

Van der Staay, D. L., "Composite Integration-Interpolation Methods for the Solution of Stiff Differential-Difference Equations," Ph.D. Thesis, Electrical and Computer Engineering Dept., Syracuse University, Syracuse, New York (1982).

Van Veldhuizen, M., "Consistency and Stability for One-Step Discretizations of Stiff Differential Equations," *Stiff Differential Systems,* R. A. Willoughby, Ed., Plenum Press, New York (1974).

Van Veldhuizen, M., "D-Stability," *SIAM J. Num. Anal.,* **18**, 45 (1981).

Varah, J. M., "Stiffly Stable Linear Multistep Methods of Extended Order," *SIAM J. Numer. Anal.,* **15**, 1234 (1978).

Varah, J. M., "On the Efficient Implementations of Implicit Runge–Kutta Methods," *Math. Comp.,* **33**, 557 (1979).

Varah, J. M., "Stability Restrictions on Second Order, Three Level Finite Difference Schemes for Parabolic Equations," *SIAM J. Numer. Anal.,* **17**, 300 (1980).

Varma, A., Georgakis, Amundson, N. R., and R. Aris, "Computational Method for the Tubular Chemical Reactor," *Comput. Methods Appl. Mech. & Eng.,* **8**, 319 (1976).

Verner, Y. H., "Families of Imbedded Runge–Kutta Methods," *SIAM J. Num. Anal.,* **16** No. 5, 857 (1979).

Verwer, J. G., "S-Stability Properties for Generalized Runge–Kutta Methods," *Numer. Math.,* **27**, 359 (1977).

Verwer, J. G., "An Analysis of Rosenbrock Methods for Nonlinear Stiff Initial Value Problems," *SIAM J. Numer. Anal.,* **19**, 155 (1981).

Von Pragenau, G. L., "Large Step Integration for Linear Dynamic Systems," *IEEE Southeast Con.* (1981).

Wait, R., "Parabolic Partial Differential Equations," Chapt. 14, *Modern Numerical Methods for Ordinary Differential Equations* Wall, G., and J. M. Watt, Eds., Clarendon Press, Oxford, 197 (1976).

Wall, S. N. and G. E. A. Anlansson, "Numerical Calculations on the Kinetics of Stepwise Micelle Association," *J. Phys. Chem.,* **84**, 727 (1980).

Wallace, C. S. and G. K. Gupta, "General Linear Multistep Methods to Solve Ordinary Differential Equations," *Austral. Comput. J.,* **5**, 62 (1973).

Wallace, J. S., Springer, G. S., and D. H. Stedman, "Photochemical Ozone and Nitric Oxide Formation in Air-Nitrogen Dioxide Mixtures Containing Sulfur Dioxide or Chlorine," **14**, 1147 (1980).

Walsh, J. L., *Amer. Math. Soc. Coll. Publ.,* **XX**, New York (1935).

Wambecq, A., "Rational Runge–Kutta Method for Solving Systems of Ordinary Differential Equations," *Computing,* **20**, 333 (1978).

Wanner, G., "A Short Proof on Non-Linear A-Stability," *BIT,* **16**, 226 (1976).

Wanner, G., "On the Integration of Stiff Differential Equations," *Int. Series in Numerical Math, Vol. 37,* Descloux, J. and J. Marti, Eds., Birkhauser Verlag, Basel and Stuttgart (1977).
Wanner, G., "On the Choice of Gamma for Singly Implicit R–K or Rosenbrock Methods," *BIT,* **20**, 102 (1980a).
Wanner, G., "Characterization of All A-Stable Methods of Order $2M - 4$," *BIT,* **20**, 367 (1980b).
Wanner, G., Presentation Given During the Rutischauser Symposium (unpublished) (1980c).
Wanner, G., Hairer, E. and S. P. Norsett, "Order Stars and Stability Theorems," *BIT,* **18**, 475 (1978a).
Wanner, G., Hairer, E., and S. P. Norsett, "When I-Stability Implies A-Stability," *BIT,* **18**, 503 (1978b).
Ward, R. A. and W. M. A. Fowler, "Thermalization of Long-Lived Nuclear Isomeric States Under Stellar Conditions," *Astrophysical Journal,* **238** 266 (1980).
Warming, R. F. and R. M. Beam, "An Extension of A-Stability to Alternating Direction Implicit Methods," *BIT,* **19**, 395 (1979a).
Warming, R. F. and R. M. Beam, "Factored, A-Stable Linear Multistep Methods —An Alternative to the Method of Lines for Multidimensions," *SIGNUM Newsl.,* **14**, 17 (1979b).
Warner, D. D., "The Numerical Solution of the Equations of Chemical Kinetics," *J. Phys. Chem.,* **81**, 25 (1977).
Warner, D. D. and C. L. Wilson, "Two Dimensional Concentration Dependent Diffusion," *Bell System Technical J.,* **59**, 1 (1980).
Wasow, W., *Asymptotic Expansions for Ordinary Differential Equations,* Wiley-Interscience, New York (1965).
Watanabe, D. S., "Block Implicit One-Step Methods," *Math. Comp.,* **32**, 405 (1978).
Watanabe, D. S. and M. G. Roth, "The Stability of Difference Formulas for Delay Differential Equations," Submitted to *SIAM J. Num. Anal.*
Watkins, D. S., "Determining Initial Values for Stiff Systems of Ordinary Equations," *SIAM J. Numer. Anal.,* **18**, 13 (1981a).
Watkins, D. S., "Efficient Initialization of Systems with One Unknown Initial Condition," *SIAM J. Numer. Anal.,* **18**, 794 (1981b).
Watson, H. D. D., "Implicit Integration for CSMP III and the Problem of Stiffness," *Simulation,* **26**, 57 (1976).
Weaver, L., Shultis, J. K., and R. E. Faw, "Analytical Solutions of a Model for Radiation-Induced Conductivity in Insulators," *J. App. Phys.,* **48**, 2762 (1977).
Wei, J. and C. D. Prater, "The Structure of an Analysis of Complex Reaction Systems," *Adv. in Catal.,* **13**, Academic Press, New York (1962).
Weimer, A. W. and D. E. Clough, "A Critical Evaluation of the Semi-Implicit Runge–Kutta Methods for Stiff Systems," *AICHE J.,* **25**, 730 (1979).
Wendt, J. O. L., Martinez, C. H., Lilley, D. G. and T. L. Corley, "Numerical Solution of Stiff Boundary Value Problems in Kinetics and Diffusion," *Chem. Eng. Sci.,* **34**, 527 (1979).
Werner, W., *Adv. Nucl. Sci. & Tech.,* **10**, Lewins, J. and M. Becker, Eds., Plenum Press (1981).
Westbrook, C. K., "A Generalized ICE Method for Chemically Reactive Flows in Combustion Systems," *J. Comp. Phys.,* **29**, 67 (1978).
Westlake, W. J., "Use of Statistical Methods in Evaluation of *In Vivo* Performance of Dosage Forms," *J. Pharm. Sci.,* **62(10)**, 1579 (1973).
White, III, C. W. and W. D. Seider, "Analysis of Chemical Reaction System," *Chem. Eng. Commun.,* **9**, 159 (1981).
Wiederholt, L. F., "Stability of Multistep Methods for Delay Differential Equations," *Math. Comput.,* **30**, 283 (1976).
Williams, J. and F. De Hoog, "A Class of A-Stable Advanced Multistep Methods," *Math. Comp.,* **28**, 163 (1974).

Willoughby, R. A., "International Symposium on Stiff Differential Systems: Introduction," *Stiff Differential Systems,* R. A. Willoughby, Ed., Plenum Press, New York (1974).

Wolfbrandt, A., "A Study of Rosenbrock Processes with Respect to Order Conditions and Stiff Stability," Chalmers University of Technology, Sweden (1977).

Womble, M. E., "Approximations to Ricatti Equations Having Slow and Fast Modes," *IEEE Trans. Autom. Control,* **AC-21**, 846 (1976).

Wood, W. L., "Control of Crank-Nicolson Noise in the Numerical Solution of the Heat Conduction Equations," *Int. J. Numer. Methods Eng.,* **11**, 1059 (1977).

Wouk, A., "Collocation for Initial Value Problems," *BIT,* **16**, 215 (1976).

Wright, J. P., "Mixed Time Integration Schemes," *Comput. & Struct.,* **10**, 235 (1979).

Wright, K., "Some Relationships Between Implicit Runge-Kutta Methods, Collocation and Lanczos Tau-Methods and Their Stability Properties," *BIT,* **10**, 217 (1970).

Yackel, R. A., "A Boundary Layer Method for the Matrix Riccati Equations," *IEEE Trans. Aut. Contr.* **AC-18**, 17 (1973).

Yanenko, N. N., Lisseikin, V. D., and V. M. Kovenia, "The Method of the Solution of Gas Dynamical Problems in Moving Meshes," 3rd Int. Symp. on Computing Methods in Appl. Sci. & Eng., Paris, France (1977).

Young, T. R. and J. P. Boris, "A Numerical Technique for Solving Stiff Ordinary Differential Equations Associated with the Chemical Kinetics of Reactive-Flow Problems," AFWL-AFOSR Workshop, p. 236 (1977).

Zadunaisky, P., "On the Estimation of Errors Propagated in the Numerical Integration of ODEs," *Numer. Math.,* **27**, 21 (1976).

Zagoruiko, A. S., "Algorithm for the Solution of Systems of Differential Equations with Automatic Selection of Order and Steplength of Integration," *Avtometriya,* 2, 92 (1981).

Zein, D. A., "On the Periodic Steady-State Problem by the Newton Method," *IEE Trans. on Circ. & Syst.,* **CAS-27**, 1263 (1980).

Zlatev, Z., "Stability Properties of Variable Stepsize Formula Methods," *Numer. Math.,* **31**, 175 (1978).

Zlatev, Z., "Zero-Stability Properties of the Three-Ordinate Variable Stepsize Variable Formula Methods," *Num. Math.,* **37**, 157 (1980).

Zlatev, Z., "Modified Diagonally Implicit Runge-Kutta Methods," *SIAM J. Sci. Stat. Comp.* 2, 321 (1981).

Zlatev, Z., Schaumberg, K., and J. Wasniewski, "Implementation of an Iterative Refinement Option in a Code for Large and Sparse Systems," *Computers & Chem.,* **4**, 87 (1980).

Zlatev, Z., Wasniewski, J., and K. Schaumburg, "Classification of the Systems of Ordinary Differential Equations and Practical Aspects in the Numerical Integration of Large Systems," *Comp. & Chem.,* **4**, 13 (1980).

Index